# The Investigation of Organic
# Reactions and Their Mechanisms

# The Investigation of Organic Reactions and Their Mechanisms

Edited by

**Howard Maskill**
Sometime lecturer
University of Newcastle upon Tyne
and visiting professor
University of Santiago de Compostela
Spain

**Blackwell**
Publishing

© 2006 by Blackwell Publishing Ltd

Editorial offices:
Blackwell Publishing Ltd, 9600 Garsington Road, Oxford OX4 2DQ, UK
    Tel: +44 (0) 1865 776868
Blackwell Publishing Professional, 2121 State Avenue, Ames, Iowa 50014-8300, USA
    Tel: +1 515 292 0140
Blackwell Publishing Asia Pty Ltd, 550 Swanston Street, Carlton, Victoria 3053, Australia
    Tel: +61 (0)3 8359 1011

First published 2006 by Blackwell Publishing Ltd

ISBN-13: 978-1-4051-3142-1
ISBN-10: 1-4051-3142-X

Library of Congress Cataloging-in-Publication Data
    The investigation of organic reactions and their mechanisms / edited by Howard Maskill.
        p.   cm.
    Includes bibliographical references and index.
    ISBN-13: 978-1-4051-3142-1 (hardback : alk. paper)
    ISBN-10: 1-4051-3142-X (hardback : alk. paper)
    1. Chemistry, Physical organic.   2. Chemical reactions.   3. Chemical processes.
    I. Maskill, Howard.
    QD476.I558 2006
    547′.2—dc22
    2006012267

A catalogue record for this title is available from the British Library

Set in 10/12pt Minion
by TechBooks Electronic Services Pvt Ltd, New Delhi, India
Printed and bound in Singapore
by Fabulous Printers Pte Ltd

The publisher's policy is to use permanent paper from mills that operate a sustainable
forestry policy, and which has been manufactured from pulp processed using acid-free and
elementary chlorine-free practices. Furthermore, the publisher ensures that the text paper and
cover board used have met acceptable environmental accreditation standards.

For further information on Blackwell Publishing, visit our website:
www.blackwellpublishing.com

# Contents

Contributors     xv

Foreword     xvii

Preface     xxi

1    Introduction and Overview    *H. Maskill*     1
    1.1   Background     1
    1.2   The nature of mechanism and reactivity in organic chemistry     1
    1.3   The investigation of mechanism and the scope of this book     2
        1.3.1   Product analysis, reaction intermediates and isotopic labelling     2
            1.3.1.1   Example: the acid-catalysed decomposition of
                     nitrosohydroxylamines     3
        1.3.2   Mechanisms and rate laws     4
        1.3.3   Computational chemistry     6
            1.3.3.1   Example: the acid- and base-catalysed decomposition
                     of nitramide     6
        1.3.4   Kinetics in homogeneous solution     7
            1.3.4.1   Example: the kinetics of the capture of pyridyl ketenes
                     by *n*-butylamine     8
        1.3.5   Kinetics in multiphase systems     9
        1.3.6   Electrochemical and calorimetric methods     10
        1.3.7   Reactions involving radical intermediates     12
        1.3.8   Catalysed reactions     13
    1.4   Summary     16
    Bibliography     16
    References     16

2    Investigation of Reaction Mechanisms by Product Studies    *T. W. Bentley*     18
    2.1   Introduction and overview – why study organic reaction mechanisms?     18
    2.2   Product structure and yield     21
        2.2.1   Quantitative determination of product yields     21

| | | |
|---|---|---|
| 2.2.2 | Product stabilities, and kinetic and thermodynamic control of product formation | 23 |
| 2.3 | Mechanistic information from more detailed studies of product structure | 25 |
| 2.3.1 | Stereochemical considerations | 25 |
| 2.3.2 | Use of isotopic labelling | 26 |
| 2.4 | Mechanistic evidence from variations in reaction conditions | 27 |
| 2.5 | Problems and opportunities arising from unsuccessful experiments or unexpected results | 32 |
| 2.6 | Kinetic evidence from monitoring reactions | 32 |
| 2.6.1 | Sampling and analysis for kinetics | 33 |
| 2.7 | Case studies: more detailed mechanistic evidence from product studies | 34 |
| 2.7.1 | Product-determining steps in $S_N1$ reactions | 34 |
| 2.7.2 | Selectivities | 36 |
| 2.7.3 | Rate–product correlations | 38 |
| Bibliography | | 43 |
| References | | 44 |

3   **Experimental Methods for Investigating Kinetics**   *M. Canle L., H. Maskill and J. A. Santaballa*   46

| | | |
|---|---|---|
| 3.1 | Introduction | 46 |
| 3.2 | Preliminaries | 46 |
| 3.2.1 | Reaction rate, rate law and rate constant | 46 |
| 3.2.2 | Reversible reactions, equilibrium and equilibrium constants | 48 |
| 3.2.3 | Reaction mechanism, elementary step and rate-limiting step | 48 |
| 3.2.4 | Transition structure and transition state | 50 |
| 3.3 | How to obtain the rate equation and rate constant from experimental data | 50 |
| 3.3.1 | Differential method | 51 |
| 3.3.1.1 | Example: reaction between RBr and $HO^-$ | 52 |
| 3.3.2 | Method of integration | 53 |
| 3.3.2.1 | Data handling | 54 |
| 3.3.2.2 | Example: decomposition of $N_2O_5$ in $CCl_4$ | 54 |
| 3.3.3 | Isolation method | 56 |
| 3.3.3.1 | Example: oxidation of methionine by HOCl | 57 |
| 3.4 | Reversible reactions and equilibrium constants | 58 |
| 3.4.1 | Rate constants for forward and reverse directions, and equilibrium constants | 58 |
| 3.4.1.1 | Example: *cis-trans* isomerisation of stilbene | 59 |
| 3.5 | Experimental approaches | 59 |
| 3.5.1 | Preliminary studies | 59 |
| 3.5.2 | Variables to be controlled | 60 |
| 3.5.2.1 | Volume | 60 |
| 3.5.2.2 | Temperature | 60 |
| 3.5.2.3 | pH | 61 |

3.5.2.4   Solvent                                                62
3.5.2.5   Ionic strength                                         63
3.5.2.6   Other experimental aspects                             64
3.6   Choosing an appropriate monitoring method                  65
3.6.1   Periodic monitoring                                      65
3.6.2   Continuous on-line monitoring                            65
3.6.3   Continuous static monitoring                             65
3.7   Experimental methods                                       66
3.7.1   Spectrometric methods                                    66
3.7.1.1   Conventional and slow reactions                        67
3.7.1.2   Fast reactions                                         69
3.7.1.3   Very fast and ultrafast reactions                      70
3.7.1.4   Magnetic resonance spectroscopy                        71
3.7.2   Conductimetry                                            71
3.7.3   Polarimetry                                              73
3.7.4   Potentiometry                                            73
3.7.5   Dilatometry                                              74
3.7.6   Pressure measurements                                    75
3.7.7   Chromatographic methods                                  75
3.7.8   Other techniques                                         76
Bibliography                                                     76
References                                                       76

4   The Relationship Between Mechanism and Rate Law   *J. A. Santaballa,*
    *H. Maskill and M. Canle L.*                                 79
4.1   Introduction                                               79
4.2   Deducing the rate law from a postulated mechanism          80
4.2.1   Single-step unidirectional reactions                     80
4.2.2   Simple combinations of elementary steps                  81
4.2.2.1   Consecutive unimolecular (first-order) reactions       81
4.2.2.2   Reversible unimolecular (first-order) reactions        83
4.2.2.3   Parallel (competitive) unimolecular (first-order)
          reactions                                              83
4.2.2.4   Selectivity in competing reactions                     86
4.2.3   Complex reaction schemes and approximations              86
4.2.3.1   The steady-state approximation (SSA)                   88
4.2.3.2   The pre-equilibrium approximation                      89
4.2.3.3   The rate-determining step approximation                89
4.2.3.4   The steady-state approximation, and solvolysis of alkyl
          halides and arenesulfonates                            90
4.3   Case studies                                               91
4.3.1   Chlorination of amino compounds                          91
4.3.2   The Aldol reaction                                       95
4.3.2.1   At low concentrations of aldehyde                      96
4.3.2.2   At high concentrations of aldehyde                     97

|  |  |  |
|---|---|---|
| 4.3.3 | Hydrogen atom transfer from phenols to radicals | 98 |
| 4.3.3.1 | Via pre-equilibrium formation of the phenolate | 100 |
| 4.3.3.2 | Via rate-limiting proton transfer to give the phenolate | 100 |
| 4.3.4 | Oxidation of phenols by Cr(VI) | 100 |
| Bibliography | | 103 |
| References | | 103 |

5   Reaction Kinetics in Multiphase Systems   *John H. Atherton*   104

| | | |
|---|---|---|
| 5.1 | Introduction | 104 |
| 5.2 | Background and theory | 105 |
| 5.2.1 | Mass transfer coupled to chemical reaction | 105 |
| 5.2.1.1 | Reaction too slow to occur within the diffusion film | 106 |
| 5.2.1.2 | Reaction fast relative to the film diffusion time | 107 |
| 5.2.1.3 | Interfacial reactions | 109 |
| 5.2.2 | Phase-transfer catalysis (PTC) | 110 |
| 5.2.3 | System complexity and information requirements | 112 |
| 5.3 | Some experimental methods | 113 |
| 5.3.1 | The stirred reactor for the study of reactive dispersions with a liquid continuous phase | 113 |
| 5.3.1.1 | Gas–liquid reactions | 113 |
| 5.3.1.2 | Dispersed liquid–liquid systems | 114 |
| 5.3.1.3 | Liquid–solid reactions in a stirred reactor | 115 |
| 5.3.2 | Techniques providing control of hydrodynamics | 116 |
| 5.3.2.1 | Techniques based on the Lewis cell | 116 |
| 5.3.2.2 | The rotated disc reactor | 117 |
| 5.3.2.3 | Rotated diffusion cell | 118 |
| 5.3.2.4 | Channel flow techniques | 119 |
| 5.3.2.5 | The jet reactor | 120 |
| 5.3.2.6 | Expanding drop methods | 121 |
| 5.3.2.7 | Confluence microreactor | 122 |
| 5.3.2.8 | Microelectrode techniques | 122 |
| 5.3.3 | Use of atomic force microscopy (AFM) | 123 |
| 5.4 | Information requirements and experimental design | 123 |
| 5.5 | Summary | 124 |
| Bibliography | | 125 |
| References | | 125 |

6   Electrochemical Methods of Investigating Reaction Mechanisms *Ole Hammerich*   127

| | | |
|---|---|---|
| 6.1 | What is organic electrochemistry? | 127 |
| 6.2 | The relationship between organic electrochemistry and the chemistry of radical ions and neutral radicals | 130 |
| 6.3 | The use of electrochemical methods for investigating kinetics and mechanisms | 131 |

6.4    Experimental considerations                                          132
        6.4.1    Two-electrode and three-electrode electrochemical cells     132
        6.4.2    Cells for electroanalytical studies                         133
        6.4.3    Electrodes for electroanalytical studies                    134
                 6.4.3.1    The working electrode (W)                        134
                 6.4.3.2    The counter electrode (C)                        135
                 6.4.3.3    The reference electrode (R)                      135
        6.4.4    The solvent-supporting electrolyte system                   135
        6.4.5    The electronic instrumentation                             135
6.5    Some basics                                                          136
        6.5.1    Potential and current                                       137
        6.5.2    The electrochemical double layer and the charging current   138
        6.5.3    Mass transport and current                                  139
6.6    The kinetics and mechanisms of follow-up reactions                   141
        6.6.1    Nomenclature                                                141
        6.6.2    Mechanisms and rate laws                                    141
        6.6.3    The theoretical response curve for a proposed mechanism     142
6.7    The response curves for common electroanalytical methods             142
        6.7.1    Potential step experiments (chronoamperometry and double
                 potential step chronoamperometry)                          143
        6.7.2    Potential sweep experiments (linear sweep voltammetry and
                 cyclic voltammetry)                                        147
                 6.7.2.1    CV conditions                                    151
                 6.7.2.2    LSV conditions                                   154
                 6.7.2.3    Fitting simulated voltammograms to experimental
                            voltammograms                                   154
        6.7.3    Potential sweep experiments with ultramicroelectrodes       155
        6.7.4    Concluding remarks                                          159
Appendix                                                                    159
        A.1    The preliminary experiments                                  159
        A.2    Preliminary studies by cyclic voltammetry                    160
        A.3    Determination of the number of electrons, $n$ (coulometry)   162
        A.4    Preparative or semi-preparative electrolysis, identification
                 of products                                                164
References                                                                  165


7    Computational Chemistry and the Elucidation of Mechanism
     Peter R. Schreiner                                                     167
     7.1    How can computational chemistry help in the elucidation of reaction
            mechanisms?                                                     167
            7.1.1    General remarks                                        167
            7.1.2    Potential energy surfaces, reaction coordinates and transition
                     structures                                            168
            7.1.3    Absolute and relative energies; isodesmic and homodesmotic
                     equations                                             170

7.2   Basic computational considerations                              172
      7.2.1   Molecular mechanics                                     172
      7.2.2   Wave function theory                                    173
      7.2.3   Semiempirical methods                                   173
      7.2.4   Hartree–Fock theory                                     175
      7.2.5   Electron-correlation methods                            176
      7.2.6   Density functional theory                               179
      7.2.7   Symmetry                                                181
      7.2.8   Basis sets                                              181
      7.2.9   Validation                                              182
7.3   Case studies                                                    182
      7.3.1   The ethane rotational barrier and wave function analysis  182
      7.3.2   The nonclassical carbocation problem and the inclusion
              of solvent effects                                      187
7.4   Matching computed and experimental data                        192
7.5   Conclusions and outlook                                         193
7.6   List of abbreviations                                           194
References                                                            194


8   Calorimetric Methods of Investigating Organic Reactions    U. Fischer
    and K. Hungerbühler                                               198
    8.1   Introduction                                                198
    8.2   Investigation of reaction kinetics and mechanisms using calorimetry and
          infrared spectroscopy                                       199
          8.2.1   Fundamentals of reaction calorimetry                200
          8.2.2   Types of reaction calorimeters                      200
                  8.2.2.1   Heat-flow calorimeters                    201
                  8.2.2.2   Power-compensation calorimeters           201
                  8.2.2.3   Heat-balance calorimeters                 202
                  8.2.2.4   Peltier calorimeters                      202
          8.2.3   Steady-state isothermal heat-flow balance of a general type of
                  reaction calorimeter                                202
          8.2.4   Infrared and IR-ATR spectroscopy                    205
          8.2.5   Experimental methods for isothermal calorimetric and infrared
                  reaction data                                       206
                  8.2.5.1   Experimental methods for isothermal calorimetric
                            reaction data                             206
                  8.2.5.2   Experimental methods for isothermal infrared
                            reaction data                             209
                  8.2.5.3   Methods for combined determination of isothermal
                            calorimetric and infrared reaction data   211
    8.3   Investigation of reaction kinetics using calorimetry and IR-ATR
          spectroscopy – examples of application                      211
          8.3.1   Calorimetric device used in combination with IR-ATR
                  spectroscopy                                        211

8.3.2    Example 1: Hydrolysis of acetic anhydride                           213
     8.3.2.1    Materials and methods                              213
     8.3.2.2    Results and discussion                             213
8.3.3    Example 2: sequential epoxidation of 2,5-di-tert-butyl-
     1,4-benzoquinone                                           216
     8.3.3.1    Materials and methods                              217
     8.3.3.2    Results and discussion                             217
8.3.4    Example 3: Hydrogenation of nitrobenzene                    222
     8.3.4.1    Materials and methods                              222
     8.3.4.2    Results and discussion                             223
8.4    Conclusions and outlook                                              223
References                                                                     225

9    The Detection and Characterisation of Intermediates in Chemical Reactions
*C. I. F. Watt*                                                                227
9.1    Introduction: What is an intermediate?                           227
    9.1.1    Potential energy surfaces and profiles                   227
    9.1.2    From molecular potential energy to rates of reaction     229
9.2    A systematic approach to the description of mechanism          231
    9.2.1    Reaction classification                                 231
    9.2.2    Consequences of uncoupled bonding changes                232
    9.2.3    Sequences of basic reactions                            233
9.3    Evidence and tests for the existence of intermediates           234
    9.3.1    Direct observation                                      234
    9.3.2    Deductions from kinetic behaviour                       238
    9.3.3    Trapping of intermediates                               242
    9.3.4    Exploitation of stereochemistry                         246
    9.3.5    Isotopic substitution in theory                         249
    9.3.6    Isotopic substitution in practice                       252
    9.3.7    Linear free energy relationships                        256
References                                                                     258

10    Investigation of Reactions Involving Radical Intermediates    *Fawaz*
*Aldabbagh, W. Russell Bowman and John M. D. Storey*                  261
10.1    Background and introduction                                      261
    10.1.1    Radical intermediates                                  261
    10.1.2    Some initial considerations of radical mechanisms and
         chapter overview                                    262
10.2    Initiation                                                       264
10.3    Radical addition to alkenes                                      266
10.4    Chain and non-chain reactions                                    268
10.5    Nitroxides                                                       268
    10.5.1    Nitroxide-trapping experiments                         269
    10.5.2    Alkoxyamine dissociation rate constant, $k_d$          270
    10.5.3    The persistent radical effect (PRE)                    273

|  |  | 10.5.4 | Nitroxide-mediated living/controlled radical polymerisations (NMP) | 275 |
|  | 10.6 | Radical clock reactions |  | 276 |
|  | 10.7 | Homolytic aromatic substitution |  | 280 |
|  | 10.8 | Redox reactions |  | 284 |
|  |  | 10.8.1 | Reductions with samarium di-iodide, $SmI_2$ | 284 |
|  |  | 10.8.2 | $S_{RN}1$ substitution | 287 |
|  | Bibliography |  |  | 291 |
|  | References |  |  | 292 |

11 Investigation of Catalysis by Acids, Bases, Other Small Molecules and Enzymes
   *A. Williams*                                                                      293

|  | 11.1 | Introduction |  | 293 |
|  |  | 11.1.1 | Definitions | 293 |
|  | 11.2 | Catalysis by acids and bases |  | 294 |
|  |  | 11.2.1 | Experimental demonstration | 294 |
|  |  | 11.2.2 | Reaction flux and third-order terms | 297 |
|  |  | 11.2.3 | Brønsted equations | 298 |
|  |  |  | 11.2.3.1  Brønsted parameters close to $-1$, 0 and $+1$ | 299 |
|  |  | 11.2.4 | Kinetic ambiguity | 299 |
|  |  |  | 11.2.4.1  Cross-correlation effects | 299 |
|  |  |  | 11.2.4.2  The diffusion-controlled limit as a criterion of mechanism | 301 |
|  |  |  | 11.2.4.3  Scatter in Brønsted plots | 302 |
|  |  |  | 11.2.4.4  Solvent kinetic isotope effects | 302 |
|  |  | 11.2.5 | Demonstrating mechanisms of catalysis by proton transfer | 302 |
|  |  |  | 11.2.5.1  Stepwise proton transfer (trapping) | 302 |
|  |  |  | 11.2.5.2  Stabilisation of intermediates by proton transfer | 304 |
|  |  |  | 11.2.5.3  Preassociation | 306 |
|  |  |  | 11.2.5.4  Concerted proton transfer | 307 |
|  |  |  | 11.2.5.5  Push–pull and bifunctional acid–base catalysis | 307 |
|  | 11.3 | Nucleophilic and electrophilic catalysis |  | 308 |
|  |  | 11.3.1 | Detection of intermediates | 308 |
|  |  | 11.3.2 | Non-linear free energy relationships and transient intermediates | 310 |
|  | 11.4 | Enzyme Catalysis |  | 311 |
|  |  | 11.4.1 | Technical applications | 312 |
|  |  | 11.4.2 | Enzyme assay | 312 |
|  |  | 11.4.3 | Steady-state kinetics | 313 |
|  |  |  | 11.4.3.1  Active-site titration | 313 |
|  |  |  | 11.4.3.2  Active-site directed irreversible inhibitors | 315 |
|  |  | 11.4.4 | Kinetic analysis | 316 |
|  |  | 11.4.5 | Reversible inhibitors | 317 |
|  |  | 11.4.6 | Detection of covalently bound intermediates | 318 |
|  |  |  | 11.4.6.1  Direct observation | 318 |
|  |  |  | 11.4.6.2  Structural variation | 319 |

| | | 11.4.6.3 | Stereochemistry | 320 |
| | | 11.4.6.4 | Kinetics | 320 |
| | | 11.4.6.5 | Trapping | 321 |
| | Bibliography | | | 322 |
| | References | | | 322 |

12 Catalysis by Organometallic Compounds    *Guy C. Lloyd-Jones*    324

| 12.1 | Introduction | 324 |
| | 12.1.1 | The challenges inherent in the investigation of organic reactions catalysed by organometallics | 324 |
| | 12.1.2 | Techniques used for the study of organometallic catalysis | 326 |
| | 12.1.3 | Choice of examples | 327 |
| 12.2 | Use of a classical heteronuclear NMR method to study intermediates 'on cycle' directly: the Rh-catalysed asymmetric addition of organoboronic acids to enones | 328 |
| | 12.2.1 | Background and introduction | 328 |
| | 12.2.2 | The $^{31}P\{^{1}H\}$ NMR investigation of the Rh-catalysed asymmetric phenylation of cyclohexenone | 330 |
| | 12.2.3 | Summary and key outcomes from the mechanistic investigation | 333 |
| 12.3 | Kinetic and isotopic labelling studies using classical techniques to study intermediates 'on cycle' indirectly: the Pd-catalysed cyclo-isomerisation of dienes | 334 |
| | 12.3.1 | Background and introduction | 334 |
| | 12.3.2 | Kinetic studies employing classical techniques | 335 |
| | 12.3.3 | 'Atom accounting' through isotopic labelling | 338 |
| | 12.3.4 | Observation of pro-catalyst activation processes by NMR spectroscopy | 341 |
| | 12.3.5 | Summary and mechanistic conclusions | 342 |
| 12.4 | Product distribution analysis, and kinetics determined by classical and advanced NMR techniques: the transition-metal-catalysed metathesis of alkenes | 343 |
| | 12.4.1 | Background and introduction | 343 |
| | 12.4.2 | Early mechanistic proposals for the alkene metathesis reaction | 344 |
| | 12.4.3 | Disproving the 'pairwise' mechanism for metathesis | 345 |
| | 12.4.4 | Mechanistic investigation of contemporary metathesis catalysts | 348 |
| | 12.4.5 | NMR studies of degenerate ligand exchange in generation I and generation II ruthenium alkylidene pro-catalysts for alkene metathesis | 351 |
| | 12.4.6 | Summary and mechanistic conclusions | 352 |
| | References | | 353 |

*Index*    354

# Contributors

Fawaz Aldabbagh

Department of Chemistry, National University of Ireland, Galway, Ireland

John H. Atherton

The School of Applied Sciences, University of Huddersfield, Huddersfield HD1 3DH (15, Prestwich Drive, Fixby Park, Huddersfield, HD2 2NU)

T. William Bentley

Department of Chemistry, University of Wales Swansea, Singleton Park, Swansea SA2 8PP

W. Russell Bowman

Department of Chemistry, Loughborough University, Loughborough, Leics. LE11 3TU

Moisés Canle López

Chemical Reactivity and Photoreactivity Group, Department of Physical Chemistry and Chemical Engineering I, University of A Coruña, Rúa Alejandro de la Sota 1, E-15008 A Coruña, Galicia, Spain

Ulrich Fischer

Swiss Federal Institute of Technology, Institute for Chemical and Bioengineering, ETH-Hoenggerberg HCI G137, CH-8093 Zurich, Switzerland.

Ole Hammerich

Department of Chemistry, University of Copenhagen, The H. C. Ørsted Institute, Universitetsparken 5, DK-2100 Copenhagen Ø, Denmark

Konrad Hungerbühler

Swiss Federal Institute of Technology, Institute for Chemical and Bioengineering, ETH-Hoenggerberg HCI G137, CH-8093 Zurich, Switzerland.

Guy C. Lloyd-Jones

The Bristol Centre for Organometallic Catalysis, School of Chemistry, University of Bristol, Cantock's Close, Bristol, BS8 1TS

Howard Maskill

School of Natural Sciences, University of Newcastle, Newcastle upon Tyne NE1 7RU

**Juan Arturo Santaballa López**   Chemical Reactivity and Photoreactivity Group, Department of Physical Chemistry and Chemical Engineering I, University of A Coruña, Rúa Alejandro de la Sota 1, E-15008 A Coruña, Galicia, Spain

**Peter R. Schreiner**   Institute of Organic Chemistry, Justus-Liebig University, Heinrich-Buff-Ring 58, D-35392 Giessen, Germany

**John M. D. Storey**   Department of Chemistry, University of Aberdeen, Meston Walk, Aberdeen, AB24 3UE

**C. Ian F. Watt**   School of Chemistry, University of Manchester, Brunswick Street, Manchester M13 9PL

**Andrew Williams**   University of Kent at Canterbury (Maple Cottage, Staithe Road, Hickling, Norfolk NR12 0YJ)

# Foreword

Physical organic chemistry is a field with a long and established tradition. Most chemists would probably identify the late 1920s through the 1930s and 1940s as the beginnings of what one might call classical physical organic chemistry. The two pioneers most often mentioned are Sir Christopher Ingold and Louis Hammett; Ingold's mechanistic studies of $S_N1$, $S_N2$ and other reactions, and the publication of Hammett's book *Physical Organic Chemistry* in 1940 indeed played key roles in shaping this emerging discipline. However, as we are reminded by John Shorter in his 1998 *Chemical Society Reviews* article, some of the groundwork had been laid by a number of less well known chemists who preceded Ingold and Hammett, among others James Walker, Arthur Lapworth, N. V. Sidgwick, J. J. Sudborough, K. J. P. Orton and H. M. Dawson. One other name one should add to this list of early pioneers is J. N. Brønsted.

What is physical organic chemistry? Jack Hine wrote in the preface of his 1962 classic book on the subject, "A broad definition of the term physical organic chemistry might include a major fraction of existing chemical knowledge and theory." Indeed, the impact of the intellectual and experimental approaches used by physical organic chemists on our understanding of chemical reactions has been profound. As pointed out by Edward Kosower in his 1968 book *Physical Organic Chemistry*, "there is scarcely a branch of organic chemistry, including that concerned with synthesis, that could not be treated within the context of physical organic chemistry." This is most clearly seen in how modern organic chemistry textbooks for undergraduate students approach the subject.

The study of reactions from the point of view of their mechanism and the relationship between structure and reactivity has always been at the core of this field, and this is what we call classical physical organic chemistry. Even though the importance of determining the products of a reaction as the starting point of any mechanistic investigation can never be emphasised enough, everything which happens between reactants and products is the domain of physical organic chemistry. This includes not only the formation of intermediates and transition structures, the mapping out of reaction trajectories and the free energy changes that occur along the reaction path, but also our attempts at understanding why a reaction 'chooses' a particular mechanism. One physical organic chemist who has probably contributed more than anyone else to current notions of how reactions choose their mechanisms is William Jencks.

Over the years, the scope of physical organic chemistry has continually evolved and expanded, and now includes an ever increasing number of new topics and subfields. Besides the more established newer disciplines of photochemistry, electrochemistry, bioorganic and bioinorganic chemistry, and computational chemistry, it presently includes developing areas such as supramolecular chemistry, combinatorial chemistry, transition metallo-organic chemistry, nanochemistry, materials science, biomimetic chemistry, femtochemistry, the building of molecular machines such as molecular switches and motors, the use of ionic liquids as reaction media, etc.

A revealing snapshot of recent and current activities physical organic chemists are engaged in is provided by the titles of some of the invited and plenary lectures to be presented at the 18th IUPAC Conference on Physical Organic Chemistry to be held in Warsaw, Poland, in August 2006. Here is a sample.

- Biosensors
- Molecular Machines in Biology
- Structural Biology
- De Novo Design Approach Based on Nanorecognition: Functional Molecules/Materials and Nanosensors/Nanodevices
- New Developments of Electron Transfer Catalytic Systems
- Time-resolved Synchrotron Diffraction Studies of Molecular Excited States
- Molecular Motions in New Catenanes
- From Crystal Engineering to Supramolecular Green Chemistry: Solid–Solid and Solid–Gas Reactions with Molecular Crystals
- Unusual Weak Interactions – Theoretical Considerations
- On the Chemical Nature of Purpose

Despite this intense and growing diversification which is a testament to the versatility and adaptability of physical organic chemistry and its practitioners, there is continued vitality at the core of this important field of study which aims at continually enhancing and refining our understanding of chemical reactions. Such understanding not only satisfies our curiosity about the world surrounding us but also helps the synthetic and industrial chemist in designing better or more practical ways of creating new compounds. In fact, the present book is mainly aimed at the chemist who wants to investigate reaction mechanisms.

In the mid-1980s, I was the editor of the 4th edition of *Investigation of Rates and Mechanisms of Reactions,* which was part of the *Techniques in Chemistry* series initiated by Arnold Weissberger. For almost half a century, this treatise, along with its earlier editions, has been a pre-eminent source of guidance for physical organic chemists as well as physical, biophysical and inorganic chemists interested in reaction mechanisms. Its scope was quite broad and dealt with both the whole spectrum of experimental techniques and numerous conceptual topics. Regarding experimental techniques, it provided a comprehensive discussion of how to measure 'slow' as well as 'fast' reactions. The latter methods included flow techniques, relaxation techniques such as the temperature jump and pressure jump, electrical field and ultrasonic methods, flash and laser photolysis, NMR/ESR, ICR and pulse radiolysis. The conceptual topics included rate laws, transition state theory, solution versus gas phase reactions, kinetic isotope effects, enzyme kinetics, catalysis, linear free energy relationships and others.

The present book is not meant to be as comprehensive as *Investigation of Rates and Mechanisms of Reactions* but nevertheless provides a rich source of information covering the most important topics necessary for chemists who want to study reaction mechanisms without having to become true experts in physical organic chemistry and kinetics. It thus fills an important need in the current literature. The fact that this book is not as comprehensive and detailed but still teaches the basics is actually a plus in terms of its intended audience. And, especially with respect to applications and coverage of the literature, it is of course more current than the now somewhat dated treatise I edited over 20 years ago.

Claude F. Bernasconi
Santa Cruz, April 2006

# Preface

This book is to help chemists who do not have a strong background in physical/mechanistic organic chemistry but who want to characterise an organic chemical reaction and investigate its mechanism. They may be in the chemical or pharmaceutical manufacturing industry and need reaction data to help identify reaction conditions for an improved yield or a shorter reaction time, or to devise safer reaction conditions. Another potential user could be a synthetic chemist who wants to investigate the mechanism of a newly discovered reaction in order, for example, to optimise reaction conditions and avoid troublesome side reactions.

The book is not primarily intended to be a review of selected current topics for expert physical organic chemists, although it may serve to some degree in this respect. Nor is the book a compendium of mechanisms of organic reactions (although many are necessarily described) or a bench manual for experimental methods (although some practical aspects of less familiar techniques are covered). Our aim was to provide a *guidebook* for the trained chemist who, for reasons of curiosity or practical need, wants to investigate an organic reaction and its mechanism. The investigator may subsequently want a more detailed exposition of a subject than we provide, so bibliographies of more advanced texts and reviews are given at the ends of most chapters, as well as selected references to the original literature, as appropriate.

The book was planned as a single coherent account of the principal methods currently used in mechanistic investigations of organic chemical reactions at a level accessible to graduate chemists in industry as well as academic researchers. Although any chapter can stand alone, we have included many links and cross-references between chapters. We have also tried to show how a particular reaction should be investigated by as wide a range of techniques as is necessary for the resolution of the issues involved. Some chapters include basic material which one may find in descriptive texts on reaction mechanisms or on kinetics, but presented in the context of *how* an organic chemical reaction is investigated, and related to the content of other chapters. The coverage is not comprehensive, and the reader will not find separate chapters on some important methods such as NMR spectroscopy and kinetic isotope effects; examples of the use of such techniques to clarify particular mechanistic problems, however, will be found in several chapters. Correspondingly, solvent and substituent effects upon organic reactivity (for example) are not included as separate topics, but discussions of such matters illuminate a number of case studies. Such topics, which are well covered in the specialist literature but in which there have not been significant recent developments, have been left out to make space for topics in which there have been significant recent developments, e.g. computational chemistry and calorimetry, or are particularly timely because of their current industrial application, e.g. reactions in multiphase systems, synthetically useful reactions involving free radicals and catalysis by organometallic compounds.

Although organic chemistry is a mature subject, different energy units, different abbreviations for metric units of volume and different systems of nomenclature, for example, are still commonly used. We have been consistent within chapters in these respects and follow common usage within each particular area, but uniformity has not been imposed upon the book as a whole.

Contributing authors are an international group of expert practitioners of the techniques covered and have varied industrial and academic backgrounds; they have illustrated their contributions by examples from their own research as well as from the wider chemical literature. To improve the prospects of a coherent book, authors shared their chapter manuscripts with other members of the team wherever connections were identified. I am very grateful to them for all their efforts, cooperation and enthusiasm. Additionally, I am most grateful to Dr Paul Sayer who initiated the project, and to his colleagues at Blackwell for their help in producing the book.

Much of my editing was carried out whilst I was a visiting professor at the University of Santiago de Compostela in Spain during several happy months in 2005 and 2006. My most sincere thanks are due to my principal host, Dr Juan Crugeiras, who made these visits possible.

H. Maskill
Santiago de Compostela, April 2006

# Chapter 1
# Introduction and Overview

## H. Maskill

## 1.1   Background

Descriptive organic reaction mechanisms are covered in every modern university chemistry degree and good current general organic chemistry books approach the subject from a mechanistic viewpoint. In addition, there are specialised texts on organic reaction mechanisms and more advanced monographs on particular techniques used in the investigation of mechanisms of organic reactions and cognate matters. However, there are few books with our aims (see Preface) which bridge the gap between the advanced single-topic monographs on techniques of physical organic chemistry and texts which describe mechanisms of organic reactions. The excellent 1986 text, *Investigation of Rates and Mechanisms of Reactions*, edited by Bernasconi is such a book [1], but is now out of print and, in some respects, out of date. Some of its chapters, however, still provide excellent coverage of certain aspects of physical organic chemistry, especially the underlying principles.

## 1.2   The nature of mechanism and reactivity in organic chemistry

Sometimes, the rebonding in a chemical transformation occurs in just a single step; this will be unimolecular (if we ignore the molecular collisions whereby the reactant molecule gains the necessary energy to react) or bimolecular (if we ignore the initial formation of the encounter complex). Otherwise, a mechanism is a sequence of *elementary* reactions, each being indivisible into simpler chemical events.

Devising possible molecular mechanisms to account for the formation of identified products from known starting materials is often routine; this is principally because newly discovered reactions are generally closely related to previously known ones. However, it is not always so; for example, before the importance of orbital symmetry was discovered, some reactions were unhelpfully said to proceed by 'no mechanism' pathways [2]. The notion that a reaction could occur without a mechanism is clearly absurd yet, at the time, reactions were known which did not appear to follow any known mechanism, i.e. those involving homolytic or heterolytic bonding processes. Furthermore, although devising possible alternative mechanisms is seldom a challenge, identifying the 'correct' one may not be easy. A necessary preliminary is to have clear ideas about the nature of mechanism.

The abbreviated representations of mechanisms introduced by Ingold and Hughes and their colleagues are still widely used [3]; they are concise, easy to understand and describe adequately the mechanisms of many reactions. There have been major subsequent developments in our mechanistic understanding, however, including the discovery of the importance of orbital symmetry considerations, and an improving appreciation of synchroneity and concertedness in the making and breaking of bonds. Although the nomenclature recommended by Guthrie and Jencks to describe a mechanism allows greater precision than the earlier system and accommodates virtually all mechanistic subtleties [4], it is not simple and has not been universally adopted; their article, however, will help any organic chemist to clarify ideas about mechanism.

## 1.3   The investigation of mechanism and the scope of this book

The structure of this book reflects to a degree the developmental nature of the subject, i.e. the progression from how one characterises an organic chemical reaction to the formulation of a molecular mechanism. Some chapters focus on *methods* of investigating reactions (product analysis, kinetics, electrochemistry, computational chemistry and calorimetry); other chapters cover particular *types of reactions* (those involving intermediates, especially radicals, and catalysed reactions) or *special reaction conditions* (multiphase systems) and the methods that have been developed for their investigation. The chapters on particular types of reactions and reaction conditions have been included because of their importance in modern synthetic and manufacturing chemistry. Throughout, examples and case studies have been included (i) to strengthen links between methodologies and reaction types and (ii) to illustrate the synergism between different techniques employed to address mechanistic problems.

In order that all chapters be self-contained and comprehensible without detailed knowledge of the content of others, some topics (e.g. the steady state approximation and kinetic versus thermodynamic control) crop up in several places. The coverage is not the same in different chapters, however, and is developed in each according to the context and the perspectives of different authors.

### 1.3.1   *Product analysis, reaction intermediates and isotopic labelling*

First, a reaction has to be characterised, i.e. identities and yields of products must be determined, and these aspects are covered generally in Chapter 2. Once these are known, alternative possible 'paper mechanisms' can be devised. Each may take the form of a sequence of linear chemical equations, or a composite reaction scheme replete, perhaps, with 'curly arrows'. The next stage is to devise strategies for distinguishing between the alternatives with a view to identifying the 'correct' one or, more realistically, eliminating the incorrect ones; wherever possible, one seeks positive rather than negative evidence.

Scheme 1.1 includes alternative *concerted* and *stepwise* routes for the transformation of one molecule into another; for present purposes, we describe any transformation which is not stepwise as concerted. The formation of the intermediate from the reactant molecule in the stepwise route may be reversible, as shown, or irreversible.

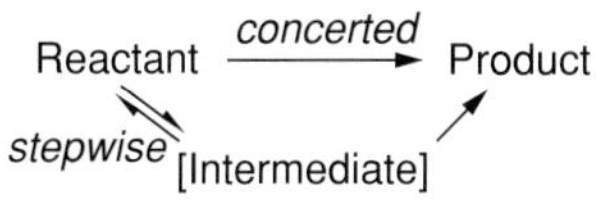

**Scheme 1.1** Alternative concerted and stepwise transformations of one molecule into another.

Positive identification or other unequivocal evidence for the involvement of the intermediate confirms the stepwise route (see Chapter 9); however, failure to detect the intermediate (i.e. negative evidence) is seldom solid proof of its non-involvement – the intermediate may be just too short lived to be detected by the methodology employed. On the other hand, positive evidence of a concerted route can be problematical and some ingenious experimental strategies have been developed to address this issue, see Chapters 9 and 11 [5].

If a comparison of alternative mechanisms indicates that products additional to those already detected should be formed by just one of the possible routes, e.g. the putative intermediate in Scheme 1.1 may be known (or reasonably expected) to give more than one product, then return to a more detailed analysis of the products is indicated. However, in this event, one needs to establish beforehand that a technique is available for the detection and (preferably) quantification of the additional product(s) being sought. In general, if alternative mechanisms can be distinguished by product analysis, it is essential that the analytical technique and experimental protocol to be used be validated and shown to be capable of giving unambiguous results.

### 1.3.1.1 Example: the acid-catalysed decomposition of nitrosohydroxylamines

Following the discovery of the wide-ranging physiological effects of nitric oxide [6], compounds which liberate it under mild conditions, including a range of nitrosohydroxylamine derivatives, were investigated. The hydroxydiazenium oxide (1) in Scheme 1.2 (which is closely related to the isomeric nitrosohydroxylamine, 2, by a proton transfer) and related *N*-nitroso-*N*,*O*-dialkylhydroxylamines undergo acid-catalysed decomposition in aqueous solution. However, preliminary mass spectrometric analysis indicated that the gaseous product was nitrous oxide, $N_2O$, rather than nitric oxide [7].

The alternative mechanisms shown to the right and to the left of 1 in Scheme 1.2 both account for the kinetics results and the initial product analysis, and both have literature analogies. However, isotope labelling experiments (the asterisk indicates the site of $^{17}O$ or $^{18}O$ incorporation) allowed a distinction between the two. In the path to the left, protonation of the hydroxyl of 1 with loss of labelled water as nucleofuge would lead to the evolution of unlabelled $N_2O$, and the residual adamantyl cation would be intercepted either by the liberated labelled water molecule or by an unlabelled solvent water molecule. In this event,

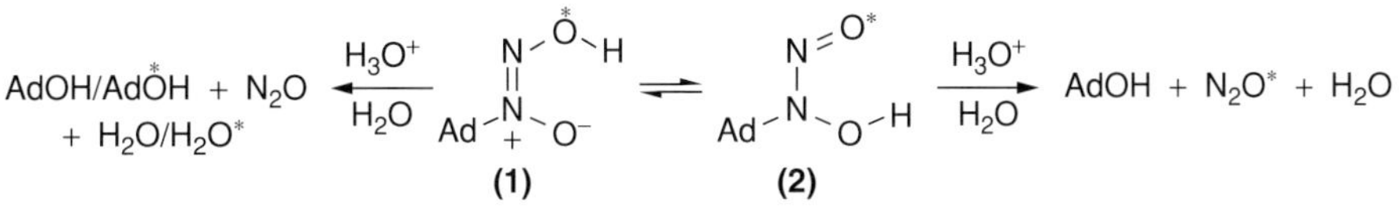

**Scheme 1.2** Alternative mechanisms for the acid-catalysed decomposition of $N^1$-adamantyl-$N^2$-hydroxydiazenium oxide (1) distinguished by an oxygen labelling study [7, 8].

the isolated adamantanol would contain some degree of incorporation of the labelled oxygen. In contrast, protonation of the hydroxyl of 2 with departure of water would lead to liberation of isotopically enriched $N_2O$ and no possibility of the label being incorporated into the adamantanol.

First, the reaction was carried out using starting material enriched with $^{17}O$ specifically as indicated in Scheme 1.2. The $^{17}O$ NMR spectrum of the isolated 2-adamantanol showed that it contained no more than natural abundance $^{17}O$; this represents negative evidence supportive of the mechanism to the right. The reaction was then carried out using starting material labelled with $^{18}O$ as indicated in Scheme 1.2, and the millimetre wavelength rotational spectrum of the nitrous oxide evolved unambiguously established that it contained the $^{18}O$ label [8]; this is positive evidence required by the mechanism to the right. These results rule out the path directly from 1, shown to the left in Scheme 1.2, and support other evidence that the reaction proceeds by protonation and fragmentation of 2 [7].

### 1.3.2 Mechanisms and rate laws

The nature of a transformation obtained by analytical methods (see Chapter 2) provides the basis for mechanistic speculation, and devising one or more possible pathways is seldom problematical. Consider the reaction of Equation 1.1 where AH is a catalytic acid, X contains an electrophilic residue and B is a base/nucleophile – it could be the acid-catalysed addition of a nucleophile to a carbonyl compound, for example:

$$B + X \xrightarrow[H_2O]{AH} Y \tag{1.1}$$

The reaction involving the three molecules in a single termolecular event is improbable, so the reaction via a pre-equilibrium is a reasonable initial hypothesis. Two possibilities are given in Scheme 1.3 where $K$ and $K'$ are equilibrium constants, and $k$ and $k'$ are *mechanistic* second-order rate constants of *elementary* steps (see below).

$$AH + X \overset{K}{\rightleftharpoons} A^- + HX^+ \qquad B + X \overset{K'}{\rightleftharpoons} B\text{-}X$$

$$B + HX^+ \overset{k}{\longrightarrow} HY^+ \qquad B\text{-}X + AH \overset{k'}{\longrightarrow} HY^+ + A^-$$

$$HY^+ + A^- \rightleftharpoons Y + AH \qquad HY^+ + A^- \rightleftharpoons Y + AH$$

**Scheme 1.3** Possible mechanisms involving pre-equilibria for the reaction of Equation 1.1.

The rate of the reaction by the mechanism on the left is given by

$$\text{rate} = k[B][HX^+],$$

but this is not a legitimate rate law as one concentration term is of a proposed intermediate, i.e. not of a reactant. However, we can substitute for $[HX^+]$ using the expression describing the prior equilibrium to give

$$\text{rate} = k\,[B]\,[X]\,[AH]\,K/[A^-],$$

so from the definition of the dissociation constant of AH,

$$K_a = [H_3O^+]\,[A^-]/[AH],$$

we may write

$$\text{rate} = k\,[\text{B}]\,[\text{X}]\,[\text{H}_3\text{O}^+]\,K/K_a$$

or

$$\text{rate} = k_{\text{exp}}\,[\text{B}]\,[\text{X}]\,[\text{H}_3\text{O}^+], \tag{1.2}$$

where $k_{\text{exp}}$ is the experimental third-order rate constant corresponding to this mechanism *if the reaction turns out to be specific acid catalysed*, and is related to the mechanistic parameters by $k_{\text{exp}} = kK/K_a$.

In a similar manner, it is easily shown that the mechanism on the right of Scheme 1.3 leads to the rate law of Equation 1.3,

$$\text{rate} = k'_{\text{exp}}\,[\text{B}]\,[\text{X}]\,[\text{AH}], \tag{1.3}$$

where $k'_{\text{exp}}$ is the experimental third-order rate constant corresponding to this mechanism *if the reaction turns out to be general acid catalysed*, and is related to the mechanistic parameters by $k'_{\text{exp}} = k'K'$.

We see here that the mechanism with a pre-equilibrium proton transfer leads to a specific acid catalysis rate law whereas that with a rate-determining proton transfer leads to general acid catalysis. It follows that, according to which catalytic rate law is observed, one of these two mechanisms may be excluded from further consideration. Occasionally, however, different mechanisms lead to the same rate law and are described as *kinetically equivalent* (see Chapters 4 and 11) and cannot be distinguished quite so easily.

In examples such as the above, the rate law establishes the composition of the activated complex (transition structure), but not its structure, i.e. not the atom connectivity, and provides no information about the sequence of events leading to its formation. Thus, the rate law of Equation 1.2 (if observed) for the reaction of Equation 1.1 tells us that the activated complex comprises the atoms of one molecule each of B and X, plus a proton and an indeterminate number of solvent (water) molecules, but it says nothing about how the atoms are bonded together. For example, if B and X *both* have basic and electrophilic sites, another mechanistic possibility includes a pre-equilibrium proton transfer from AH to B followed by the reaction between $\text{HB}^+$ and X, and this also leads to the rate law of Equation 1.2. Observation of this rate law, therefore, allows transition structures in which the proton is bonded to a basic site in either B or X, and distinguishing between the kinetically equivalent mechanisms requires evidence additional to the rate law.

We have seen above that the rate law of a reaction is a consequence of the mechanism, so the protocol is that (i) we propose a mechanism, (ii) deduce the rate law required by the mechanism and (iii) check experimentally whether it is observed. If the experimental result is *not* in agreement with the prediction, the mechanism is defective and needs either refinement or rejection. Clearly, the ability to deduce the rate law from a proposed mechanism is a necessary skill for any investigator of reaction mechanisms (see Chapter 4).

It is self-evident that a unimolecular mechanism, e.g. an isomerisation, will lead to a first-order rate law, and a bimolecular mechanism, e.g. a concerted dimerisation, to a second-order rate law. However, whilst it is invariably true that a simple mechanism leads to a simple rate law, the converse is not true – a simple rate law does not necessarily implicate a simple mechanism.

### 1.3.3   Computational chemistry

Computational chemistry has been especially helpful in distinguishing concerted and step-wise alternatives. Molecular properties are now calculable to a high degree of reliability (see Chapter 7), even for compounds too unstable to allow direct measurements. Consequently, putative intermediates in hypothetical reaction schemes can be scrutinised and their via-bility investigated. An intermediate corresponds to a local minimum in a potential energy hypersurface and should be contrasted with a *transition structure* which corresponds to a saddle point, i.e. a maximum in one dimension (the *reaction coordinate*, see Chapter 7) and minima in others. In principle, any intermediate in a proposed mechanism may be investi-gated theoretically. Initially, this will give the energy and structure of the molecular species in the gas phase; if the postulated species does not correspond to a stable bonded structure, i.e. an energy minimum, the proposed mechanism is not viable. Additionally, there are pro-tocols which allow consideration of the effect of the environment (e.g. a counter-ion in the case of charged species, proximate solvent molecules or the effect of the medium considered as a continuum). The reaction may also be converted computationally from the molecular to the molar scale, and entropy included.

Since the availability of inexpensive high-speed computers and spectacular advances in computational chemistry methodology, mechanistic subtleties may now be investigated which are not amenable to experimental scrutiny. Of course, validation of a particular computational technique is essential and a comparison with a sound *relevant* experimental result is one method (see Chapter 7).

### 1.3.3.1   Example: the acid- and base-catalysed decomposition of nitramide

The base-catalysed decomposition of nitramide (**3** in Scheme 1.4) is of special historical importance as it was the reaction used to establish the Brønsted catalysis law. The reaction has been studied over many years and considerable evidence indicates that the decomposition

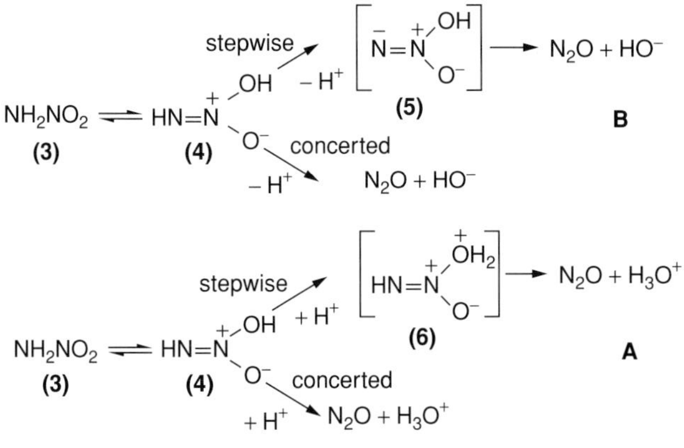

**Scheme 1.4**   Decomposition of nitramide (**3**) in aqueous solution via its *aci*-form (**4**): **B**, catalysed by bases; **A**, catalysed by acids [9, 10].

proceeds via the *aci*-form (**4**) [9], and a stepwise process via **5** was generally accepted (upper path in Scheme 1.4B). However, recent computational work has established that anion **5**, formed by deprotonation from the nitrogen of **4** by the catalytic base, does not correspond to an energy minimum [10]. This base-catalysed reaction, therefore, occurs by an enforced concerted mechanism, i.e. departure of hydroxide from **4** is concerted with the proton abstraction (lower path in Scheme 1.4B).

Correspondingly, in the acid-catalysed decomposition of nitramide (Scheme 1.4A), protonation of **4** on the hydroxyl also leads to an ion (**6**) which spontaneously fragments, i.e. **6** does not correspond to an energy minimum, so the upper stepwise path in Scheme 1.4A is not viable. The acid-catalysed decomposition of **3** via **4**, therefore, also involves concerted proton transfer and fragmentation (the lower path in Scheme 1.4A).

In the above example, the computational investigation followed experimental work. Computational chemistry may be exploited to assist in designing experimental investigations, and occasionally leads to predictions which may be tested experimentally. Arenediazonium ions, $ArN_2^+$, are well known, and dediazoniation reactions are important in preparative aromatic chemistry [11]. In contrast, alkanediazonium ions, $RN_2^+$, are known only as unstable reactive intermediates, e.g. in deamination of primary amines induced by nitrous acid. Recently, alkaneoxodiazonium ions, $(RN_2O)^+$, have been implicated as intermediates in acid-catalysed decompositions of *N*-nitroso-dialkylhydroxylamines [7], which led to the interesting possibility that the arene analogues, $(ArN_2O)^+$, may also be viable species. This was explored computationally and, indeed, $(PhN_2O)^+$ and $PhON_2^+$ correspond to energy minima [12], so salts (perhaps of substituted analogues) may be preparable. Here, as in general, selection of the most appropriate computational method for the particular application is critical, and the relative strengths and weaknesses of various packages are discussed in Chapter 7.

For many years, transition state theory [13] has been the foundation of most studies of mechanism and reactivity in organic chemistry, and collision theory the preserve of physical chemists dealing with simple small molecules in the gas phase. Following the ever-increasing availability of inexpensive powerful computers, the development of molecular dynamics has allowed new insight into mechanisms of gas phase reactions of organic molecules. This is presently an expanding area and its anticipated application to reactions in solution will surely lead to revision of many cherished notions. Also, the implications of ongoing developments at the interface between high-level theory and femtosecond experimental gas phase studies remain to be explored for reactivity studies in solution.

### 1.3.4   *Kinetics in homogeneous solution*

Once a transformation has been characterised, rate laws can be investigated. Sometimes, the kinetic study is simply to obtain rate data for technological reasons, and empirical rate laws may be sufficient. Fundamental knowledge of the reaction mechanism, however, generally offers better prospects for process optimisation. A simple kinetics study seldom allows identification of a single mechanism because different mechanisms may lead to the same rate law (see *kinetic equivalence* above and in Chapters 4 and 11). A mechanistic possibility may be rejected, however, if its predicted rate law is not in accord with what is observed experimentally.

Discussions of *results* of rate studies permeate this book because kinetics investigations are the single most important group of techniques in mechanistic determinations. However, kinetics *results* have to be derived from *measurements* which are the outcome of *experiments*. Chapter 3 on conventional kinetics methods includes techniques which are generally applicable, and also current procedures for extracting rate constants (and, in some cases, equilibrium constants) from raw experimental data.

In principle, any property of a reacting system which changes as the reaction proceeds may be monitored in order to accumulate the experimental data which lead to determination of the various kinetics parameters (rate law, rate constants, kinetic isotope effects, etc.). In practice, some methods are much more widely used than others, and UV–vis spectrophotometric techniques are amongst these. Often, it is sufficient simply to record continuously the absorbance at a fixed wavelength of a reaction mixture in a thermostatted cuvette; the required instrumentation is inexpensive and only a basic level of experimental skill is required. In contrast, instrumentation required to study very fast reactions spectrophotometrically is demanding both of resources and experimental skill, and likely to remain the preserve of relatively few dedicated expert users.

A major recent development is the increasing exploitation of time-resolved IR spectrophotometry for kinetics which has a major advantage over UV methods – in addition to kinetic data, it also provides readily interpretable IR spectroscopic information which allows some degree of structural characterisation of reactive intermediates.

### 1.3.4.1 Example: the kinetics of the capture of pyridyl ketenes by n-butylamine

Laser flash photolysis of 3-pyridyl diazomethyl ketone in acetonitrile containing *n*-butylamine generates 3-pyridyl ketene as a reactive intermediate, as shown in Scheme 1.5. Subsequent sequential IR measurements at 2125 cm$^{-1}$ on the microsecond time scale allowed determination of the pseudo-first-order rate constant for its capture by the amine. From the rate law for the capture of the ketene,

$$\text{rate} = k_q[\text{ketene}][n\text{-butylamine}]$$

$$\text{or rate} = k_{obs}[\text{ketene}] \text{ where } k_{obs} = k_q[n\text{-butylamine}],$$

i.e. $k_{obs}$ is the pseudo-first-order rate constant and $k_q$ is the second-order rate constant, we see that the plot of $k_{obs}$ against [$n$-butylamine] gives $k_q$. Results for 2-, 3- and 4-pyridyl ketenes are shown in Fig. 1.1 [14].

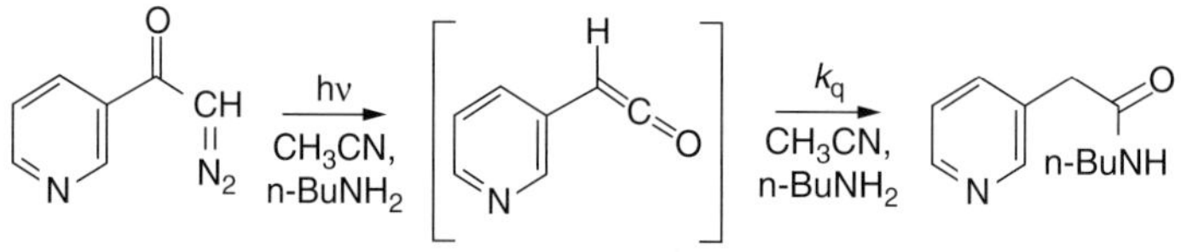

**Scheme 1.5** 3-Pyridyl ketene generated by laser flash photolysis in acetonitrile, and trapping by *n*-butylamine investigated by time-resolved IR measurements [14].

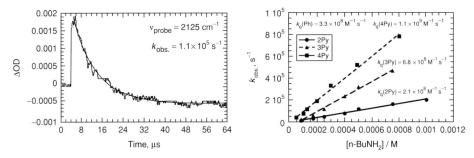

**Fig. 1.1** (a) Absorbance versus time measurements taken after the laser flash which generates 3-pyridyl ketene from 3-pyridyl diazomethyl ketone; (b) second-order plots for capture of 2-, 3- and 4-pyridyl ketene by *n*-butylamine. Data taken from reference [14].

### 1.3.5 Kinetics in multiphase systems

Multiphase reactions have long been used (though not always recognised as such) in organic chemistry, e.g. the acylation of amines and phenols (the Schotten–Baumann reaction), and a summary of synthetically useful examples has been collected by Atherton [15]. In particular, phase transfer catalysis (PTC) is now a familiar technique [16a], which is included in many undergraduate organic chemistry practical courses [16b], and may be used to facilitate (for example) substitutions, oxidations and dihalocyclopropane formation by dihalocarbene addition to alkenes (e.g. Scheme 1.6) [16c]. This technique is effective and reasonably well understood, but it involves reactants distributed between just two liquid phases. PTC and more complicated multiphase reactions are already employed in chemical manufacturing and other processes such as heterogeneous photocatalysis; they may involve reactions between gases, liquid reactants, solutions of reactants and solids (sometimes sparingly soluble). In even the simplest of these systems, an appreciation of the partitioning of materials between phases, and of the transport dynamics of materials within liquids and across phase boundaries, is required.

Recovery of metals such as copper, the operation of batteries (cells) in portable electronic equipment, the reprocessing of fission products in the nuclear power industry and a very wide range of gas-phase processes catalysed by condensed phase materials are applied chemical processes, other than PTC, in which chemical reactions are coupled to mass transport within phases, or across phase boundaries. Their mechanistic investigation requires special techniques, instrumentation and skills covered here in Chapter 5, but not usually encountered in undergraduate chemistry degrees. Electrochemistry generally involves reactions at phase boundaries, so there are connections here between Chapter 5 (Reaction kinetics in multiphase systems) and Chapter 6 (Electrochemical methods of investigating reaction mechanisms).

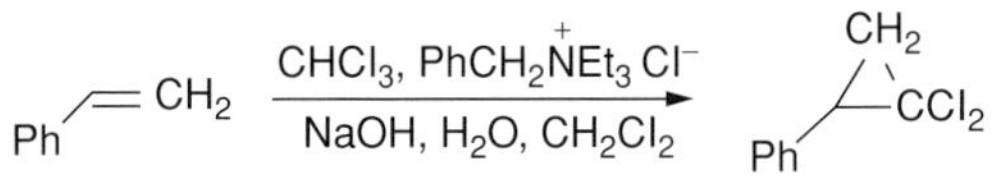

**Scheme 1.6** Addition of dichlorocarbene to styrene by phase transfer catalysis [16c].

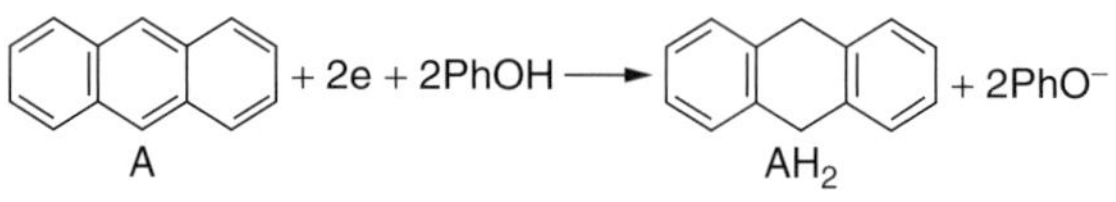

**Scheme 1.7**   Electrochemical reduction of anthracene with phenol as proton donor.

## 1.3.6   Electrochemical and calorimetric methods

Electrochemistry may be exploited for the analysis of extremely low concentrations of electrochemically active species, for laboratory scale preparations and in manufacturing processes on a massive scale [17]. In the context of the investigation of reaction mechanisms in solution, the present focus is on electroanalytical techniques; these are included in Chapter 6, but there are obvious connections with Chapter 5 (see above) and Chapter 10 on free radicals.

The basic methods of electrochemistry (electrolysis and the use of anodes and cathodes as reducing and oxidising agents, respectively) have been generally accessible since the earliest days of chemistry; however, following the development of increasingly sophisticated experimental techniques, electrochemical methodologies became the preserve of specialist investigators. More recently, however, increased interest in electron transfer processes (e.g. in reactions involving organometallic compounds [18] and in biological processes [19]) and increasing use of synthetic reactions involving radicals and radical ions [20] have led to a wider use of electrochemical methods of investigation of mechanism. And whilst such methods are limited to processes involving electron transfer (so are not as widely applicable as spectroscopy, for example), when applicable, they can provide information in considerable detail on intermediates implicated, e.g. in the electrochemical reduction of anthracene in the presence of phenol, Scheme 1.7 and Chapter 6. In this reduction of anthracene (A), the first-formed intermediate is the anthracene radical anion ($A^{\bullet-}$), which undergoes rate-determining protonation by phenol to give the neutral radical, $AH^{\bullet}$; this is reduced by further $A^{\bullet-}$ to give the anion ($AH^{-}$), protonation of which then yields 9,10-dihydroanthracene ($AH_2$).

The electrohydrodimerisation of acrylonitrile to give adiponitrile (a one-electron process at high substrate concentrations, Scheme 1.8A and Chapter 6) is an example of how an industrially important electrosynthetic process has been investigated following recent instrumental developments, viz. the application of ultramicroelectrodes at low-voltage sweep rates. Use of conventional electrodes would have required substrate concentrations in the mM range but, under these conditions, acrylonitrile undergoes a different reaction – a two-electron electrochemical reduction of the alkene residue (Scheme 1.8B). The switchover between the two reactions occurs at about 1 mol dm$^{-3}$ substrate concentration.

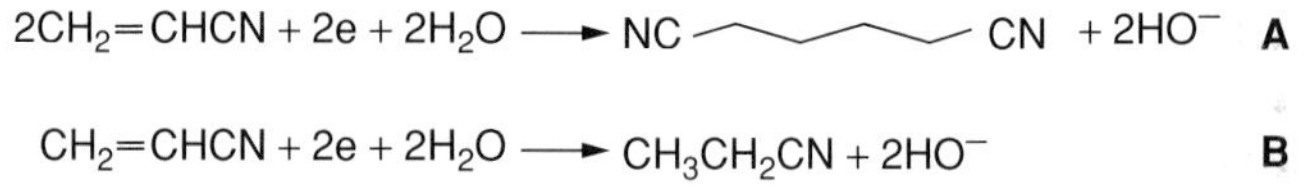

**Scheme 1.8**   Different electrochemical reductions of acrylonitrile according to reaction conditions (**A**, high concentration; **B**, low concentration).

Calorimetry, one of the experimental aspects of thermodynamics, has long been an area of physical chemistry practised largely by specialists and used principally for the acquisition of thermochemical data. Thermodynamics in general, of course, dictates the maximum conversion that may be obtained in a chemical reaction and, in a reversible process, is quantified by the equilibrium constant; if this is very large, or if the reaction is in practice irreversible, the maximum conversion is controlled simply by the initial amount of the limiting reagent. Calorimetric data more specifically may relate to individual compounds, e.g. enthalpies of formation, or they may relate to chemical reactions as most chemical and physical processes are accompanied by heat effects. Thermochemical characterisation of both compounds and chemical reactions is vitally important in the context of process safety to avoid the potentially disastrous consequences of a 'thermal runaway' when insufficient cooling is provided for an exothermic reaction; many industrial laboratories have calorimetric equipment for such purposes.

Several methods have been developed over the years for the thermochemical characterisation of compounds and reactions, and the assessment of thermal safety, e.g. differential scanning calorimetry (DSC) and differential thermal analysis (DTA), as well as reaction calorimetry. Of these, reaction calorimetry is the most directly applicable to reaction characterisation and, as the heat-flow rate during a chemical reaction is proportional to the rate of conversion, it represents a differential kinetic analysis technique. Consequently, calorimetry is uniquely able to provide kinetics as well as thermodynamics information to be exploited in mechanism studies as well as process development and optimisation [21].

More significantly, when calorimetry is combined with an integral kinetic analysis method, e.g. a spectroscopic technique, we have an expanded and extremely sophisticated method for the characterisation of chemical reactions. And when the calorimetric method is linked to FTIR spectroscopy (in particular, *attenuated total reflectance IR spectroscopy*, IR-ATR), structural as well as kinetic and thermodynamic information becomes available for the investigation of organic reactions. We devote much of Chapter 8 to this new development, and the discussion will focus on reaction calorimeters of a size able to mimic production-scale reactors of the corresponding industrial processes.

Reaction characterisation by calorimetry generally involves construction of a model complete with kinetic and thermodynamic parameters (e.g. rate constants and reaction enthalpies) for the steps which together comprise the overall process. Experimental calorimetric measurements are then compared with those simulated on the basis of the reaction model and particular values for the various parameters. The measurements could be of heat evolution measured as a function of time for the reaction carried out isothermally under specified conditions. Congruence between the experimental measurements and simulated values is taken as the support for the model and the reliability of the parameters, which may then be used for the design of a manufacturing process, for example. A *reaction model* in this sense should not be confused with a *mechanism* in the sense used by most organic chemists–they are different but equally valid descriptions of the reaction. The model is empirical and comprises a set of chemical equations and associated kinetic and thermodynamic parameters. The mechanism comprises a description of how at the molecular level reactants become products. Whilst there is no necessary connection between a useful model and the mechanism (known or otherwise), the application of sound mechanistic principles is likely to provide the most effective route to a good model.

$$Ac_2O + H_2O \xrightarrow[k]{H_3O^+} 2AcOH; \Delta H$$

**Scheme 1.9**   Acid-catalysed hydrolysis of acetic anhydride.

The kinetics of the hydrolysis of acetic anhydride in dilute hydrochloric acid, Scheme 1.9, may be described by a single pseudo-first-order rate constant, $k$, and the investigation by calorimetry combined with IR spectroscopy, as we shall see in Chapter 8, provides a clear distinction between the heat change due to mixing of the acetic anhydride into the aqueous solution and that due to the subsequent hydrolysis. This model of the reaction is sufficient for devising a safe and efficient large-scale process. We know from other evidence, of course, that the reaction at the molecular level is not a single-step process – it involves tetrahedral intermediates – but this does not detract from the validity or usefulness of the model for technical purposes.

$$Ar\text{-}NO_2 + 3H_2 \xrightarrow[\text{solvent}]{\text{catalyst}} Ar\text{-}NH_2 + 2H_2O$$

**Scheme 1.10**   Catalytic hydrogenation of nitro-arenes to give arylamines.

The catalytic hydrogenation of nitro-arenes (Scheme 1.10) to give arylamines is another important industrial process. A calorimetric investigation of the hydrogenation of nitrobenzene to give aniline was coupled with *simultaneous* reaction monitoring by IR and hydrogen uptake measurements, see Chapter 8. The same kinetics description was obtained by all three monitoring methods, which provides mutual support for their validity, and the enthalpy of reaction was obtained from the integrated heat-flow measurements. In accord with a one-step kinetic model, only minor traces of phenylhydroxylamine (a potential intermediate in a stepwise reaction) were detected in the IR spectrum. This multi-instrumental investigation was aided by a sound appreciation of gas–liquid phase transfer phenomena.

### 1.3.7   *Reactions involving radical intermediates*

Unimolecular heterolyses of neutral molecules in solution are viable only if the resultant ions, and the associated developing dipolarity in the transition structure, are stabilised by solvation. Such reactions are rare in the gas phase and nonpolar solvents because, under such conditions, there are no mechanisms for stabilising high-energy polar states; so, if a reaction occurs, it usually proceeds by an alternative lower energy non-polar route. This could be either a concerted mechanism, bypassing high-energy polar intermediates, or a path via homolysis and uncharged radical intermediates. In contrast to concerted mechanisms, whose elucidation required the discovery of the importance of orbital symmetry considerations, radical reactions in solution and the gas phase were amongst the earliest to be investigated. Indeed, studies of radical reactions predate the classic work by Ingold's school on polar mechanisms of substitution and elimination, and have continued to the present. Interestingly, most investigators of mechanism have either focussed wholly on radical reactions, or have avoided them altogether. Nevertheless, one notes that the so-called azide clock technique for measuring rate constants for capture of carbocations by nucleophiles (see Chapter 2) [22] follows closely (in methodology if not in time) the radical clock protocol for estimating rate constants of radical reactions (see Chapter 10) [23].

Since the heroic early mechanistic investigations, there have been two developments of major significance in radical chemistry. The first was the advent of *electron spin resonance* (ESR) spectroscopy (and the associated technique of *chemically induced dynamic nuclear polarisation*, CIDNP) [24], which provided structural as well as kinetic information; the second is the more recent development of a wide range of synthetically useful radical reactions [20]. Another recent development, the combination of the pulse radiolysis and laser-flash photolysis techniques, is enormously powerful for the study of radicals but beyond the scope of this book.

Radical reactions have been used for many years on an industrial scale, especially in the polymer industry in continuous and batch processes, and their often complex mechanisms have been rigorously investigated and described elsewhere [25]. In Chapter 10, we include discussion of some 'living' mechanisms established in the 1990s; these have increased the number and complexity of macromolecular structures produced by radical reactions. Another significant recent development, fuelled by recent advances in the area of radical mediators, has been the exploitation of radical reactions in a wide range of functional group transformations, and it is the investigation of these types of reaction which are the main thrust of Chapter 10 here. This chapter, separate from our consideration of reaction intermediates more generally in Chapter 9, is warranted by the distinctive strategies involved and the cohesion of the subject matter.

## 1.3.8   Catalysed reactions

The systematic investigation of catalysis in homogeneous solution over the past half century and more has been driven by the desire to understand how enzymes bring about massive rate enhancements in biological processes, and aspects of enzyme catalysis remain amongst the most regularly (and often repetitively) reviewed areas of chemistry. It was appreciated at an early stage that a relatively small number of molecular processes are involved and, amongst these, those involving proton transfer have been the most intensely and rigorously investigated. Much more recently, catalysis by organometallic species has become extremely important to synthetic organic chemists, previously having long been a specialist interest of organometallic chemists. The reasons are obvious – organometallic catalysis has made possible new transformations, especially in C—C bond-forming reactions, of great value in organic synthesis. In this book, we have not attempted descriptive expositions on the *nature* of catalysis; our focus in Chapter 11 is on illustrating the methods of investigating catalysis by small ions and molecules, and by enzymes; the methodologies and techniques for investigating organometallic catalysis are illustrated by several case studies in Chapter 12. Reaction intermediates are necessarily involved in catalysis, so there are connections with Chapter 9, and unravelling problems associated with *kinetic equivalence* involves material covered in Chapter 4.

The distinction between stepwise and concerted processes is clear in principle even though establishing whether an overall transformation is one or the other may be problematical [5]. Once it has been established that two aspects of an overall transformation are concerted, the

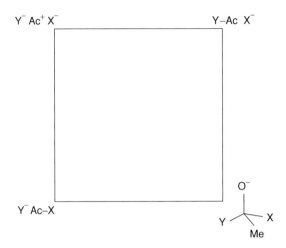

**Fig. 1.2**  Template for a two-dimensional reaction map for the overall transformation $Y^- + Ac-X \rightarrow$ Y–Ac $+ X^-$.

degree of coupling between them remains to be considered. Such matters are of paramount importance in catalytic processes, and included in Chapter 11. The issue may be introduced in a simpler context, e.g. a substitution occurring by an $S_N2$ mechanism ($A_N D_N$), Equation 1.4:

$$Y^- + R-X \rightarrow Y-R + X^- \tag{1.4}$$

It is the coupling of the bond making between $Y^-$ and C with the unbonding of $X^-$ from C in the transition structure which makes the process possible; and the poorer the nucleofuge, the more necessary the coupling. Discussion of these matters was enormously facilitated by the introduction of two-dimensional reaction maps by More O'Ferrall [26]. A simple case for substitution at unsaturated carbon, i.e. acetyl transfer from one Lewis base to another, is shown in Fig. 1.2.

In this, reactants and products are represented by the bottom left and top right of the diagram, respectively; the top left corresponds to a configuration in which the Ac—X bond has broken heterolytically with no associated bonding by $Y^-$, and the bottom right corresponds to full bonding by $Y^-$ with no C—X bond breaking. The diagram allows descriptions of mechanisms by either of two stepwise possibilities (associative via a tetrahedral intermediate, or dissociative via an acylium cation intermediate), or by concerted alternatives. A diagonal path across the centre of the diagram corresponds to a *synchronous* concerted process whereas paths strongly curved towards the top left or bottom right represent *asynchronous* concerted paths bypassing the respective intermediates in conceivable stepwise processes. Of course, which path best describes the real reaction depends upon the outcome of theoretical and experimental investigations. More elaborate versions of these reaction maps, frequently called Albery–More O'Ferrall–Jencks diagrams, are employed in Chapter 11.

Whereas many reactions catalysed by organometallic compounds are characteristically different from those catalysed by Brønsted acids and bases, and to a degree their investigators have developed distinctive terminologies, the strategies employed in mechanistic studies are

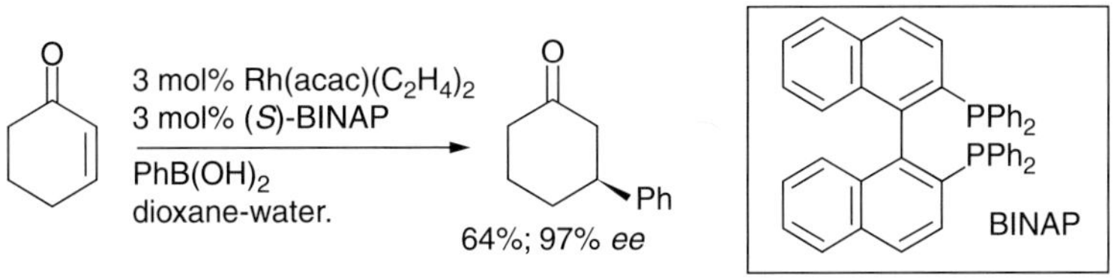

**Scheme 1.11** Enantioselective Rh-catalysed phenylation of cyclohexenone [27].

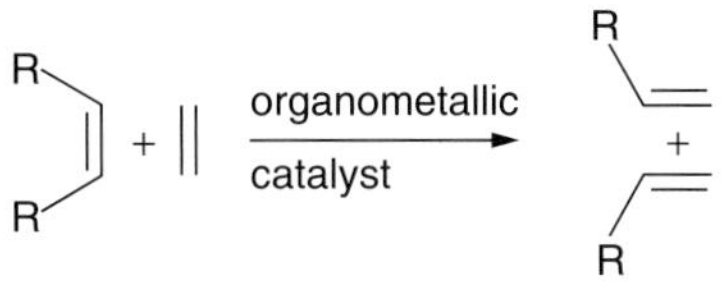

**Scheme 1.12** Metathesis of alkenes with organometallic catalysis.

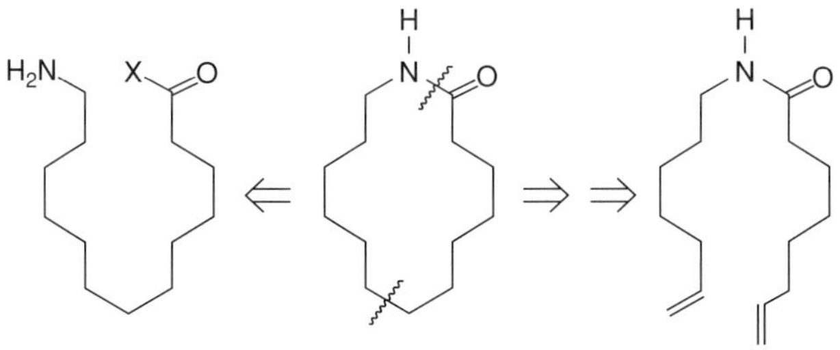

**Scheme 1.13** Alternative disconnections of a macrocyclic amide.

closely similar to those of more conventional physical organic chemistry (see Chapter 12). The asymmetric phenylation reaction of cyclohexenone in Scheme 1.11 clearly relates to the familiar Michael reaction, but the mechanism, whose elucidation is described in Chapter 12, is very different [27]. The most important feature of this reaction, of course, is the asymmetric induction which is brought about by use of a chiral ligand for the catalytically active metal centre, a recurrent and very important theme in catalysis by organometallics.

Transition-metal-catalysed metathesis of alkenes (Scheme 1.12) is more removed from conventional organic chemistry than the above Michael-like reaction, and its investigation has been a major challenge (see Chapter 12). The novelty and enormous value of these reactions have been recognised by the award of the 2005 Nobel Prize for Chemistry to Chauvin, Schrock and Grubbs for their seminal investigations in this area [28].

Metathesis reactions may be intramolecular and *ring-closing diene metathesis* (RCM, implicated in Scheme 1.13, see Chapter 12) allows disconnections in retro-synthetic analysis otherwise of little use. The normal disconnection of the macrocyclic amide in Scheme 1.13 would be at the amide but, because of the ready reduction of alkenes to alkanes, the alternative disconnection now becomes a viable option. And since *any* of the C—C linkages could be formed by RCM, such a disconnection allows far greater synthetic flexibility than the conventional disconnection at the functional group.

## 1.4 Summary

There are different aspects to the characterisation of an organic chemical reaction. It could involve determination of thermodynamic parameters such as the equilibrium constant at a specified temperature (if the transformation is reversible), and the enthalpy and entropy of reaction; still on the same macroscopic scale, it could also involve measurements of rate constants and activation parameters. Alternatively, it could involve learning how reactants are transformed into products at the molecular level; although most of the evidence about the molecular reaction mechanism is gleaned from experimental measurements on the macroscopic scale, computational chemistry is becoming increasingly important. As the foregoing overview indicates, we shall cover macroscopic and molecular descriptions of organic reactions in this book, the relationships between the two and how they are investigated.

## Bibliography

Anslyn, E.V. and Dougherty, D.A. (2004) *Modern Physical Organic Chemistry*. University Science Books, Mill Valley, CA.

Carpenter, B.K. (1984) *Determination of Organic Reaction Mechanisms*. Wiley-Interscience, New York.

Espenson, J.H. (1995) *Chemical Kinetics and Reaction Mechanisms* (2nd edn). McGraw-Hill, New York.

Isaacs, N.S. (1995) *Physical Organic Chemistry* (2nd edn).Longman, Harlow.

Lowry, T.H. and Richardson, K.S. (1987) *Mechanism and Theory in Organic Chemistry* (3rd edn). Harper Collins, New York.

Maskill, H. (1985) *The Physical Basis of Organic Chemistry*. Oxford University Press, Oxford.

Moore, J.W. and Pearson, R.G. (1981) *Kinetics and Mechanism* (3rd edn). Wiley, New York.

## References

1. Bernasconi, C.F. (Ed.) (1986) *Investigation of Rates and Mechanisms of Reactions, Part 1, General Considerations and Reactions at Conventional Rates* (4th edn). Wiley-Interscience, New York.
2. Rhoads, S.-J. (1963) Chapter 11 in: P. de Mayo (Ed.) *Molecular Rearrangements, Part 1*. Wiley-Interscience, New York and London.
3. Ingold, C.K. (1969) *Structure and Mechanism in Organic Chemistry* (2nd edn). Cornell University Press, Ithaca, NY.
4. Guthrie, R.D. and Jencks, W.P. (1989) *Accounts of Chemical Research*, **22**, 343.
5. Williams, A. (2000) *Concerted Organic and Bio-organic Mechanisms*. CRC Press, Boca Raton, FL; Williams, A. (2003) *Free-Energy Relationships in Organic and Bio-organic Chemistry*. Royal Society of Chemistry, Cambridge.
6. Butler, A.R. (1990) *Chemistry in Britain*, **26**, 419; Butler, A.R. and Williams, D.L.H. (1993) *Chemical Society Reviews*, **22**, 233.
7. Maskill, H., Menneer, I.D. and Smith, D.I. (1995) *Journal of the Chemical Society, Chemical Communications*, 1855; Bhat, J.I., Clegg, W., Elsegood, M.R.J., Maskill, H., Menneer, I.D. and Miatt, P.C. (2000) *Journal of the Chemical Society, Perkin Transactions 2*, 1435.
8. Haider, J., Hill, M.N.S., Menneer, I.D., Maskill, H. and Smith, J.G. (1997) *Journal of the Chemical Society, Chemical Communications*, 1571.
9. Arrowsmith, C.H., Awwal, A., Euser, B.A., Kresge, A.J., Lau, P.P.T., Onwood, D.P., Tang, Y.C. and Young, E.C. (1991) *Journal of the American Chemical Society*, **113**, 172.

10. Eckert-Maksić, M., Maskill, H. and Zrinski, I. (2001) *Journal of the Chemical Society, Perkin Transactions 2*, 2147.
11. Zollinger, H. (1994) *Diazo Chemistry I, Aromatic and Heteroaromatic Compounds*. VCH, Weinheim, Germany.
12. Eckert-Maksić, M., Glasovac, Z., Maskill, H. and Zrinski, I. (2003) *Journal of Physical Organic Chemistry*, **16**, 491.
13. Kreevoy, M.M. and Truhlar, D.G. (1986) Chapter 1 in: C.F. Bernasconi (Ed.) *Investigation of Rates and Mechanisms of Reactions, Part 1, General Considerations and Reactions at Conventional Rates* (4th edn). Wiley-Interscience, New York. See also Albery, W.J. (1993) *Advances in Physical Organic Chemistry*, **28**, 139.
14. Acton, A.W., Allen, A.D., Antunes, L.M., Fedorov, A.V., Najafian, K., Tidwell, T.T. and Wagner, B.D. (2002) *Journal of the American Chemical Society*, **124**, 13790.
15. Atherton, J.H. (1999) *Process Development: Physicochemical Concepts*. Oxford Science Publications, Oxford University Press, Oxford, Chapter 8.
16. (a) Dehmlow, E.V. (1977) *Angewandte Chemie*, **89**, 521; (b) Mohrig, J.R., Hammond, C.N., Morrill, T.C. and Neckers, D.C. (1998) *Experimental Organic Chemistry*. Freeman, New York; (c) *Organic Syntheses*, Coll. Vol. 7, 1990, p. 12.
17. Lund, H. and Hammerich, O. (Eds) (2001) *Organic Electrochemistry* (4th edn). Dekker, New York.
18. Kochi, J.K. (1994) *Advances in Physical Organic Chemistry*, **29**, 185; Evans, D.H. (1990) *Chemical Reviews*, **90**, 739.
19. Gray, H.B. and Winkler, J.R. (1996) *Annual Review of Biochemistry*, **65**, 537.
20. (a) Renaud, P. and Sibi, M.P. (Eds) (2001) *Radicals in Organic Synthesis* (vols 1 and 2). Wiley-VCH, Weinheim, Germany; (b) Zard, S.Z. (2003) *Radical Reactions in Organic Synthesis*. Oxford University Press, Oxford.
21. Zogg, A., Stoessel, F., Fischer, U. and Hungerbühler, K. (2004) *Thermochimica Acta*, **419**, 1.
22. Richard, J.P., Rothenberg, M.E. and Jencks, W.P. (1984) *Journal of the American Chemical Society*, **106**, 1361.
23. Griller, D. and Ingold, K.U. (1980) *Accounts of Chemical Research*, **13**, 317; Newcomb, M. (2001) Kinetics of radical reactions: radical clocks. In: P. Renaud and M.P. Sibi (Eds) *Radicals in Organic Synthesis* (vols 1 and 2). Wiley-VCH, Weinheim, Germany, p. 317.
24. Gerson, F. and Huber, W. (2003) *Electron Spin Resonance Spectroscopy of Organic Radicals*. Wiley-VCH, Weinheim, Germany.
25. Moad, G. and Solomon, D.H. (1995) *The Chemistry of Free Radical Polymerization*. Pergamon, Bath.
26. More O'Ferrall, R.A. (1970) *Journal of the Chemical Society B*, 274; Jencks, W.P. (1972) *Chemical Reviews*, **72**, 705.
27. Takaya, Y., Ogasawara, M., Hayashi, T., Sakai, M. and Miyaura, N. (1998) *Journal of the American Chemical Society*, **120**, 5579.
28. http://nobelprize.org/chemistry/laureates/2005/index.html.

# Chapter 2

# Investigation of Reaction Mechanisms by Product Studies

## T. W. Bentley

## 2.1 Introduction and overview – why study organic reaction mechanisms?

Reaction mechanisms have been an integral part of the teaching of organic chemistry and in the planning of routes for organic syntheses for about 50 years [1]. Prior to the 1962 *Annual Report of the Chemical Society*, kinetics and reaction mechanisms were presented in a chapter entitled *Theoretical organic chemistry*, and a separate section on physical organic chemistry did not appear in *Chemical Abstracts* until 1963. Less well established is the use of kinetic and mechanistic principles in the optimisation of particular steps of organic syntheses (both laboratory preparations and commercial manufacturing processes). The main aim of the first sections of this chapter is to show that optimisations can be achieved, using the standard spectroscopic and chromatographic techniques of preparative organic chemistry, through an improved understanding of the factors which control the yields of products of organic reactions. As the experimental techniques are well known, relatively few comments on *how to investigate* need to be made – instead, the emphasis will be on the design of suitable experiments, i.e. *what to investigate*.

A well-known example of the application of mechanistic understanding to help to control product yields is also of commercial significance – the addition of HBr to alkenes which may occur via cationic or radical mechanisms, Scheme 2.1 [2a]. Very pure alk-1-enes (1), in the absence of peroxides, react to give the 2-bromo-products (2) by Markovnikov addition. In the presence of peroxides or other radical sources, anti-Markovnikov addition gives the 1-bromo-products (3).

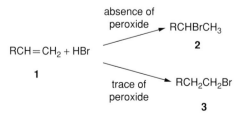

**Scheme 2.1** Radical and ionic additions of HBr to an alk-1-ene.

The first sentence of Hammett's influential book, *Physical Organic Chemistry*, states, "A major part of the job of the chemist is the prediction and control of the course of chemical reactions" [3a]. Hammett then explains that one approach is the application of broadly ranging principles, but another involves "bit-by-bit development of empirical generalizations, aided by theories of approximate validity whenever they seem either to rationalize a useful empirical conclusion or to suggest interesting lines of experimental investigation".

Mechanistic considerations based on product studies form the early stages of the *bit-by-bit development*. The structure of the product, along with that of the starting material, defines the chemical reaction and is the starting point for mechanistic investigations. After the structures of the products of a reaction have been determined, 'paper mechanisms' can be drawn; these are mechanistic hypotheses, based as much as possible on precedents and analogies, on which further experimental investigations can be based. Unexpected results are of particular interest.

Initial curiosity about unusual observations is one of the main reasons for undertaking mechanistic investigations, and such studies may lead to new chemistry. Sometimes (with luck!), the eventual outcome may be of great significance. For example, the rapid addition of diborane to alkenes (hydroboration) in ether solvents opened up new areas of organoboron chemistry. Brown and Subba Rao discovered hydroboration by systematic investigations into why the unsaturated carboxylic ester, ethyl oleate, consumed more than the quantity of reducing agent expected simply for reduction of the carboxylic ester function [4]. As this observation was the only anomalous result within a series of reducible compounds, it would have been tempting to have ignored it. Other examples where mechanistic explanations of unexpected results have led to important new chemistry are given in Section 2.5.

Discussion of a reaction mechanism usually involves an individual reaction, often extended to include the effects of substituents or solvents on that particular reaction, i.e. a single reaction or one of several closely related compounds may be considered within a mechanistic study. In contrast, reaction yields depend on combinations of kinetic and/or thermodynamic factors for several different reactions of just one compound. An understanding of these factors, additional mechanistic information and other matters all contribute to a greater understanding of a particular reaction. This understanding can then be used to optimise the yields of desired products and ensure that an organic synthesis (by a laboratory reaction or a commercial manufacturing process) is relatively robust, i.e. it is possible to obtain good yields without the need to reproduce exactly very particular reaction conditions.

The acquired understanding of *mechanism* (using the term in a broader sense than usual) can then be used by synthetic or process development chemists to design systematically other changes to the reaction conditions (in addition to achieving increases in product yields), such as shorter reaction times, lower product costs and/or less waste for disposal. However, alternative approaches to the design of experiments should not be excluded. Diverse, but less systematic, investigations could introduce valuable new leads by intuition, or simply by good luck (serendipity, see Chapter 12). Statistical methods of yield optimisation are also available, and studies of the factors which control product stabilities may also be important (Section 2.2.2). Helpful information may even be obtained from 'unsuccessful' experiments (Section 2.5).

Spectroscopic techniques may provide stereochemical information, from which useful mechanistic deductions can be made (Section 2.3.1). Some of the most surprising (but also convincing) mechanistic evidence has arisen from mechanistic studies involving isotopic labelling (e.g. evidence for benzyne intermediates) and the use of isotopic labels provides ways to identify which bonds are broken or formed during a reaction (Section 2.3.2).

Knowledge of reaction intermediates (see Chapter 9) is also important for mechanistic understanding and yield optimisation. Relatively stable intermediates (intermediate products) can sometimes be detected simply by monitoring reactions as they proceed, or by minor changes to the reaction conditions such as lowering the temperature or omitting one or more of the reagents (see Section 2.4 and Chapter 12). Even if intermediate products are formed in such small concentrations that they cannot be observed directly, their formation during rearrangement reactions may be investigated by *crossover* experiments (see Section 2.4). Knowledge of minor or intermediate products is particularly helpful in designing reaction conditions to optimise the yield of a desired final product.

The formation of relatively stable intermediate products can be of direct commercial significance. Details of commercial processes are not usually considered in academic texts for several reasons, including their commercial sensitivity and the financial, technical and legal complexity. Details of processes may be patented to try to protect information, or it may be decided that publication of the patent might reveal more information than it protects, so details of a process may be kept secret. Also, patent laws vary from country to country. Although academic scientists may act as expert witnesses in courts, consultancy activities are usually confidential. Recognising the difficulties in discussing this topic, the following paragraph is an attempt to give an indication of the relevance of mechanistic studies to the complex world of patents and their defence.

Consider Scheme 2.2. Suppose that company X has patented a process leading from starting material **B** to product **A**, via pathway 1. Also suppose that a second company, Y, devises an alternative process to manufacture product **A** from starting material **C**, and they claim that the reaction of **C** to **A** occurs directly (pathway 2). If company X can establish that the mechanism of the reaction of **C** to **A** proceeds via compound **B**, by pathway 3 followed by pathway 1, then company Y may be infringing the patent of company X. If cases such as this are disputed, the reaction mechanisms could be the subject of legal proceedings. In marked contrast to the usual academic debates, mechanisms of reaction could then be considered in a law court, with a yes/no (proven or not proven) decision after the scientific evidence has been presented. The outcome of such a case could have major financial implications, e.g. legal costs, compensation, and possibly licensing agreements.

Reaction monitoring is usually beneficial. Even for well-established syntheses, the progress of organic reactions is often monitored *qualitatively* by chromatographic techniques, most simply by TLC, to determine the 'reaction time', i.e. when the starting materials have been consumed. Application of standard *quantitative* chromatographic methods, e.g. GC or

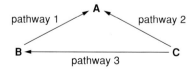

**Scheme 2.2**   Direct and indirect industrial preparations of a compound A.

HPLC, should lead to improved monitoring, and Section 2.6 includes an account of how kinetically useful information may be obtained by good monitoring under controlled reaction conditions (especially temperature). Applications of chromatographic and/or spectroscopic techniques help to identify minor products in the final product mixture, and these may provide information about the structure of reactive intermediates (Section 2.4 and Chapter 9). The yields of minor or intermediate products may be changed by variations in the reaction conditions (e.g. lower temperatures), and examples are given in Section 2.4.

A note of caution is also warranted. It is well established that reaction mechanisms depend on structures of reactants, so extrapolation of mechanistic deductions from one reaction to another of a 'similar' reactant should not be automatic. Mechanistic changes could also arise through changes in the reaction conditions (including solvent, temperature, concentrations of reagents and presence of catalysts), and impurities in starting materials or solvents could be catalysts or inhibitors, e.g. acid, base, water or metal ions (see Chapter 11).

A wide range of mechanisms are mentioned in this chapter, but a comprehensive discussion of any particular mechanism requires the consideration of additional independent evidence. Consequently, references to original literature sources and cross references to later chapters in this book are given.

To summarise so far, mechanistic principles established over the past 100 years or more are already integrated into the teaching of organic chemistry, and ongoing mechanistic investigations make important current contributions to

(1) optimising yields of organic syntheses in building up a 'bit-by-bit' understanding of what factors control a particular reaction;
(2) new chemistry through curiosity-driven research – why/how does an unexpected product form?
(3) defence of patents for manufacturing processes.


## 2.2   Product structure and yield

Before more detailed mechanistic studies begin, a reaction must be defined by the structures of the starting materials and products. In some cases, one may be limited to the study solely of the reactants and the products (see Chapter 12). With the availability of a wide range of spectroscopic techniques (IR, MS, NMR, UV–vis), incorrect assignments of the structures of pure organic compounds are very rare nowadays. Uncertainties about structure can often be resolved by X-ray crystallography. Some incorrect assignments of mechanistic interest from the older literature were summarised by Jackson [2b].


### 2.2.1   *Quantitative determination of product yields*

Initial analyses of product mixtures can be carried out by GC-MS, LC-MS or NMR. Using GC or LC, components present in very small amounts (<1%) can be identified. MS is very useful for the analysis of mixtures of known compounds, but usually does not provide sufficient information on which to base structural assignments for new compounds. Because of the

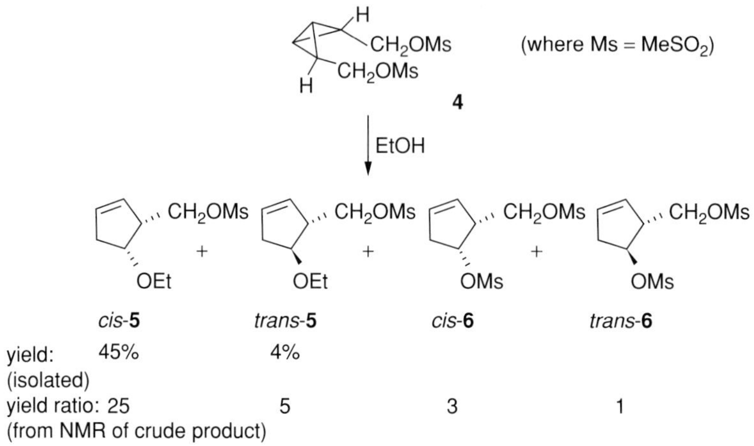

Scheme 2.3   Products from ethanolysis of *endo, endo*-dimesylate (**4**).

high sensitivity and resolution of high field NMR instruments, several components in crude mixtures of low MW organic products can be identified. In Scheme 2.3, for example, four products (due to a 1,2-alkyl rearrangement leading to ring expansion, followed by a ring opening) were identified after ethanolysis of the *endo, endo*-bicyclic dimesylate (**4**), including the *cis*- and *trans*-monocyclic dimesylates (*cis*-**6** and *trans*-**6**) [5a]. A detailed product analysis from the corresponding *exo,exo*-bicyclic isomeric substrate required a fully proton-coupled 151 MHz $^{13}$C NMR spectrum (the coupling assists in making the assignments), and an in-depth knowledge of the NMR spectra of related compounds.

Quantitative analyses by integration can be made using FT $^1$H NMR if, as is usually the case, nuclei relax to their equilibrium distributions between successive pulses. However, in $^{13}$C NMR, peaks for individual carbon atoms may be of markedly different intensities. Appropriate protocols are available to improve the accuracy of NMR integrations for both $^1$H and $^{13}$C nuclei [6]. The ratios of yields of products from (**4**) are also shown in Scheme 2.3; these were obtained by integration of the $^1$H NMR spectra, using 90° observation pulses [5b]. As might be expected, isolated yields for two of the four components shown in Scheme 2.3 reveal losses of material and a significant change in the ratio of yields of *cis*-**5**:*trans*-**5** from 5:1 in the crude product to 11:1 for isolated products. These results emphasise the importance of obtaining product yields directly from crude products, with minimal work-up (see also Section 2.6 and Chapter 10, Section 10.6). If a work-up procedure is used, solutions for disposal such as aqueous washings could be analysed directly by reverse phase HPLC to check for minor products or the presence of the desired product.

More accurate quantitative analyses can be carried out by GC or LC. A suitable *internal standard* is usually required; it must be chemically stable and involatile under the conditions of the experiment (prior to injection into the chromatography apparatus), and must be resolved from the other signals in the chromatogram. It is usually added to the product mixture after the reaction has been completed but before any extraction, purification or analysis steps are undertaken. Saturated hydrocarbons of $C_{10}$ or above are typically used as internal standards for GC [7]. Response factors are obtained for each component of the mixture by prior quantitative analysis of standard mixtures of the components and the internal standard [7].

If a pure sample of a component of the mixture is not available, a response factor can be estimated, e.g. in GC from the number of carbon atoms in the molecule [7]. However, caution should be exercised; for example, UV responses for LC detection are determined by $\varepsilon$ values, which can vary by orders of magnitude – the area of a signal (or worse, the peak height) may be a very misleading indicator of the molar concentration of the corresponding analyte. As well as calibrations to obtain response factors, additional checks to show a linear relationship between integrated peak areas and concentrations should be made, all calibrations being in the same range of concentrations as the product mixture under investigation.

The yield and purity of each product depend in part on competing reactions, which can be investigated by the methods discussed below. Other important aspects are methods of purification or extraction and the required specification of product purity.

## 2.2.2 Product stabilities, and kinetic and thermodynamic control of product formation

The initial products of organic reactions are formed under conditions of kinetic control – the products are formed in proportions governed by the relative rates of the parallel (forward) reactions leading to their formation. Subsequently, product composition may become thermodynamically controlled (equilibrium controlled), i.e. when products are in proportions governed by the equilibrium constants for their interconversion under the reaction conditions. The reaction conditions for equilibrium control could involve longer reaction times than those for kinetic control, or addition of a catalyst. The mechanism of equilibrium control could simply involve reversal of the initial product-forming reactions (as in Scheme 2.4, see below), or the products could interconvert by another process (e.g. hydrolysis of an alkyl chloride could produce a mixture of an alcohol and an alkene, and the $H_3O^+$ by-product could catalyse their interconversion).

When a mixture of 0.01 mol of each of cyclohexanone (**7**), furfural (**8**) and semicarbazide (Scheme 2.4) reacted in aqueous ethanol at 25°C, the product isolated after a few seconds was cyclohexanone semicarbazone (**9**); but, after a few hours, the product was the semicarbazone

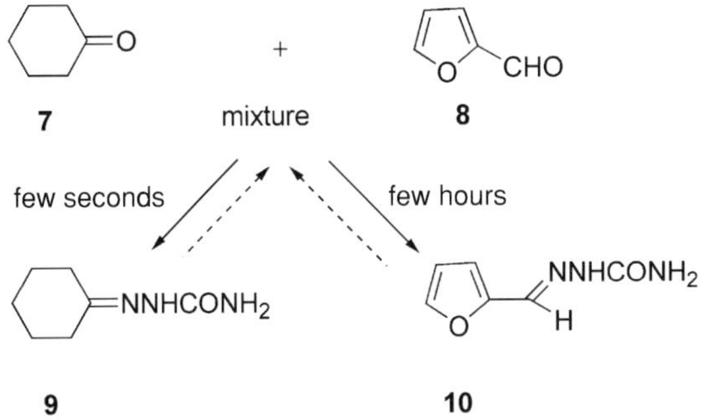

**Scheme 2.4** Kinetic and thermodynamic products from equimolar amounts of cyclohexanone (**7**), furfural (**8**) and semicarbazide.

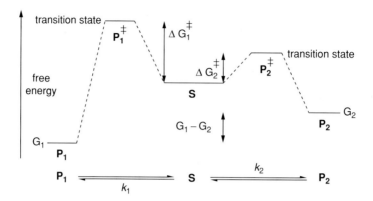

**Fig. 2.1**   Free energy diagram showing the reaction of **S** to give the products $P_1$ favoured by thermodynamic control and $P_2$ favoured by kinetic control.

(**10**) of furfural [**8**]. The relative rates of formation of products favour **9** rather than **10** by a ratio of 50:1, so formation of **9** is favoured kinetically. However, in the reverse reactions, **9** hydrolyses faster than **10** by a ratio of 300:1, and **10** is favoured thermodynamically.

The free energy diagram in Fig. 2.1 shows a general case. Products $P_1$ and $P_2$ have free energies of $G_1$ and $G_2$, respectively. The thermodynamically favoured product is $P_1$, having a lower free energy than $P_2$. The starting material (**S**) reacts to give $P_1$ with a free energy of activation $\Delta G_1^{\ddagger}$, which is greater than the free energy of activation $\Delta G_2^{\ddagger}$ leading to $P_2$, so formation of $P_2$ is favoured kinetically. A change from kinetic to thermodynamic control depends on the relative magnitudes of the free energy differences $(G_1 - G_2)$, $\Delta G_1^{\ddagger}$, and $\Delta G_2^{\ddagger}$, so will not always involve a sudden switchover. In practice, higher temperatures may be required to achieve thermodynamic control within a convenient time (aromatic substitutions are well-known examples) [9, 10]. Because equilibrium constants depend on temperature, the thermodynamically controlled product composition after long reaction times will not be exactly the same as that achieved by increasing the temperature [10].

Product compositions can depend on many other kinetic factors. Even 'stable' organic compounds will slowly decompose. Yields of many desired products may be affected by decomposition caused, for example, by acid, base, oxygen or water; the method of work-up and purification will also reduce the yield of an isolated product. Consequently, prompt analysis of the whole reaction mixture, or a representative aliquot, is recommended (see Section 2.6).

Product compositions under conditions of kinetic control depend on the relative energies of the transition states leading to their formation (Fig. 2.1). If the two products are initially formed in equal amounts, the free energies of activation $(\Delta G^{\ddagger} = -RT \ln k)$ leading to the two transition states are the same $(\Delta G_1^{\ddagger} = \Delta G_2^{\ddagger}$, so $\delta \Delta G^{\ddagger} = 0)$. If, by changing the reaction conditions, a $\delta \Delta G^{\ddagger}$ of 1.4 kcal mol$^{-1}$ between the two transition states is achieved, the product ratio (at 25°C) will change from 50:50 to 90:10 because $\delta \Delta G^{\ddagger} = -RT \ln(k_1/k_2)$. Such a small free energy change can often be achieved by changes in reaction conditions (e.g. temperature or solvent). The same free energy change (1.4 kcal mol$^{-1}$) will increase the yield of a product from 1% to 10%, from 10% to 50% or from 90% to 99%. It is encouraging that an even smaller energy change is required to change the yield from, say, 50 to 80%. However, whilst such improvements are very important synthetically, it is not usually appropriate to try to explain or interpret such small differences in $\Delta G^{\ddagger}$ mechanistically.

## 2.3 Mechanistic information from more detailed studies of product structure

The more detailed product analytical techniques to be covered under this heading are stereochemical and isotopic labelling probes.

### 2.3.1 Stereochemical considerations

A reaction mechanism must account for the product stereochemistry, and investigations of stereochemistry have played a major role in developing our current understanding of organic reaction mechanisms (e.g. Walden inversion and $S_N2$ mechanisms, neighbouring group participation and reactions controlled by orbital symmetry). The historical background and many examples have been discussed in detail in other texts [2c, 11], and will not be presented again here. Nowadays, the stereochemical course of many reactions is known, although the mechanistic explanation may still be under debate.

Two general problems should be appreciated. First, the inadvisability of putative 'explanations' of small effects (noted above) commonly applies to issues of stereoselectivity. If, through changes in reagents and reaction conditions, a non-stereoselective reaction (ratio of stereoisomeric products = 50:50) becomes stereoselective to some degree (e.g. stereoisomeric product ratio = 90:10), the improvement may be highly significant synthetically, but the energy change involved is small and any 'explanation' is likely to be tenuous at best. Second, concerted reactions are difficult to distinguish from reactions in which a reactive intermediate (e.g. a biradical) is formed, but is so short lived that there is insufficient time for bond rotation, for example [12].

Knowledge of the expected stereochemistry is an integral part of the planning of organic syntheses, and commercial pressures for greater stereoselectivity are increasing. The most useful reactions for controlling product stereochemistry are usually kinetically controlled, and much current research involves investigations of suitable reagents or catalysts [13] (including enzymes/biocatalysts [14]), see Chapters 11 and 12. Under more vigorous reaction conditions, thermodynamic control may take over, e.g. conversion of maleic acid to fumaric acid by boiling with concentrated hydrochloric acid [15], or epimerisation of alkoxides formed by reduction of ketones, i.e. under strongly basic conditions [16].

If there is no change in bonding to the stereogenic centre, there can be no configurational change. Thus, in the hydrolysis of carboxylic esters by acyl-oxygen cleavage, the alkyl-oxygen bond is unaffected, so the configuration of the alcohol product is the same as in the ester starting material. Similarly, the configurational stability of migrating groups during rearrangement reactions is an important indicator of mechanism. For 1,2-shifts to electron-deficient atoms, Scheme 2.5, the group R usually retains configuration when it migrates from atom X to the electron-deficient atom Y in cation (**11**) following departure

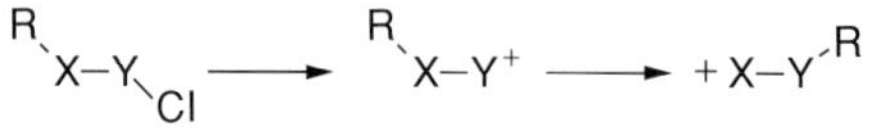

$$R-\underset{Cl}{X-Y} \longrightarrow \underset{R}{X-Y^+} \longrightarrow + X-Y^{R}$$

**11**              **12**

**Scheme 2.5**   1,2-Migration of alkyl group (R) with bonding electron pair with retention of configuration.

of the nucleofuge Cl$^-$ to give the rearranged cation (**12**) [17]. Such reactions usually involve migration of R from X to Y without detachment/reattachment of R.

If the migrating group R in **11** became fully unbonded from X before rebonding to Y, the detached R group would not retain its configuration, unless detachment/reattachment were very rapid. The Stevens rearrangement of ylid **13** in Scheme 2.6 is known to be intramolecular from *crossover* experiments (see section 2.4), and the configuration of the migrating group R is retained. A mechanism corresponding to a front-side $S_N2$ reaction could be drawn, but later CIDNP results and detection of a small amount of a radical-coupling product (R—R) were explained by a reaction via radical pairs in a solvent cage; the intermediate (**14**) can be drawn in two resonance forms differing only in the location of one electron, either on nitrogen or carbon [18].

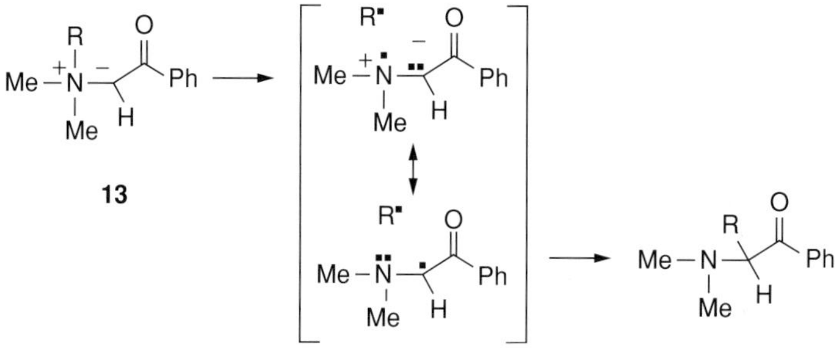

**Scheme 2.6**   1,2-Migration via a radical pair in a Stevens rearrangement of ylid **13**.

### 2.3.2   Use of isotopic labelling

The commercial availability of a wide range of non-radioactive isotopically enriched organic compounds, and convenient analyses by NMR or MS, provide many opportunities to investigate reaction mechanisms using isotopic labelling (see also Chapter 9). Such experiments are much less demanding than some of the important labelling experiments previously carried out with radioactive isotopes (e.g. $^{14}$C or $^{3}$H), which often involved laborious selective degradations to locate labels. A more detailed account is available elsewhere [19], and examples of isotopic labelling of solvents are included in Section 2.4.

The benefits of initial product studies, followed by a more detailed examination using isotopic labelling, are well illustrated by the amination of halobenzenes, Scheme 2.7. The reaction of *p*-bromomethoxybenzene (**15**) with lithium diethylamide in ether gives a 1:1 mixture of *m*- and *p*-diethylamino-substituted products (**16** and **17**), with no trace of the *o*-isomer. One possible mechanism for these and many related reactions was that a 'normal' direct displacement reaction to give **17** was in competition with an 'abnormal' displacement,

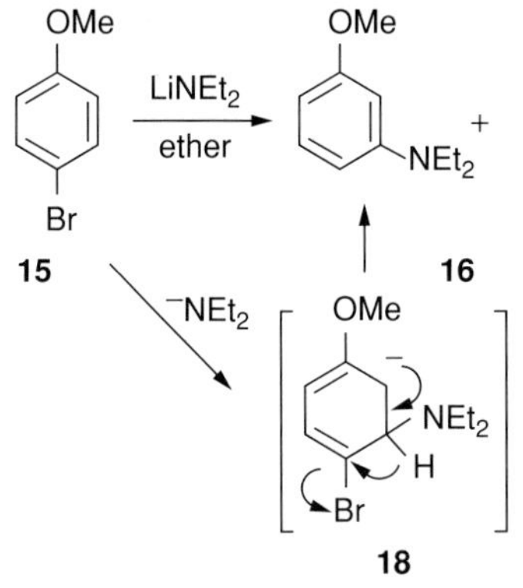

**Scheme 2.7**  Non-benzyne mechanism to account for rearrangement in the reaction of 4-bromomethoxy-benzene (**15**) with lithium diethylamide.

involving an initial addition to give **18**, followed by a 1,2-hydride shift and loss of bromide to give **16**[20].

The alternative benzyne mechanism was supported by a study of rearrangements in aminations of iodobenzene-1-$^{14}$C and chlorobenzene-1-$^{14}$C (**19** in Scheme 2.8), which gave unrearranged and rearranged anilines **21** and **22**. The slightly greater preference for rearrangement was explained by a 10% kinetic isotope effect on the rate of addition of ammonia to the benzyne (**20**). It was convincingly argued that fortuitous combinations of normal and abnormal pathways (Scheme 2.7) were highly unlikely to explain the observations that equal amounts of unrearranged and rearranged products (**21** and **22**) were formed from both chlorobenzene and iodobenzene [20].

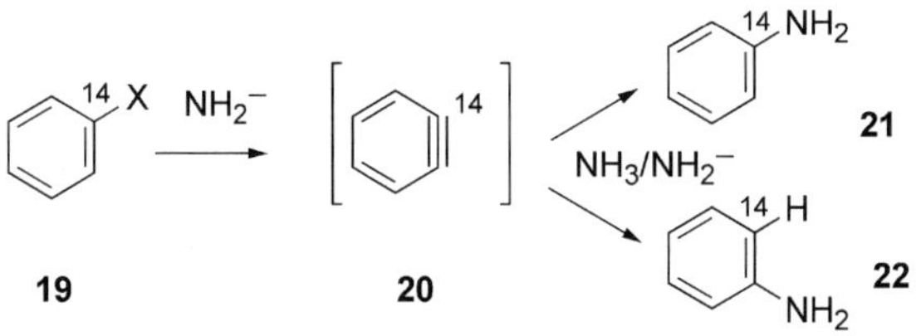

**Scheme 2.8**  Benzyne mechanism to account for rearrangement during amination of halobenzenes.

## 2.4  Mechanistic evidence from variations in reaction conditions

The following examples illustrate the strong evidence obtainable by product analysis under different reaction conditions. The simplest variations are reaction time and temperature. Monitoring reactions qualitatively as they proceed is also included here, because reactions are investigated at shorter than normal reaction times. More quantitative aspects of reaction monitoring are discussed in Section 2.6. First, however, some limitations need to be considered. If reaction conditions are varied, it cannot be established with certainty that the reaction mechanism does not change. Also, it needs to be established that any reaction

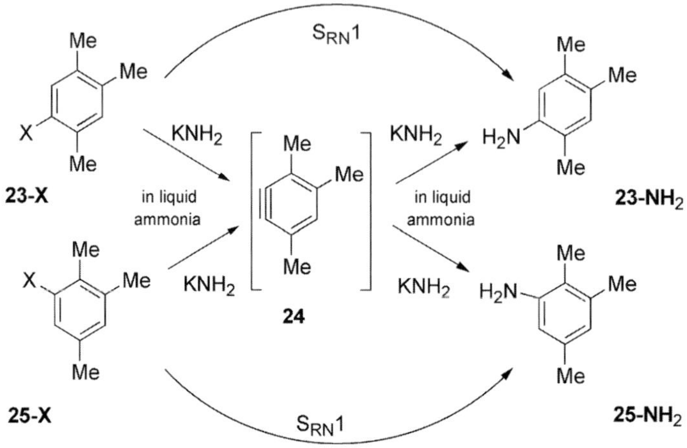

**Scheme 2.9** Competing benzyne and $S_{RN}1$ mechanisms for halotrimethylbenzenes with potassium amide in liquid ammonia.

intermediates observed under changed reaction conditions would proceed to the main product under normal reaction conditions.

A variation on the aryne mechanism for nucleophilic aromatic substitution (discussed above, Scheme 2.8) is the $S_{RN}1$ mechanism (see also Chapter 10). Product analysis, with or without radical initiation or radical inhibition, played a crucial role in establishing a radical anion mechanism [21]. The four isomeric bromo- and chloro-trimethylbenzenes (23-X and 25-X, Scheme 2.9) reacted with potassium amide in liquid ammonia, as expected for the benzyne mechanism, giving the same product ratio of 25-NH$_2$/23-NH$_2$ = 1.46. As the benzyne intermediate (24) is unsymmetrical, a 1:1 product ratio is not observed.

However, when the corresponding iodo compounds were examined under the same reaction conditions, 23-I gave a 25-NH$_2$/23-NH$_2$ product ratio of 0.63:1, and 25-I gave a ratio of 5.9:1. The preference for direct substitution was largely suppressed by addition of a radical scavenger (tetraphenylhydrazine), whereas addition of potassium metal gave entirely unrearranged substitution. The proposed $S_{RN}1$ mechanism (Equations 2.1–2.5 in Scheme 2.10) involves a radical chain. The reaction is initiated when solvated electrons (from potassium in liquid ammonia) add to the substrate (ArX) to give a radical anion

initiation:    $e^- + ArX \longrightarrow [ArX]^{\bullet-}$          (2.1)

propagation:
$$[ArX]^{\bullet-} \longrightarrow Ar^\bullet + X^- \quad (2.2)$$
$$Ar^\bullet + Y^- \longrightarrow [ArY]^{\bullet-} \quad (2.3)$$
$$[ArY]^{\bullet-} + ArX \longrightarrow ArY + [ArX]^{\bullet-} \quad (2.4)$$

overall:    $ArX + Y^- \longrightarrow ArY + X^-$          (2.5)

**Scheme 2.10** The $S_{RN}1$ mechanism of nucleophilic aromatic substitution.

(Equation 2.1). Three propagation steps then follow, including dissociation of the radical anion to an aryl radical and $X^-$ (Equation 2.2). In contrast, the corresponding alternative $S_N1$ reaction would lead to the much less stable aryl cation (the empty p-orbital is part of the $\sigma$-framework, and so cannot be stabilised by the $\pi$-electrons). The aryl radical then reacts rapidly with another nucleophile ($Y^-$ in general or $NH_2^-$ in this case) to give another radical anion (Equation 2.3); then electron transfer from one radical anion to another reactant molecule (Equation 2.4) initiates another chain. The overall consequence of the three propagation steps is nucleophilic aromatic substitution (Equation 2.5).

The above example (Schemes 2.9 and 2.10) illustrates a potentially general procedure: (i) identify unexpected products; (ii) consider mechanistic explanations; (iii) modify reaction conditions (in this case using catalysts or inhibitors) to optimise the yield of desired products, and to test mechanistic deductions. Additional and more direct supporting evidence for radical intermediates is given in Chapter 10.

Another example is the Hofmann reaction in which an amide is converted to an amine having one fewer carbon atoms (Equation 2.6 and Scheme 2.11). The reaction is usually initiated by adding the amide to cold alkaline sodium hypobromite solution (bromine and aqueous NaOH), followed by rapid warming [22]:

$$RCONH_2 + Br_2 + 4OH^- = RNH_2 + CO_3^{2-} + 2Br^- + 2H_2O. \qquad (2.6)$$

Several reaction intermediates can be observed by varying the reaction conditions. If equimolar amounts of bromine and hydroxide are added to acetamide (**26**, R = Me, Scheme 2.11), the product is *N*-bromoacetamide (**27**, R = Me). Further reaction with base gives unstable salts (**28**), which rearrange to isocyanates (**29**); reaction with water and an excess of hydroxide finally leads to the amine product (**30**), but in alcoholic solutions urethanes (**31**) are formed.

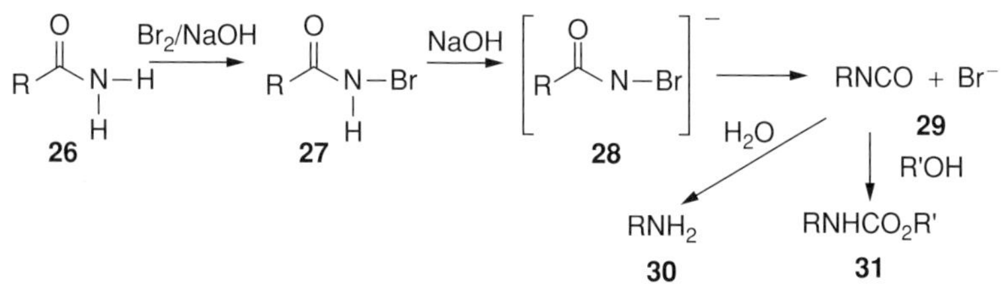

**Scheme 2.11**   Effect of variations in reaction conditions in the Hofmann rearrangement of amides **26**.

Crossover experiments, often used to investigate whether rearrangement reactions are intramolecular or intermolecular, involve modified reaction conditions. Consider a rearrangement of R—X—Y to X—Y—R. Instead of a pure starting material, a mixture of two starting materials (R—X—Y and R'—X—Y') is used, where R' and Y' differ slightly from R and Y (e.g. an Et instead of an Me substituent, or an isotopic label is incorporated). The reaction is carried out under otherwise normal reaction conditions, and the products are examined (Scheme 2.12). If the rearrangement is intramolecular (i.e. R or R' migrate without becoming detached from X—Y), only two products will be formed (X—Y—R and X—Y'—R'); but, if the reaction is intermolecular, two additional products could be formed (X—Y'—R and X—Y—R').

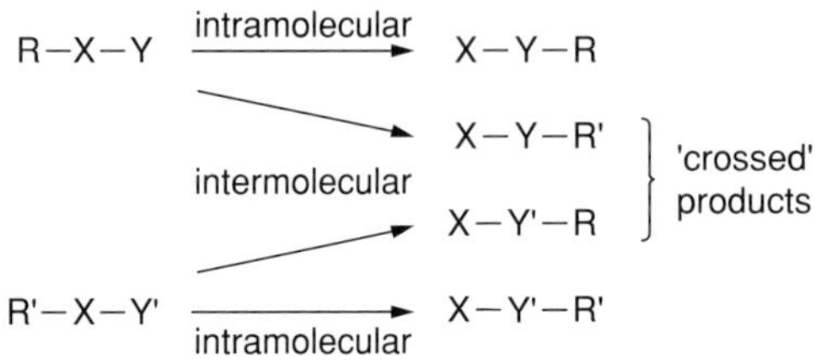

Scheme 2.12  Crossover experiments to distinguish between intermolecular and intramolecular mechanisms.

Reaction intermediates can be detected by reaction monitoring (i.e. analyses at several reaction times), and their presence may be inferred or even observed more readily at low temperatures. In a Wittig reaction, the ylid **32** in Scheme 2.13 was produced from ethyltriphenylphosphonium bromide and butyl lithium, and reacted with a small excess of cyclohexanone in THF at −70°C; the initial product, the oxaphosphetane **33**, was identified by $^{31}$P NMR and converted to the alkene product and triphenylphosphine oxide (**34**) above −15°C (see also Chapter 9). These results provide relatively direct experimental evidence for the mechanism shown in Scheme 2.13 [23].

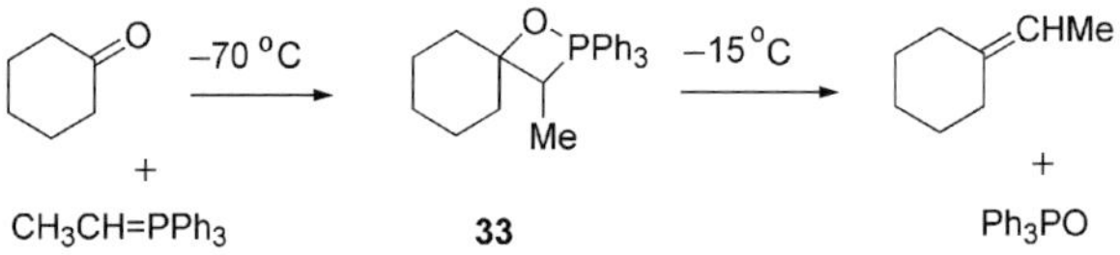

Scheme 2.13  Mechanism of the Wittig reaction examined by NMR at low temperatures.

On treatment with aluminium chloride in pentane at room temperature, 2,2-dichloronorbornane (**35** in Scheme 2.14) is converted to 1-chloronorbornane (**36**) in about 50% yield. Many carbocations can be investigated under weakly nucleophilic conditions (e.g. 'magic acid', a mixture of SbF$_5$ and FSO$_3$H). Investigations by $^1$H and $^{13}$C NMR methods showed that **35** may be ionised to the cation **37** in SbF$_5$/SO$_2$ClF at −100°C, and, on warming to −20°C, the rearranged cation **38** was detected [24]. When the synthesis was

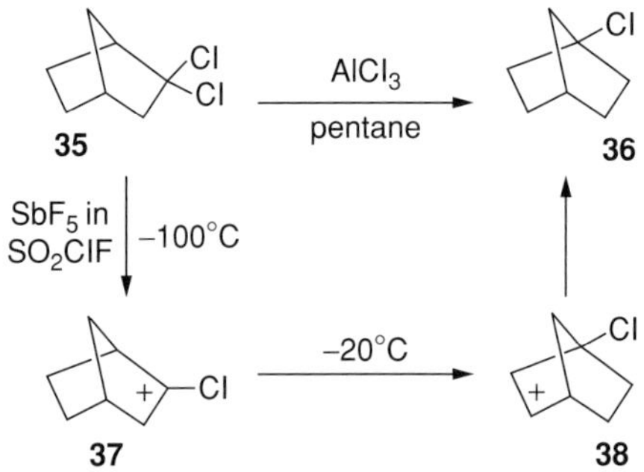

Scheme 2.14  Effect of temperature on the conversion of 2,2-dichloronorbornane into 1-chloronorbornane induced by Lewis acids.

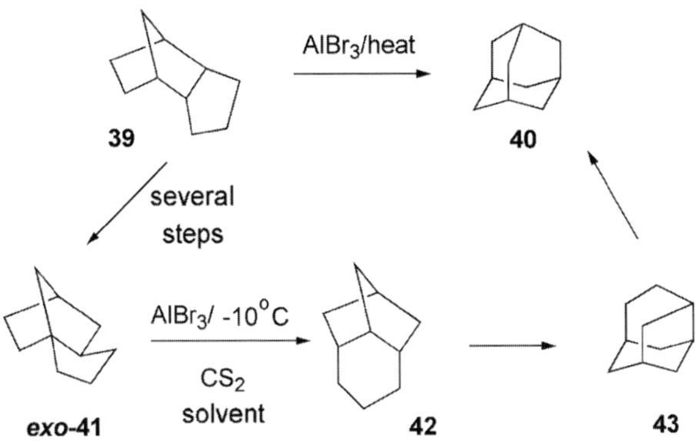

**Scheme 2.15**  Detection of intermediates **42** and **43** by GC in the isomerisation of tetrahydrodicyclopentadiene (**39**) to adamantane (**40**) induced by Lewis acids.

carried out in deuterated solvent (pentane–d$_{12}$), the product (**36**) was shown by GC-MS to contain either one deuterium atom or none, indicating that the source of hydride needed to convert **38** to **36** is both solvent and substrate/product. The GC-MS study also identified undeuterated dichloronorbornanes amongst the products [24].

Monitoring by GC provided evidence [25] for several intermediates in the isomerisation of tetrahydrodicyclopentadiene (**39**) to adamantane (**40**), Scheme 2.15, a thermodynamically controlled process (discovered serendipitously) [26]. Molecular mechanics calculations indicated that the slow step of the rearrangement could occur prior to formation of *exo*-**41**, which rearranges in a few minutes to **40** on treatment with AlBr$_3$ in CS$_2$ at 25°C. When the reaction of *exo*-**41** was conducted at −10°C and monitored by GC over a period of 100 minutes, the presence of two additional intermediates (**42** and **43**) was revealed [25].

Re-emphasis of a cautionary note is appropriate here: when reaction conditions are varied, mechanisms may change. Variations in pH often lead to mechanistic changes because substrates may be protonated by H$_3$O$^+$, or attacked by HO$^-$, much more rapidly than by water (see Chapter 11). Dediazoniation of methylbenzenediazonium ions (**44** in Scheme 2.16) in aqueous ethanol is heterolytic at low pH; typical products are ArOH or ArOEt (**45**) derived from attack on a cationic intermediate by nucleophiles. Above pH 6, however, the process is homolytic and toluene (**46**) is formed from a radical intermediate by reduction of

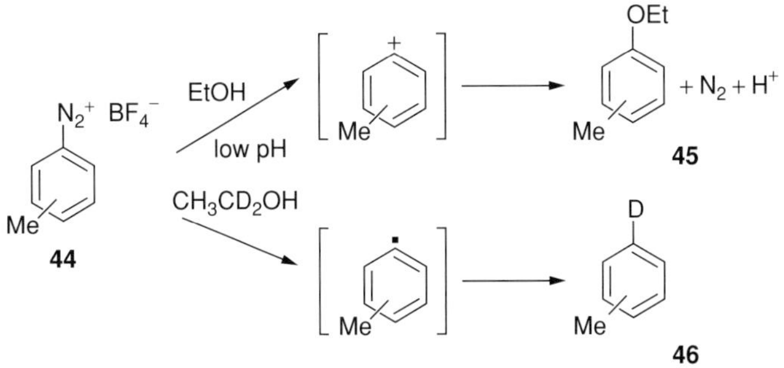

**Scheme 2.16**  Dediazoniation by ionic and radical mechanisms.

methylbenzenediazonium ions [27]. In specifically deuterated ethanol ($CH_3CD_2OH$), the reduction product contained deuterium, which identified the source of the hydrogen atom involved in the reduction [28].

## 2.5    Problems and opportunities arising from unsuccessful experiments or unexpected results

Unexpected results can lead to new opportunities, but the provenance of the discovery is not always clear from research publications. Bunnett [21] admitted that the discovery of the $S_{RN}1$ mechanism (Scheme 2.9) was serendipitous, and the discoveries of hydroboration [4] and of the conservation of orbital symmetry in concerted reactions (Woodward was working on the synthesis of vitamin $B_{12}$) [29] originated in unexpected results. In another context, monitoring by NMR of reaction products as they were formed led to the chance observation of negative signals (emission of radiation instead of the usual absorption), explained by CIDNP in radical pairs (Scheme 2.6 earlier; see also Chapter 10) [30]. However, as Pasteur remarked in the nineteenth century [31], "chance favours only the prepared mind".

Unsuccessful reactions, leading to little or none of the desired product, occur for many reasons, and an understanding of the cause of a low product yield can be informative. A zero yield of the desired product is considerably worse than a yield of 1% (see discussion of $\delta \Delta G^{\ddagger}$ in Section 2.2.2). Consequently, a yield of a few % is an opportunity for investigation of alternative reagents or catalysts, and improved reaction conditions. However, a zero yield usually provides more of a problem and less of an opportunity!

One of the most common reasons for low yields is an incomplete reaction. Rates of organic reactions can vary enormously; some are complete in a few seconds whereas rates of others are measured on a geological timescale. Consequently, to ensure that the problem of low yields is not simply due to low reactivity, reaction conditions should be such that some or all of the starting material does actually react. If none of the desired product is obtained, but similar reactions of related compounds are successful, the mechanistic implications should be considered. This situation has been referred to as *Limitation of Reaction*, and several examples have been given [32]; the Hofmann rearrangement, for example, does not proceed for secondary amides (RCONHR') because the intermediate anion **28** cannot form (Scheme 2.11). Sometimes, a substrate for a mechanistic investigation may be chosen deliberately to exclude particular reaction pathways; for example, unimolecular substitution reactions of 1-adamantyl derivatives have been studied in detail in the knowledge that rear-side nucleophilic attack and elimination are not possible and hence not complications (see Section 2.7.1).

## 2.6    Kinetic evidence from monitoring reactions

The use of standard spectroscopic and chromatographic techniques of organic chemistry for mechanistic studies has been emphasised in this chapter. Monitoring of organic syntheses is usually desirable, not only to ensure a suitable extent of reaction before work-up, but also to avoid the apocryphal situation where a reaction claimed to be occurring 'overnight' was actually complete in a few minutes! With relatively small changes in experimental procedures, worthwhile kinetic evidence (i.e. profiles of changes in concentrations of starting materials and/or products with time) can be obtained. This additional evidence can be obtained at the same time as experiments designed to investigate the effects of variations in reaction conditions (Section 2.4) or mechanistic evidence from product mixtures (Section 2.7). The aim is to obtain realistic but imperfect preliminary kinetic data.

Instead of obtaining excellent fits to theoretical rate laws by studying reactions in very dilute solutions ($10^{-2}$ to $10^{-5}$ M), reactions could be monitored at the same concentrations as those used for syntheses (e.g. $>1$ M ). Although this is far from ideal because the reaction medium changes as the reaction occurs, the important point is that relatively few changes in the procedure (requiring little extra effort) are required. Methods of handling the raw kinetic data are considered in Chapter 3.

### 2.6.1   Sampling and analysis for kinetics

When data are collected automatically and calculations are also automated, hundreds of data points are often used to obtain rate constants (see Chapter 3). Readers contemplating their first foray, using a manual method of sampling, will be relieved to learn that reliable rate constants can be obtained from only about ten data points, collected at approximately equal extents of reaction, for a well-thermostatted, well-behaved reaction. Even fewer data points are adequate, especially if an 'infinity' reading is available; if the disappearance of starting material is being monitored, an 'infinity' reading of zero may be assumed.

The most convenient way to monitor very slow reactions is in a well-sealed NMR tube, using deuterated solvents or by suppression of solvent signals. If a reaction is proceeding at $20°C$ for several days, several NMR spectra could be obtained over a 24-hour period, without greatly inconveniencing other users of the instrument – relatively small errors would be introduced by the temperature changes which occur when the NMR tube is removed for short periods of time from the thermostat for the measurements. Alternatively, small aliquots of the reaction mixture could be removed from the reaction mixture and analysed by NMR, GC or HPLC. If chromatographic methods are used, chromatography equipment dedicated to the project is highly desirable to avoid delays in analysing the samples and to ensure that calibrations are reliable.

For faster reactions, it may be necessary to monitor the progress of the reaction in an NMR tube in a thermostatted NMR probe. If an aliquot method is used, it should be possible to remove an aliquot rapidly and quench it by adding cold solvent – reaction rates are lower under colder, more dilute conditions. In reactions involving bases or acids, reactions may be quenched by addition of acidic or basic buffer solutions. Of course, whether or not the reaction can be quenched efficiently, the aliquots should be analysed as expeditiously as possible to reduce the possibility of further reaction or degradation.

Many different types of reactions can be monitored by reverse phase HPLC, in some cases by automated injections from a thermostatted HPLC sample tray [33], or alternatively after quenching aliquots in methanol. No additional work-up is required, and a wide range of product polarities can be accommodated. If all of the products dissolve in the methanol solution, all of the products should be detected, unless their retention times are exceedingly long. If reaction volumes are controlled accurately either by automated injections [33] or by controlling solvent volumes (of aliquots withdrawn, volumes of solvent used to quench and solutions injected onto the HPLC column), internal standards may not be required. If more accurate data are required, corrections can be made mathematically for small changes in the volumes injected [34].

## 2.7   Case studies: more detailed mechanistic evidence from product studies

In this section, mechanistic investigations of details of particular reaction pathways will be considered. Such studies have fundamental long-term aims: (i) understanding the factors influencing organic reactivity, e.g. quantitative relationships between structure and reactivity; (ii) correlating data from diverse reactions by mechanistic classification, e.g. the $S_N2$ mechanism which unifies a wide range of reactions such as hydrolysis of alkyl halides and quaternisation of amines; (iii) obtaining increasingly detailed knowledge about reaction intermediates and pathways.

Usually, mechanistic proposals based on product studies are supplemented by independent evidence from similar reactions (i.e. arguments based on analogies). Such proposals might also be supported by a knowledge of characteristic reactions of functional groups and properties of reactive intermediates (e.g. acidities of C—H groups, or relative stabilities of carbocations), and the proposals are a basis for the more detailed investigations by methods discussed in later chapters.

Natural product chemistry has played an important role in mechanistic developments. For example, some of the earliest work on what is now known as *neighbouring group participation* involved steroids [35a], and some of the first mechanistic investigations from product studies were by characterising the structures of terpenes by chemical degradation (prior to the era of spectroscopic methods). Product studies, in which minor and/or rearranged components were detected, provided some of the earliest evidence for carbocationic intermediates [35b]. Initial mechanistic proposals (e.g. a positive charge on carbon) were speculative and controversial. Gradually over time, and with improved methods of investigation, such mechanisms became more firmly established.

In these later sections, interpretations of quantitative data for product mixtures are emphasised, and the relationship between kinetics and product analysis will be developed. Mechanistic applications of kinetic data are limited to steps of reactions prior to and including the rate-determining step. As separate later steps often determine the reaction products, detailed product studies and investigations of reactive intermediates are important supplements to kinetic studies. Examples of solvolytic and related ($S_N$) reactions have been chosen first because they provide a consistent theme, and second because $S_N$ reactions provide an opportunity to assess critically many of the mechanistic concepts of organic chemistry. Product composition in solvolytic reactions will be discussed next followed by product selectivities (Section 2.7.2) and rate–product correlations (Section 2.7.3).

### 2.7.1   *Product-determining steps in $S_N1$ reactions*

Mechanistic arguments based on product yields have been used to attempt to decide whether a free carbocation is the reaction intermediate in $S_N1$ reactions. If a free carbocation is formed during the solvolysis of an alkyl chloride, added chloride anion may decrease the reaction rate through the *common ion (rate depression) effect*. This phenomenon occurs when there is a reversible initial ionisation/dissociation – see later discussion of Table 2.5 in Section 2.7.3 [3b]. Solvolyses leading to relatively unstable cations, such as the *t*-butyl cation, do not show clear kinetic evidence for common ion rate depression. Alternative evidence apparently supporting formation of a free *t*-butyl cation (rather than the ion pair **48** in Scheme 2.17) was obtained from product studies for various *t*-butyl-X substrates, **47**; it was claimed to show (reasonably at the time) that the yield of alkene **49** did not depend on the leaving group X (Table 2.1) [36]. The 100-fold variation in rate constants at 25°C shows that the first step of the reaction (heterolysis of the C—X bond) depends strongly on the leaving group X, but the product-determining steps appeared to be independent of X. However, it was later shown that solvolyses of esters of optically active tertiary alcohols did not give racemic products, as would be expected for reaction through a 'free' carbocation [37]. Also, in less polar solvents (ethanol and acetic acid), the product ratios did depend on the leaving group X (Table 2.2) [38].

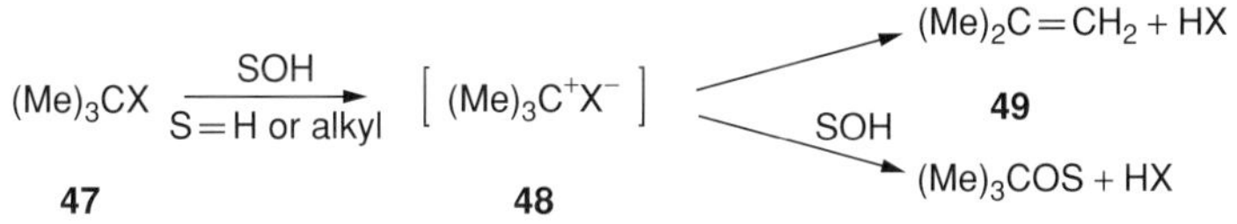

$$(Me)_3CX \xrightarrow[S=H \text{ or alkyl}]{SOH} \left[ (Me)_3C^+X^- \right] \begin{array}{c} \nearrow (Me)_2C=CH_2 + HX \\ \xrightarrow{SOH} \quad \textbf{49} \\ \searrow (Me)_3COS + HX \end{array}$$

**47**　　　　　　　　　　　**48**

**Scheme 2.17** Formation of substitution and elimination product in the solvolysis of *t*-butyl substrates.

The explanation of the later data (Table 2.2) involved making a distinction between ionisation and dissociation [38]. Heterolysis (ionisation) of the C—X bond in a substrate RX (**50** in Scheme 2.18) leads initially to a contact or tight ion pair (**51**). Insertion of one solvent molecule between the cation and the anion leads to a solvent-separated ion pair (**52**), and further dissociation leads to a free or dissociated cation (**53**).

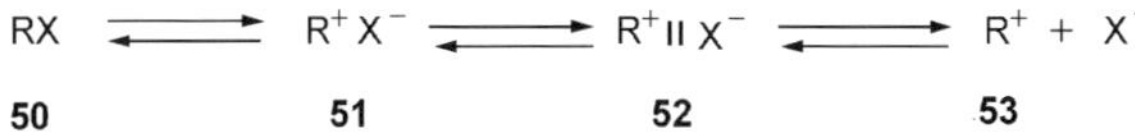

$$RX \rightleftharpoons R^+X^- \rightleftharpoons R^+ \| X^- \rightleftharpoons R^+ + X^-$$

**50**　　　　　**51**　　　　　**52**　　　　　**53**

**Scheme 2.18** Extended ion pair mechanism of solvolysis of RX.

The results in Table 2.2 could be explained most simply by product formation via a contact ion pair (**51**), in which the leaving group X$^-$ removes a proton from the cation to give the alkene. When **51** reacts with solvent to give substitution products, the leaving group X$^-$

**Table 2.1** Early data for mole% alkene from solvolyses of *t*-butyl-X substrates in 80% ethanol–water (Scheme 2.17) [36].

| X in *t*-Bu-X | 25°C | 65°C | $10^5 k_{25}$ (s$^{-1}$) |
|---|---|---|---|
| Cl | 17 | 36 | 0.85 |
| Br | 13 | – | 37 |
| I | 13 | – | 90 |
| Me$_2$S$^+$ | – | 36 | – |

**Table 2.2**   Later, more extensive data on mole% alkene from solvolyses of *t*-butyl-X substrates in Scheme 2.17 in various solvents.*

| X in *t*-Bu-X | $H_2O$, 25°C | $H_2O$, 75°C | EtOH, 75°C | AcOH, 75°C |
|---|---|---|---|---|
| Cl | 5.0 | 7.6 | 44.2 | 73 |
| Br | 5.0 | 6.6 | 36.0 | 69.5 |
| I | 4.0 | 6.0 | 32.3 | – |
| $Me_2S^+$ | – | 6.5 | 17.8 | 11.7 |
| $H_2O^+$ | 3 | 4.7 | – | – |

*Data from reference 38; typical errors, ±1.0.

will be in such a close proximity that the product ratio and the stereochemistry will not be that expected of a free planar carbocation. These results (Tables 2.1 and 2.2) illustrate the evolving nature of mechanistic understanding, and the evolutionary process has continued. More recent evidence supports a concerted elimination process for elimination of HCl from tertiary alkyl substrates [39].

The convenient synthesis of adamantane [26] led to several significant developments. 1-Adamantyl substrates (**54**, Scheme 2.19) are tertiary alkyl compounds for which the caged structure prevents rear-side nucleophilic attack, and elimination does not occur because adamantene (**55**) is too highly strained. The following question arises: when does product formation occur in the solvolytic process? Product studies from competing nucleophilic substitutions in mixed alcohol–water solvent mixtures have provided an answer. To explain the background to this work, we first need to discuss *product selectivities.*

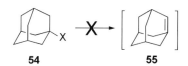

**54**                              **55**

**Scheme 2.19**   Non-formation of adamantene (**55**) from 1-adamantyl substrates.

### 2.7.2   Selectivities

Although the term 'selectivity' is often used, there is no widespread agreement yet on an exact definition. If two competing reactions are 'unselective', a 50:50 product ratio would be obtained from an unbiased competition between the two. It might then be considered to have no selectivity. To convert this product ratio of unity into a *numerical value* of selectivity equal to zero, we take the logarithm of unity [40]. There are various other definitions of selectivity, and caution is needed when obtaining data from the published literature; often, a logarithmic term is not used [41].

We will define *selectivity* using Scheme 2.20 and will assume that products from a substrate (RCl) in a mixed aqueous alcohol solvent are formed by kinetic control in two parallel competing second-order reactions. Attack by alcohol (R′OH), with a rate constant $k_a$, yields an ether product (ROR′), and attack by water, with a rate constant $k_w$, yields an alcohol product (ROH). The product ratio is then given by Equations 2.7 and 2.8, and the selectivity

($S$) can be defined by rearranging the equation as in Equations 2.9 and 2.10:

$$\text{product mole ratio} = \text{mol\% ether product/mol\% alcohol product} \qquad (2.7)$$
$$= [\text{ether product}]/[\text{alcohol product}]$$
$$= k_a\,[\text{alcohol solvent}]/k_w\,[\text{water}] \qquad (2.8)$$
$$S = k_a/k_w = [\text{ether}]/[\text{alcohol product}] \times [\text{water}]/[\text{alcohol solvent}] \quad (2.9)$$
$$= \text{product mole ratio/solvent mole ratio.} \qquad (2.10)$$

The above definition of $S$ is employed independently of the actual kinetic order of the product-determining step. Solvolytic reactions may be of higher kinetic order, or it could be argued that their kinetic order is ill-defined. There is no need to derive Equation 2.9 – it can stand alone as the definition of $S$. To illustrate how the equation works, suppose that the solvent is an equimolar mixture of alcohol and water, so the solvent mole ratio is 1.0. If the product mole ratio is also 1.0, then $S = 1.0$, and the product-forming reactions are unselective (so log $S = 0$ for an unselective reaction). If twice as much ether is formed as alcohol in an equimolar solvent mixture, then $S = 2$. In practice, many solvent compositions are employed, and the equation corrects automatically for variations in solvent composition.

Selectivities for alcohol–water mixtures depend strongly on the pH of the solution. If the pH is sufficiently high that the alcohol and water are partially deprotonated, then alkoxide and hydroxide will be effective nucleophiles, and their concentrations will depend on the $pK_a$ values of the alcohol and water. For relatively acidic alcohols, calculation of $S$ using Equation 2.9 could then give $S$ values over 1000, e.g. nucleophilic aromatic substitutions, including the dyeing of cotton using fibre-reactive dyes [42].

For only weakly acidic alcohols (e.g. methanol–water and ethanol–water mixtures), typical values of $S$ are often close to unity, but can vary from ca. 0.1 to ca. 10 [41]. When $S < 1$, selectivity may be said to be 'inverse'; in this event, if selectivity were defined using log $S$, the selectivity value would become negative. Note also that, if the reactions were simply competing second-order bimolecular processes with the solvent, $S$ would not depend on the solvent composition, so mechanistic information can be obtained from variations in $S$.

Data for solvolyses of 1-bromoadamantane (**54**, X = Br in Scheme 2.19) in ethanol–water in Table 2.3 show that $S$ is approximately independent of solvent composition, but the selectivity is inverse. Why is $S < 1$ for competing nucleophilic substitutions when ethanol is normally more nucleophilic than water? A credible explanation is that the products are formed by front-side collapse of a solvent-separated ion pair (**52** in Scheme 2.18) – the caged structure prevents rear-side approach, so attack must occur from the front-side, and the proportion of water in solvent-separated ion pairs must be greater than in the bulk solvent.

Selectivities for solvolyses of *p*-methoxybenzoyl chloride (**56** in Fig. 2.2) in ethanol–water are similar to those for 1-bromoadamane (Table 2.3), and $S$ is almost independent of solvent

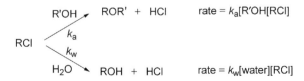

**Scheme 2.20**   Selectivity in the solvolysis of RCl in an aqueous alcohol.

**Table 2.3** Product selectivities (Equation 2.11) for solvolyses of 1-bromoadamantane (**54**, X = Br) in ethanol–water at 75°C and 100°C [41].

| Solvent % (v/v) | 50 | 70 | 80 | 90 | 95 |
|---|---|---|---|---|---|
| Selectivity at 75°C | 0.61 | 0.57 | 0.53 | 0.56 | 0.58 |
| Selectivity at 100°C | 0.49 | – | 0.48 | 0.45 | 0.45 |

composition – as expected from the definition of S. Constant, but higher values of S are observed for methanol–water mixtures. However, for solvolyses of p-nitrobenzoyl chloride (**57**, Fig. 2.2), S increases as water is added to methanol. Selectivities for solvolyses of p-chlorobenzoyl chloride (**58**, Fig. 2.2) initially increase, and then decrease. The results (and independent kinetic evidence) are consistent with a mechanistic change for solvolyses of p-chlorobenzoyl chloride (**58**, Fig. 2.2), which arises simply from a change in the solvent composition [43].

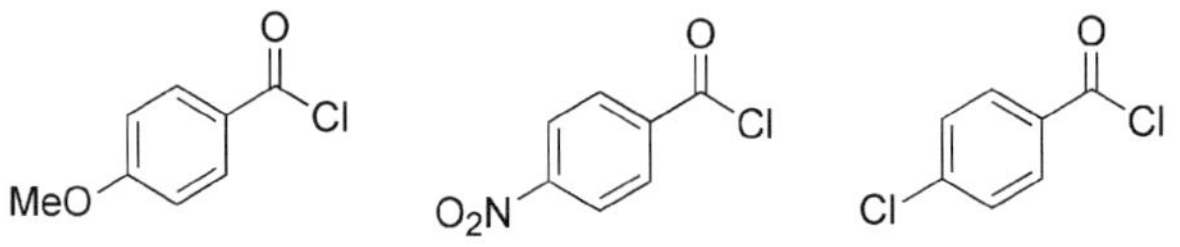

| **56** | **57** | **58** |

**Fig. 2.2** Substituted benzoyl chlorides.

The above results emphasise the value of product selectivities in mechanistic studies, but also show that reaction mechanisms can change as substituents and/or solvents are varied. Consequently, when the kinetic effects of substituents and solvents are used to probe reaction mechanisms, the possibility that the probe induces a mechanistic change should be considered.

## 2.7.3 Rate–product correlations

Combinations of product studies with kinetic data provide particularly powerful indications of reaction mechanisms. Either the presence or absence of a rate–product correlation may be of mechanistic significance. First, we have an explanation of rate–product correlations using the example of competing methanolysis (second-order rate constant, $k_{MeOH}$) and aminolysis (second-order rate constant, $k_{am}$) of benzoyl chloride (**59**) in Scheme 2.21. The mechanism is initially assumed to involve independent competing pathways, as shown, so that the equations of correlation can be derived.

As the methanol solvent is in large excess, the observed solvolysis rate constant ($k_{solv}$) is pseudo-first order, and is related to the second-order rate constant of the bimolocular attack, $k_{MeOH}$, by Equation 2.11. If the concentration of benzoyl chloride is very low ($<10^{-3}$ M), and the amine concentration is at least ten times higher, the observed *overall* rate constant ($k_{obs}$) is pseudo-first order, and is related to the competing solvolysis and aminolysis rate constants by Equation 2.12. The validity of this equation was established by plotting $k_{obs}$ against [ArNH$_2$]; a straight line was obtained with slope $k_{am}$ and intercept $k_{solv}$ [44]. A

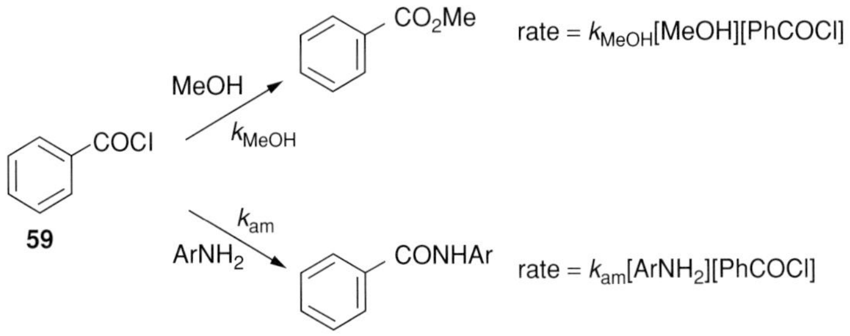

$$\text{rate} = k_{\text{MeOH}}[\text{MeOH}][\text{PhCOCl}]$$

$$\text{rate} = k_{\text{am}}[\text{ArNH}_2][\text{PhCOCl}]$$

**Scheme 2.21** Competing pathways in the reaction of benzoyl chloride (**59**) with methanol containing an arylamine.

further check was provided by the direct determination of $k_{\text{solv}}$, the solvolysis rate constant observed in the absence of added amine. Division of both sides of Equation 2.12 by $k_{\text{solv}}$ gives Equation 2.13, which quantifies the rate enhancement due to the addition of the amine to the methanol:

$$k_{\text{solv}} = k_{\text{MeOH}}[\text{MeOH}] \tag{2.11}$$

$$k_{\text{obs}} = k_{\text{solv}} + k_{\text{am}}[\text{ArNH}_2] \tag{2.12}$$

$$k_{\text{obs}}/k_{\text{solv}} = 1 + (k_{\text{am}}/k_{\text{solv}})[\text{ArNH}_2]. \tag{2.13}$$

The product ratio is equal to the ratio of rates of the second-order processes (Scheme 2.21) leading to their formation (Equation 2.14); rearranging Equations 2.11, 2.13 and 2.14 gives Equation 2.15. Alternatively, the mole fraction of amide product is obtained from Equation 2.16:

$$[\text{PhCONHAr}][\text{PhCO}_2\text{Me}] = k_{\text{am}}[\text{ArNH}_2]/k_{\text{MeOH}}[\text{MeOH}] = k_{\text{am}}[\text{ArNH}_2]/k_{\text{solv}} \tag{2.14}$$

$$\text{rate enhancement} = k_{\text{obs}}/k_{\text{solv}} = 1 + [\text{PhCONHAr}]/[\text{PhCO}_2\text{Me}] \tag{2.15}$$

$$k_{\text{am}}[\text{ArNH}_2]/k_{\text{obs}} = [\text{PhCONHAr}]/([\text{PhCONHAr}] + [\text{PhCO}_2\text{Me}]). \tag{2.16}$$

The left-hand side of Equation 2.15 refers to a ratio of experimental rate constants; $k_{\text{obs}}$ and $k_{\text{solv}}$ are obtained entirely from kinetic data (Equation 2.12) and the ratio $k_{\text{obs}}/k_{\text{solv}}$ is the rate enhancement for a particular concentration of amine; for example, if the ratio is 2.0, addition of the amine has doubled the rate, corresponding to a rate enhancement of 100%. The right-hand side refers to the observed product ratio, which is obtained by independent measurements (e.g. by HPLC). A fit to Equation 2.15, as observed for *m*-nitroaniline in methanol at 25°C, exemplifies a rate–product correlation (Table 2.4) [44]. From the agreement between calculated and observed yields of amide shown in Table 2.4, we conclude that there are indeed competing second-order reactions, as assumed in Scheme 2.21. If the mechanism proceeded via an intermediate formed in a slow step, which was trapped by the amine in a subsequent rapid step, then more amide product would be formed than predicted by Equation 2.15.

An added nucleophile may contribute a *medium effect* which could complicate interpretations of rate–product correlations (Equation 2.15); to avoid this, only low concentrations of nucleophiles should be used ($<10^{-2}$ M). Allowances can be made (at least partially) for the medium effect of added electrolytes by conducting reactions at a constant *ionic strength* (*I*); as the concentration of a reactive anionic nucleophile such as chloride or bromide

**Table 2.4** Rate constants and % amide product for solvolysis of benzoyl chloride (**59**) in methanol with added *m*-nitroaniline at 25°C [44].

| Initial concentrations (mol dm$^{-3}$) | | | % PhCONHAr | |
|---|---|---|---|---|
| [PhCOCl] | [ArNH$_2$] | $10^3 k_{obs}$ (s$^{-1}$) | Observed | Calculated* |
| $5 \times 10^{-5}$ | – | 4.50 | – | – |
| $5 \times 10^{-5}$ | $6.3 \times 10^{-4}$ | 5.47 | 20.2 | 17.7 |
| $1 \times 10^{-4}$ | $3.0 \times 10^{-3}$ | 9.05 | 48.4 | 50.3 |
| $1 \times 10^{-3}$ | $1.0 \times 10^{-2}$ | 18.8 | 77.0 | 76.1 |

*From Equation 2.15, $k_{solv} = 4.5 \times 10^{-3}$ s$^{-1}$ (first entry in the table); the amide:ester product ratio ($P_R$) was determined experimentally, so % amide is $100 P_R/(1 + P_R)$.

is reduced, a non-nucleophilic anion such as perchlorate or nitrate is added to maintain *I* constant. Medium effects of non-electrolytes have also been observed. For solvolyses of benzoyl chloride (**59**) in 50% acetone–water, 0.1 M *o*-nitroaniline unexpectedly caused a significant kinetic effect; when an inert model nitro compound (nitrobenzene) was used to maintain a constant solute concentration of nitroarene, Equation 2.15 was successful [45]. This use of nitrobenzene is exactly analogous to the use of an inert electrolyte to maintain a constant ionic strength.

The results in Table 2.4 are also relevant to preparative scale reactions. Adapting the definition of selectivities (Equation 2.10) for the reaction shown in Scheme 2.21 by rearranging Equation 2.14 gives a definition of *S* for competing methanolysis and aminolysis (Equation 2.17). Although the molar concentration of methanol in almost pure solvent is high (24.7 M), the major product is amide even when the concentration of *m*-nitroaniline is only $10^{-2}$ M (Table 2.4), and *S* is calculated from Equation 2.17 to be over 8000; very high yields are predicted for reactions in more concentrated solutions of *m*-nitroaniline. In contrast, under the same conditions, the less basic amine *o*-nitroaniline has an *S* value of only 6 [44]:

$$S = k_{am}/k_{MeOH} = [\text{PhCONHAr}][\text{MeOH}]/[\text{PhCO}_2\text{Me}][\text{ArNH}_2]. \qquad (2.17)$$

Observations of rate–product correlations for competing solvolysis without rearrangement ($k_s$ pathway) and solvolysis with rearrangement ($k_\Delta$ pathway) of 3-aryl-2-butyl sulfonates (**60** in Scheme 2.22) showed that the two pathways are independent, i.e. there is no crossover between the two pathways [46]. At that time (up to mid-1960s), it was thought that the solvolytic reactions of simple secondary alkyl sulfonates proceeded via cationic intermediates (carbenium ions as ion pairs or otherwise, as in Scheme 2.18). The following question was then asked. If a cationic intermediate (e.g. **61**) is formed, why are the reaction pathways independent? Alternatively, what prevents the aryl group in the cationic intermediate (**61**) from rearranging to give (**62**) or (**63**)? The answer to these questions was thought to be that any cationic intermediates formed during the direct substitution process ($k_s$ pathway) must be strongly solvated by a solvent molecule acting as a nucleophile [46].

Supporting evidence was later obtained from rate–product correlations for solvolyses of 2-propyl and 2-octyl sulfonates in the presence of added azide ion (N$_3^-$). The observed rate–product correlation [47] is consistent with competing S$_N$2 reactions of the covalent substrates (Scheme 2.23), rather than the trapping of a cationic intermediate by azide ion (Scheme 2.24). Although the medium effect of the added electrolyte complicates the interpretation of

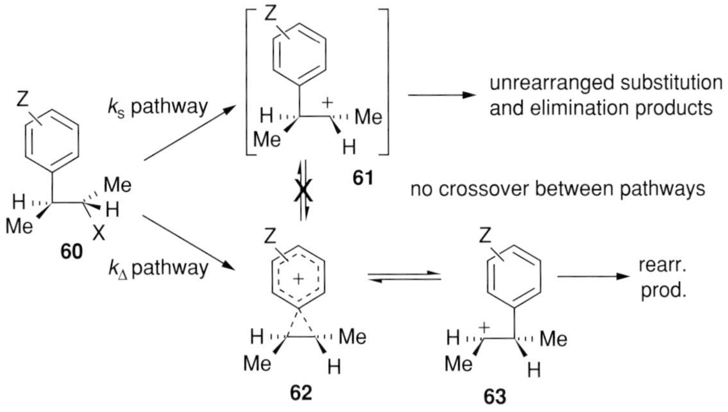

**Scheme 2.22**  Independent pathways in the solvolysis of 3-aryl-2-butyl sulfonates (**60**) established by the absence of crossover product.

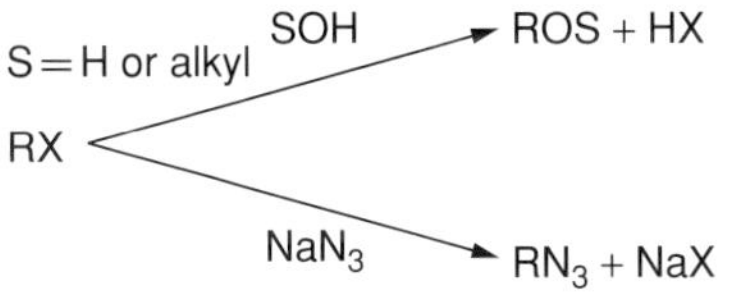

**Scheme 2.23**  Competing $S_N2$ pathways in solvolysis of 2-alkyl sulfonates in alcohol containing sodium azide.

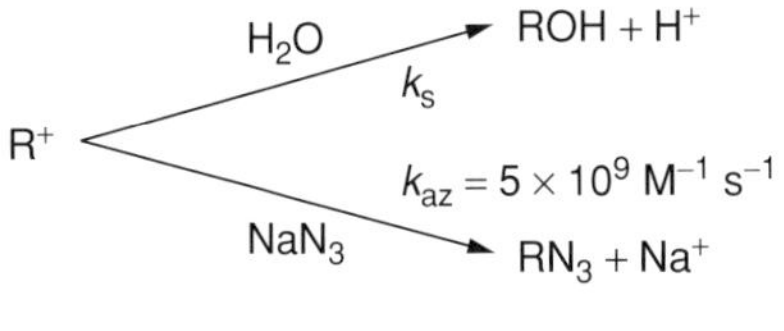

**Scheme 2.24**  The 'azide clock' involving competitive trapping of a carbenium ion by water and azide.

the rate–product correlation [47], independent kinetic and stereochemical results are also consistent with an $S_N2$ mechanism for solvolyses of simple secondary sulfonates.

The reaction of azide ions with carbocations is the basis of the 'azide clock' method for estimating carbocation lifetimes in hydroxylic solvents (lifetime $= 1/k_s$, where $k_s$ is the first-order rate constant for attack of water on the carbocation); this is analogous to the radical clock technique discussed in Chapter 10. In the present case, a rate–product correlation is assumed for the very rapid competing product-forming steps of $S_N1$ reactions (Scheme 2.24). Because the slow step of an $S_N1$ reaction is formation of a carbocation, typical kinetic data do not provide information about this step. Furthermore, the rate constant for the reaction of azide ion with a carbocation ($k_{az}$) is assumed to be diffusion controlled (ca. $5 \times 10^9\ M^{-1}\ s^{-1}$). The rate constant for attack by water can then be obtained from the mole ratio of azide product/solvolysis product, and the molar concentrations of azide (Equation 2.18, equivalent to Equation 2.14) [48]. The reliability of the estimated lifetimes was later confirmed independently by generating the cations very rapidly by laser flash photolysis, and then measuring their relatively slow decay [49]. This example clearly illustrates the general point noted previously – initial arguments based on product data and analogies may later be confirmed independently.

$$k_{az}[\text{azide}]/k_s = [\text{azide product}]/[\text{solvolysis products}]. \tag{2.18}$$

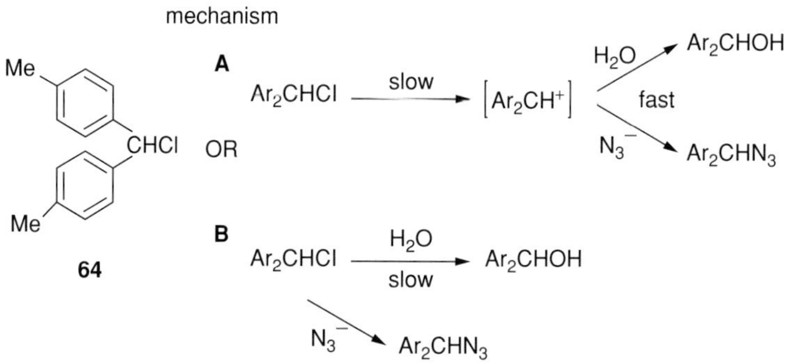

**Scheme 2.25** Alternative $S_N1$ and $S_N2$ mechanisms for the reaction of $p,p'$-dimethylbenzhydryl chloride (**64**) in aqueous acetone containing sodium azide.

Sometimes the absence of a rate–product correlation is mechanistically significant. If reactions of a substrate (e.g. **64**) occur via free carbocations, mechanism A in Scheme 2.25 is applicable. Solvolysis data are given in Table 2.5 [50], where the solvolysis rate constants now refer to ionisation/dissociation. In the presence of a strong nucleophile (such as azide), a relatively large amount of azide product is obtained (Table 2.5, entry 4). If the rate enhancement of 1.65 (70.5/42.7) due to the added azide were due to a competing process in which azide reacted directly with $Ar_2CHCl$ (Scheme 2.25, path B), a rate–product correlation would be expected. Then, as in Equation 2.16, the mole fraction of azide product would be 0.65/1.65 = 0.39 (i.e. 39%); the higher observed yield of azide product (65%) shows the absence of a rate–product correlation and supports a trapping mechanism for a reactive intermediate in which azide is reacting very rapidly with a reactive intermediate cation formed in an $S_N1$ reaction (Scheme 2.25, mechanism A), rather than reacting with the covalent substrate (mechanism B).

**Table 2.5** Solvolyses of $p,p'$-dimethylbenzhydryl chloride (**64**) in 85% acetone–water at 0°C in the presence of added salts [50].

| Entry no. | Salt | Salt conc (mol dm$^{-3}$) | $10^5 k_{solv}$ (s$^{-1}$)* | % Azide product |
|---|---|---|---|---|
| 1 | None | – | 42.7 | – |
| 2 | LiBr | 0.0504 | 57.0 | – |
| 3 | Me$_4$NNO$_3$ | 0.0506 | 72.0 | – |
| 4 | NaN$_3$ | 0.0512 | 70.5 | 65 |
| 5 | LiCl | 0.0555 | 23.0 | – |

*Initial rate constant.

Addition of non-nucleophilic electrolytes (e.g. nitrates, Table 2.5, entry 3) shows that the consequent increased rate is due to the increase in the ionic strength of the medium, further supporting the trapping mechanism. A comparison of the effects of added bromide and chloride (entries 2 and 5) illustrates the *common ion effect*, i.e. trapping $Ar_2CH^+$ by chloride to regenerate starting material retards solvolyses of $Ar_2CHCl$.

In general, mechanistic evidence for a reactive intermediate from trapping experiments needs to be linked to arguments against the introduction of an alternative pathway from the reactant, i.e. to show that an intermediate really has been trapped, not the reactant. A classic case is the hydrolysis of 4-nitrophenyl acetate catalysed by imidazole. The mechanism is nucleophile catalysis and the intermediate (*N*-acetylimidazolium cation) was trapped by aniline (to give acetanilide) with no kinetic effect, i.e. the aniline does not react directly with the substrate [51].

# Bibliography

### *Texts on structure elucidation by spectroscopic methods*

Field, L.D., Sternell, S. and Kalma, J.R. (2002) *Organic Structures from Spectra* (3rd edn). Wiley, New York.
Pavia, D.L., Lampman, G.M. and Kriz, G.S. (2003) *Introduction to Spectroscopy* (3rd edn). Harcourt Brace, Fort Worth, TX.
Silverstein, R.M. and Webster, F.X. (2005) *Spectrometric Identification of Organic Compounds* (7th edn). Wiley, New York.
Williams, D.H. and Fleming, I. (1995) *Spectroscopic Methods in Organic Chemistry* (5th edn). McGraw-Hill, New York.

### *Texts on quantitative chromatographic analyses*

Christian, G.D. (2004) *Analytical Chemistry* (6th edn). Wiley, New York.
Harris, D.C. (2003) *Quantitative Chemical Analysis* (6th edn). Freeman, New York.
Skoog, D.A., West, D.M., Holler, J.F. and Holler, F.J. (2003) *Fundamentals of Analytical Chemistry* (8th edn). Brooks/Cole, Pacific Grove, CA.

### *Texts giving more of the commercial background to the optimisation of organic syntheses*

Atherton, J.H. and Carpenter, K.J. (1999) *Process Development: Physicochemical Concepts*, Oxford Chemistry Primers 79. Oxford University Press, Oxford.
Lee, S. and Robinson, G. (1995) *Process Development – Fine Chemicals from Grams to Kilograms*, Oxford Chemistry Primers 30. Oxford University Press, Oxford.
Repic, O. (1998) *Principles of Process Research and Chemical Development in the Pharmaceutical Industry*. Wiley, New York.

# References

1. Influential early teaching texts include Hine, J. (1956) *Physical Organic Chemistry*. McGraw-Hill, New York; Gould, E.S. (1959) *Mechanism and Structure in Organic Chemistry*. Holt, New York; Sykes, P. (1961) *A Guidebook to Mechanism in Organic Chemistry*. Longman, London.
2. Jackson, R.A. (1972) *Mechanism, an Introduction to the Study of Organic Reactions*. Clarendon, Oxford: (a) p. 2; (b) pp. 12–15; (c) Chapter 5.
3. Hammett, L.P. (1970) *Physical Organic Chemistry* (2nd edn). McGraw-Hill, New York, (a) p. 1; (b) pp. 152–8.
4. Brown, H.C. (1972) *Boranes in Organic Chemistry*. Cornell University Press, Ithaca, NY, pp. 258–61.
5. (a) Bentley, T.W., Llewellyn, G., Kottke, T., Stalke, D., Cohrs, C., Herberth, E., Kunz, U. and Christl, M. (2001) *European Journal of Organic Chemistry*, 1279; (b) in association with the research described in reference 5(a), the NMR results were obtained by E. Ruckdeschel and M. Grüne, University of Wurzburg.
6. Claridge, T.D.W. (1999) *High Resolution NMR Techniques in Organic Chemistry*. Pergamon-Elsevier Science, Oxford, pp. 114–6.
7. Maskill, H. and Wilson, A.A. (1984) *Journal of the Chemical Society, Perkin Transactions 2*, 119.
8. Conant, J.B. and Bartlett, P.D. (1932) *Journal of the American Chemical Society*, 54, 2881.
9. Carroll, F.A. (1998) *Perspectives on Structure and Mechanism in Organic Chemistry*. Brooks/Cole, Pacific Grove, CA, pp. 344–9.
10. Snadden, R.B. (1985) *Journal of Chemical Education*, 62, 653.
11. Billups, W.E., Houk, K.N. and Stevens, R.V. (1986) Chapter 10 in: C.F. Bernasconi (Ed.) *Investigations of Rates and Mechanisms of Organic Reactions (Part 1)*. Wiley, New York.
12. Carpenter, B.K. (1984) *Determination of Organic Reaction Mechanisms*. Wiley, New York, Chapter 3; see also A. Williams, *Concerted Organic and Bio-organic Mechanisms*. CRC Press, Boca Raton, FL, 2000; A. Williams, *Free-Energy Relationships in Organic and Bio-organic Chemistry*. Royal Society of Chemistry, Cambridge, 2003.
13. Labinger, J.A. and Bercaw, J.E. (2002) *Nature*, 417, 507.
14. Roberts, S.M. (2004) *Tetrahedron*, 60, 499, and other articles in the same issue.
15. Furniss, B.S., Hannaford, A.J. Smith, P.W.G. and Tatchell, A.R. (1989) *Vogel's Textbook of Practical Organic Chemistry* (5th edn). Longman, New York, p. 808.
16. House, H.O. (1972) *Modern Synthetic Reactions* (2nd edn). Benjamin, Menlo Park, CA, p. 154.
17. Sykes, P. (1972) *The Search for Organic Reaction Pathways*. Longman, London, pp. 165–7.
18. Smith, J.B. and March, J. (2001) *Advanced Organic Chemistry* (5th edn). Wiley, New York, pp. 1419–21.
19. Wentrup, C. (1986) Chapter 9 in: C.F. Bernasconi (Ed.) *Investigations of Rates and Mechanisms of Organic Reactions (Part 1)*. Wiley, New York.
20. Roberts, J.D., Semenow, D.A., Simmons, Jr., H.E. and Carlsmith, L.A. (1956) *Journal of the American Chemical Society*, 78, 601.
21. Bunnett, J.F. (1978) *Accounts of Chemical Research*, 11, 413.
22. Wallis, E.S. and Lane, J.F. (1946) *Organic Reactions* (vol 3). Wiley, New York, Chapter 7.
23. Vedejs, E. and Snoble, K.A.J. (1973) *Journal of the American Chemical Society*, 95, 5778.
24. Smith, K., Conley, N., Hondrogiannis, G., Glover, L., Green, J.F., Mamantov, A. and Pagni, R.M. (2004) *Journal Organic Chemistry*, 69, 4843.
25. Engler, E.M., Farcasiu, M., Sevin, A., Cense, J.M. and Schleyer, P.v.R. (1973) *Journal of the American Chemical Society*, 97, 5769.
26. Schleyer, P.v.R. (1957) *Journal of the American Chemical Society*, 79, 3292.
27. Pazo-Llorente, R., Bravo-Diaz, C. and Gonzalez-Romero, E. (2004) *European Journal of Organic Chemistry*, 3221.

28. Canning, P.J.S., Maskill, H., McCrudden, K. and Sexton, B. (2002) *Bulletin of the Chemical Society of Japan*, **75**, 789.
29. Woodward, R.B. (1967) *Aromaticity, an International Symposium*, Special Publication No. 21. The Chemical Society, London, pp. 217–49.
30. Pine, S.H. (1972) *Journal of Chemical Education*, **49**, 664.
31. Brock, W.H. (1992) *The Fontana History of Chemistry*. Fontana Press, London, p. 259.
32. Gould, E.S. (1959) *Mechanism and Structure in Organic Chemistry*. Holt, New York, pp. 150–2.
33. Bentley, T.W., Llewellyn, G., Norman, S.J., Kemmer, R., Kunz, U. and Christl, M. (1997) *Liebigs Annals of Chemistry*, 229.
34. Bentley, T.W. and Gream, G.E. (1985) *Journal of Organic Chemistry*, **50**, (1776).
35. Ingold, C.K. (1953) *Structure and Mechanism in Organic Chemistry* (1st edn). Bell and Sons, London, (a) p. 392; (b) pp. 482–94.
36. Cooper, K.A., Hughes, E.D., Ingold, C.K. and MacNulty, B.J. (1948) *Journal of the Chemical Society*, 2038.
37. Doering, W.v.E. and Zeiss, H.H. (1953) *Journal of the American Chemical Society*, **75**, 4733.
38. Cocivera, M. and Winstein S. (1963) *Journal of the American Chemical Society*, **85**, 1702.
39. Toteva, M.M. and Richard, J.P. (1996) *Journal of the American Chemical Society*, **118**, 11434.
40. Pross, A. (1977) *Advances in Physical Organic Chemistry*, **14**, 69.
41. Ta-Shma, R. and Rappoport, Z. (1992) *Advances in Physical Organic Chemistry*, **27**, 239.
42. Bentley, T.W., Ratcliff, J., Renfrew, A.H.M. and Taylor, J.A. (1995) *Journal of the Society of Dyers and Colourists*, **111**, 288.
43. Bentley, T.W. and Koo, I.S. (1989) *Journal of the Chemical Society, Perkin Transactions 2*, 1385.
44. Bentley, T.W. and Freeman, A.E. (1984) *Journal of the Chemical Society, Perkin Transactions 2*, 1115.
45. Bentley, T.W., Carter, G.E. and Harris, H.C. (1985) *Journal of the Chemical Society, Perkin Transactions 2*, 983.
46. Brown, H.C., Kim, C.J., Lancelot, C.J. and Schleyer, P.v.R. (1970) *Journal of the American Chemical Society*, **92**, 5244.
47. Raber, D.J., Harris, J.M., Hall, R.E. and Schleyer, P.v.R. (1971) *Journal of the American Chemical Society*, **93**, 4821.
48. Richard, J.P., Rothenberg, M.E. and Jencks, W.P. (1984) *Journal of the American Chemical Society*, **106**, 1361.
49. McClelland, R.A. (1996) *Tetrahedron*, **52**, 6823.
50. Bateman, L.C., Hughes, E.D. and Ingold, C.K. (1940) *Journal of the Chemical Society*, 974.
51. Maskill, H. (1985) *The Physical Basis of Organic Chemistry*. Oxford University Press, Oxford, pp. 348–9.

# Chapter 3
# Experimental Methods for Investigating Kinetics

## M. Canle L., H. Maskill and J. A. Santaballa

## 3.1 Introduction

This chapter covers experimental approaches to the investigation of chemical kinetics. Well-established techniques in the field include spectroscopy, titrimetry, polarimetry, conductimetry, etc., but the wide range of circumstances of experimental studies of reaction mechanisms makes it impossible to include in a limited space all the techniques potentially available and the different ways in which they may be applied. Consequently, we limit ourselves to those which are more readily available and commonly used, and even here we shall not always go into detail; our aim is to indicate what is possible and to explore the underlying relationships between what is observed experimentally and the chemical phenomenon under investigation; more specialised texts provide greater detail for particular methods [1]. After covering some necessary concepts, we shall concentrate on practicalities, and on how one proceeds from experimental data to rate constants.

There are different justifications for the measurement of rate constants, but they generally fall into one of two categories: the need for quantitative information about the reactivity of a particular compound (or a family of compounds) for practical or technological reasons, and the wish to elucidate the mechanism of a particular reaction. The first objective is, usually, more easily fulfilled and requires less experimental work; investigating in detail the mechanism of a process usually involves much more than quantification of rate constants and will invariably require the application of other techniques, covered in other chapters of this book.

## 3.2 Preliminaries [2]

### 3.2.1 Reaction rate, rate law and rate constant

A chemical reaction is typically the transformation of one compound into another by one or more reagents (reversible reactions which proceed towards equilibrium are considered later). The rate of reaction ($r$) is defined as the change in the extent of the reaction ($\xi$) with time, Equation 3.1,

$$r = \frac{d\xi}{dt} = \frac{1}{v_i}\frac{dn_i}{dt},\qquad(3.1)$$

where $n_i$ is the number of moles of any reactant or product $i$, and $v_i$ is the stoichiometric coefficient for species $i$ in the chemical equation describing the reaction; by convention, $v_i$ is positive for products and negative for reactants.

Usually, two conditions are imposed: (i) the process is isothermal, and (ii) the process takes place at constant volume. These restrictions, which are relatively easily fulfilled in the overwhelming majority of reactions we might wish to study, allow us to rewrite Equation 3.1 for the reaction rate as Equation 3.2 where $[i]$ is the molar concentration of species $i$, and $V$ is the volume of the reaction:

$$r = \frac{d\xi}{dt} = \frac{V}{v_i} \left( \frac{d[i]}{dt} \right)_{V,T} . \tag{3.2}$$

Thus, for the reaction of Equation 3.3, we may write Equation 3.4:

$$v_a A + v_b B \quad \rightarrow \quad v_c C \tag{3.3}$$

$$\text{rate} \propto \left( \frac{-1}{v_a} \frac{d[A]}{dt} = \frac{-1}{v_b} \frac{d[B]}{dt} = \frac{+1}{v_c} \frac{d[C]}{dt} \right), \tag{3.4}$$

where the plus and minus signs indicate whether the species is a reactant (negative) or product (positive). Of course, reactants *decrease* in concentration, so the differential term for a reactant, e.g. $d[A]/dt$, is negative, and products *increase* in concentration, so $d[C]/dt$ is positive. It follows that all terms in Equation 3.4 are positive when real values are introduced, and the equation as a whole is mathematically sound.

We note from Equation 3.4 that the changing concentrations of the components in the reaction with time are, strictly speaking, only *proportional to* the rate of reaction, as rigorously defined in Equations 3.1 and 3.2. Experimentally, the instantaneous rates of changes of the concentrations of the components in a chemical equation are usually proportional to the instantaneous concentrations of the components themselves, each raised to some power. Such a relationship is called the *rate law*, or *rate equation*, of the reaction. For example, the rate law for the reaction of Equation 3.3 might be Equation 3.5:

$$\left( \frac{-1}{v_a} \frac{d[A]}{dt} = \frac{-1}{v_b} \frac{d[B]}{dt} = \frac{+1}{v_c} \frac{d[C]}{dt} \right) = k \, [A]^\alpha \, [B]^\beta, \tag{3.5}$$

where $k$ is the *rate coefficient* [2], or *rate constant*, and $\alpha$ and $\beta$ are the *order of the reaction* in components A and B (*rate constant* is much the more common name for $k$, and we shall use it, although it is less satisfactory since $k$ depends on many factors, as will be seen later).

When the stoichiometric coefficients, $v_a$, $v_b$, etc., are included in the rate law, as in Equation 3.5, the reaction has a unique rate constant ($k$) under specified conditions regardless of whether the rate is measured by monitoring the changing concentration of A, B or C. It also follows from Equation 3.5 that (except for zero-order reactions) the *instantaneous* rate of a reaction changes as the reaction proceeds, as will be illustrated later in Fig. 3.1. Thus, $k$ is the parameter which measures whether the reaction (imprecisely expressed) is 'fast' or 'slow'. In any case, it follows that any property of a reacting system which relates (preferably directly) to the concentration of any component in the chemical reaction may be monitored to measure the rate and, hence, to investigate the rate law and quantify the rate constant.

### 3.2.2   Reversible reactions, equilibrium and equilibrium constants

Equilibrium is reached in a reversible chemical process at constant temperature and pressure when the free energy of the system becomes minimal with respect to the extent of reaction, Equation 3.6:

$$\left(\frac{\partial G}{\partial \xi}\right)_{P,T} = 0 \text{ and } \left(\frac{\partial^2 G}{\partial \xi^2}\right)_{P,T} > 0. \tag{3.6}$$

By standard thermodynamics, a dimensionless *equilibrium constant* can be derived from this condition and, for a general reversible chemical reaction, Equation 3.7, we may write Equation 3.8 which defines the equilibrium condition:

$$\nu_A A + \nu_B B + \cdots \rightleftharpoons \nu_X X + \nu_Y Y + \cdots \tag{3.7}$$

$$K_a^o = \frac{a_X^{\nu_X} a_Y^{\nu_Y} \cdots}{a_A^{\nu_A} a_B^{\nu_B} \cdots} (a^o)^{-\sum \nu_i}, \tag{3.8}$$

where $a_i$ is the *activity* of species $i$ at equilibrium, $\nu_i$ is the stoichiometric coefficient for $i$ and $a^o$ is the standard state activity (with units). Taking into account the relationship between activity and concentration ($a_i = \gamma_i[i]$ where $\gamma_i$ is an activity coefficient), the term comprising activities in Equation 3.8 can be dissected into a term comprising concentrations ($K_c$) and another comprising activity coefficients ($K_\gamma$), Equation 3.9:

$$K_a^o = \frac{[X]^{\nu_X}[Y]^{\nu_Y} \cdots}{[A]^{\nu_A}[B]^{\nu_B} \cdots} \frac{\gamma_X^{\nu_X} \gamma_Y^{\nu_Y} \cdots}{\gamma_A^{\nu_A} \gamma_B^{\nu_B} \cdots} (a^o)^{-\sum \nu_i} = K_c K_\gamma (a^o)^{-\sum \nu_i}. \tag{3.9}$$

However, for most reactions in solution, a less rigorously defined *practical equilibrium constant*, derived from Equation 3.9 and expressed in terms of molar concentrations of species at equilibrium, may be sufficient, and we note that, defined in Equation 3.10, $K_c$ is not dimensionless, but has units (unless $[\nu_X + \nu_Y + \cdots] = [\nu_A + \nu_B + \cdots]$):

$$K_c = \frac{[X]^{\nu_X} \cdot [Y]^{\nu_Y} \cdots}{[A]^{\nu_A} \cdot [B]^{\nu_B} \cdots}. \tag{3.10}$$

Thus, any physical property that relates (preferably directly) to concentration, e.g. pH or spectrophotometric absorbance, may be used to investigate a chemical equilibrium and to measure the corresponding equilibrium constant, $K_c$.

### 3.2.3   Reaction mechanism, elementary step and rate-limiting step

For our present purposes, we use the term *reaction mechanism* to mean a set of simple or elementary chemical reactions which, when combined, are sufficient to explain (i) the products and stoichiometry of the overall chemical reaction, (ii) any intermediates observed during the progress of the reaction and (iii) the kinetics of the process. Each of these elementary steps, at least in solution, is invariably unimolecular or bimolecular and, in isolation, will necessarily be kinetically first or second order. In contrast, the kinetic order of each reaction component (i.e. the exponent of each concentration term in the rate equation) in the observed chemical reaction does not necessarily coincide with its stoichiometric coefficient in the overall balanced chemical equation.

An elementary step cannot be further divided into simpler steps, i.e. elementary steps are the indivisible components of chemical reaction mechanisms. One of the elementary steps may sometimes be identified as the so-called *rate-limiting* or *rate-determining step*, i.e. the one which principally limits the rate of the overall process. This can be clarified by a real example.

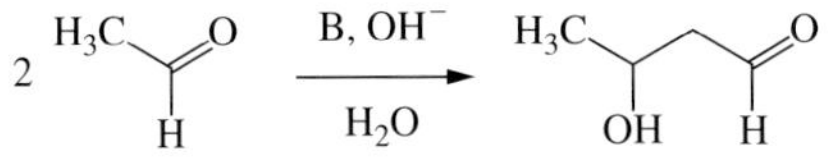

$$(3.11)$$

The base-catalysed Aldol reaction is shown in Equation 3.11 [3, 4], and a mechanism to account for the global process in Scheme 3.1. At low concentrations of acetaldehyde, reverse of the proton-abstraction steps is fast compared with the forward bimolecular enolate capture ($k_3[CH_3CHO] \ll k_{-1}$ and $k_{-2}[BH^+]$), and the $k_3$ step is rate limiting. Under these conditions, the kinetics are second order in $[CH_3CHO]$ and show specific base catalysis, i.e. the reaction is first order in $[HO^-]$ and, even though B is involved in the mechanism, it does not appear in the rate law [5]. According to this mechanism, therefore, the overall rate law is given by Equation 3.12:

$$\text{rate} = (k_3 k_1 / k_{-1})[CH_3CHO]^2[HO^-]. \qquad (3.12)$$

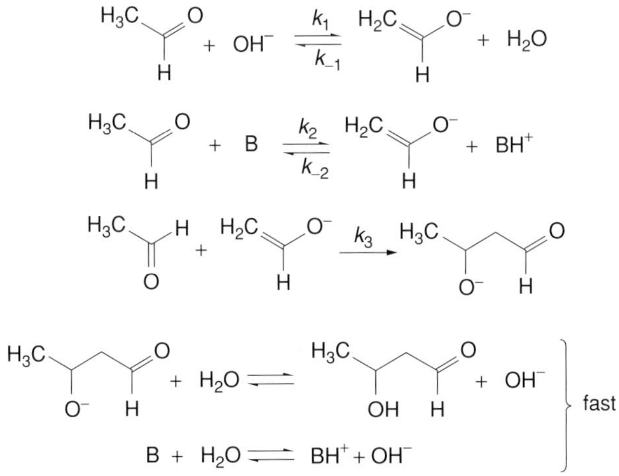

**Scheme 3.1** A mechanism for the base-catalysed Aldol reaction.

On the other hand, at high concentrations of acetaldehyde, when the intermediate enolate carbanion is rapidly captured by another molecule of aldehyde, reverse of the initial parallel proton-abstraction steps is prevented ($k_3[CH_3CHO] \gg k_{-1}$ and $k_{-2}[BH^+]$), and the rate of the overall reaction is effectively limited by the initial proton abstractions; these then constitute (parallel) rate-limiting steps. The overall process is now first order in acetaldehyde and shows general-base catalysis [5], i.e. the rate law is given by Equation 3.13:

$$\text{rate} = (k_1[HO^-] + k_2[B])[CH_3CHO]. \qquad (3.13)$$

During the course of an Aldol reaction, the concentration of acetaldehyde inevitably decreases, so a reaction may have the rate law of Equation 3.13 in the early stages when

[CH$_3$CHO] is high, and the rate law of Equation 3.12 towards the end when [CH$_3$CHO] is low; in the middle stages of the reaction, there will not be a clean simple-order rate law.

How a rate law is derived from a postulated mechanism will be dealt with in the following chapter, including a more detailed consideration of the Aldol reaction.

### 3.2.4    Transition structure and transition state

The most widely accepted theoretical model to describe the progress from reactant to product(s) in an elementary step is *transition state theory*. Various aspects of transition state theory are invoked in various chapters of this book, but the reader is referred to other texts specifically on kinetics for a detailed exposition [6]. Fundamental precepts, however, are that reactant molecules are in statistical thermodynamic equilibrium with *activated complexes* (also called *transition structures*) and that these species then partition between forward progress to the product of the elementary step, and reversion to the reactant. The activated complex, therefore, is the molecular species corresponding to the saddle point which separates the reactant and product on the potential energy hypersurface of the elementary reaction. It has an imaginary vibrational frequency corresponding to the main molecular event taking place along the reaction coordinate at the saddle point (e.g. a bond being broken).

If the reaction is considered on the molar scale, the activated complexes are the molecular species at the hypothetical free energy maximum which separates reactant and product molecules. The distinction between an activated complex (transition structure), which is a real molecular species, though of exceedingly short lifetime, and transition state, which is a hypothetical thermodynamic state on the molar scale, is important though frequently confused [7].

## 3.3    How to obtain the rate equation and rate constant from experimental data

As seen above in Section 3.2.1, the rate equation for the chemical process in Equation 3.3 was given as Equation 3.5. We shall now use the word 'rate' not as rigorously defined in Section 3.2.1, but more loosely to mean any of the differential terms in Equation 3.5. Consequently, the rate law for the reaction of Equation 3.3 is now written as Equation 3.14:

$$\nu_a A + \nu_b B \rightarrow \nu_c C \tag{3.3}$$

$$\text{rate}(r) = k[A]^\alpha [B]^\beta, \tag{3.14}$$

where $k$ is the rate constant, $\alpha$ and $\beta$ are the orders of the reaction with respect to reactants A and B, and which have to be determined experimentally; two possible methods follow immediately from the form of Equation 3.14: the *differential method* and the *method of integration*.

### 3.3.1 Differential method

Taking logarithms of Equation 3.14, we obtain Equation 3.15 – another relationship between the rate at any instant, the rate constant, and the instantaneous concentrations of reactants:

$$\ln r = \ln k + \alpha \ln[A] + \beta \ln[B]. \tag{3.15}$$

There are alternative methods of proceeding. When [B] is kept constant (e.g. in large excess), a plot of [A] against time allows the rate $(r)$ to be determined at any instant from the tangent at any value of [A], as in Fig. 3.1a. The gradient of the linear plot of $\ln r$ against $\ln$ [A] then gives $\alpha$, the reaction order with respect to A. In another experiment, [A] is kept constant, and a plot of [B] against time gives the rate $(r)$ at any stage; the gradient of a new plot of $\ln r$ against $\ln$ [B] then gives $\beta$, the reaction order with respect to B. Reaction orders obtained in this way are called 'with respect to concentration'. In Fig. 3.1c, the gradients at time $= 0$ for three separate reactions give the rates at time $= 0$ for the different initial concentrations of the limiting reactant, i.e. in these cases, the *initial rates* $(r_0)$ at the known *initial concentrations*; Fig. 3.1d shows the plot of $\ln r_0$ against $\ln$ [reactant]$_0$. This, the so-called *method of initial rates*, is a special case of the differential method and is used for very slow reactions, or for when it would be impracticable for other reasons to monitor the reaction until completion.

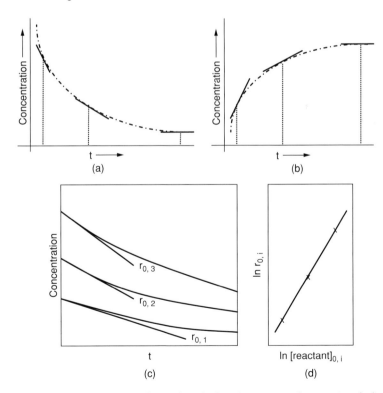

**Fig. 3.1** Illustration of how to use the differential method: (a) for a reagent disappearing; (b) for a product appearing; (c) [reactant] versus $t$ for different initial concentrations showing the slopes at $t = 0$; (d) plot of the logarithms of the different slopes obtained in (c) versus logarithms of the initial concentrations, i.e. $\ln r_{0,i}$, versus $\ln$[reactant]$_{0,i}$.

If only about the first 1% of a slow reaction is monitored, a plot of the concentration versus time plot is virtually linear, so the initial rate can be measured with little of the uncertainty associated with fitting a tangent to a curve.

An alternative way of employing the differential method is to use only one concentration versus time curve for each reactant in turn, all other reactants in each experiment being kept constant. For the reaction of Equation 3.3, for example, the plot of [A] against time at constant [B] allows tangents at different values of $[A]_t$ to generate a set of $r_t$ values. The slope of the linear plot of ln $r_t$ against ln $[A]_t$ gives $\alpha$. Then, from a new experiment at constant [A], [B] is first plotted against time to give a set of $r_t$ values; then the linear plot of ln $r_t$ against ln$[B]_t$ gives $\beta$. When obtained in this way, reaction orders are called 'with respect to time'. In a simple process, orders *with respect to concentration* should be the same as when determined *with respect to time*, but this will not be the case if an intermediate or a product inhibits or catalyses the reaction (in the latter case, we have *autocatalysis*). Inhibition by an intermediate or a product will lead to the order with respect to time being lower than the order with respect to concentration, while autocatalysis will lead to the order with respect to time being higher than the order with respect to concentration.

Determination of the rate constant by the differential method is straightforward; either of the above-mentioned plots used to determine $\alpha$ and $\beta$ gives $k$ from the intercept.

The mathematical relationships using the differential method to determine $\alpha$, $\beta$ and $k$ are linear, and any commercial spreadsheet or scientific software package will have a suitable fitting program. It is desirable that the fitting software gives the standard deviations of the fitted parameters, so that the statistical errors of reaction orders and rate coefficients are known.

Although not very commonly used (with the exception of the initial rate procedure for slow reactions), the differential method has the advantage that it makes no assumption about what the reaction order might be (note the contrast with the method of integration, Section 3.3.2), and it allows a clear distinction between the order with respect to concentration and order with respect to time. However, the rate constant is obtained from an intercept by this method and will, therefore, have a relatively high associated error. The initial rates method also has the drawback that it may miss the effect of products on the global kinetics of the process.

### 3.3.1.1    Example: reaction between RBr and HO⁻

The initial rate of reaction between an alkyl bromide and $HO^-$ was measured as a function of the initial concentrations of the reagents, with the following results [8]:

| $[RBr]_0$ (mol dm$^{-3}$) | 0.1 | 0.2 | 0.2 |
|---|---|---|---|
| $[HO^-]_0$ (mol dm$^{-3}$) | 0.01 | 0.01 | 0.02 |
| $r_0$ (mol dm$^{-3}$ s$^{-1}$) | $2.1 \times 10^{-5}$ | $4.2 \times 10^{-5}$ | $8.1 \times 10^{-5}$ |

The initial rate is given by Equation 3.16:

$$r_0 = k[RBr]_0^{\alpha}[HO^-]_0^{\beta}. \tag{3.16}$$

Taking logarithms, log $r_0$ = log $k + \alpha$ log $[RBr]_0 + \beta$ log $[HO^-]_0$, and using the data in the first two columns (for which $[HO^-]_0$ is constant), we obtain a pair of simultaneous equations from which we calculate $\alpha = 1.0$. Then, from the second and third columns (for which $[RBr]_0$ is constant), we obtain two other simultaneous equations, from which we calculate $\beta = 0.97$. Realistically, these are both 1, so the reaction is second order overall, i.e. the rate law is $r = k\,[\text{RBr}][\text{HO}^-]$. Although it is also possible to calculate the rate constant, it would have a high error, and is better determined by the method of integration (see below).

## 3.3.2 Method of integration

This method allows accurate rate constants to be determined once the reaction order with respect to each reactant is known. It consists of integrating the differential rate equation to obtain a mathematical expression for the relationship between the concentration of some component and time; fitting concentration and time data to the integrated rate equation then allows determination of the rate constant. In a sense, it is complementary rather than alternative to the differential method, but often it is used alone on a trial and error basis using different *assumed* reaction orders, i.e. different integrated rate equations are used in turn to see which gives the best fit. This is satisfactory if the reaction order turns out to be simple (0, 1 or 2), but is unsound if non-integral orders are involved, which is not uncommon. Table 3.1 compiles the differential and integrated rate equations for the commonest rate laws, i.e. those of orders 0, 1 and 2; the reader is referred to the specialised literature for treatments of more complicated rate laws [9].

   This method has the important advantage that it only requires knowing the way the concentration (or some property that is proportional to it) changes with time, which may greatly simplify the work by avoiding time-consuming calibrations of the instrumentation used. Too commonly, however, the method of integration is used without a preliminary analysis of the reaction using the differential method. As a consequence, a reaction order is assumed without proper consideration of possible alternatives (including fractional orders) or of deviations that may result from inhibition or autocatalysis (see Section 3.3.1). Additionally, it is not always easy to distinguish between alternative integrated rate equations simply by scrutiny of the concentration versus time curves. For example, first-order and second-order reactions of compound A in Table 3.1 lead to superficially similar [A] versus time plots (see

**Table 3.1** Differential and integrated rate equations for irreversible processes.*

| Reaction | Order in A | Differential rate equation | Integrated rate equation |
|---|---|---|---|
| $A \rightarrow X + Y + \cdots$ | 0 | $\dfrac{dx}{dt} = k$ | $[A]_t = [A]_0 - kt,$  or  $x = kt$ |
| $A \rightarrow X + Y + \cdots$ | 1 | $\dfrac{dx}{dt} = k([A]_0 - x)$ | $[A]_t = [A]_0 e^{-kt}$  or  $\ln \dfrac{[A]_0}{([A]_0 - x)} = kt$ |
| $2A \rightarrow X + Y + \cdots$ | 2 | $\dfrac{dx}{dt} = k([A]_0 - x)^2$ | $[A]_t = \dfrac{[A]_0}{(1 + [A]_0 kt)}$  or  $\dfrac{1}{[A]_0}\dfrac{x}{([A]_0 - x)} = kt$ |
| $A + B \rightarrow X + Y + \cdots$ | 2 | $\dfrac{dx}{dt} = k([A]_0 - x)([B]_0 - x)$ | $\dfrac{1}{([A]_0 - [B]_0)} \ln \dfrac{[B]_0([A]_0 - x)}{([A]_0([B]_0 - x))} = kt$ |

* In all cases, $[A]_t$ is the concentration of the (limiting) reactant A at any time ($t$), $x$ is the concentration of A that has been consumed in the reaction at time $t$, which, from the stoichiometry, is equal to the concentration of the product X that has been formed; $[A] = [A]_0$ when time $= 0$; correspondingly (when B is involved), $[B] = [B]_0$ when $t = 0$.

the example in Section 3.3.3.1). Of course, the possibility of a wrong attribution of reaction order will be minimised by monitoring the reaction for the greatest practicable extent of reaction and by using data of the best possible quality.

### 3.3.2.1   Data handling

It was previously normal practice to use linear forms of rate equations to simplify determination of rate constants by graphical methods. For example, the logarithmic version of the first-order rate law (Table 3.1), Equation 3.17a, allows $k$ to be determined easily from the gradient of a graph of $\ln C_t$ against time, by fitting the data to the mathematical model, $y = a + bx$:

$$\text{(a) } \ln C_t = \ln C_0 - kt \qquad \text{(b) } C_t = C_0 e^{-kt}. \qquad (3.17)$$

However, mathematical manipulations (in this case, taking logarithms of the observed concentrations or of related physical properties) increase the uncertainty in the result by expanding errors when using imprecise data. This is specially true of data collected towards the end of a reaction when the differences in concentration between successive readings are small and decreasing.

With the development of inexpensive powerful computers and appropriate software, it has become increasingly possible to use non-linear optimisation procedures with direct experimental readings as the input. Thus, concentration–time data may be fitted directly to the exponential version of the first-order rate law, Equation 3.17b.

In both versions of Equation 3.17, $C_0$ and $k$ are the parameters which are optimised (even though an experimental value of $C_0$ may be available) using an iterative method starting from estimated approximate values. All the popular scientific fitting packages include these and other common equations of correlation, but only some allow the kineticist to define the mathematical equation for a non-standard rate law. This is a critical point to check when choosing fitting software. It is also important that the software allows a graphical representation of the residuals of the fit, i.e. the differences between experimentally measured values and the calculated ones. A good fit to the mathematical model should produce random residuals, and the better the data, the smaller the residuals. But if the mathematical model is not appropriate, the residuals show a clear systematic pattern [10]. Fig. 3.2 shows a comparison of the residuals obtained for first-order and second-order fits of the same experimental data from a process A + B $\rightarrow$ products, with $[A]_0 = [B]_0$. The random nature of the residuals in (b) compared with the non-random pattern in (a) allows a clear differentiation: it is a second-order reaction. This criterion frequently gives a better idea about the quality of the fit than the regression coefficient, and there are nice examples in the literature [11].

### 3.3.2.2   Example: decomposition of $N_2O_5$ in $CCl_4$

The following data were obtained at 45°C for the decomposition of dinitrogen pentoxide in $CCl_4$ according to the equation, $2N_2O_5 \rightarrow 4NO_2 + O_2$.

| $t$ (s) | 0 | 319 | 867 | 1198 | 1877 | 2315 |
|---|---|---|---|---|---|---|
| $[N_2O_5]$ (mol dm$^{-3}$) | 2.33 | 1.91 | 1.36 | 1.11 | 0.72 | 0.55 |

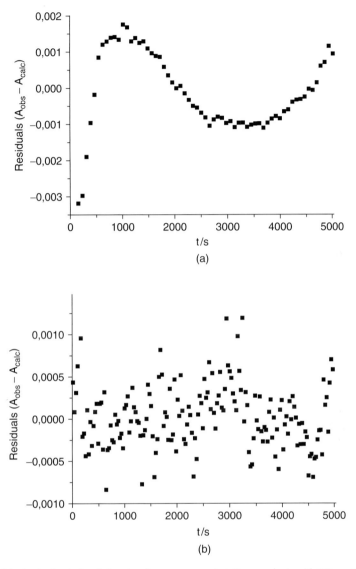

**Fig. 3.2**    Residuals obtained after fitting data from a process A + B → products with $[A]_0 = [B]_0$ to (a) the first-order rate equation and (b) the second-order rate equation [11].

To test whether the reaction is first order, we simply fit the data to the exponential integrated first-order rate equation (Table 3.1) using a non-linear optimisation procedure and the result is shown in Fig. 3.3. The excellent fit shows that the reaction follows the mathematical model and, therefore, that the process is first order with respect to $[N_2O_5]$, i.e. the rate law is $r = k_{obs}[N_2O_5]$. The rate constant is also obtained in the fitting procedure, $k_{obs} = (6.10 \pm 0.06) \times 10^{-4}\,\mathrm{s}^{-1}$. We see that, even with such a low number of experimental points, the statistical error is lower than 1%, which shows that many data points are not needed if

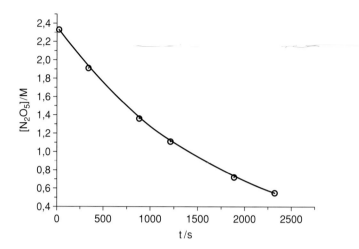

**Fig. 3.3** Fit of the experimental data to the integrated first-order rate equation for the process $2N_2O_5 \rightarrow 4NO_2 + O_2$ at 45°C in $CCl_4$.

their quality is good. Note also that, although $N_2O_5$ has a stoichiometric coefficient of 2 in the balanced equation, the reaction is still first order in this reagent. This tells us that the overall reaction cannot be an elementary bimolecular step, so we have to find a mechanism comprising elementary steps that together allow derivation of the observed rate law and explain the products and stoichiometry. Such matters are covered in the next chapter.

### 3.3.3   Isolation method

This is not by itself a kinetic method. It must be combined with either the differential or the integration method and involves keeping all the reactants but one in large excess so that their concentration does not vary through the reaction; under these conditions, the observed reaction order is that of the limiting reagent. For example, the simple second-order reaction of Equation 3.18,

$$A + B \rightarrow C, \tag{3.18}$$

could have a rate law of the form shown in Equation 3.19,

$$r = k[A]^{\alpha}[B]^{\beta}. \tag{3.19}$$

If B is in large excess, [B] will not change appreciably as A is consumed, so the rate law becomes $r = k'[A]^{\alpha}$, i.e. the observed reaction now has a pseudo-$\alpha$-order rate law, and the relationship between $k$ and $k'$ is shown in Equation 3.20:

$$k' = k[B]^{\beta}. \tag{3.20}$$

Correspondingly, [A] could be in large excess allowing the reaction order with respect to [B] to be isolated, so the overall rate law is established part by part in turn.

Most commonly, this method is used to transform complex rate laws into much more manageable pseudo-first-order versions. However, there are traps for the unwary; as we saw for the Aldol reaction in Section 3.2.3, changing concentrations of reactants may lead to changes in the rate law if a different step becomes rate limiting.

### 3.3.3.1  Example: oxidation of methionine by HOCl

The essential amino acid methionine (Met) acts intracellularly as an antioxidant, scavenging HOCl that is generated by the enzymatic system, myeloperoxidase/$Cl^-$/$H_2O_2$ [12]. During the reaction, the corresponding sulfoxide (MetSO) and $Cl^-$ are generated. Results by monitoring $[Cl^-]$ against time in one reaction are plotted in Fig. 3.4 using the integrated second-order rate equation (see Table 3.1) and a non-linear optimisation procedure. As shown, the data nicely fit the integrated second-order rate equation, indicating that the process is second order overall, with an observed rate constant $k_{obs} = (40 \pm 7)$ dm$^3$ mol$^{-1}$ min$^{-1}$.

However, we are still not ready to write the complete rate equation, as it is necessary to establish the order with respect to each of the reagents (Met and HOCl) separately – so far we know only that the reaction is second order overall. This matter can be resolved using the isolation method. By taking logarithms of Equation 3.20, we obtain $\ln k' = \ln k + \beta \ln[B]$. Applying this to the present problem, we may write $\log k'_{obs} = \log k_{obs} + \beta \log[Met]$, where $\beta$ is the reaction order with respect to methionine. Thus, $k'_{obs}$, the experimental pseudo-first-order rate constant, is measured at different excess concentrations of methionine, and the gradient of a plot of $\log k'_{obs}$ against $\log [Met]$ gives $\beta$, the order with respect to methionine.

The log–log plot shown in Fig. 3.5 is indeed linear with gradient $= 1.1 \pm 0.1$, which shows that the reaction is first order with respect to [Met] and, therefore (since it is second order overall), it must be first order with respect to [HOCl]. Thus, the complete rate equation is $r = k_{obs}[HOCl][Met]$, and $k_{obs} = (40 \pm 7)$ dm$^3$ mol$^{-1}$ min$^{-1}$ under the conditions of the reaction.

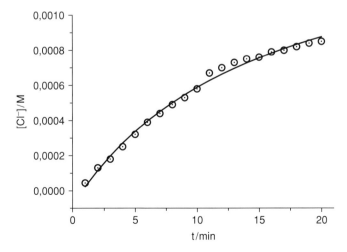

**Fig. 3.4**  Rate of formation of chloride in the oxidation of methionine (Met) with HOCl; $[Met]_0 = [HOCl]_0 = 1$ mM, 298.0 K, $I = 0.10$ M.

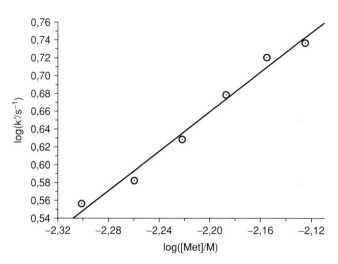

**Fig. 3.5**   Determination of the kinetic order with respect to methionine (Met) for the reaction between Met and HOCl (see the example in Section 3.3.3.1); [HOCl] = 0.5 mM, pH = 12, $I$ = 0.50 M, $T$ = 298.0 K.

## 3.4   Reversible reactions and equilibrium constants

The main feature about reversible reactions is that the reactants are not fully converted into products – only to a concentration determined by the equilibrium constant. If the reversibility is not recognised, it may just seem that the forward reaction is not quantitative.

### 3.4.1   Rate constants for forward and reverse directions, and equilibrium constants

Consider the generic case of a reversible isomerisation, $A \rightleftharpoons B$; if we define $x$ as the concentration of A that has reacted, i.e. the instantaneous concentration of A during the reaction is $([A]_0 - x)$, the rate law is given by Equation 3.21:

$$A \rightarrow B \qquad \text{rate constant} = k_f$$

$$B \rightarrow A \qquad \text{rate constant} = k_b$$

$$\frac{dx}{dt} = k_f([A]_0 - x) - k_b x. \tag{3.21}$$

When the system is at equilibrium, we denote the equilibrium value of $x$ by $x_e$, and the rates of the forward and back reactions are equal, i.e.

$$k_f([A]_0 - x_e) = k_b x_e,$$

so using this to substitute for $[A]_0$ in Equation 3.21, we obtain Equation 3.22:

$$\frac{dx}{dt} = (k_f + k_b)(x_e - x). \tag{3.22}$$

Since $x = 0$ when $t = 0$, integration of Equation 3.22 gives Equation 3.23:

$$\ln\left(\frac{x_e}{x_e - x}\right) = (k_f + k_b)t, \qquad (3.23)$$

which has the form of a first-order rate equation with $k_{obs} = (k_f + k_b)$. Consequently, $(k_f + k_b)$ may be obtained from $x$ versus $t$ data using Equation 3.23. The equilibrium constant $(K = [B]_e/[A]_e)$ may be determined from the measured equilibrium concentrations, but $K$ is also given by $K = k_f/k_b$. Consequently, the value of $(k_f + k_b)$ from the rate data and that of $k_f/k_b$ from the equilibrium allow determination of individual rate constants, $k_f$ and $k_b$.

### 3.4.1.1   Example: cis-trans *isomerisation of stilbene*

The following data are for the reversible isomerisation:

*cis*-stilbene $\rightleftharpoons$ *trans*-stilbene at 298 K[13].

| $t$ (min) | 0 | 20 | 50 | 80 | 120 | 170 | $\infty$ |
|---|---|---|---|---|---|---|---|
| % *cis*-stilbene | 100 | 92.5 | 82.3 | 73.6 | 63.7 | 53.8 | 17.1 |

From these data, $x_e = 82.9$ and, with this value, a plot $\ln\{x_e/(x_e - x)\}$ versus $t$ can be constructed. The data fit Equation 3.23 very well and, from the gradient, $(k_f + k_b) = 4.8 \times 10^{-3}$ min$^{-1}$. From the infinity concentrations at equilibrium, $K = 82.9/17.1$, so $k_f/k_b = 4.85$. With values now for the sum and ratio of $k_f$ and $k_b$, we obtain $k_f = 3.98 \times 10^{-3}$ min$^{-1}$ and $k_b = 8.2 \times 10^{-4}$ min$^{-1}$.

The forward and reverse directions of an elementary reaction represent a particular case of a reversible process. The *principle of microscopic reversibility* states that, for such a reaction, there is a single common pathway, i.e. a single transition structure for forward and reverse directions [14]. For the thermal interconversion of *cis*- and *trans*-stilbenes, the single transition structure comprises two PhCH moieties, singly bonded with a dihedral angle of 90°. (In contrast, the forward and reverse paths in the more complicated photochemical interconversion of *cis*- and *trans*-stilbenes are different, and the principle of microscopic reversibility does not apply.)

## 3.5   Experimental approaches

As indicated earlier, the kinetics of a reaction are characterised by a rate constant under specified conditions, and it is important that the conditions be specified as comprehensively as possible. In this section, we cover preliminary studies and identify the principal features that need to be kept constant with an indication of how the kinetics might be affected if they are not kept constant.

### 3.5.1   *Preliminary studies*

Before getting involved in any kind of detailed study on kinetics, the structures of products and the stoichiometry of the chemical process must be established (see Chapter 2). It is also prudent to consider possible side reactions leading to by-products, as these may interfere with the process under study. Most of these processes can be classified into a few categories: parallel, consecutive and reversible reactions.

As many physicochemical characteristics as possible of reactants, possible intermediates and products should be considered. These might include spectral features (IR, UV–vis, NMR), ion conductivities (if ions participate in the reaction), optical activity, etc. If such data are not available in the literature, they should be investigated early on as they may lead to a monitoring procedure for the kinetic study. Although physicochemical properties which are directly proportional to concentration are most convenient, others such as pH or electrode potentials may be used as their relationship to concentration is well understood. When the relationship between an observed property or measured signal and concentration of a component in the reaction mixture is not theoretically derived, e.g. GLC signals from analysed samples of the reaction mixture, calibration curves may be used. These are constructed by analysis of standard solutions of a reaction component (see Chapter 2).

The conditions under which the study is to be performed must be established, and these include the volume, temperature, pH for aqueous solutions, solvent composition, concentration and ionic strength. Almost invariably, it is helpful to perform some exploratory rate measurements at an early stage in the investigation to identify the most favourable conditions.

### 3.5.2    *Variables to be controlled*

#### 3.5.2.1    *Volume*

As indicated above, studying the reaction at constant volume in a single homogeneous phase greatly simplifies the analysis of the results (see also Chapter 4). There are, however, many industrial processes carried out in flow or stirred reactors in which the volume cannot be assumed constant. In such cases, the rigorous definition of the rate equations in terms of the extent of reaction must be used.

Unwanted changes in solution volume may be due to changes in the density of the solution or to evaporation. The former effect may be reduced by working at lower concentrations of reagents; problems due to evaporation will be reduced at lower reaction temperatures and eliminated by using sealed reaction vessels.

Reactions in solution are often initiated by mixing at least two solutions. Mixing may not be efficient if they are of very different viscosities or densities. In order to have a homogeneous medium, of course, mixing must be efficient, and this requires a minimal 'dead time' – the period during which monitoring is not possible. Ideally, mixing should cause as little turbulence as possible; this is especially important if the reaction is to be followed using a spectroscopic technique as eddies lead to optical complications.

#### 3.5.2.2    *Temperature*

It is common knowledge that rates of chemical reactions depend upon the temperature, and this is a critical variable that must be kept under rigorous control. There are two issues: temperature stability during a reaction and the accuracy of the measurement of the temperature of the reaction. The temperature dependence of the rate of a reaction is overwhelmingly due to the effect of temperature upon the rate constant, which is described by the Arrhenius equation where all terms have their usual meanings:

$$k = A\,e^{\frac{-E_a}{RT}}.$$

As a rule of thumb, an increase of $10°C$ leads to a doubling of the reaction rate, or a change of $1°C$ leads to a change of about 10% in the rate constant at normal temperatures. Even if a temperature can be controlled to a high degree of stability, if its value is not known accurately, it cannot be reproduced reliably between different laboratories. There is no point, therefore, in quoting a rate constant with a precision of less than 1% if the temperature is not accurate to, say, $<0.1°C$. And if the temperature stability during a reaction is not better than $\pm0.1°C$, the rate constant cannot possibly be precise to better than 1%.

Nowadays, however, temperature control can be very precise, especially near room temperature, and accurate temperature measurements are possible though seldom routine. A wide range of thermostats is commercially available, and the selection of an appropriate model will depend on the type of experiments to be carried out. Usually, it is desirable that a thermostatting device includes a cooling facility to allow temperatures below $0°C$ to be maintained, and that it be equipped with a circulating pump to allow efficient thermostatting of an external reactor. If water is the circulating medium, it is important to add either an algaecide or some substance that avoids algae proliferation, and to check periodically that the flow is not obstructed. If the thermostatting device is outside the reaction vessel itself, the temperature should be controlled by a probe inside the reaction vessel as the heat loss from the connecting tubing might be appreciable. If the reaction is slow and thermostatting has to be maintained for long periods, the temperature should be checked at the beginning, during and at the end of the experiment.

We could face two limiting situations: an inconveniently slow reaction [15] and a process that is too fast to monitor [16]. One may choose to run experiments at a higher temperature for reactions that are too slow at around the usual 298 K. In this case, a limitation is imposed by the boiling point and vapour pressure of the solvent which may lead to evaporation and, consequently, volume change (see above). On the other hand, it may be necessary to carry out studies at low temperatures if the reaction is too fast at 298 K. In this case, potential problems include solutions freezing and condensation of atmospheric water vapour on the reaction vessel and optical surfaces, which may affect spectrophotometric measurements.

For a reaction initiated by mixing reagents and/or solutions, they should all be stored at the required temperature to minimise temperature changes upon mixing (see Chapter 8). Even then, it is prudent to leave out the first few readings when fitting data to the rate law.

### 3.5.2.3 pH

The presence of $H_3O^+$ or $HO^-$ may alter drastically the observed reaction rate either because they catalyse the reaction (acid or base catalysis, see Section 3.2.3 for the Aldol reaction, and Chapter 11) or because of ionic strength effects. Proper pH control in an aqueous solution will require a buffer system which is described by the appropriate version of the Henderson–Hasselbach equation, according to whether the acid or base is the charged species:

$$pH = pK_a + \log\frac{[B]}{[BH^+]} \quad \text{or} \quad pH = pK_a + \log\frac{[B^-]}{[BH]}.$$

According to these expressions, the pH can be adjusted and controlled by the *buffer ratio*, $[B]/[BH^+]$ or $[B^-]/[BH]$, and there are excellent manuals on the subject [17]. The choice of reaction conditions will affect the selection of an appropriate buffer system, e.g. there may be solubility or ionic strength considerations. For the latter reason, buffer components with low ionic charges are usually preferred. For example, in their overlapping buffering range, acetate/acetic acid is preferred to citrate/citric acid, other matters being equal. The buffering capacity of the selected buffer must also be tested by measuring the pH both at the beginning and at the end of the reaction. To ensure effective buffer control, the concentration of the buffer component that is present in lower concentration should be at least 20 times higher than that of any of the reactants, and the targeted pH value should be within $\pm2$ pH units of the $pK_a$ of the conjugate acid of the buffer system. Special care must be taken when using high proportions of co-solvents as they may cause appreciable changes of $pK_a$. For example, the $pK_a$ of $CH_3CO_2H$ has been shown to increase by ca. 3 units between 0 and about 70% $CH_3CN/H_2O$ (v/v) [18].

If a reactant is acidic or basic, it is important to establish which is the reactive form under the reaction conditions. This might involve measuring the acid–base properties of the reactant under non-reacting (but otherwise similar) reaction conditions and then measuring the pH dependence of the reaction rate (see also Chapter 11).

### 3.5.2.4   Solvent

The effect of solvents upon reactions is a large and important topic and has been extensively discussed [6, 19]. The solvent is unlikely to change during a reaction except in concentrated solutions as reactants become products, or with mixed solvents in the event of selective evaporation. Although our present purpose is not to give an account of solvent effects upon rates or of how they contribute towards an understanding of mechanism, it is useful for the kineticist to have an appreciation of how the nature of the medium affects the rate constant. Correlation analysis has been applied to this field and there are several equations for the prediction of the effect of a solvent change on a given reaction using tabulated literature parameters for wide-ranging solvents [19].

The polarity of the solvent will influence different types of reactions in different ways, depending upon whether they involve ions, dipoles or polarisable molecules. At the simplest level, we can analyse the effects of the solvent in terms of the different degrees of solvation of species in the initial state and the transition state. For example, in the reaction between pyridine and methyl iodide (Equation 3.24) the reactants are separate neutral molecules, the products are separate fully formed ions, but the transition structure is a single molecular entity with an appreciable degree of polarity.

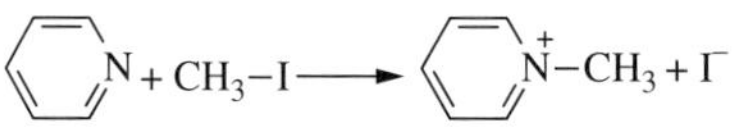

$$\text{C}_5\text{H}_5\text{N} + CH_3-I \longrightarrow C_5H_5\overset{+}{N}-CH_3 + I^-$$

$$(3.24)$$

The increase in solvation between the initial state and the transition state by polar solvents will be greater than that by nonpolar solvents, so the free energy of activation ($\Delta G^{\ddagger}$) will be lower and the rate constant greater. For the same change of nonpolar to polar solvent, a similar line of reasoning leads one to predict a lower rate constant for the reaction between ions of opposite charge where there is a reduction in the polarity in the transition state,

Equation 3.25:

$$Y^- + X^+ \rightleftharpoons (Y^{\delta -} \cdots X^{\delta +})^{\ddagger} \rightarrow Y-X. \tag{3.25}$$

In general terms, solvent effects are better understood in reactions between ions because electrostatic forces between ions are much stronger than non-electrostatic forces. Thus, if the solvent is treated as a continuum of dielectric constant $\varepsilon$ and the ions as rigid spheres that come in contact in an encounter complex, the rate constant is given by Equation 3.26:

$$\ln k = \ln k_0 - \frac{z_A z_B e^2}{4\pi \varepsilon_0 \varepsilon d_{AB} k_B T}, \tag{3.26}$$

where $k_0$ is the rate constant when the dielectric constant is infinite (i.e. with no electrostatic forces between ions), $z_A$ and $z_B$ are the charge numbers of the ions (these may be positive or negative), $e$ is the charge of the electron ($1.602 \times 10^{-19}$ C), $\varepsilon_0$ is the permittivity of vacuum ($8.85419 \times 10^{-12}$ F m$^{-1}$), $d_{AB}$ is the distance between the centres of the ions in the encounter complex, $k_B$ is the Boltzmann constant ($1.38066 \times 10^{-23}$ JK$^{-1}$) and $T$ is the absolute temperature [6]. This relationship is rather generally obeyed, with deviations for solvents of low dielectric constants. A linear graph of $\ln k$ versus ($1/\varepsilon$) affirms that the relationship holds and allows determination of $d_{AB}$.

Co-solvents are commonly used to increase the solubility of reactants and reagents, and ensure a homogeneous solution. The choice of co-solvent must be considered carefully, bearing in mind that it should be inert in the reaction in question, but will have an effect on the rate of reactions and (for reversible reactions) the equilibrium. It is generally desirable that co-solvents have similar boiling points (or volatilities) to minimise selective evaporation during a reaction. Sometimes, two solvents that are perfectly miscible when pure may separate into distinct phases at high ionic strengths (see Section 3.5.2.5), e.g. water and acetonitrile. It is important to take this possibility into account and check that the final reaction medium is homogeneous, rather than biphasic, or microheterogeneous [20, 21].

### 3.5.2.5 Ionic strength

Since an equilibrium is assumed between the transition state and the reactant(s), and because the corresponding equilibrium constant can be expressed in terms of activity coefficients and concentrations to account for the non-ideality of the medium, it follows that there should be an activity coefficient effect upon reaction rates. This is observed as a dependence of the rate constant upon ionic strength – *the kinetic electrolyte effect* [2]. Thus, for a bimolecular reaction,

$$A^{z_A} + B^{z_B} \rightarrow \text{Products},$$

the rate constant can be expressed as Equation 3.27:

$$\log k_{obs} = \log k_0 + 2B z_A z_B I^{1/2}, \tag{3.27}$$

where $k_0$ is the rate constant extrapolated to zero ionic strength, $B$ is a constant obtained from the Debye–Hückel limiting law ($\log \gamma_1 = -B z_i^2 I^{1/2}$, with $B \sim 0.51$ dm$^{-3/2}$ mol$^{-1/2}$ for aqueous solutions at 298.0 K), $z_A$ and $z_B$ are the charge numbers of the ions (which may be positive or negative) and $I$ is the ionic strength of the medium, which can be calculated by Equation 3.28, where $C_i$ is the concentration of component $i$:

$$I = \frac{1}{2} \sum_i C_i z_i^2. \tag{3.28}$$

Note that while the ionic strengths of 1:1 salts coincide with their concentration, $I$ becomes larger than $C$ as the absolute values of the ionic charges increase; for example, while $I = 1$ M for a 1 M solution of NaCl, it would be 15 M for a 1 M solution of $Al_2(SO_4)_3$! The above dependence of the rate constant on the ionic strength holds for low ionic strengths, i.e. the conditions under which the Debye–Hückel limiting law is valid. For medium or high ionic strengths, more sophisticated treatments of ionic interactions are needed [6].

In a bimolecular process between ions of the same charge, an increase in the reaction rate will be observed as the ionic strength increases. Conversely, if the ions are of opposite charges, the reaction rate decreases. If one of the reactants is a neutral species (or if the reaction is unimolecular), the reaction rate becomes essentially independent of the ionic strength, according to this model, and this is approximately true in practice. These effects have been studied in detail and summarised graphically [22].

It follows from the above that maintenance of constant ionic strength is important in two contexts. First, reactions such as those of Equations 3.24 and 3.25 remove or generate ions, so progress of the reaction alters the ionic strength of the medium, and the rate of the latter would be expected to be sensitive to the ionic strength. Maintenance of relatively high ionic strength, i.e. much higher than the concentrations of reactants, will minimise the perturbation as the reaction proceeds. Second, if the effect of ionic acids or bases is being investigated, e.g. in an investigation of acid or base catalysis, it is important to distinguish between the effect of the electrolyte as an acid or a base, and its kinetic electrolyte effect. Thus, it is normal good practice to investigate the effect of the changing concentration of an acid or a base at constant ionic strength, i.e. changes in buffer concentration should be made at constant ionic strength, see Chapter 11.

The choice of electrolyte for ionic strength control is not always straightforward, and the following points need to be considered.

(i) The electrolyte should be inert and soluble.
(ii) 1:1-Electrolytes are preferred, as weighing or volumetric errors associated with smaller amounts of more highly charged electrolytes will lead to a poorer control of the ionic strength.
(iii) Electrolytes with cation and anion of similar mobilities are preferred, as this will reduce liquid-junction potentials in glass electrodes when measuring pH.
(iv) The type of pH electrode to be used may affect the choice of salt to be used; for example, the use of $NaClO_4$ may lead to precipitation of $KClO_4$ in the salt bridge of glass electrodes if the internal electrolyte of these is KCl.

With these matters in mind, $NaClO_4$, NaCl and KCl are usually the most appropriate electrolytes for maintenance of ionic strength.

### 3.5.2.6    Other experimental aspects

Less commonly, other experimental conditions may need to be controlled. For example, it may be that the presence of $O_2$ in solution affects the reaction, in which case stock solutions and the reaction mixture should be flushed and then kept saturated with an inert gas (nitrogen or, preferably, argon). For reactions in nonaqueous solvents, of course, water may need to be rigorously excluded. And sometimes, a chemical process is affected by light if any of the species involved is light sensitive. In this event, stock solutions and the reaction mixture should be kept in the dark or illuminated only by light of non-interfering wavelengths.

## 3.6 Choosing an appropriate monitoring method

Once preliminary studies have been carried out on the process under study, and suitable general reaction conditions have been identified, the next task is to consider the most appropriate kinetic method and the experimental set-up for measuring rate constants. Usually, the first decision to be taken is whether the reaction is to be monitored periodically or continuously.

### 3.6.1 Periodic monitoring

It is quite easy to get sufficient data for a preliminary kinetic analysis from a few experimental measurements of the change in the concentration of starting materials or products with time (see Chapter 2). To do this, we can use a simple periodic method, withdrawing samples from our reactor from time to time and determining the relevant concentrations by some chromatographic, spectroscopic or other analytical method, e.g. HPLC, GLC, UV–vis, NMR or titrimetry. Ideally, to facilitate stability of reaction conditions, studies of this kind are performed in dilute solutions (about 10 μM to 10 mM) and under mild conditions of temperature, pH, solvent, salt concentration, etc. These are not typical synthetic conditions but the experiments will be *easy* for gaining preliminary kinetic data. Good kinetic data should be reproducible to within 3%, but acceptable data for preliminary considerations need be reliable to only within about 10%. For example, this level of precision is normally adequate for studies of the effect of solvent composition or ionic strength. On the other hand, reproducibility and precision must be better if we need to measure, for example, secondary deuterium kinetic isotope effects, and very much better for heavy atom isotope effects [23, 24].

### 3.6.2 Continuous on-line monitoring

There are cases in which monitoring a flowing medium is needed. This may be required for the investigation of a reaction in an industrial flow system, for example. Ideally, the rate of flow should be known precisely; then a number of observation chambers placed at known distances along the flow system allow measurements to be made at calculable reaction times. The observation chambers may consist of a sampling device with a septum or, more simply, a device to measure some physical property in situ. However, continuous monitoring in flow systems is a specialised topic usually requiring dedicated experimental set-ups and will not be covered here.

### 3.6.3 Continuous static monitoring

In this much more common method, the reaction takes place in a closed or sealed vessel, with or without stirring and with or without gas bubbling through, depending on the characteristics of the experiment. The reaction may be followed by withdrawing samples (or using the full content of each of a series of sealed vessels initially and simultaneously prepared) in order to measure the concentration(s) of reactant(s), intermediate(s) or product(s). This procedure has the disadvantage that appreciable initial amounts of reactants may be required as there must be sufficient to allow the desired number of samples to be withdrawn. More simply, a physical property of a single reaction may be measured, preferably avoiding opening the reaction vessel. This could be spectrophotometric absorbance, for example, in which case the reaction would be carried out in a thermostatted spectrophotometric cell, or conductivity using a thermostatted conductivity cell. Alternatively, the measurement could be by a probe inserted into the reaction, e.g. an electrochemical device of some sort. In both types of cases, measurements can be made at either very short time intervals or very long ones, depending upon the nature of the reaction and the device in use. And the measuring device may remain in place and measurements be taken over longer time intervals, but still without opening or otherwise disturbing the reaction vessel. Most reactions nowadays are investigated by a continuous static monitoring method.

## 3.7   Experimental methods

Once a decision has been taken about which property is most suitable for monitoring the reaction, an appropriate practical method must be chosen, and an obvious general requirement is that the response time (or analysis time) must be short compared with the reaction time. The most obvious *direct* property is the concentration of a reactant or product, and component analysis is covered in Chapter 2. In the remainder of this chapter, we shall focus on various experimental methods directly applied to measuring rate constants which relate (sometimes indirectly) to concentration.

### 3.7.1   *Spectrometric methods*

These are now probably the most widely used methods in kinetic and mechanistic studies, and include a wide range of spectral frequencies: radio frequencies (NMR, ESR), IR and UV–vis. Appropriate instrumentation which is easily adapted for kinetics is readily available in most research laboratories; it is usually easy to use, and the output easily interpreted. Spectrophotometric methods are also widely used for the determination of equilibrium constants [25]. However, before deciding upon a spectrophotometric technique, the following experimental aspects must be considered carefully.

(i) The compound must have an absorption/emission peak or band whose intensity changes as the reaction proceeds, and these changes must be accurately measurable.
(ii) Effective mixing of the reagents and the homogeneity of the medium must be ensured.
(iii) It is convenient (although not essential) that the spectroscopic signal due to a component be linearly proportional to its concentration in the range of concentrations to be used. If so, it is possible to use spectrophotometric signal versus time curves directly to obtain the kinetics, but the signal versus concentration relationship must be investigated to establish the range of concentrations where the linear dependence

holds. With UV–vis radiation, for example, it is normal practice to identify the upper limit of the concentration range over which the Beer–Lambert law is obeyed. Working with any physical property in a region where there are deviations from direct proportionality with concentration leads to poorer results. This is mainly because the onset of non-linearity coincides with higher values of the physical property, which are either close to the resolution limits of the experimental device or so high that they contain a large intrinsic error. When the signal is not directly proportional to the concentration, calibration curves of spectrophotometric signal versus concentration are needed; these are easily constructed from the spectra of standard solutions of known concentration.

(iv) The reaction medium/solvent must be sufficiently transparent to the radiation in the region of absorbance by the targeted species. It is possible to correct for some extent of absorption by the medium by using a blank cell containing just the reaction medium in a double beam spectrophotometer, but this appreciably worsens the signal:noise ratio and may lead to imprecise results.

(v) The extent of overlap of the targeted signal with other signals due to reagents, intermediates or products should be minimal. Although this is not an essential condition, it certainly facilitates the experimental work. If there is considerable overlap, the extinction coefficients (molar absorptivities) of the different species involved must be measured to make sure that it is possible to follow the process with confidence.

(vi) The required signal:noise ratio must be considered carefully, and the procedure for data acquisition may need to be changed as the reaction proceeds. For example, if changes in an $^{17}O$ NMR signal are to be followed, the quality of the data will depend upon the number of spectral acquisitions which constitute each measurement, and these may need to be changed as the reaction proceeds.

(vii) The timescale of the signal changes and of the data acquisition is a fundamental point. Nowadays, it is possible to follow chemical processes taking place on the femtosecond timescale ($1\,s = 10^{-15}$ fs), but this is far from routine. In general, the faster the process to be followed, the more sophisticated (and expensive) the equipment is, the more difficult it is to use and the more difficult it is to interpret the data. The different techniques are usually classified according to their time resolution, and are compiled in Fig. 3.6 see below.

### 3.7.1.1 Conventional and slow reactions

The most widely used routine spectroscopic methods are those based on the absorption/emission of UV–vis light. The UV–vis method is generally preferred over IR and NMR spectroscopy because (a) it is more sensitive, so more dilute solutions may be used, and (b) water and other common solvents present fewer problems with UV–vis than with IR and NMR methods. It is rare, but not impossible, that the light source used initiates some photochemical process, and this should be checked.

A reaction which is complete within, say, a day may be monitored for most of its course whereas a slower reaction would usually be studied using the initial rates method (see Section 3.3.1) using data collected during the initial stages – perhaps just the first 1% of the reaction.

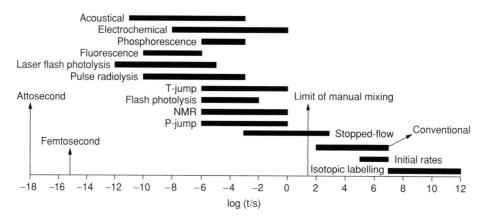

**Fig. 3.6**  Some of the experimental methods available for the study of kinetics and mechanisms in solution, classified according to their approximate time resolution.

Laboratory spectrophotometers and fluorimeters which allow monitoring of reactions that take place within a few seconds are fairly routine now, and time resolutions of tenths of a second are available; detailed descriptions are available in texts on spectroscopy [26]. Solution cells may be glass or plastic for light in the visible range, but quartz cells are needed for UV work. In an absorption spectrophotometer, the light source and photodetector are in line. In a fluorimeter, the detector is at 90° to the incident light, so cells must have four optical faces.

It is highly informative to study time-resolved spectra of a reaction, not just time-absorbance readings at a single wavelength. Diode array detectors are ideal for this purpose due to their fast speed of acquisition, although they are less precise than conventional photodetectors. If a band is due to a reactant, its intensity will decrease while, if it is due to a product, it will increase. An intermediate species will be first formed and then disappear, i.e. the corresponding band will first increase and then decrease. Isosbestic points, i.e. where the absorbance does not change with time because extinction coefficients of reactants and products are the same, are significant. A well-defined isosbestic point indicates that there is no accumulation of intermediates (or an accumulating intermediate has an extinction coefficient identical with that of reactants and products). An example of the usefulness of time-resolved spectral recording is the identification of $N$-(2-imino, 1-oxo-propyl)-glycine as an intermediate in the decomposition of $N$-halo-dipeptides, which are products of intracellular oxidation by the enzymatic system, myeloperoxidase/$Cl^-/H_2O_2$ [27].

Although IR monitoring would be desirable in many cases, it is not at all straightforward. All common solvents absorb to some degree, and often strongly, within the conventional IR range, and there are only few suitable materials for cells which are transparent to even moderate ranges of IR radiation. Frequently, one has to identify a solvent and a cell material which have only weak absorption within a useful range, and then ensure that there is no incompatibility; for example, water cannot be used in NaCl cells. In principle, the problem with using water may be solved by using Raman spectroscopy, but the extreme weakness of the signals obtained is a handicap. However, there are nice examples of IR spectroscopic applications. Simard used IR kinetics to identify ketene, formaldehyde and acetaldehyde as intermediates in the pyrolysis of ethylene oxide [28], and time-resolved IR spectroscopy was used recently to study the formation of pyridylketenes from 2-, 3- and 4-diazoacetylpyridines [29].

### 3.7.1.2   Fast reactions

These are reactions that take place within the $10^{-1}$–$10^{-3}$ s range, and stopped-flow methods are most widely used [30]; a typical set-up is shown in Fig. 3.7. The technique involves mixing the flows from two separate syringes into an already filled mixing chamber by compressed air. The solution that is displaced triggers a digitiser where the trace showing the change in absorption or emission at a given wavelength is registered for analysis. It is essential that the viscosities of the reacting solutions are similar to keep the mixing time as low as possible. This limits not only the nature of the solvents, but also the ionic strengths of the two solutions. Ideally, both solutions should have similar solvent compositions and ionic strengths. Both mixing syringes and the mixing chamber must be thermostatted, and the mixing chamber designed to avoid turbulence inside it (see Section 3.5.2.1).

There are two kinds of stopped-flow apparatus, according to the geometry of the mixing unit: horizontal and vertical. The vertical arrangement is preferred as it facilitates exclusion of air bubbles from the syringes and therefore promotes more efficient mixing. Frequently, the thermostatting bath has components made from Teflon which has the major disadvantage of undergoing severe dilations/contractions with temperature changes. This must be carefully controlled as it may lead to leaks in the syringes and/or damage to the system due to blocking after dilation. This problem also limits the temperature range that can be used with this kind of apparatus to an interval of about 35°C, which sometimes is not sufficient for a proper determination of the enthalpy and entropy of activation/reaction.

The intracellular oxidation of thiols by HOCl was investigated in a stopped-flow investigation and shown to take place via formation of sulfenyl chlorides with elementary rate constants of ca. $10^9$ M$^{-1}$ s$^{-1}$ for the rate-limiting step [12]. For the oxidation of amines by HOCl [31], activation parameters of $\Delta H^{\ddagger} \sim 10$ kJ mol$^{-1}$ and $\Delta S^{\ddagger} \sim -60$ J K$^{-1}$ mol$^{-1}$ were determined by measuring second-order rate constants at different temperatures. These results were fundamental to the later characterisation of the mechanism as a Cl transfer accompanied by several proton transfers along a chain of hydrogen-bonded

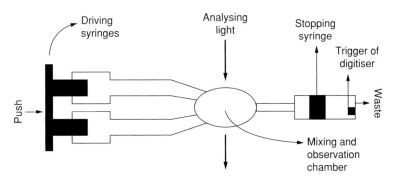

**Fig. 3.7**   Generic layout of a single-mixing stopped-flow system.

water molecules [31, 32]. Nowadays, so-called multi-mixing stopped-flow systems (in contrast to the single-mixing versions described above) allow the sequential mixing of several solutions.

### 3.7.1.3    Very fast and ultrafast reactions

There are a number of techniques that have been designed specifically for the study of processes taking place in the millisecond (ms) to femtosecond (fs) range. These include relaxation methods (temperature jump, pressure jump, pH jump) for equilibrium processes [33], acoustical methods, the use of electrical pulses and waves, pulse radiolysis, flash photolysis and laser flash photolysis (µs, ns, ps and fs). In all of them, a chemical system at equilibrium is instantaneously perturbed by some physical method; then its return to equilibrium is observed (the final equilibrium state may not coincide with the original). All these techniques have been discussed in the literature [30, 34] and tend not to be very widely applicable, i.e. any one technique is usually applicable to a rather narrow range of reaction type. It generally follows that considerable effort has to be invested in the design of an experiment for the investigation of a particular type of very fast reaction, and help by an expert in each technique is essential.

Flash photolysis and laser flash photolysis are probably the most versatile of the methods in the above list; they have been particularly useful in identifying very short-lived intermediates such as radicals, radical cations and anions, triplet states, carbenium ions and carbanions. They provide a wealth of structural, kinetic and thermodynamic information, and a simplified generic experimental arrangement of a system suitable for studying very fast and ultrafast processes is shown in Fig. 3.8. Examples of applications include the ketonisation of acetophenone enol in aqueous buffer solutions [35], kinetic and thermodynamic characterisation of the aminium radical cation and aminyl radical derived from *N*-phenylglycine [36] and phenylureas [37], and the first direct observation of a radical cation derived from an enol ether [38].

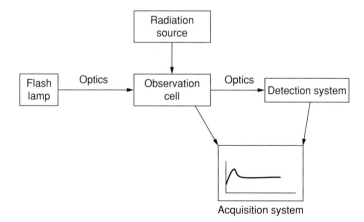

**Fig. 3.8**  Generic layout of a system suitable for studying very fast and ultrafast processes. Appropriate radiation sources may be a flash lamp, a laser or an electron accelerator, while optical, conductivity, or ESR detection systems may be employed.

### 3.7.1.4 *Magnetic resonance spectroscopy*

As indicated previously, NMR may be used simply as an analytical technique for monitoring the decomposition of a reactant or formation of a product. In addition, NMR and ESR merit a special mention due to their importance in studying the dynamics of systems at equilibrium; these so-called equilibrium methods do not alter the dynamic equilibrium of the chemical process under study. They have been used to study, for example, $^1$H-transfer reactions, valence isomerisations, conformational interconversions, heteronuclear isotopic exchange processes (NMR) and electron-transfer reactions (ESR). These techniques can be applied to the study of fast or very fast reactions by analysis of spectral line broadening [16, 39].

Two objectives are fulfilled when using NMR: (a) the functional groups involved in the chemical change are characterised and (b) kinetic parameters are determined. For a system $A \rightleftharpoons B$, a single rate constant, $k = (k_f + k_b)$, is determined from line shape analysis, $k_f$ and $k_b$ being the rate constants for the forward and back reactions, respectively. Individual values of $k_f$ and $k_b$ can then be calculated from the equilibrium constant, $K = k_f /k_b$, if that is independently measured from the equilibrium concentrations of A and B [40].

The higher concentrations of the sample needed for the NMR method compared with other physical methods is a drawback, as also is the lower precision in the determination of rate constants. The latter is usually because the temperature of the sample in the NMR probe is controlled by a flow of heated or cooled nitrogen which does not normally provide highly accurate temperature control and measurement. Sometimes, the need for isotopically labelled substrates and solvents can be an additional drawback.

The acid-catalysed proton exchange in secondary amides, with estimation of the corresponding rate constants [41], and, more recently, the analysis of the relative base strengths of N, O and S within sulfenamides using $^{14}$N NMR are examples of the application of the NMR technique [42]. The conformational interconversion of 2-methoxy-1,3-dimethylhexahydropyrimidine has been scrutinised by $^1$H and $^{13}$C NMR spectroscopy in order to investigate the origins of the anomeric effect [43], a topic considered in detail in Chapter 7. Also, an extremely high water exchange rate ($k = 8.8 \times 10^8$ s$^{-1}$ at 298.0 K) was observed recently for a Gd$^{3+}$ chelate using $^{17}$O relaxation rates and $^1$H relaxivities as a function of the magnetic field at different temperatures [44].

Electron spin resonance (ESR) spectroscopy is of application to organic species containing unpaired electrons radicals, radical ions and triplet states, and is much more sensitive than NMR; it is an extremely powerful tool in the field of radical chemistry (see Chapter 10). Highly unstable radicals can be generated in situ or, if necessary, trapped into solid matrices at very low temperatures. Examples of the application of this techniques include study of the formation of radical cations of methoxylated benzenes by reaction with different strong oxidants in aqueous solution [45], and the study of the photodissociation of N-trityl-anilines [46].

## 3.7.2 *Conductimetry*

Conductance measurements may be useful when there is a significant contribution to the total conductivity of the medium by ions formed or consumed during a chemical reaction. Of the processes that could in principle be monitored by this method, those in which $H_3O^+$

and $HO^-$ are consumed or released are, in practice, the most favourable. The technique has requirements regarding the initial presence of ions in the medium, and on its ionic strength and buffer concentration. If the concentrations of several ions change in the course of the reaction, the information obtained gives a global balance for all the ions, and not specific information for one of them (unless the ion mobilities are very different). Furthermore, conductance measurements in nonpolar solvents are complicated, which limits the types of media that can be used.

The medium must be homogeneous, and it is convenient if the conductance of the medium is directly proportional to the concentration of the ions that are responsible for the conductivity. In such cases, kinetic results are obtained directly from conductance versus time curves. The measured physical property is the specific conductance ($\kappa$) with units Siemen per metre ($S\ m^{-1}$), which is directly proportional to the concentration,

$$\kappa = 10^{-3}\Lambda C,$$

where $\Lambda$ (the constant of proportionality) is the *equivalent conductance* and $C$ is the concentration of the conducting species. For a very dilute solution and small changes of concentration, $\Lambda$ can be considered a constant that is characteristic of the ionic species. If weak electrolytes are involved in the reaction, it becomes necessary to find the empirical relationship between $\kappa$ and the concentration.

The specific conductance ($\kappa$) is related to the solution's resistance ($R$) to the passage of electric current (the measured physical property) by the equation

$$\frac{1}{R} = \kappa\frac{A}{l},$$

where $A$ is the area of the electrodes and $l$ is the distance between them. The term ($l/A$) is a temperature-dependent parameter that is characteristic of the conductivity cell and must be evaluated prior to use of the cell using a standard solution of known specific conductance if absolute conductances are to be measured. A calibration is not necessary for routine kinetic work when only changes in conductance are measured.

The timescale over which the conductance of the medium changes is a fundamental issue; standard conductivity cells are designed for use with alternating current (AC), but the period of this current imposes a limit on rates of reactions that can be followed. To investigate reactions faster than the AC conductivity cell can handle, it is necessary to build and calibrate appropriate direct current (DC) conductivity cells, which is not a routine business. Conductivity meters that record continuously are uncommon. Nowadays, however, it is easy to interface a simple apparatus to a computer and collect the data with ad hoc software.

The hydrolysis of ethanoic anhydride [47] and the reaction between phenacyl bromide and pyridine to give phenacylpyridinium bromide (Equation 3.29, an example of the Menschutkin reaction) are reported cases of kinetics studies by conductance monitoring [48].

$$(3.29)$$

Sometimes, conductance detection may serve to distinguish between possible reaction mechanisms. For example, when hydroxyl radical ($HO^\bullet$) reacts as an oxidant yielding $HO^-$, an appreciable increase in conductance should be observed, but this will not be the case if it reacts by hydrogen atom abstraction or by addition to an unsaturated carbon [49].

### 3.7.3  Polarimetry

If reactants and products have different optical rotation properties, it is possible to study the transformation by monitoring the optical rotation using a polarimeter. When more than one optically active substance is present in the reaction, their combined optical rotation effect upon the plane of polarised light is observed. The angle of rotation ($\alpha$) caused by a solution of a single pure compound is given by:

$$\alpha = [\alpha]_\lambda^T lC$$

where $[\alpha]_\lambda^T$ is the *specific rotation* of the compound for radiation of wavelength $\lambda$ at temperature $T$, $l$ is the optical path length (dm) and $C$ is the concentration (g cm$^{-3}$). Because the effect depends strongly on the temperature, the polarimeter cell must be thermostatted and have a temperature-measuring device; optical rotation also depends on the solvent and (if it is aqueous) could depend on the pH of the solution if the compound is acidic or basic and the different forms have different $[\alpha]_\lambda^T$ values.

Typically, standard polarimeters operate at a single wavelength, usually the D line of sodium (589 nm) or the green line of mercury (546 nm), although more sophisticated polarimeters have monochromators that allow the recording of optical rotatory dispersion (ORD) curves which show how $\alpha$ depends on $\lambda$. Such curves are useful in exploratory studies to select the best wavelength to be used in a kinetic investigation. It is now possible to use wavelengths that lie in the UV region of the spectrum, although the cost of the polarimeter and of the cells (which must be made of quartz) is much higher. Measurements in the visible range are much less sensitive, so higher concentrations are needed.

It is difficult to take more than about three readings a minute by visual monitoring using a standard polarimeter, which limits rates to be followed to relatively slow ones. Nowadays, however, a digital polarimeter can be interfaced to a computer, allowing continuous recording and monitoring of faster reactions.

Polarimetry is extremely useful for monitoring reactions of optically active natural products such as carbohydrates which do not have a useful UV chromophore, and samples for study do not need to be enantiomerically pure. Nevertheless, compared with spectrophotometry, the technique has been applied to relatively few reactions. It was, however, the first technique used for monitoring a chemical reaction by measuring a physical property when Wilhemy investigated the mutarotation of sucrose in acidic solution and established the proportionality between the rate of reaction and the amount of remaining reactant [50]. The study of a similar process, the mutarotation of glucose, served to establish the well-known Brønsted relationship, a fundamental catalysis law in mechanistic organic chemistry.

### 3.7.4  Potentiometry

It is possible nowadays to determine with very high accuracy the concentration of various ions and gases in solution using combined-glass or selective electrodes; these provide an inexpensive straightforward procedure for monitoring chemical reactions [51–55].

The measured property is the electrical potential which is related to the concentration of the target species by the Nernst equation (see Chapter 6)

$$E = E^0 - \frac{RT}{nF} \ln \frac{[\text{Red}]}{[\text{Ox}]},$$

where $E^0$ is the standard reduction potential, $R$ is the gas constant ($8.3144 \text{ JK}^{-1} \text{ mol}^{-1}$), $T$ is the absolute temperature, $n$ is the number of electrons involved in the process, $F$ is the Faraday constant ($9.64846 \times 10^{-12} \text{ C mol}^{-1}$) and [Red] and [Ox] are the concentrations of the reduced and oxidised species, respectively. As is evident, this is not a linear relationship between the measured physical property and the concentration, so the change in $E$ (or pH) with time does not lead directly to rate constants – calibration curves are needed to convert $E$ into concentration.

An important advantage of these techniques is that the measurements of electrical potential can be very accurate, which allows monitoring until almost complete conversion, or until equilibrium in a reversible process. In fact, potentiometry is extremely powerful for obtaining equilibrium constants [25]. However, there are also restrictions and limitations: (a) the solution must be conducting; (b) the response time of the electrode can be relatively long, so there is a limit to the speed of a reaction which can be monitored and (c) there can be appreciable interference from impurities, or intermediates and products.

Formation of $Cl^-$ was monitored potentiometrically in the reaction of methionine and HOCl shown earlier in the example in Section 3.3.3.1 and plotted in Fig. 3.4 [12]. The technique has also been used to study the elimination of HCl from DDT [56] and various bromination reactions [57].

### 3.7.5 Dilatometry

Many reactions involve volume changes that can be followed by the rise or fall of a meniscus in a capillary dilatometer. This is a readily available, simple, inexpensive technique that has fallen out of use due to the availability of alternative instrumental methods that require a lower degree of experimental skill and less effort in the way of double-checks. Its main disadvantage is the need for very rigorous temperature control. Since a dilatometer works very much as a thermometer, the variation in the meniscus due to a tiny temperature change may hide any volume change due to the reaction under study. Moreover, dilatometry also requires high initial concentrations in order to achieve readily observable volume changes, and this leads to complications such as the accumulation of appreciable concentrations of intermediates, e.g. the accumulation of intermediate hemiacetals in the acid hydrolysis of concentrated solutions of acetals [58].

It is prudent at this point to note that dilatometry does not fulfil the initial requirement of a reaction at constant volume (see Sections 3.2.1 and 3.5.2.1), so the definition of reaction rate needs to be revised to take account of the possibility that the volume may change as the reaction proceeds. Thus, for a first-order reaction, for example,

$$\nu_a A \rightarrow \nu_b B + \nu_c C + \cdots,$$

the integrated rate equation can be expressed as [59]

$$kt = \ln \left( \frac{[A]_0}{([A]_0 - x)} \right) + \Delta V^0 x,$$

where $k$ is the rate constant, $t$ is time, $[A]_0$ is the initial concentration of A, $x$ is the amount of A that has been transformed and $\Delta V^0$ is the change in molar volume for the process $(\Delta V^0 = (\nu_b V_B^0 + \nu_c V_C^0 + \cdots) - \nu_a V_A^0$, where $V_i^0$ is the molar volume of each participating species). As is evident, the above equation differs from the usual integrated first-order rate equation by the presence of the term $\Delta V^0 x$. Since $\Delta V^0$ may be very low in solution, $\Delta V^0 x$ could be within experimental error, in which case this term may be neglected, and the usual integrated first-order rate equation used.

Dilatometry has been applied, for example, in the study of polymerisation or depolymerisation reactions that can lead, respectively, to a decrease or increase in volume [59]. Recently, Mayr and colleagues used this technique to study the kinetics of hetero Diels–Alder reactions, namely $[2^+ + 4]$ cycloadditions of iminium ions with 1,3-dienes [60].

### 3.7.6 Pressure measurements

Gas phase reactions at constant volume and temperature are routinely monitored by pressure measurements when the gas is assumed to behave ideally and a direct proportionality between pressure and concentration can be assumed, but the technique is much less commonly used for reactions in solution. If a gas is evolved or consumed in a reaction, it can be monitored continuously by the change in pressure when the reaction is investigated in a closed reactor at constant volume. An early example is the thermal decomposition of arenediazonium salts with evolution of $N_2$ [61]:

$$ArN_2^+ + 2H_2O \rightarrow ArOH + N_2 + H_3O^+.$$

The technique has drawbacks associated with the establishment of equilibrium between the gas and the other phase, typically a liquid. If the gas is a reactant, its rate of dissolving in the liquid phase should be faster than the chemical rate-determining step. If it is a product, its release should be sufficiently fast that supersaturation in the liquid phase does not become a problem. (The relationships between rates of transport between phases and chemical processes are explored in Chapter 5.)

### 3.7.7 Chromatographic methods

Various chromatographic techniques may be used to monitor the progress of a reaction (see Chapter 2). In some cases, like thin layer chromatography (TLC), it is difficult to go beyond a merely qualitative analysis of how reagents disappear, intermediates come and go, and products are formed, all observed by the increasing or decreasing intensity of the TLC spots. Other techniques, such as gas chromatography (GC) or high performance liquid chromatography (HPLC), allow not only following the reaction components, but also a relatively easy quantification of rate constants, since the chromatographic signal is proportional to the concentration of the species in the analysed sample. In order to quantify rate constants, it is not strictly necessary to perform calibrations, or use known reactants and products, i.e. a rate constant can be estimated by monitoring an unidentified species whose concentration reflects the extent of reaction.

Techniques of chromatographic analysis continue to develop and for up-to-date methods, the specialist literature should be consulted [62, 63]. In all cases, reaction samples have to be taken at known time intervals and quenched by an appropriate method (sudden cooling, change of pH, dilution, etc.) before chromatographic analysis. It is important to check the stability of the reaction component to the chromatographic and work-up conditions. For example, are the compounds to be analysed thermally stable to the GC conditions? (Conditions inside a GC injection port and, indeed, within the column are not unlike those of a heterogeneous catalytic reactor!) Are they stable to the pH of the HPLC eluent? An obvious restriction is that chromatographic component analysis does not lend itself to the study of fast reactions.

Examples of the use of chromatographic techniques to study reaction rates and mechanisms from the recent literature are the photodegradation of aniline derivatives in suspensions of $TiO_2$ which acts as a photocatalyst [64], and a similar study on the mechanism of photodegradation of phenylurea herbicides [37].

### 3.7.8   Other techniques

Essentially, any technique applicable to the measurement of physicochemical properties of compounds may be considered for the study of reaction rates, and it is up to the imagination of the researcher to exploit the principles behind the technique and devise an experimental method. A number of extremely useful electrochemical techniques are covered in Chapter 6, and Chapter 8 includes a very promising new method combining calorimetric and FTIR measurements. Mass spectrometry, a field in constant development and with an abundant literature, is ideally suited for the study of wide-ranging reaction types in the gas phase, including those related to atmospheric investigations, but is beyond the scope of this chapter.

## Bibliography

Bernasconi, C.F. (Ed.) (1986) *Investigation of Rates and Mechanisms of Reactions, Parts I and II, in Techniques of Chemistry* (vol 6, 4th edn). Wiley-Interscience, New York.
Eyring, H., Lin, S.H. and Lin, S.M. (1980) *Basic Chemical Kinetics*, Wiley, New York.
Laidler, K.J. (1987) *Chemical Kinetics.* Harper & Row, New York.
Strehlow, H. (1992) *Rapid Reactions in Solution,* VCH, Weinheim.
Zuman, P. and Patel, R.C. (1984) *Techniques in Organic Reaction Kinetics,* Krieger, New York.

## References

1. Bernasconi, C.F. (Ed.) (1986) *Investigation of Rates and Mechanisms of Reactions, Parts I and II, Techniques of Chemistry* (vol 6, 4th edn). Wiley-Interscience, New York.
2. Müller, P. (1994) *Pure and Applied Chemistry*, **66**, 1077.
3. Bell, R.P. (1973) *The Proton in Chemistry*, Chapman & Hall, London.
4. Bell, R.P. and McTigue, P.T. (1960) *Journal of the Chemical Society*, 2983.

5. Maskill, H. (1985) *The Physical Basis of Organic Chemistry.* Oxford University Press, Oxford.
6. Laidler, K.J. (1987) *Chemical Kinetics*, Harper & Row, New York.
7. Williams, I.H. (1993) *Chemical Society Reviews*, **22**, 277.
8. Pilling, M.J. and Seakins, P.W. (1995) *Reaction Kinetics*, Oxford Science Publications, Oxford.
9. Margerison, D. (1969) The treatment of experimental data. In: C.H. Bamford and C.F.H. Tipper (Eds) *Comprehensive Chemical Kinetics* (vol 1). Elsevier, Amsterdam, p. 343.
10. Draper, N.R. and Smith, H. (1981) *Applied Regression Analysis.* Wiley, New York.
11. Armesto, X.L., Canle L., M., Losada, M. and Santaballa, J.A. (1993) *International Journal Chemical Kinetics*, **25**, 331.
12. Armesto, X.L., Canle L., M., Fernández, M.I., García, M.V. and Santaballa, J.A. (2000) *Tetrahedron*, **56**, 1103.
13. Logan, S.R. (1996) *Fundamentals of Chemical Kinetics.* Addison-Wesley/Longman, Essex.
14. Krupka, R.M., Kaplan, H. and Laidler, K.J. (1966) *Transactions of the Faraday Society*, **62**, 2754.
15. Batt, L. (1969) Experimental methods for the study of slow reactions. In: C.H. Bamford and C.F.H. Tipper (Eds) *Comprehensive Chemical Kinetics* (vol 1). Elsevier, Amsterdam, p. 1.
16. Hague, D.N. (1969) Experimental methods for the study of fast reactions. In: C.H. Bamford and C.F.H. Tipper (Eds) *Comprehensive Chemical Kinetics* (vol 1). Elsevier, Amsterdam, p. 112.
17. Perrin, D.D. and Dempsey, B. (1974) *Buffers for pH and Metal Ion Control.* Chapman & Hall, London.
18. Espinosa, S., Bosch, E. and Rosés, M. (2000) *Analytical Chemistry*, **72**, 5193.
19. Reichardt, C. (2003) *Solvents and Solvent Effects in Organic Chemistry* (3rd edn). Wiley-VCH, Weinheim.
20. Gratzel, M. and Kalyanasundaram, K. (1991) *Kinetics and Catalysis in Microheterogeneous Systems.* Dekker, New York.
21. Thomas, J.M. and Thomas, J.W. (1997) *Principles and Practice of Heterogeneous Catalysis.* VCH, Weinheim.
22. Steinfeld, J.I., Francisco, J.S. and Hase, W.L. (1999) *Chemical Kinetics and Dynamics.* Prentice-Hall, Englewood Cliffs, NJ.
23. Buncel, E. and Saunders, W.H. Jr. (Eds), (1992) *Heavy Atom Isotope Effects.* Elsevier, Amsterdam.
24. Collins, C.J. and Bowman, N.S. (1970) *Isotope Effects in Chemical Reactions.* Van Nostrand Reinhold, New York.
25. Albert, A. and Serjeant, E.P. (1962) *Ionization Constants of Acids & Bases. A Laboratory Manual.* Wiley, New York.
26. Banwell, C.N. and McCash, E.M. (1994) *Fundamentals of Molecular Spectroscopy.* McGraw-Hill, London.
27. Armesto, X.L., Canle L., M., García, M.V. and Santaballa, J.A. (1997) *Tetrahedron*, **53**, 12615.
28. Simard, G.L. (1948) *Journal of Chemical Physics*, **16**, 836.
29. Acton, A.W., Allen, A.D., Antunes, L.M., Fedorov, A.V., Najafian, K., Tidwell, T.T. and Wagner, B.D. (2002) *Journal of the American Chemical Society*, **124**, 13790.
30. Strehlow, H. (1992) *Rapid Reactions in Solution.* VCH, New York.
31. Abia, L., Armesto, X.L., Canle L., M., Garcia, M.V. and Santaballa, J.A. (1998) *Tetrahedron*, **54**, 521.
32. Andrés, J., Canle L., M., García, M.V., Vázquez, L.F.R. and Santaballa, J.A. (2001) *Chemical Physics Letters*, **342**, 405.
33. Czerlinski, G.H. (1966) *Chemical Relaxation.* Dekker, New York.
34. Crooks, J.E. (1982) Kinetics of reactions in solution: Part II. Fast reactions. *Annual Reports in the Progress of Chemistry. Section C: Physical Chemistry* (vol 79). Royal Society of Chemistry, UK, Chapter 3, p. 41.
35. Chiang, Y., Kresge, A.J., Santaballa, J.A. and Wirz, J. (1988) *Journal of the American Chemical Society*, **110**, 5506.

36. Canle L., M., Santaballa, J.A. and Steenken, S. (1999) *Chemistry – A European Journal*, **5**, 1192.
37. Canle L., M., Fernández, M.I., Rodríguez, S., Santaballa, J.A., Steenken, S. and Vulliet, E. (2005) *ChemPhysChem*, **6**, 2064.
38. Bernhard, K., Geimer, J., Canle-Lopez, M., Reynisson, J., Beckert, D., Gleiter, R. and Steenken, S. (2001) *Chemistry – A European Journal*, **7**, 4640.
39. Dwek, R.A. (1973) *Nuclear Magnetic Resonance in Chemistry*, Claredon, London.
40. Zuman, P. and Patel, R.C. (1984) *Techniques in Organic Reaction Kinetics*, Wiley, New York.
41. Perrin, C.L., Lollo, C.P. and Johnston, E.R. (1984) *Journal of the American Chemical Society*, **106**, 2749.
42. Bagno, A., Eustace, S.J., Johansson, L. and Scorrano, G. (1994) *Journal of Organic Chemistry*, **59**, 232.
43. Perrin, C.L., Armstrong, K.B. and Fabian, M.A. (1994) *Journal of the American Chemical Society*, **116**, 715.
44. Mato-Iglesias, M., Platas-Iglesias, C., Djanashvili, K., Peters, J.A., Tóth, E., Balogh, E., Muller, R.N., Elst, L.V., Blas, A.d. and Rodríguez-Blas, T. (2005) *Chemical Communications*, 4729.
45. O'Neill, P., Steenken, S. and Schulte-Frohlinde, D. (1975) *Journal of Physical Chemistry*, **79**, 2773.
46. Siskos, M.G., Zarkadis, A.K., Steenken, S., Karakostas, N. and Garas, S.K. (1998) *Journal of Organic Chemistry*, **63**, 3251.
47. Greenberg, D.B. (1962) *Journal of Chemical Education*, **39**, 140.
48. Barnard, P.W.C. and Smith, B.V. (1981) *Journal of Chemical Education*, **58**, 282.
49. Azenha, M.E.D.G., Burrows, H.D., Canle L., M., Coimbra, R., Fernández, M.I., García, M.V., Rodrígues, A.E., Santaballa, J.A. and Steenken, S. (2003) *Chemical Communications*, 112.
50. Wilhelmy, L.F. (1850) *Annalen der Physik*, **81**, 413.
51. Bockris, J.O.M. and Reddy, A.K.N. (1970) *Modern Electrochemistry; an Introduction to an Inter-disciplinary Area*, Plenum, New York.
52. Buck, R.P. and Cosofret, V. (1993) *Pure and Applied Chemistry*, **65**, 1849.
53. Meyerhoff, M.E. and Fraticelli, Y.M. (1982) *Analytical Chemistry*, **54**, 27R.
54. Moody, G.J. and Thomas, J.D.R. (1978) Applications of ion-selective electrodes. In: H. Freiser (Ed.) *Ion-Selective Electrodes in Analytical Chemistry*. Plenum, New York.
55. Veselý, J. and Stulik, K. (1987) Potentiometry with ion-selective electrodes. In: R. Kalveda (Ed.) *Electroanalytical Methods in Chemical and Environmental Analysis*. Plenum, Prague.
56. England, B.D. and McLennan, D.J. (1966) *Journal of the Chemical Society B*, 696.
57. Smith, R.H. (1973) *Journal of Chemical Education*, **50**, 441.
58. Schubert, W.M. and Brownawell, D.W. (1982) *Journal of the American Chemical Society*, **104**, 3487.
59. Senent Pérez, S. (1987) *Química Física* (vol 1). UNED, Madrid.
60. Mayr, H., Ofial, A.R., Sauer, J. and Schmied, B. (2000) *European Journal of Organic Chemistry*, 2013.
61. DeTar, D.F. (1956) *Journal of the American Chemical Society*, **78**, 3911.
62. Stahl, E. (1990) *Thin Layer Chromatography. A Laboratory Handbook*. Springer, Berlin.
63. Sherma, J. (Ed.), (2000) *CRC Handbook of Chromatography*. CRC Press, Boca Raton, FL.
64. Canle L., M., Santaballa. J.A. and Vulliet, E. (2005) *Journal of Photochemistry and Photobiology A*, **175**, 172.

# Chapter 4

# The Relationship Between Mechanism and Rate Law

*J. A. Santaballa, H. Maskill and M. Canle L.*

## 4.1 Introduction

The kinetic investigation of an organic chemical reaction involves

(i) application of appropriate experimental techniques for the collection of reliable experimental data; and
(ii) proper data handling of the experimental results to deduce a kinetic model for the reaction which describes the observed kinetics and can predict results under conditions not used in the study.

The kinetic study assists in the development of a credible reaction mechanism which describes all aspects of the reaction – not just the kinetics [1]. The complete exercise involves empirical and theoretical considerations which run in parallel; they are complementary and feedback between them is essential [2]. Aspects (i) and (ii) above were covered in the previous chapter, and we now focus first on the derivation of the rate law (rate equation) from a mechanistic proposal (the *mechanistic* rate law) for comparison with the experimental finding. In simple cases, the derivation is usually straightforward but can be mathematically challenging for complex reaction mechanisms. Once derived, the mechanistic rate law is compared with the experimental, and the quality of the agreement is one test of the applicability of the mechanism. Different mechanisms may lead to the same rate law (they are *kinetically equivalent*), and, whilst agreement between mechanistic and experimental rate laws is required, this alone is not a sufficient proof of the validity of the mechanism [3–7]. We conclude the chapter by working through several case histories.

Many chemical processes are initiated simply by mixing the appropriate reagents, and (usually) the higher the temperature, the faster the reaction rate; such reactions are classified as thermally activated or thermal reactions. Sometimes, thermal activation is not enough to initiate the reaction or, in orbital-symmetry-controlled concerted processes, initiates the 'wrong' reaction, and photochemical activation is necessary. Although the procedure to obtain a *mechanistic* rate law also applies for photochemical reactions, we shall not consider them specifically in this chapter.

In the following, the subscript 'obs' is appended to rate constants obtained from experiments, i.e. $k_{obs}$ is an experimentally determined reaction parameter. A *mechanistic* rate constant, i.e. the one in a rate law deduced from a proposed reaction mechanism, has no such subscript, but it may have a numerical subscript as in $k_3$, for example, indicating the elementary step in the proposed mechanism to which it relates.

## 4.2  Deducing the rate law from a postulated mechanism

An early objective in a mechanistic investigation is to establish the rate law (see Chapter 3) which is an algebraic equation describing the instantaneous dependence of the rate on concentrations of compounds or other properties proportional to concentrations (e.g. partial pressures). Rate laws cannot be reliably deduced from the stoichiometry of the overall balanced chemical equation – they have to be determined experimentally. The functional dependence of rates on concentrations may be simple or complicated, and concentrations may be of reactants, products or even materials not appearing in the overall chemical equation, as in the case of catalysis (see Chapters 11 and 12) [3–7].

A credible mechanism and the rate law deduced from it must obey the following three general requirements.

(i)  The algebraic sum of all mechanistic steps must give the balanced overall chemical equation, although this statement should be taken with care [8].
(ii)  The rate equation deduced from the mechanism must not include concentrations of reaction intermediates.
(iii)  The rate equation deduced from the mechanism must agree with the experimental one.

Sometimes, a complex mechanistic rate law is obtained which is not obviously in agreement with the experimental finding; often, consideration of the experimental conditions and/or probable relative magnitudes of rate constants of individual steps will allow simplification of the mechanistic rate law, which leads to correspondence with what is observed experimentally.

Of course, a mechanistic rate law which corresponds to the one determined experimentally (i.e. has exactly the same form) indicates no more than that the mechanism is not wrong – it is insufficient evidence that the mechanism is correct. Commonly, more than one mechanism is consistent with the observed rate equation, and further experimental work is required to allow rejection of the wrong ones. And, although only the overall chemical change is usually directly observed for most chemical reactions, kinetic experiments can sometimes be designed to detect reaction intermediates (see Chapter 9), and the possible sequence of steps in the overall proposed mechanism [3–7].

### 4.2.1  Single-step unidirectional reactions

The simplest reactions have the one-step unimolecular or bimolecular mechanisms illustrated in Table 4.1 along with their differential rate equations, i.e. the relationships between instantaneous reaction rates and concentrations of reactants. That simple unimolecular reactions are first order, and bimolecular ones second order, we take as self-evident. The integrated rate equations, which describe the concentration–time profiles for reactants, are also given in Table 4.1. In such simple reactions, the order of the reaction coincides with the molecularity and the stoichiometric coefficient.

**Table 4.1** Differential and integrated rate equations for single-step, unimolecular and bimolecular, irreversible processes.*

| Reaction | Molecularity | Differential rate equation | Integrated rate equation |
|---|---|---|---|
| $A \rightarrow C^{\dagger}$ | 1 | $-\dfrac{d[A]_t}{dt} = k[A]_t$ | $[A]_t = [A]_0\, e^{-kt}$ |
| | | | or $\ln \dfrac{[A]_t}{[A]_0} = -kt$ |
| | | | or $[C]_t = [A]_0\,(1 - e^{-kt})$ |
| $2A \rightarrow C^{\ddagger}$ | 2 | $-\dfrac{d[A]_t}{dt} = k[A]_t^2$ | $\dfrac{1}{[A]_t} - \dfrac{1}{[A]_0} = kt$ |
| $A + B \rightarrow C^{\S}$ | 2 | $-\dfrac{d[A]_t}{dt} = -\dfrac{d[B]_t}{dt} = k[A]_t[B]_t$ | $\dfrac{1}{([A]_0 - [B]_0)} \ln \dfrac{[B]_0[A]_t}{[A]_0[B]_t} = kt$ |

*In all cases, $[i]_t$ is the concentration of $i$ at any time, $t$; $[i] = [i]_0$ when $t = 0$, and $i =$ A or B.
†e.g., an isomerisation.
‡e.g., a dimerisation.
§e.g., a cycloaddition.
Throughout, there is no requirement that the product be a single compound.

Because simple rate laws follow necessarily from these simple mechanisms, a postulated simple unimolecular or bimolecular mechanism must have the corresponding rate law and this is easily tested (see Chapter 3) [9]. However, observation of a first- or second-order rate law is not sufficient evidence that the mechanism is unimolecular or bimolecular, respectively.

## 4.2.2 Simple combinations of elementary steps

In this section, we derive the expected rate laws from selected possible mechanisms involving simple combinations of elementary steps which, individually, are unimolecular or bimolecular, i.e. simple combinations of first- and second-order reactions.

### 4.2.2.1 Consecutive unimolecular (first-order) reactions

This corresponds to the formation of a product via a single intermediate in irreversible steps, usually written as Equation 4.1:

$$A \xrightarrow{k_1} B \xrightarrow{k_2} C \tag{4.1}$$

The corresponding system of rate equations, and their exact solution, assuming that only A is present initially, is shown in Scheme 4.1; the expressions for $[B]_t$ and $[C]_t$ do not apply when the two rate constants are identical ($k_1 = k_2$), an improbable situation in chemical processes. The concentration versus time curves are plotted in Fig. 4.1 (the same concentration–time dependences would be observed for any two consecutive first-order processes A → B → C even if they are not unimolecular, e.g. if pseudo-first-order steps are involved)[10].

The integrated equations in Scheme 4.1 assume simpler forms when one of the rate constants is much higher than the other. When $k_1 \gg k_2$, the reactions can be treated separately

$$-\frac{d[A]}{dt} = k_1 [A] \qquad\qquad [A]_t = [A]_0 e^{-k_1 t}$$

$$-\frac{d[B]}{dt} = k_1 [A] - k_2 [B] \qquad [B]_t = \frac{k_1 [A]_0}{k_2 - k_1}(e^{-k_1 t} - e^{-k_2 t})$$

$$-\frac{d[C]}{dt} = k_2 [B] \qquad\qquad [C]_t = [A]_0[1 - \frac{1}{k_2 - k_1}(k_2 e^{-k_1 t} - k_1 e^{-k_2 t})]$$

**Scheme 4.1**   Differential and integrated rate equations for the reaction of Equation 4.1.

and independently. In the time taken for A to react completely to give B, B undergoes negligible conversion to C and may be detectable or even isolable. Then, B *subsequently* converts into C in a much slower process uncomplicated by the presence of A.

On the other hand, if $k_2 \gg k_1$, B is a reactive intermediate (see Chapter 9) and the expression for the formation of C may be approximated to Equation 4.2:

$$[C]_t = [A]_0(1 - e^{-k_1 t}), \tag{4.2}$$

which is also the rate law for the *direct* conversion of A to C (Equation 4.3 and Table 4.1),

$$A \xrightarrow{k_1} C \tag{4.3}$$

We see, therefore, that observation of a first-order rate law could be due to a single-step first-order reaction, or a first-order initial step followed by very rapid subsequent reactions of all intermediates. This is true when the fast subsequent steps are other than first order.

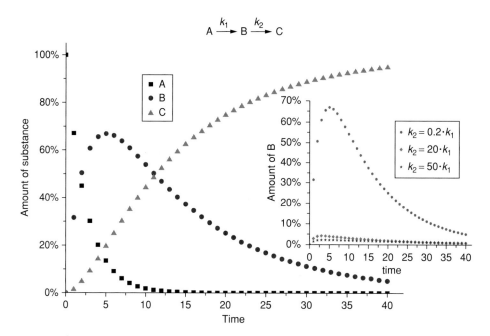

**Fig. 4.1**   Profiles of the species involved in the consecutive first-order reactions A → B → C for $k_2 = 0.2k_1$. The inset shows the changes in the amount of B for other ratios of the two rate constants.

Strategies for detecting reactive intermediates are required to distinguish between these possibilities (Chapter 9). We also see in the inset to Fig. 4.1 that the concentration of the reactive intermediate B in the mechanism of Equation 4.1 builds up to a value determined by the ratio $k_1/k_2$; this is relevant to our later consideration of the *steady-state approximation* (Section 4.2.3.1).

### 4.2.2.2 Reversible unimolecular (first-order) reactions

This could be considered as a particular case of Equation 4.1 above when A = C, and we re-label $k_2$ as $k_{-1}$ as in Equation 4.4:

$$A \xrightarrow{k_1} B \xrightarrow{k_{-1}} A \tag{4.4}$$

$$\text{or } A \underset{k_{-1}}{\overset{k_1}{\rightleftarrows}} B$$

The rate law for Equation 4.4 is

$$-\frac{d[A]}{dt} = k_1[A] - k_{-1}[B],$$

which, assuming $[B]_0 = 0$, leads to

$$[A]_t = [A]_{eq} + ([A]_0 - [A]_{eq})e^{-(k_1 + k_{-1})t},$$

where $[A]_{eq}$ is the equilibrium concentration of A, or

$$\ln\left(\frac{([A]_t - [A]_{eq})}{([A]_0 - [A]_{eq})}\right) = -(k_1 + k_{-1})t. \tag{4.5}$$

Equation 4.5 shows that, for this kind of reversible reaction, a plot of $\ln([A]_t - [A]_{eq})$ against time is linear with the gradient $-(k_1 + k_{-1})$, i.e. the reaction approaches equilibrium as a typical first-order reaction with an observed constant $k_{obs} = k_1 + k_{-1}$.

The equilibrium constant ($K$ = (activity of B)$_{eq}$/(activity of A)$_{eq} \approx [B]_{eq}/[A]_{eq}$) is also given by $K = k_1/k_{-1}$, see Fig. 4.2. Consequently, values are available for the sum and ratio of $k_1$ and $k_{-1}$, so they are individually determinable.

A particular case of this kinetic system is the racemisation of an enantiomeric compound, i.e. Equation 4.4 where A and B are enantiomers, and $k_1 = k_{-1}$, the rate constant for enantiomerisation. The rate constant for the approach to equilibrium, measurable by polarimetry, is the rate constant for racemisation, which is twice the rate constant for enantiomerisation, $k_{rac} = 2k_1$ [11].

### 4.2.2.3 Parallel (competitive) unimolecular (first-order) reactions

Scheme 4.2 shows competing unidirectional first-order reactions of a single reactant to give different products. The overall rate of disappearance of A is the sum of the rates of the two routes:

$$-\frac{d[A]}{dt} = k_1[A] + k_2[A]$$

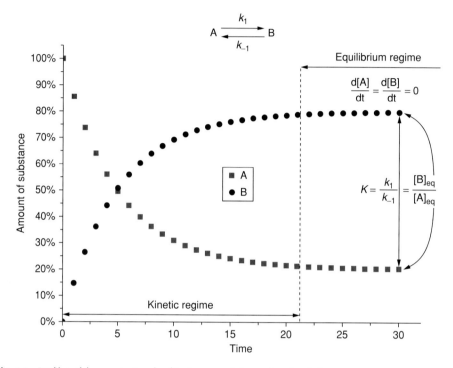

**Fig. 4.2**   Profiles of the species involved in the reversible reaction A $\rightleftharpoons$ B.

or

$$-\frac{d[A]}{dt} = k[A] \qquad \text{where} \quad k = (k_1 + k_2).$$

We see, therefore, that the reactant A disappears in an overall first-order process with $k = k_1 + k_2$, and the products B and C appear together, as shown in Fig. 4.3. The reaction proceeds mainly by the pathway with the higher rate constant, but it is important to appreciate (though surprising to the uninitiated) that the same rate constant $(k)$ is obtained regardless of whether one monitors the declining concentration of A, or the increasing concentration of *either* of the two products. In other words, fitting $[B]_t$ against time data, for example, to the integrated first-order rate law

$$[B]_t = \frac{k_1 [A]_0}{k}(1 - e^{-kt}) = [B]_\infty (1 - e^{-kt})$$

gives $k = (k_1 + k_2)$ and not just $k_1$. Evaluation of $k_1$ and $k_2$ separately requires a product analysis, see below.

**Scheme 4.2**   Parallel (competitive) first-order unidirectional reactions of a single reactant.

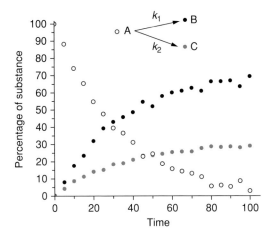

**Fig. 4.3** Concentration–time profiles for two parallel first-order unidirectional reactions of a single compound A.

The ratio of the products formed by the mechanism of Scheme 4.2 is given by $[B]/[C] = k_1/k_2$ and this remains true throughout when the reaction is *kinetically controlled* (see later and also Chapter 2). Thus, at any time during the reaction, product analysis allows determination of the *ratio* of the two elementary rate constants, $k_1/k_2$. As the rate study gives $(k_1 + k_2)$, the two elementary rate constants are individually determinable.

Two cases are included in Fig. 4.4, with the rate of disappearance of A the same in both; in one, $k_1/k_2 = 7/3$ (case 1) and in the other $k_1/k_2 = 9/1$ (case 2). In the figure, the amounts

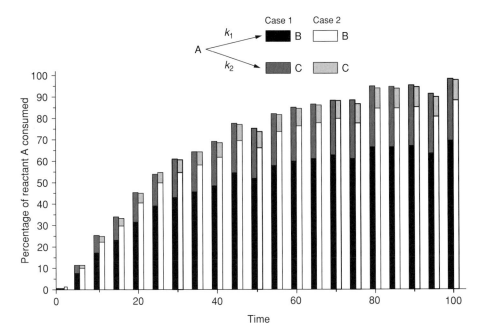

**Fig. 4.4** Relative amounts of the products B and C in the reaction of Scheme 4.2 for two ratios of $k_1$ and $k_2$. In both, $k = (k_1 + k_2)$ is the same, but $k_1/k_2 = 7/3$ in one (case 1), and $k_1/k_2 = 9/1$ in the other (case 2).

of the two products in each case have been stacked, and the sums in the two cases are equal to each other, and equal to the amount of the reactant A consumed.

If the reaction includes a mechanism for equilibration of the products, e.g. by direct interconversion following their formation (B $\rightleftharpoons$ C), the ratio [B]/[C] will not remain constant but drifts towards the equilibrium value. This is easily detected by product analysis during the course of the reaction. Scheme 4.2 with no product interconversion corresponds to *kinetic control* and the alternative corresponds to *thermodynamic* (or *equilibrium*) *control*, see Chapter 2.

### 4.2.2.4   Selectivity in competing reactions

In general, relative values of rate constants for all pathways from a single reactant are obtained by product analysis, which relates to the concept of *selectivity* (see Chapter 2). Selectivity in the present context may be defined as the ratio of the rate of formation of one component to the rate of formation of another and, in the simple example of Scheme 4.2, the selectivity of A for the formation of B rather than C is given by Equation 4.6:

$$\text{Selectivity} = \frac{[B]}{[C]} = \frac{\text{rate of formation of B}}{\text{rate of formation of C}} = \frac{r_1}{r_2} = \frac{k_1}{k_2}. \tag{4.6}$$

It is essential to establish that the alternative reactions from the reactant are kinetically controlled; otherwise product ratios could reflect relative stability rather than relative rates of formation.

The concept of selectivity is most commonly encountered (and most useful in mechanistic investigations, see Chapter 2) when a reactant or a reactive intermediate has alternative *bimolecular* routes; it is then also very useful in yield optimisation in chemical process development [12]. The reaction in Scheme 4.3 involves an electrophilic intermediate (X) which is captured by nucleophilic reagent C (which could be solvent). If another nucleophilic species (D) is added to the reaction mixture, the additional product D—X is formed in competition with C—X. If $k_D$ is known (e.g. if D is known to react with electrophiles at the diffusion limit), then values for [D], [C] and the product ratio [C—X]/[D—X] allow determination of $k_C$, i.e. quantitative information about the reactivity of X with C, and information about the selectivity of X in reactions with nucleophiles.

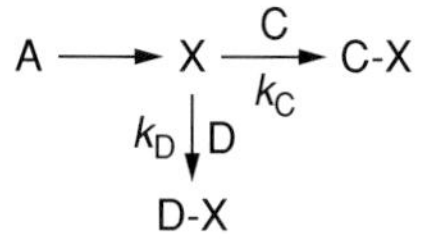

$$
\begin{array}{ll}
A \longrightarrow X \xrightarrow[k_C]{C} C\text{-}X & \text{rate}_C = k_C\,[X]\,[C] \\
\qquad\; k_D \downarrow D & \text{rate}_D = k_D\,[X]\,[D] \\
\qquad\quad D\text{-}X & \dfrac{[C\text{-}X]}{[D\text{-}X]} = \dfrac{\text{rate}_C}{\text{rate}_D}
\end{array}
$$

**Scheme 4.3**   Selectivity of an electrophilic intermediate X in a bimolecular reaction.

### 4.2.3   Complex reaction schemes and approximations

Many reactions with mechanisms like those discussed in the previous section have been investigated, i.e. reactions with relatively simple experimental rate laws which are congruent with rate laws derived from credible mechanisms. However, there are also many other reactions whose kinetics cannot be modelled by such simple combinations of elementary steps. They require schemes comprising complex combinations of elementary steps and lead to sets of coupled differential equations which may not have exact analytical solutions, or the extraction of rate constants from the experimental data involves cumbersome numerical methods [3–7].

Fortunately, introduction of *chemical* approximations often leads to *mathematical* simplification of complex rate equations, which allows the reaction to be modelled. The following are the three most commonly used approximations.

- The steady-state approximation
- The pre-equilibrium approximation
- The rate-determining step approximation

We shall use the representative mechanism in Equation 4.7 comprising three unimolecular steps to illustrate their application; each elementary step is assigned a *mechanistic* rate constant:

$$A \underset{k_{-1}}{\overset{k_1}{\rightleftarrows}} B \overset{k_2}{\longrightarrow} C \tag{4.7}$$

Qualitatively, this mechanism involves the reversible formation of an intermediate (B) which partitions between return to starting material (A) and irreversible forward reaction to product (C). Because these are coupled first-order processes, the differential equations can be solved exactly and, after rather involved but standard mathematical manipulations, the analytical solutions for the dependences of the concentrations of A, B and C upon time (assuming no B or C is initially present) are shown in Equations 4.8 [6]:

$$-\frac{d[A]}{dt} = k_1[A] - k_{-1}[B] \qquad [A]_t = \frac{k_1[A]_0}{\lambda_2 - \lambda_3}\left(\frac{\lambda_2 - k_2}{\lambda_2}e^{-\lambda_2 t} - \frac{\lambda_3 - k_2}{\lambda_3}e^{-\lambda_3 t}\right)$$

$$\frac{d[B]}{dt} = k_1[A] - k_{-1}[B] - k_2[B] \qquad [B]_t = \frac{k_1[A]_0}{\lambda_2 - \lambda_3}(e^{-\lambda_3 t} - e^{-\lambda_2 t}) \tag{4.8}$$

$$\frac{d[C]}{dt} = k_2[B] \qquad [C]_t = [A]_0\left(1 + \frac{\lambda_3}{\lambda_2 - \lambda_3}e^{-\lambda_2 t} - \frac{\lambda_2}{\lambda_2 - \lambda_3}e^{-\lambda_3 t}\right)$$

where $\quad \lambda_2 = \frac{1}{2}(p+q); \quad \lambda_3 = \frac{1}{2}(p-q); \quad p = k_1 + k_{-1} + k_2; \quad q = (p^2 - 4k_1 k_2)^{\frac{1}{2}}.$

The mechanism of Equation 4.7 is not especially complicated, yet the rigorous derivation of the rate equations is mathematically challenging, and the concentration–time expressions in Equations 4.8 are complex. It will be clear that when more unimolecular steps are involved in a mechanism, or if bimolecular elementary steps intervene, derivation of analytical solutions may become a formidable task. If the magnitudes of the elementary rate constants are similar, mathematical simplifications are not feasible, so the difficult rigorous methods have to be used. However, approximations become possible when the elementary rate constants are appreciably unequal in magnitude. This allows considerable mathematical simplification of the concentration–time relationships. Fortunately, the approximations are valid for many reactions of interest to organic chemists as we shall demonstrate.

### 4.2.3.1   The steady-state approximation (SSA)

The essence of the steady-state assumption is that the concentration of a reactive intermediate (X) is assumed to build up during a brief initial induction period, but then, during most of the rest of the reaction, its formation and decomposition are balanced (i.e. $d[X]/dt = 0$) so that its concentration remains essentially constant. Obviously, during the final stages of the reaction, the concentration of the intermediate decreases to zero.

The SSA is illustrated for the reaction $A \rightarrow B \rightarrow C$ by the inset of Fig. 4.1 (p. 82); as the ratio $k_2/k_1$ increases, the concentration of B becomes smaller and more closely approximates a constant value during the middle part of the reaction. Under these conditions, i.e. $k_2 \gg k_1$, B is a reactive intermediate, and the SSA is valid. We may generalise these conditions of validity of the SSA: the rate constants for loss of the reactive intermediate must be greater than those for its formation. The value of the SSA will be demonstrated by application to the reaction of Equation 4.7 for which the conditions of applicability are $(k_2$ and/or $k_{-1}) \gg k_1$.

Applying the steady-state approximation $(d[B]/dt = 0)$ to Equations 4.8,

$$\frac{d[B]}{dt} = k_1[A] - k_{-1}[B] - k_2[B] = 0 \quad \text{so} \quad [B]_{SS} = \frac{k_1[A]}{k_{-1} + k_2},$$

where $[B]_{SS}$ is the (very low) steady-state concentration of B during the central part of the reaction of Equation 4.7. Using this expression for [B] to substitute in

$$-\frac{d[A]}{dt} = k_1[A] - k_{-1}[B],$$

we obtain (after simplification) Equation 4.9:

$$-\frac{d[A]}{dt} = \frac{k_1 k_2[A]}{k_{-1} + k_2}. \tag{4.9}$$

Thus, the conditions for the applicability of the SSA to the mechanism of Equation 4.7 lead to the prediction of a simple first-order rate law for the disappearance of the reactant A. If the assumptions are sound, the experimental rate law shown in Equation 4.10 will be observed:

$$-\frac{d[A]}{dt} = k_{obs}[A], \tag{4.10}$$

i.e. the reaction will be first order in reactant concentration, and the experimental rate constant, $k_{obs}$, may be equated with the combination of the elementary rate constants in the proposed mechanism as follows:

$$k_{obs} = \frac{k_1 k_2}{k_{-1} + k_2}.$$

Integration of Equation 4.10 gives

$$[A]_t = [A]_0 e^{-k_{obs} t},$$

an equation much simpler than that in Equations 4.8 for $[A]_t$!

In a similar manner, substitution for [B] in

$$\frac{d[C]}{dt} = k_2[B]$$

in Equations 4.8 leads to

$$\frac{d[C]}{dt} = \frac{k_1 k_2 [A]}{k_{-1} + k_2}.$$

Since $[B]_{SS}$ is extremely small, mass conservation allows us to write $[A]_t = [A]_0 - [C]_t$, so we may now substitute for $[A]$, whereupon we obtain as our prediction,

$$\frac{d[C]}{dt} = \frac{k_1 k_2 ([A]_0 - [C])}{k_{-1} + k_2}.$$

Again, if the assumptions are sound, a first-order increase in $[C]$ will be observed experimentally:

$$-\frac{d[C]}{dt} = k_{obs}([A]_0 - [C]) \qquad \text{where} \quad k_{obs} = \frac{k_1 k_2}{k_{-1} + k_2}$$

and, since $[A]_0$ is mathematically a constant, integration now gives

$$[C]_t = [A]_0 (1 - e^{-k_{obs}t}).$$

Here also, then, we see that application of the steady-state approximation to the concentration of B when $(k_2$ and/or $k_{-1}) \gg k_1$ leads to a simple first-order concentration–time relationship for the formation of the product C, i.e. very much simpler than that in Equations 4.8.

Of course, experimental observations of clean first-order relationships for the concentrations of A and/or C with time are not *proof* that the mechanism for conversion of A into C is that of Equation 4.7 with $(k_2$ and/or $k_{-1}) \gg k_1$; they are simply compatible with it. But they do allow rejection of other more complex possibilities, e.g. Equation 4.7 *without* the rate constant inequalities required by the SSA.

### 4.2.3.2  The pre-equilibrium approximation

There are two limiting forms of Equation 4.9. If B reverts to A in the mechanism of Equation 4.7 much faster than it proceeds to give the product C, i.e. $k_{-1} \gg k_2$, the initial reversible step becomes a *pre-equilibrium* with $k_1/k_{-1} = K_1$. The predicted rate equation of Equation 4.9 then becomes

$$-\frac{d[A]}{dt} = k_2 K_1 [A].$$

If the assumptions are sound, therefore, a first-order reaction will be observed experimentally,

$$-\frac{d[A]}{dt} = k_{obs}[A],$$

and we may now write $k_{obs} = k_2 K_1$, where $K_1$ is the equilibrium constant for the initial pre-equilibrium. This is the so-called *pre-equilibrium approximation* when the intermediate is in equilibrium with reactant(s).

### 4.2.3.3  The rate-determining step approximation

When $k_2 \gg k_{-1}$ in the mechanism of Equation 4.7, i.e. the intermediate B proceeds to the product C much faster than it reverts to the reactant A, the predicted rate law of Equation 4.9

may be approximated to

$$-\frac{d[A]}{dt} = k_1[A].$$

Thus, if the assumptions are sound, a first-order rate law will be observed and the experimentally observed first-order rate constant may be equated with the *mechanistic* rate constant of the first step, $k_{obs} = k_1$. In this event, the overall rate of reaction is effectively controlled by the first step, and this is known as the *rate-determining* (or *rate-limiting*) *step* of the reaction.

As long as the SSA is valid for the mechanism in Equation 4.7, but regardless of whether it either involves a pre-equilibrium or proceeds via an initial rate-limiting step (or neither), the same prediction is obtained – a first-order rate law will be observed. However, the correspondence between the measured first-order rate constant, $k_{obs}$, and mechanistic rate constants is different, and additional evidence is required to distinguish between the alternatives.

### 4.2.3.4    *The steady-state approximation, and solvolysis of alkyl halides and arenesulfonates*

Many alkyl halides and arenesulfonates (R—X) undergo first-order solvolysis in ionising solvents (SOH) and the essential experimental features may be accommodated by the generic mechanism shown in Scheme 4.4. The mechanism includes reversible ionisation of the covalent substrate to give a contact (intimate) ion pair, $R^+ \cdot X^-$, which undergoes internal return or nucleophilic capture by the solvent (in some cases, there is evidence for the intervention of solvent-separated ion pairs, and even fully dissociated ion pairs, but the minimalist mechanism of Scheme 4.4 serves our present purpose adequately) [13].

When R—X is enantiomerically enriched *exo*-2-norbornyl 4-bromobenzenesulfonate (brosylate), 1 in Fig. 4.5, and SOH is aqueous ethanol, for example, solvolysis is accompanied by racemisation. However, the rate of racemisation is *faster* than the rate of product formation, and starting material isolated before completion of the solvolysis is partially racemised [14]. The most economical interpretation of these results (and much other evidence, see Chapter 7) is that $R^+ \cdot X^-$ includes the achiral nonclassical carbenium ion (3 in Fig. 4.5), and the ion pair undergoes internal return faster than nucleophilic capture. In other words, $k_{-1} > k_2$ in Scheme 4.4, the precise value of the ratio $k_{-1}/k_2$ depending upon the particular solvent, and the greater this ratio, the closer the initial reversible process approaches a pre-equilibrium.

In contrast, when R—X in Scheme 4.4 is enantiomerically enriched *endo*-2-norbornyl brosylate (2 in Fig. 4.5) under the same reaction conditions, solvolysis is still accompanied by racemisation, but the rate of racemisation is *identical* with that of product formation [14]. The achiral carbocation in this reaction, therefore, is captured by solvent much faster than it can undergo internal return, i.e. $k_2 \gg k_{-1}$ in Scheme 4.4. In this case, therefore, the initial ionisation is the rate-determining step.

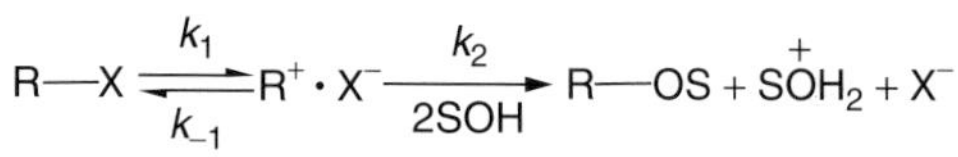

$$R—X \underset{k_{-1}}{\overset{k_1}{\rightleftharpoons}} R^+ \cdot X^- \xrightarrow[2SOH]{k_2} R—OS + S\overset{+}{O}H_2 + X^-$$

**Scheme 4.4**   Generic mechanism for the first-order solvolysis of alkyl halides and arenesulfonates.

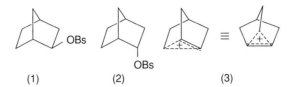

(1)     (2)     (3)

**Fig. 4.5** *exo*- and *endo*-2-Norbornyl brosylates (**1** and **2**) and the achiral nonclassical 2-norbornyl carbenium ion, **3**.

## 4.3 Case studies

In the following, representative examples have been selected to illustrate diverse features including

- widening mechanistic possibilities when acid–base forms of a reactant are possible;
- environmentally or commercially significant reactions;
- how a mechanism not in accord with the observed rate law can be discarded;
- changes in the rate equation when the concentration of a reactant is changed; and
- mechanistic deductions based on a combination of product analysis and kinetics.

### 4.3.1 Chlorination of amino compounds

Chlorination by aqueous chlorine or its derivatives is a widespread oxidation technique used in the disinfection of natural and waste waters. Oxidation by hypochlorous acid is also used in vivo as part of an unspecific defence system in mammals.

Chlorine in aqueous solution can be present as different species depending on the acidity (Scheme 4.5). At 25°C and below pH $\approx$ 4, the predominant species is $Cl_2(aq)$, and $Cl_3^-(aq)$ when $Cl^-$ is present; in the range pH $\approx$ 4–7.5, hypochlorous acid (HOCl) is the major species, which is in acid–base equilibrium with its conjugate base, hypochlorite ($ClO^-$); hypochlorite is the predominant species above pH $\approx$ 7.5.

$$Cl_2(aq) + 2H_2O(l) \rightleftharpoons HClO(aq) + Cl^-(aq) + H_3O^+(aq)$$

$$Cl_2(aq) + Cl^-(aq) \rightleftharpoons Cl_3^-(aq)$$

$$HClO(aq) + H_2O(l) \rightleftharpoons ClO^-(aq) + H_3O^+(aq)$$

**Scheme 4.5** Equilibria involving chlorine in aqueous solution.

Kinetic runs carried out at constant pH > 5 and constant ionic strength indicate that chlorination of amines ($R_1R_2NH$) is a second-order reaction, Equation 4.11 [15],

$$r_{obs} = k_{obs}[R_1R_2NH]_T\,[Cl_2]_T, \tag{4.11}$$

where the subscript 'T' means the total concentration of all forms of the specified material, amine or aqueous chlorine, i.e.

$$[Cl_2]_T = [Cl_2] + [Cl_3^-] + [HOCl] + [ClO^-]$$

$$[R_1R_2NH]_T = [R_1R_2NH] + [R_1R_2NH_2^+].$$

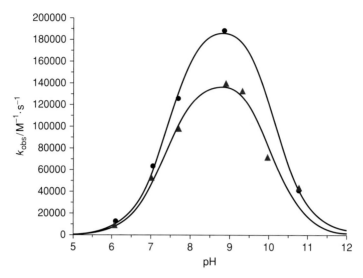

**Fig. 4.6** Dependence of the observed second-order rate constant ($k_{obs}$) upon acidity in the chlorination of amino acids, glycine (●) and isoleucine (▲); $T = 298$ K. The solid lines correspond to non-linear fitting of the experimental data to Equation 4.12.

It was found that $k_{obs}$ shows a complex dependence on acidity as exemplified in Fig. 4.6 for two amino acids. This acid dependence is satisfactorily described by Equation 4.12 where $m$, $n$ and $o$ are empirical parameters,

$$k_{obs} = m \left( \frac{[H^+]}{(n + [H^+])(o + [H^+])} \right), \tag{4.12}$$

so Equation 4.11 may be expanded as shown in Equation 4.13:

$$r_{obs} = k_{obs}[R_1R_2NH]_T [Cl_2]_T = \frac{m[H^+][R_1R_2NH]_T[Cl_2]_T}{(n + [H^+])(o + [H^+])}. \tag{4.13}$$

Note that, in this equation and elsewhere, we use $H^+$ in place of $H_3O^+$ for brevity.

The experimental rate law expressed in Equation 4.13 has only one term, admittedly a complex one, which suggests only one reaction path, and the form of the denominator indicates two acid–base equilibria prior to the rate-determining step. It is common to find sigmoidal dependences on acidity when acid–base pre-equilibria are involved, and a sudden change in $k_{obs}$ usually takes place within just 2–3 pH units (when broader sigmoids are observed, results should be analysed with care). The pH corresponding to the half-height of the sigmoid gives an estimate of the $pK_a$ of the conjugate acid involved (see also Chapter 11), so the bell-shaped curve in Fig. 4.6 indicates that, in the present case, two acid–base pre-equilibria are involved with acids of $pK_a$ values around 7.5 and 10.

The proposal of a mechanism obviously requires identification of the species able to react; in this case, at first sight, there are up to six species: four from chlorine ($Cl_2$, $Cl_3^-$, HOCl and $ClO^-$) and two from the amine ($R_1R_2NH_2^+$ and $R_1R_2NH$). The eight elementary bimolecular steps shown in Scheme 4.6, therefore, are possible in principle. For the chlorination of amino acids, of course, there are additional acid–base equilibria to be taken into account, which expands the number of mechanistic possibilities.

$$R_1R_2NH_2^+ + ClO^- \rightarrow R_1R_2NCl + H_2O \quad k_1 \Rightarrow r_1 = k_1\,[R_1R_2NH_2^+][ClO^-]$$
$$R_1R_2NH \quad + ClO^- \rightarrow R_1R_2NCl + HO^- \quad k_2 \Rightarrow r_2 = k_2\,[R_1R_2NH][ClO^-]$$
$$R_1R_2NH_2^+ + HClO \rightarrow R_1R_2NCl + H_3O^+ \quad k_3 \Rightarrow r_3 = k_3\,[R_1R_2NH_2^+][HClO]$$
$$R_1R_2NH \quad + HClO \rightarrow R_1R_2NCl + H_2O \quad k_4 \Rightarrow r_4 = k_4\,[R_1R_2NH][HClO]$$

$$R_1R_2NH_2^+ + Cl_2 \quad \rightarrow R_1R_2NCl + 2H^+ + Cl^- \quad k_5 \Rightarrow r_5 = k_5\,[R_1R_2NH_2^+][Cl_2]$$
$$R_1R_2NH \quad + Cl_2 \quad \rightarrow R_1R_2NCl + H^+ \quad + Cl^- \quad k_6 \Rightarrow r_6 = k_6\,[R_1R_2NH][Cl_2]$$
$$R_1R_2NH_2^+ + Cl_3^- \rightarrow R_1R_2NCl + 2H^+ + 2Cl^- \quad k_7 \Rightarrow r_7 = k_7\,[R_1R_2NH_2^+][Cl_3^-]$$
$$R_1R_2NH \quad + Cl_3^- \rightarrow R_1R_2NCl + H^+ \quad + 2Cl^- \quad k_8 \Rightarrow r_8 = k_8\,[R_1R_2NH][Cl_3^-]$$

**Scheme 4.6** Possible elementary bimolecular steps in the chlorination of amines.

For the sake of brevity, we shall not consider further the possible reactions involving $Cl_2$ and $Cl_3^-$ except to state that the observed rate dependence on acidity is not predicted by any mechanism involving these species. In addition, their concentrations in the working pH range would require second-order rate constants for $k_5$ to $k_8$ above the diffusion-controlled limit, so their contribution as the chlorinating species, in the working pH range, can be rejected. So, because virtually all chlorine in the working pH range is going to be reacting as HOCl or $ClO^-$, $[Cl_2]_T$ in previous rate equations will now be redesignated $[HOCl]_T$, and we rewrite the previous experimental rate law (Equation 4.13) as Equation 4.14:

$$r_{obs} = k_{obs}[R_1R_2NH]_T[HClO]_T = \frac{m[H^+][R_1R_2NH]_T[HClO]_T}{(n+[H^+])(o+[H^+])}. \tag{4.14}$$

The mechanism to be proposed is now fairly simple; it comprises the acid–base equilibria of hypochlorous acid and the amine, whose equilibrium constants are known [15],

$$HOCl + H_2O \rightleftharpoons ClO^- + H_3O^+; \quad K_{a1}$$
$$R_1R_2NH_2^+ + H_2O \rightleftharpoons R_1R_2NH + H_3O^+; \quad K_{a2}$$

and (as there is only one term in the observed rate law) just one of the first four elementary steps in Scheme 4.6.

Table 4.2 shows the rate laws predicted by mechanisms involving the acid–base equilibria above and one of the elementary steps 1–4 of Scheme 4.6 as the rate-determining step. The forms of the experimental rate laws which would correspond to those derived from the mechanisms are also shown in Table 4.2.

We have already seen from Equation 4.11 that it is not possible to distinguish between the first four reactions of Scheme 4.6 from kinetic data at constant pH. But when the influence of acidity is taken into account, only two possibilities are compatible with the observed rate law – the first and the fourth ($r_1$ and $r_4$). These two are *kinetically equivalent* mechanisms, i.e. they cannot be distinguished from the form of the rate law alone – the distinction has to be made on other grounds.

It should be stressed that whilst these two mechanistic rate equations have the same form, the molecular processes they are derived from are different, and the respective second-order rate constants (corresponding to the mechanistic rate constants $k_1$ and $k_4$ for bimolecular processes $r_1$ and $r_4$ in Scheme 4.6) extracted from the experimental results differ by three orders of magnitude, as shown in Table 4.3. The more credible of the two alternatives, in

**Table 4.2**  Mechanistic and corresponding experimental rate laws for the chlorination of amino compounds by aqueous chlorine at pH > 5.

| Rate law predicted by mechanism | Form of the corresponding experimental rate law |
|---|---|
| $r_1 = \dfrac{k_1 K_{a1}[H^+]}{([H^+] + K_{a1})([H^+] + K_{a2})}[R_1R_2NH]_T[HClO]_T$ | $r_{obs} = \dfrac{m[H^+]}{(n + [H^+])(o + [H^+])}[R_1R_2NH]_T[HClO]_T$ |
| $r_2 = \dfrac{k_2 K_{a1} K_{a2}}{([H^+] + K_{a1})([H^+] + K_{a2})}[R_1R_2NH]_T[HClO]_T$ | $r_{obs} = \dfrac{p}{(n + [H^+])(o + [H^+])}[R_1R_2NH]_T[HClO]_T$ |
| $r_3 = \dfrac{k_3[H^+]^2}{([H^+] + K_{a1})([H^+] + K_{a2})}[R_1R_2NH]_T[HClO]_T$ | $r_{obs} = \dfrac{q[H^+]^2}{(n + [H^+])(o + [H^+])}[R_1R_2NH]_T[HClO]_T$ |
| $r_4 = \dfrac{k_4 K_{a2}[H^+]}{([H^+] + K_{a1})([H^+] + K_{a2})}[R_1R_2NH]_T[HClO]_T$ | $r_{obs} = \dfrac{m[H^+]}{(n + [H^+])(o + [H^+])}[R_1R_2NH]_T[HClO]_T$ |

accord with a substantial body of evidence of amines reacting as bases/nucleophiles, is the path via $r_4$ with $k_4 \approx 10^7$ dm$^3$ mol$^{-1}$ s$^{-1}$ at 298 K [15].

Sometimes it is helpful to analyse the mechanistic rate equation at high or low concentration of the reactants and check the prediction with the observed behaviour. In this case, the mechanistic rate equation in Table 4.3 for $r_4$ at high pH values becomes

$$r_4 = \frac{k_4 K_w}{K_{a1}} \frac{[R_1R_2NH]_T[HOCl]_T}{[HO^-]} \quad \text{corresponding to} \quad r_{obs} = k_{obs}\frac{[R_1R_2NH]_T[ClO^-]_T}{[HO^-]},$$

**Table 4.3**  Two kinetically equivalent mechanisms for the chlorination of amines by aqueous chlorine at pH > 5.

| Path with step of rate $r_1$ in Scheme 4.6 | Path with step of rate $r_4$ in Scheme 4.6 |
|---|---|
| $HClO + H_2O \rightleftharpoons ClO^- + H_3O^+; K_{a1}$ | $HClO + H_2O \rightleftharpoons ClO^- + H_3O^+; K_{a1}$ |
| $R_1R_2NH_2^+ + H_2O \rightleftharpoons R_1R_2NH + H_3O^+; K_{a2}$ | $R_1R_2NH_2^+ + H_2O \rightleftharpoons R_1R_2NH + H_3O^+; K_{a2}$ |
| $R_1R_2NH_2^+ + ClO^- \rightarrow R_1R_2NCl + H_2O; k_1$ | $R_1R_2NH + HClO \rightarrow R_1R_2NCl + H_2O; k_4$ |

*Rate equations derived from the mechanisms*

| | |
|---|---|
| $r_1 = k_1[R_1R_2NH_2^+][ClO^-]$ | $r_4 = k_4[R_1R_2NH][HClO]$ |
| $r_1 = \dfrac{k_1 K_{a1}[H^+][R_1R_2NH]_T[HClO]_T}{([H^+] + K_{a1})([H^+] + K_{a2})}$ | $r_4 = \dfrac{k_4 K_{a2}[H^+][R_1R_2NH]_T[HClO]_T}{([H^+] + K_{a1})([H^+] + K_{a2})}$ |

*Corresponding experimental rate laws*

$$r_{obs} = k_{obs}[R_1R_2NH]_T[HClO]_T = m\frac{[H^+]}{(n + [H^+])(o + [H^+])}[R_1R_2NH]_T[HClO]_T$$

*Correspondence of mechanical and experimental parameters*

| | |
|---|---|
| $k_{obs} = k_1 K_{a1}\left(\dfrac{[H^+]}{(K_{a1} + [H^+])(K_{a2} + [H^+])}\right)$ | $k_{obs} = k_4 K_{a2}\left(\dfrac{[H^+]}{(K_{a1} + [H^+])(K_{a2} + [H^+])}\right)$ |
| with $m = k_1 K_{a1}$, $n$ and $o$ being $K_{a1}$ and $K_{a2}$ | with $m = k_4 K_{a2}$, $n$ and $o$ being $K_{a1}$ and $K_{a2}$ |
| $k_1 \approx 10^4$ dm$^3$ mol$^{-1}$ s$^{-1}$ at 298 K | $k_4 \approx 10^7$ dm$^3$ mol$^{-1}$ s$^{-1}$ at 298 K |

i.e. the experimental rate law should be first order in hypochlorite ion and amine, and of order $-1$ with respect to the hydroxide ion. This is indeed the observed behaviour [15], although the kinetic study was complicated by the subsequent reaction of the $N$-chloramine product.

On the other hand, at low pH, the mechanistic rate law becomes

$$r_4 = k_4 K_{a2} \frac{[R_1R_2NH]_T[HOCl]_T}{[H^+]} \quad \text{corresponding to} \quad r_{obs} = k_{obs} \frac{[R_1R_2NH]_T[HOCl]_T}{[H^+]},$$

i.e. it is still predicted to be first order in each of total hypochlorous acid and amine, but now of order $-1$ in hydronium ion, which is not the observed behaviour; chlorination by $Cl_2$ is now taking place instead of by HOCl [15].

### 4.3.2   The Aldol reaction

The Aldol reaction (Equation 4.15) is an important synthetic prototype mentioned in Chapter 3 [16, 17]:

$$2 \quad \ce{H3C-CHO} \xrightarrow[\text{H2O}]{\text{B, OH}^-} \quad \ce{H3C-CH(OH)-CH2-CHO} \tag{4.15}$$

A possible mechanism which accounts for the general features of the reaction in aqueous solution is shown in Scheme 4.7. B represents any base, in addition to $OH^-$, present in the aqueous reaction mixture, $CO_3{}^{2-}$ or an amine, for example; and, because acetaldehyde is only very weakly acidic in aqueous solution ($pK_a = 16.7$) [18], the proposed enolate carbanion intermediate can be present at only extremely low concentrations.

Different rate laws have been observed experimentally, depending on working conditions, i.e. first order in acetaldehyde and general base catalysed, or second order in acetaldehyde and specific base catalysed [17]. We shall use the steady-state approximation to deduce rate laws based upon the mechanism of Scheme 4.7, and see how different mechanistic assumptions lead to different rate law predictions.

The final proton transfer will be extremely fast, so the rate of formation of product is effectively shown in Equation 4.16, i.e. the rate of the addition of the enolate anion to another molecule of aldehyde:

$$r = k_4[CH_3CHO][CH_2CHO^-]. \tag{4.16}$$

This, however, is not a legitimate rate law as it includes the concentration of an intermediate. We need to be able to express the concentration of the enolate in terms of the concentrations of acetaldehyde and the bases present.

Applying the steady-state approximation (SSA) to the transient enolate intermediate in Scheme 4.7, its concentration is given by

$$[CH_2CHO^-]_{SSA} = \frac{(k_1[OH^-] + k_2[H_2O] + k_3[B])[CH_3CHO]}{k_{-1}[H_2O] + k_{-2}[H^+] + k_{-3}[BH^+] + k_4[CH_3CHO]}.$$

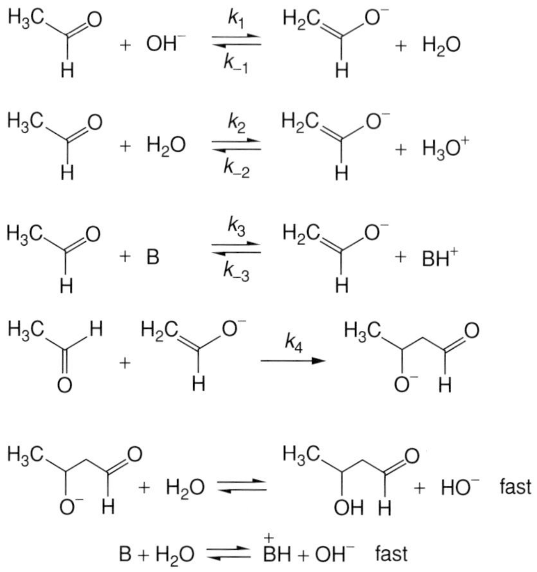

**Scheme 4.7**  A general mechanism for the Aldol reaction in aqueous solution.

Substituting the SSA concentration of $CH_2CHO^-$ into Equation 4.16, the reaction rate is now given by Equation 4.17:

$$r = \frac{k_4(k_1[OH^-] + k'_2 + k_3[B])[CH_3CHO]^2}{k'_{-1} + k_{-2}[H^+] + k_{-3}[BH^+] + k_4[CH_3CHO]}, \tag{4.17}$$

where $k'_2 = k_2[H_2O]$ and $k'_{-1} = k_{-1}[H_2O]$. According to Equation 4.17, which includes concentrations of only reactants, so is legitimate in that respect, the reaction has no simple assignable kinetic order in any of the reactants. In other words, application of the SSA to the enolate in the mechanism of Scheme 4.7 has not led to a simple rate law (cf. its application in the mechanism of Equation 4.7). However, we see below that different simple rate laws are predicted in alternative limiting situations for comparison with experiment.

### 4.3.2.1   At low concentrations of aldehyde

This limiting case of Equation 4.17 occurs when reprotonation of the enolate is much faster than its bimolecular capture by another aldehyde molecule, i.e. $(k'_{-1} + k_{-2}[H^+] + k_{-3}[BH^+]) \gg k_4[CH_3CHO]$; under these conditions, the reactions of aldehyde with the bases are pre-equilibria. The above inequality allows Equation 4.17 to be approximated by Equation 4.18:

$$r = \frac{k_4(k_1[OH^-] + k'_2 + k_3[B])[CH_3CHO]^2}{k'_{-1} + k_{-2}[H^+] + k_{-3}[BH^+]}. \tag{4.18}$$

Taking into account the equilibrium relationships

$$K_1 = \frac{k_1}{k_{-1}} = \frac{[CH_2CHO^-]}{[CH_3CHO][OH^-]} \qquad K_2 = \frac{k_2'}{k_{-2}} = \frac{[CH_2CHO^-][H^+]}{[CH_3CHO]}$$

$$K_3 = \frac{k_3}{k_{-3}} = \frac{[CH_2CHO^-][BH^+]}{[CH_3CHO][B]},$$

we may substitute for $k_1$, $k_2'$ and $k_3[B]$ in Equation 4.18 and thereby obtain

$$r = \frac{k_4(k_{-1}'K_1[OH^-] + k_{-2}K_1[H^+][OH^-] + k_{-3}K_1[BH^+][OH^-])[CH_3CHO]^2}{k_{-1}' + k_{-2}[H^+] + k_{-3}[BH^+]},$$

which, upon simplification, gives

$$r = k_4 K_1[OH^-][CH_3CHO]^2,$$

i.e. the reaction is now predicted to be second order in $[CH_3CHO]$ and *specific* base catalysed. This is indeed the observed rate law at low concentrations of acetaldehyde,

$$r_{obs} = k_{obs}[OH^-][CH_3CHO]^2,$$

i.e. even though general bases are involved in the chemical reaction, they do not appear in the rate law, and the observed third-order rate constant, $k_{obs}$, corresponds to $k_4 K_1$ of the mechanism.

### 4.3.2.2    At high concentrations of aldehyde

At higher concentrations of acetaldehyde, bimolecular trapping of the enolate in Scheme 4.7 will become faster, so, at some stage, this will compete effectively with the reprotonation of the enolate. When the bimolecular capture of enolate by another acetaldehyde molecule becomes much faster than the reprotonation of the enolate, i.e. when $k_4[CH_3CHO] \gg k_{-1}' + k_{-2}[H^+] + k_{-3}[BH^+]$, another limiting approximation to the complex rate equation predicted from the mechanism (Equation 4.17) is obtained, Equation 4.19:

$$r = (k_1[OH^-] + k_2' + k_3[B])[CH_3CHO]. \qquad (4.19)$$

Under buffered conditions at high aldehyde concentrations, the experimental rate law is

$$r_{obs} = k_{obs}[CH_3CHO] \quad \text{where} \quad k_{obs} = k_{OH}[OH^-] + k_0 + k_B[B],$$

i.e. the reaction is now first order in acetaldehyde, and first order in each of the bases present, an example of *general base catalysis* (see Chapter 11). Under these conditions, not only do we have the exact correspondence between mechanistic and experimental rate laws, but also we have the correspondence between individual experimental and mechanistic parameters:

$$k_{obs} = (k_1[OH^-] + k_2' + k_3[B]); \qquad k_{OH} = k_1; \quad k_0 = k_2'; \quad k_B = k_3.$$

Under these conditions of high concentrations of $CH_3CHO$, there is competition between the three bases present ($OH^-$, $H_2O$ and B) for the acetaldehyde, and the overall rate of the reaction is controlled by the sum of the rates of the parallel deprotonation steps, i.e. we have

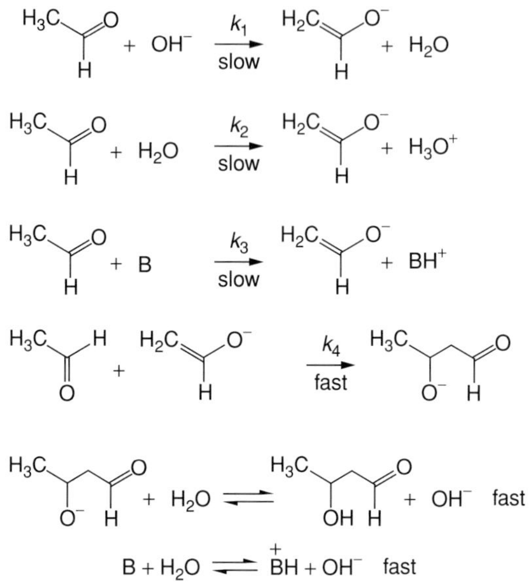

**Scheme 4.8**  Mechanism of the Aldol reaction at high aldehyde concentrations in aqueous solution.

parallel independent rate-limiting processes generating a common reactive nucleophilic intermediate which is trapped by the electrophilic second aldehyde molecule. And note that the three parallel mechanistic reaction routes imply three terms in the corresponding experimental rate law.

In addition to the rate law dependence upon the *concentration* of a general base, we might also expect a dependence of its rate constant upon its *base strength*. Indeed, for this reaction, rate constants for general bases increase with increasing base strength, with a levelling off at high values. This levelling off occurs when the deprotonation becomes diffusion controlled [4], i.e. the base has become sufficiently strong that deprotonation takes place at every molecular encounter between the base and the acetaldehyde. The rate constant for this step ($k_3$ in Scheme 4.7) is no longer dependent on the nature of the base – it is controlled simply by diffusion together of the reacting species which, in turn, is affected by the viscosity of the medium [4]. Scheme 4.7 has been rewritten as Scheme 4.8 showing the details of the general base catalysis mechanism.

### 4.3.3   Hydrogen atom transfer from phenols to radicals

The mechanism shown in Scheme 4.9 has been proposed for the hydrogen atom transfer from phenols (ArOH) to radicals (Y$^\bullet$) in non-aqueous solvents, a kinetic effect of the solvent (S) being expected when ArOH is a hydrogen bond donor and the solvent a hydrogen bond acceptor. Steps with mechanistic rate constants $k_1$, $k_{-1}$ and $k_3$ involve proton transfer (the latter two near to the diffusion-controlled limit), and $k_2$ involves electron transfer. The step with rate constant $k_0$ involves a direct hydrogen atom transfer, and the other path around the cycle involves a stepwise alternative.

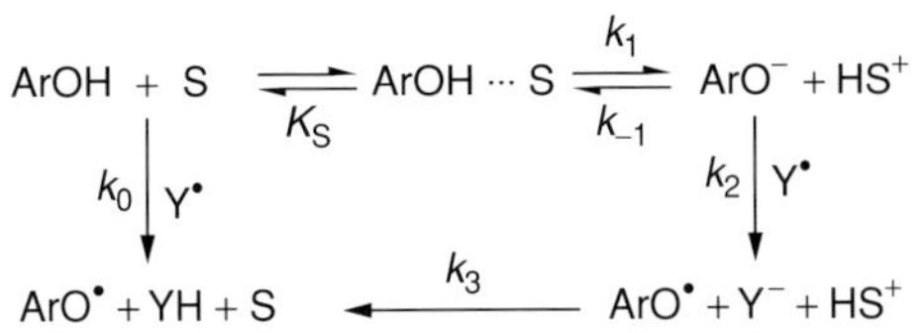

$$
\begin{array}{ccc}
\text{ArOH} + \text{S} \; \underset{K_\text{S}}{\rightleftharpoons} \; \text{ArOH} \cdots \text{S} \; \underset{k_{-1}}{\overset{k_1}{\rightleftharpoons}} \; \text{ArO}^- + \text{HS}^+ \\
k_0 \downarrow \text{Y}^\bullet \qquad\qquad\qquad\qquad\qquad k_2 \downarrow \text{Y}^\bullet \\
\text{ArO}^\bullet + \text{YH} + \text{S} \; \xleftarrow{\;\;k_3\;\;} \; \text{ArO}^\bullet + \text{Y}^- + \text{HS}^+
\end{array}
$$

**Scheme 4.9**  Proposed mechanism for hydrogen atom transfer from phenols to radicals in non-aqueous solvents [19].

According to this mechanism, the rate of formation of products is given by

$$
r = k_0[\text{ArOH}][\text{Y}^\bullet] + k_3[\text{Y}^-][\text{HS}^+],
$$

an equation which includes species other than reactants. By applying the steady-state approximation to the transient intermediates $\text{Y}^-$ and $\text{ArO}^-$, the rate law of Equation 4.20 is obtained:

$$
r = \frac{k_0}{1 + K_\text{S}[\text{S}]}[\text{ArOH}]_\text{T}[\text{Y}^\bullet] + \frac{k_1 k_2 K_\text{S}[\text{S}][\text{ArOH}]_\text{T}[\text{Y}^\bullet]}{(k_{-1}[\text{HS}^+] + k_2[\text{Y}^\bullet])(1 + K_\text{S}[\text{S}])}, \tag{4.20}
$$

in which it has been assumed that $[\text{ArOH}]_\text{T} \approx [\text{ArOH}]_\text{free} + [\text{ArOH}\cdots\text{S}]$.

An experimental rate law of the following form would be the consequence of this mechanistic rate expression:

$$
r_\text{obs} = a[\text{ArOH}]_\text{T}[\text{Y}^\bullet] + \frac{b[\text{ArOH}]_\text{T}[\text{Y}^\bullet]}{(c[\text{HS}^+] + d[\text{Y}^\bullet])} = a'[\text{Y}^\bullet] + \frac{b'[\text{Y}^\bullet]}{(c[\text{HS}^+] + d[\text{Y}^\bullet])},
$$

where the primed parameters, $a'$ and $b'$, are employed for reactions using an excess of the phenol. We see that, according to this mechanism, the reaction should be first order in the phenol and show a complex order in radical concentration.

In solvents where there is no formation of the phenolate anion $\text{ArO}^-$, only the direct $k_0$ path is possible. Although simple second-order kinetics are expected in such solvents, first order in $[\text{ArOH}]$ and in $[\text{Y}^\bullet]$, the denominator of the expected rate law (Equation 4.21 derived from Equation 4.20) includes a term acknowledging the equilibrium formation of the hydrogen-bonded complex which depletes the concentration of the free phenol:

$$
r = \frac{k_0[\text{ArOH}]_\text{T}}{1 + K_\text{S}[\text{S}]}[\text{Y}^\bullet]. \tag{4.21}
$$

Equation 4.21 corresponds to

$$
r_\text{obs} = k_\text{obs}[\text{Y}^\bullet] \quad \text{where} \quad k_\text{obs} = \frac{k_0}{1 + K_\text{S}[\text{S}]}[\text{ArOH}]_\text{T}.
$$

This model describes the observed behaviour when 2,2′-methylene-bis(4-methyl-6-*tert*-butylphenol) (**ArOH** in Fig. 4.7) reacts with 2,2-diphenyl-1-picrylhydrazyl (**Y$^\bullet$**), for example, in several solvents including $\text{CCl}_4$ and 1,4-dioxane [19].

When phenolate formation is possible, simplification of Equation 4.20 is possible in two limiting cases.

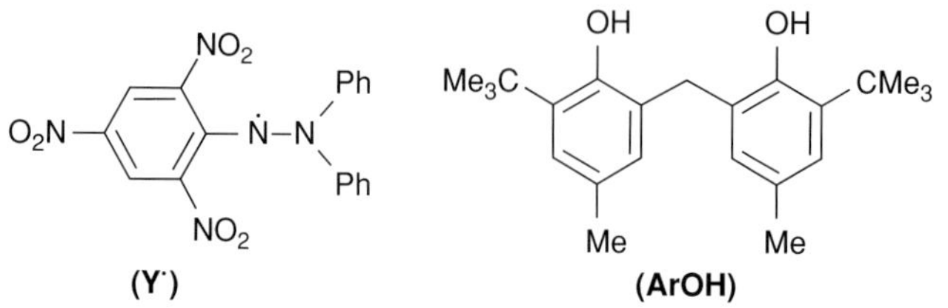

**Fig. 4.7**   Reactants in a hydrogen atom transfer reaction.

### 4.3.3.1   Via pre-equilibrium formation of the phenolate

When $k_{-1}[HS^+] \gg k_2[Y^\bullet]$, i.e. the proton transfer is another pre-equilibrium, Equation 4.20 may be simplified to give

$$r = \frac{(k_0 + k_2 K_S K_1[S]/[SH^+])}{1 + K_S[S]}[ArOH]_T[Y^\bullet] \quad \text{where} \quad K_1 = k_1/k_{-1}$$

corresponding to $r_{obs} = k_{obs}[Y^\bullet]$ where $k_{obs} = \dfrac{(k_0 + k_2 K_S K_1[S]/[SH^+])}{1 + K_S[S]}[ArOH]_T.$

In buffered solutions, the term $k_2 K_S K_1[S]/[SH^+]$ is constant, so the expected overall rate law is again second order (i.e. pseudo first order in $[Y^\bullet]$) but the correspondence of $k_{obs}$ with mechanistic rate constants is different. Of course, if the equilibrium constant $K_1$ is appreciable, the phenolate concentration must be taken into account in the mass balance for the total phenol, i.e. $[ArOH]_T \approx [ArOH]_{free} + [ArOH \cdots S] + [ArO^-]$, whereupon the mechanistic rate equation becomes more complicated.

### 4.3.3.2   Via rate-limiting proton transfer to give the phenolate

If $k_2[Y^\bullet] \gg k_{-1}[HS^+]$, i.e. the proton transfer within the hydrogen-bonded complex $(ArOH \cdots S)$ is rate determining, the mechanistic rate law becomes

$$r = \frac{(k_0[Y^\bullet] + k_1 K_S[S])[ArOH]_T}{1 + K_S[S]} \quad \text{corresponding to} \quad r_{obs} = k_{obs,1}[Y^\bullet] + k_{obs,2}$$

$$\text{where} \quad k_{obs,1} = \frac{k_0[ArOH]_T}{1 + K_S[S]} \quad \text{and} \quad k_{obs,2} = \frac{k_1 K_S[S][ArOH]_T}{1 + K_S[S]}.$$

In this case, the predicted rate law includes two reaction channels – both first order in phenol, with one first order in radical concentration and the other zero order. This behaviour has been found, for example, in the reaction between 2,2′-methylene-bis(4-methyl-6-*tert*-butylphenol) (**ArOH**) and 2,2-diphenyl-1-picrylhydrazyl ($Y^\bullet$) in Fig. 4.7 when acetonitrile, benzonitrile, acetone, cyclohexanone or DMSO are the solvents [19].

## 4.3.4   Oxidation of phenols by Cr(VI)

Both phenols and Cr(VI) are chemicals of high environmental concern. In this redox reaction, one- or two-electron transfer pathways could take place, and a proposed mechanism is shown in Scheme 4.10 [20]. In aqueous acidic media (pH $\leq$ 5), it involves fast equilibrium

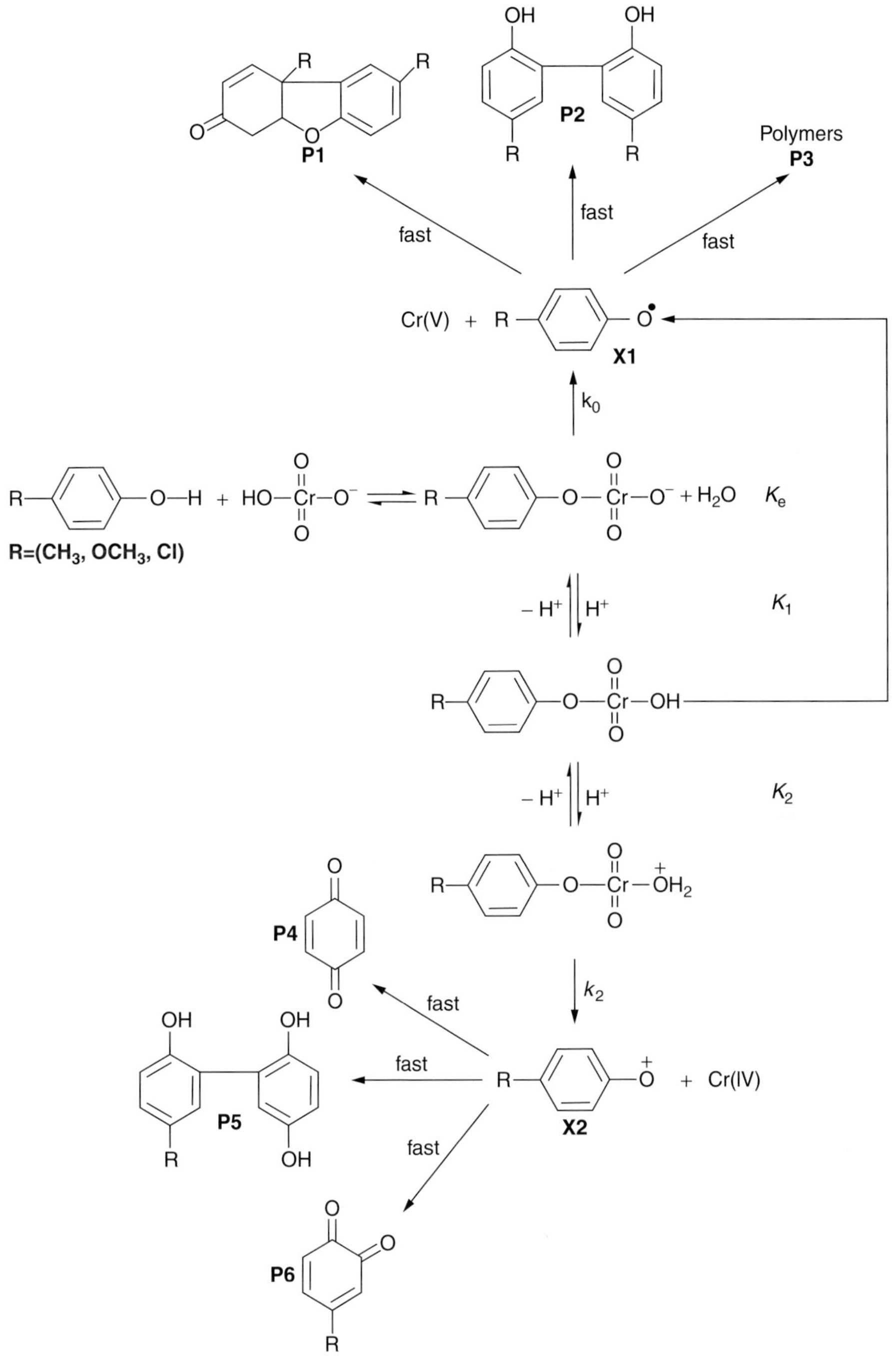

**Scheme 4.10** Proposed one-electron and two-electron transfer pathways for the oxidation of phenols by Cr(VI) [20].

formation of chromate esters, which then decompose by rate-determining electron transfer to the Cr(VI) centre with rate constants $k_0$ and/or $k_1$ (one-electron processes) and $k_2$ (two-electron process). The phenoxy radical (**X1**) and phenoxy cation (**X2**) then react rapidly to form the products (**P1–P6**).

The rate of product formation according to this mechanistic proposal is

$$r = k_0[\text{ArO–CrO}_3{}^-] + k_1[\text{ArO–Cr(O}_2)\text{–OH}] + k_2[\text{ArO–Cr(O}_2)\text{–OH}_2{}^+],$$

which, assuming all acid–base equilibria are fast, may be transformed into

$$r = \left( k_0 + \frac{k_1[\text{H}^+]}{K_1} + \frac{k_2[\text{H}^+]^2}{K_1 K_2} \right) [\text{ArOCrO}_3^-].$$

Assuming that equilibrium formation of the chromate ester is fast, this rate expression may be rewritten as Equation 4.22:

$$r = \left( k_0 + \frac{k_1[\text{H}^+]}{K_1} + \frac{k_2[\text{H}^+]^2}{K_1 K_2} \right) K_e[\text{HCrO}_4{}^-][\text{ArOH}]. \tag{4.22}$$

At pH $\leq 5$, deprotonation of phenolic groups can be neglected, so $[\text{ArOH}] \approx [\text{ArOH}]_\text{T}$, and the following mass balance holds for Cr(VI):

$$[\text{Cr(VI)}]_\text{T} = [\text{HCrO}_4{}^-] + [\text{CrO}_4{}^{2-}] + [\text{chromate ester}].$$

Because $K_e$ is very small and the formation of chromate dianion is minimal in the working pH range, this may be approximated to give

$$[\text{Cr(VI)}]_\text{T} \approx [\text{HCrO}_4^-] \quad .$$

Consequently, Equation 4.22 becomes

$$r = \left( k_0 + \frac{k_1[\text{H}^+]}{K_1} + \frac{k_2[\text{H}^+]^2}{K_1 K_2} \right) K_e[\text{Cr(VI)}]_\text{T}[\text{ArOH}]_\text{T}. \tag{4.23}$$

Accordingly, an overall second-order rate law is expected at constant pH (first order in phenol and in Cr(VI)), and there will be a complex dependence on acidity varying between kinetic order 0 and 2. In addition, the $k_2$ (two-electron transfer) pathway should be the main route at higher acidities, with predominance of the products derived from the phenoxy cation, **X2**.

Product analysis supports these deductions [20]. At lower acidities (pH $\approx$ 2–5), o-cresol gives mainly products by one-electron transfer (**P1** and **P2**). At higher acidities (pH $\leq 2$), however, 4-chlorophenol forms **P4** as the major product, with some **P5** and **P6**, but only a minor amount of **P2** and traces of polymers **P3**.

Kinetic analysis under pseudo-first-order conditions leads to the following:

$$r_\text{obs} = k_\text{obs}[\text{ArOH}]_\text{T} \quad \text{with} \quad k_\text{obs} = k'_\text{obs}[\text{Cr(VI)}]_\text{T} \quad \text{and} \quad k'_\text{obs} = a + b[\text{H}^+] + c[\text{H}^+]^2.$$

Comparison with Equation 4.23 establishes the following relationships between experimental and mechanistic parameters:

$$a = k_0 K_e; \qquad b = \frac{k_1 K_e}{K_1}; \qquad c = \frac{k_2 K_e}{K_1 K_2}.$$

In summary, product analysis and kinetic analysis support the proposed mechanism comprising three reaction channels – two of them involving a one-electron transfer pathway operating at lower acidities, and the other a two-electron transfer pathway at higher acidities.

## Bibliography

Bernasconi, C.F. (Ed.) (1986) *Investigation of Rates and Mechanisms of Reactions* (4th edn). Part I. *Techniques of Chemistry* (vol 6). Wiley, New York.

Carpenter, B.K. (1984) *Determination of Organic Reaction Mechanisms.* Wiley, New York.

Chorkendorff, I. and Niemantsverdriet, J.W. (2003) *Concepts of Modern Catalysis and Kinetics.* Wiley-VCH, Weinheim.

March, J. and Smith, M.B. (2001) *March's Advanced Organic Chemistry: Reactions, Mechanisms, and Structure* (5th edn). Wiley-Interscience, New York.

Maskill, H. (1985) *The Physical Basis of Organic Chemistry.* Oxford University Press, Oxford.

Tobe, M.L. and Burgess, J. (1999) *Inorganic Reaction Mechanisms.* Addison Wesley Longman, Harlow.

## References

1. Santacesaria, E. (1999) *Catalysis Today*, **52**, 113.
2. Szabó, A., Galambos-Faragó, Á., Mucsi, Z., Timáti, G., Vasvári-Debreczy, L. and Hermecz, I. (2004) *European Journal of Organic Chemistry*, 687.
3. Moore, J.W. and Pearson, R.G. (1981) *Kinetics and Mechanism* (3rd edn). Wiley, New York.
4. Laidler, K.J. (1987) *Chemical Kinetics* (3rd edn). Harper Collins, New York.
5. Connors, K.A. (1990) *Chemical Kinetics. The Study of Reaction Rates in Solution.* VCH, New York.
6. Espenson, J.H. (1995) *Chemical Kinetics and Reaction Mechanisms* (2nd edn). McGraw-Hill, New York.
7. Steinfeld, J.I., Francisco, J.S. and Hase, W.L. (1999) *Chemical Kinetics and Dynamics* (2nd edn). Prentice-Hall, Englewood Cliffs, NJ.
8. Toby, S. (2000) *Journal of Chemical Education*, **77**, 188.
9. Kotani, M., Shigetomi, Y., Imada, M., Oki, M. and Nagaoka, M. (1997) *Heteroatom Chemistry*, **8**, 35.
10. Celo, V. and Scott, S.L. (2005) *Inorganic Chemistry*, **44**, 2507.
11. Ceccacci, F., Mancini, G., Mencarelli, P. and Villani, C. (2003) *Tetrahedron: Asymmetry*, **14**, 3117.
12. Fogler, H.S. (2001) *Elements of Chemical Reaction Engineering* (3rd edn). Prentice-Hall, Englewood Cliffs, NJ.
13. Tsuji, Y. and Richard, J.P. (2005) *The Chemical Record*, **5**, 94.
14. Winstein, S. and Trifan, D.S. (1952) *Journal of the American Chemical Society*, **74**, 1154; Winstein, S. and Trifan, D.S. (1949) *Journal of the American Chemical Society*, **71**, 2953; Winstein, S. and Trifan, D.S. (1952) *Journal of the American Chemical Society*, **74**, 1147.
15. Armesto, X.L., Canle López, M., García Dopico, M.V. and Santaballa, J.A. (1998) *Chemical Society Reviews*, **27**, 453.
16. Bell, R.P. (1973) *The Proton in Chemistry* (2nd edn). Chapman and Hall, London.
17. Maskill, H. (1985) *The Physical Basis of Organic Chemistry.* Oxford University Press, Oxford.
18. Chiang, Y., Hojatti, M., Keeffe, J.R., Kresge, A.J., Schepp, N.P. and Wirz, J. (1987) *Journal of the American Chemical Society*, **109**, 4000.
19. Litwinienko, G. and Ingold, K.U. (2005) *Journal of Organic Chemistry*, **70**, 8982.
20. Elovitz, M.S. and Fish, W. (1995) *Environmental Science and Technology*, **29**, 1933.

# Chapter 5
# Reaction Kinetics in Multiphase Systems

## John H. Atherton

## 5.1   Introduction

Chemical reactions coupled to the transfer of reactants or products across a phase boundary are ubiquitous in nature and in many kinds of chemical processes. Examples from nature include respiration, many transport processes across cell membranes and digestive processes. Important industrial and commercial examples include the following.

- Electrochemistry, which is used to recover metals from solution and in electrical storage devices where mass transfer is a significant factor in limiting the cell performance.
- Solvent extraction in metal recovery, which is used on a large scale, particularly in the recovery of copper and in the processing of fission products in the nuclear power industry.
- Dyeing of textiles with reactive dyes.
- From the chemical manufacturing industry, catalytic cracking and catalytic hydrogenation, gas absorption or 'scrubbing' processes in which desired or waste products are removed from a waste stream, the nitration of benzene and toluene where the reactants have limited mutual solubility, and carbonylation processes using carbon monoxide.

Wide-ranging examples are to be found amongst a variety of process types in the fine chemicals and pharmaceuticals manufacturing industries, and it is in this sphere that the present author has experience. Consequently, the emphasis of this chapter will be on examples from this area of technology.

Complexity in multiphase processes arises predominantly from the coupling of chemical reaction rates to mass transfer rates. Only in special circumstances does the overall reaction rate bear a simple relationship to the limiting chemical reaction rate. Thus, for studies of the chemical reaction mechanism, for which true chemical rates are required allied to known reactant concentrations at the reaction site, the study technique must properly differentiate the mass transfer and chemical reaction components of the overall rate. The coupling can be influenced by several physical factors, and may differently affect the desired process and undesired competing processes. Process selectivities, which are determined by *relative* chemical reaction rates (see Chapter 2), can then be modulated by the physical characteristics of the reaction system. These physical characteristics can be equilibrium related, in particular to reactant and product solubilities or distribution coefficients, or may be related to the mass transfer properties imposed on the reaction by the flow properties of the system.

## 5.2   Background and theory [1]

The reactions considered in this review include at least one liquid phase. When a liquid is in contact with another phase, even in a system which is stirred, transport between the two phases is not instantaneous. Close to the interface, in the receiving phase, there is a thin region referred to as the *diffusion film*, which behaves as if it is unstirred. Transport across this film is by diffusion only. Consider a pure material (liquid or gas) in contact with a liquid in which it dissolves to some extent (Fig. 5.1). It is assumed that equilibrium is maintained across the interface, so that the concentration of the solute adjacent to the interface in the receiving phase is always at saturation.

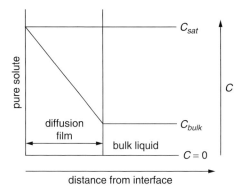

**Fig. 5.1**   Concentration profile for steady-state diffusion of a solute into a liquid receiving phase.

Application of Fick's first law gives Equation 5.1:

$$j = \frac{D}{\delta}(C_{sat} - C_{bulk}), \tag{5.1}$$

where $j$ is the flux of the solute into solution in mol cm$^{-2}$ s$^{-1}$, $D$ is the diffusion coefficient of the solute in the receiving phase in cm$^2$ s$^{-1}$, $\delta$ is the film thickness in cm and $C$ represents the concentration, as shown in Fig. 5.1.

The film thickness is not directly accessible, and the term $D/\delta$, which has units cm s$^{-1}$, is referred to as the *mass transfer coefficient*, $k_L$. In solid/liquid and liquid/liquid systems, values of $k_L$ are typically $1$–$2 \times 10^{-3}$ cm s$^{-1}$ with values of $D$ typically $5 \times 10^{-6}$ to $2 \times 10^{-5}$ cm$^2$ s$^{-1}$, corresponding to diffusion film thicknesses of $50$–$100$ μm and film diffusion times of around 5 seconds.

### 5.2.1   *Mass transfer coupled to chemical reaction*

Most reactions in two-phase systems occur in a liquid phase following the transfer of a reactant across an interface; these are commonly known as *extractive reactions*. If the transfer is facilitated by a catalyst, it is known as *phase-transfer catalysis* [2]. Unusually, reactions may actually occur at an interface (*interfacial reactions*); examples include solvolysis and nucleophilic substitution reactions of aliphatic acid chlorides [3] and the extraction of cupric ion from aqueous solution using oxime ligands insoluble in water [4], see Section 5.2.1.3(ii).

Extractive reactions can be usefully categorised into those which are too slow to occur within the diffusion film and those which occur largely within the film. Borderline cases will show intermediate behaviour.

Reactions involving completely insoluble solid surfaces are common. Processes involving heterogeneous catalysis and most electrochemical processes fall into this category.

### 5.2.1.1  Reaction too slow to occur within the diffusion film

For reactions with a first-order (or pseudo-first-order) rate constant of $<0.01\,s^{-1}$, little reaction will occur in the diffusion film.

Such a system is included in Fig. 5.2 and is treated as a pseudo-equilibrium in which the chemical reaction rate at any time during the reaction must be equal to the mass transfer rate. It is necessary to be careful in the choice of units. The chemical rate, $r$ (in mol cm$^{-3}$ s$^{-1}$), can be written as Equation 5.2:

$$r = k_1 C_{\text{bulk}}, \tag{5.2}$$

where $k_1$ is the first-order reaction rate constant, and the mass transfer rate (equal to the chemical reaction rate, and in the same units) is given by Equation 5.3:

$$r = k_L a(C_{\text{sat}} - C_{\text{bulk}}), \tag{5.3}$$

where $a$ is the interfacial area per unit volume of the receiving phase (units, cm$^{-1}$).

Following simplification of the expression obtained by substitution for $C_{bulk}$ in Equation 5.3 using Equation 5.2, we obtain Equation 5.4:

$$\frac{C_{\text{sat}}}{r} = \frac{1}{k_1} + \frac{1}{k_L a}, \tag{5.4}$$

which shows how the reaction rate and mass transfer process are coupled when the reaction occurs mainly in the bulk solution. As the interfacial area is increased, the rate asymptotically approaches the limiting chemical rate, as shown in Fig. 5.3 for a reaction with $k_1 = 10^{-2}\,s^{-1}$ and $k_L a = 10^{-3}\,s^{-1}$.

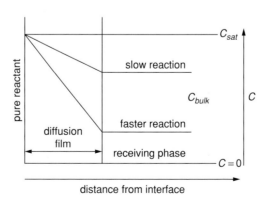

**Fig. 5.2**  Concentration profiles across the film for reactions which are slow relative to the film diffusion time.

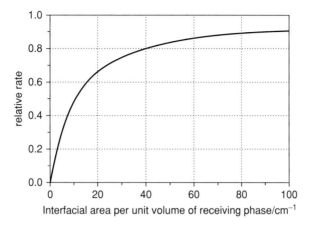

**Fig. 5.3** Reaction rate relative to the chemically limiting rate versus the interfacial area per unit volume of a receiving phase for $k_1 = 10^{-2}$ s$^{-1}$ and $k_L a = 10^{-3}$ s$^{-1}$.

### 5.2.1.2 Reaction fast relative to the film diffusion time

When a chemical reaction is fast enough to become complete within the diffusion film, the chemical rate and diffusion rates are coupled differently. Fig. 5.4 shows the basis for derivation of the rate expression for a reaction in a two-phase organic reactant/water system when the reaction is first order in solute with rate constant $k_1$, the diffusion coefficient of the organic species in water is $D$ and the saturation solubility of the organic reactant in water is $C_{sat}$. We consider the system at steady state and take a mass balance across a slab

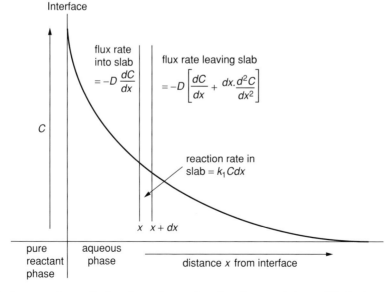

**Fig. 5.4** Concentration profile for a first-order reaction of a solute coupled to its diffusion into an infinite aqueous medium.

of unit cross-section area and thickness $dx$. The concentration gradient at $x$ is $dC/dx$. At $x + dx$, the change in the concentration gradient due to diffusion is the product of the rate of change of concentration, $d^2C/dx^2$, and the slab thickness, $dx$. Noting that

$$\text{diffusion in} = \text{diffusion out} + \text{chemical reaction},$$

and applying Fick's first law, we may write Equation 5.5:

$$-D\frac{dC}{dx} = -D\left[\frac{dC}{dx} + dx\frac{d^2C}{dx^2}\right] - k_1 C dx, \tag{5.5}$$

which may be simplified to give

$$D\frac{d^2C}{dx^2} = k_1 C. \tag{5.6}$$

Integration of Equation 5.6 once with the appropriate boundary conditions gives Equation 5.7 for the concentration gradient,

$$\frac{dC}{dx} = -C\sqrt{\frac{k_1}{D}}, \tag{5.7}$$

from which Equation 5.8 for the flux, $j$, at the interface can be derived using Fick's first law:

$$j = -D\frac{dC}{dx} = D\,C_{sat}\sqrt{\frac{k_1}{D}} = C_{sat}\sqrt{Dk_1}. \tag{5.8}$$

A further integration gives the concentration profile, Equation 5.9:

$$\frac{C}{C_{sat}} = \exp\left(-x\sqrt{\frac{k_1}{D}}\right). \tag{5.9}$$

Fig. 5.5 shows the concentration profile for such a reaction where $k_1 = 1$ s$^{-1}$ and $D = 5 \times 10^{-6}$ cm$^2$ s$^{-1}$. In this case, where the reaction is complete within the diffusion film, the

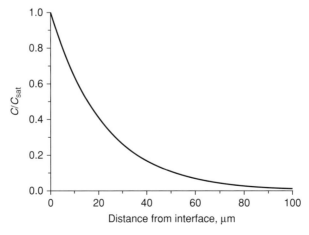

**Fig. 5.5** Concentration profile for a reaction fast enough to become complete within the diffusion film; first-order reaction rate constant $k_1 = 1$ s$^{-1}$, and reactant diffusion coefficient in the receiving phase $D = 5 \times 10^{-6}$ cm$^2$ s$^{-1}$.

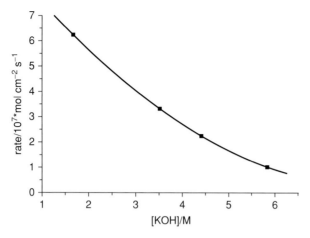

**Fig. 5.6** Plot of the rate of reaction of *n*-butyl formate with potassium hydroxide against the concentration of aqueous potassium hydroxide in a two-phase system at 30°C.

overall reaction rate is linearly dependent on the interfacial area and is independent of the volume of the phase in which the reaction occurs.

For reactions which are first order in the dissolving reactant, the rate will be proportional to the solubility of that species, regardless of whether the reaction occurs in the bulk solution or in the film (Equations 5.4 and 5.7). Inorganic salts usually depress the solubility of organic solutes in water [5], which can lead to unusual effects. Fig. 5.6 shows how the rate of reaction of *n*-butyl formate with potassium hydroxide changes with hydroxide concentration [6]. There are three reasons for the decrease in the rate as the concentration of hydroxide is increased: (i) salting out of the ester, (ii) a decrease in the diffusion coefficient of the *n*-butyl formate in water and (iii) a sharp decrease in the second-order rate constant presumably due to medium effects on $k_2$.

Reactions on or close to solid surfaces may be inhibited by deposition of insoluble or poorly soluble products on the reactant surface, a phenomenon referred to as *overgrowth*. Examples include the reaction of amines with chloranil [7], the diazotisation of poorly soluble aromatic amines in which the product diazonium salt is also insoluble and halogen exchange reactions of chloroaromatic compounds using potassium fluoride in dipolar aprotic solvents where the potassium chloride product may coat the potassium fluoride [8].

### 5.2.1.3 Interfacial reactions

*(i) At insoluble solid surfaces.* The commonest reaction involving insoluble solid surfaces is hydrogenation using a heterogeneous catalyst, Fig. 5.7. There are now two film resistances, but the analysis takes the same form as before and leads to Equations 5.10a–5.10c where terms are indicated in Fig. 5.7 and $k_{red}$ is the rate constant for reduction [9]:

$$\text{Rate of mass transfer across film (1)} = k_{GL}a_{gl}(H_{gs} - H_{bulk}) \tag{5.10a}$$

$$\text{Rate of mass transfer across film (2)} = k_{SL}a_{cat}(H_{bulk} - H_{cs}) \tag{5.10b}$$

$$\text{Rate of hydrogenation} = k_{red}a_{cat}H_{cs}. \tag{5.10c}$$

$H_{gs}$ is the solution concentration of hydrogen in equilibrium with a pressure $P_H$ of pure hydrogen in the gas phase. Equating pairs of Equations 5.10a–5.10c and eliminating the

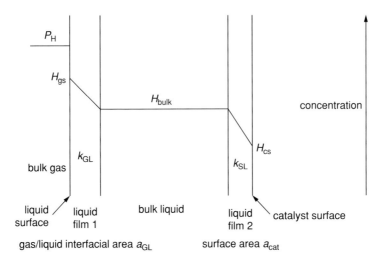

**Fig. 5.7** Film model for a heterogeneous catalytic hydrogenation.

unknown bulk and catalyst surface hydrogen concentration terms gives Equations 5.11a and 5.11b:

$$\frac{H_{gs}}{\text{rate}} = \frac{1}{k_{GL} a_{GL}} + \frac{1}{a_{cat} k_{SL}} + \frac{1}{a_{cat} k_{red}} \tag{5.11a}$$

$$\frac{H_{gs}}{\text{rate}} = \frac{1}{k_{GL} a_{GL}} + \frac{1}{a_{cat}} \left( \frac{1}{k_{SL}} + \frac{1}{k_{red}} \right). \tag{5.11b}$$

A plot of 1/rate against 1/(catalyst weight) at constant hydrogen pressure and interfacial area should give a straight line from which the intercept gives $\frac{1}{k_{GL} a_{GL}}$ and the slope gives $\frac{1}{k_{SL}} + \frac{1}{k_{red}}$. This analysis assumes that no deactivation of catalyst occurs and that there is no competitive product adsorption [10].

*(ii) At liquid–liquid interfaces.* This type of reaction is rare, although examples have been mentioned in Section 5.2.1. The rate of simple first-order interfacial reactions will be independent of the diffusion film thickness and dependent only on the surface area. For a reaction which has a rate dependence on a reactant in one phase, the rate will depend on the concentration local to the interface, which may be different from the bulk solution concentration. If transport across the diffusion film is rate limiting, i.e. slower than a chemical reaction, then the rate will depend on the film thickness (Fig. 5.8 and Equation 5.1) and hence on the hydrodynamic conditions. Should the delivery phase be a solution rather than a neat reactant, there may also be a resistance on that side of the interface.

### 5.2.2 Phase-transfer catalysis (PTC) [2]

The chemical reaction in an organic solvent involving a reagent which is only very slightly soluble in that solvent (but which is freely soluble in an immiscible aqueous phase) will necessarily be slow. However, if the concentration of this reagent in the organic phase

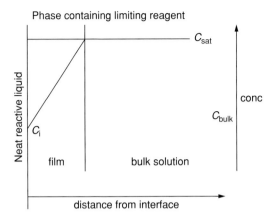

**Fig. 5.8** Concentration profile for an interfacial reaction with reagent depletion in the diffusion film.

can be increased by facilitated transfer from the aqueous phase, the reaction rate in the organic solvent will be enhanced; this is known as *phase-transfer catalysis* (PTC). The reagent whose transfer into the organic phase is facilitated is usually anionic, and PTC involves its solubilisation in the organic solvent by a lipophilic positively charged counter ion, usually a quaternary ammonium or phosphonium salt. The source of the anionic species may be an aqueous solution (as implied above) or it could be a solid salt. In the case of liquid–liquid PTC, exchange of the reactive anion can occur via partitioning of the quaternary salt between the two phases or, if the salt is insoluble in water, by interfacial exchange of the counter anion. This is illustrated in Fig. 5.9, where the nucleophilic substitution reaction,

$$A{-}X + Y^- \rightarrow A{-}Y + X^-,$$

is driven by replenishment of $Y^-$ in the organic phase and removal of $X^-$ from the organic phase by a catalytic amount of the quaternary cation, $Q^+$.

Either a chemical reaction or an anion mass transfer (which is dependent on the catalyst concentration and on the interfacial area) can be rate determining, depending on their relative rates. A recent review by Naik and Doraiswamy covers the subject comprehensively [11].

Carbanions can be generated and alkylated in a two-phase liquid–liquid system using concentrated aqueous sodium or potassium hydroxide as the base. Makosza has shown that deprotonation occurs at the interface, Fig. 5.10 [12].

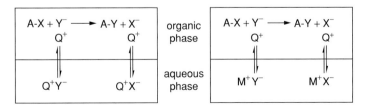

**Fig. 5.9** Phase-transfer catalysis (PTC) with and without partitioning of the cation.

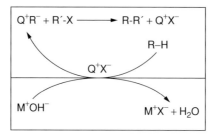

**Fig. 5.10**  Carbanion formation and alkylation via interfacial deprotonation using concentrated aqueous hydroxide solution.

## 5.2.3   *System complexity and information requirements*

The brief survey above of the coupling of the mass transfer and chemical reaction has taken simple schemes using first-order reactions to illustrate the principles involved. We have not discussed cases where there are film resistances on both sides of the interface [13]. Very complicated examples may occur in practice, especially for higher-order reactions where steep concentration gradients occur within the diffusion film. One of the more complex examples is the extraction of cupric ion from aqueous solution using a water-insoluble oxime ligand, Fig. 5.11 [4]. In this case, there are coupled adsorption, ligation and proton transfer reactions. Complete understanding of this system required around 10 years study by the methods then available.

In order to determine rate constants rigorously in interfacial reactions, methods are required which allow determination of reactant concentrations and chemical rates actually at the reactive surface. This requires a degree of control of the hydrodynamics which is not available in stirred vessels, and more sophisticated methods are used in these cases. Minimal criteria for the unambiguous determination of interfacial chemical kinetics have been enumerated elsewhere and are as follows [13].

1. Transport of the reactants to the interface should be well defined and calculable.
2. The interface between the two phases should be well defined in terms of area and topography.
3. There should be good control of the bulk solution phase concentrations, ideally with constant conditions being maintained; this is usually best achieved in a flow system.
4. Transport to the detector should be well defined.

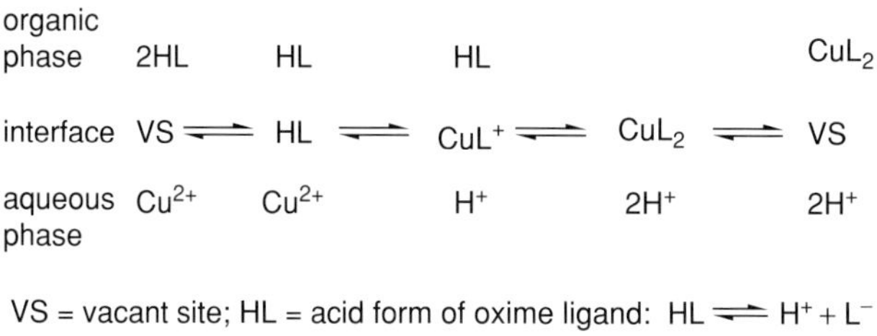

| organic phase | 2HL | HL | | HL | | | | $CuL_2$ |
|---|---|---|---|---|---|---|---|---|
| interface | $VS \rightleftharpoons$ | HL | $\rightleftharpoons$ | $CuL^+$ | $\rightleftharpoons$ | $CuL_2$ | $\rightleftharpoons$ | VS |
| aqueous phase | $Cu^{2+}$ | $Cu^{2+}$ | | $H^+$ | | $2H^+$ | | $2H^+$ |

VS = vacant site; HL = acid form of oxime ligand:  $HL \rightleftharpoons H^+ + L^-$

**Fig. 5.11**  Mechanism for the extraction of cupric ion as a complex with an oxime ligand.

## 5.3  Some experimental methods

The rest of this chapter is a discussion of selected examples of equipment used to study the kinetics of multiphase reactions. We begin with instrumentation suitable for industrial process development in the fine chemicals area and then move on to more sophisticated methods which can be used to extract true surface kinetics data even in the presence of sharp concentration gradients near the surface.

### 5.3.1   *The stirred reactor for the study of reactive dispersions with a liquid continuous phase*

A good general purpose vessel for process development work is illustrated in Fig. 5.12 [14]. It is usual to work at a height:depth ratio of about 1:1 corresponding to a working volume of around 750 cm$^3$. Baffles enforce predominantly axial flow in the vessel and ensure good mixing. Some characterisation of the mass transfer characteristics of this equipment has been carried out with both gas–liquid and liquid–liquid systems.

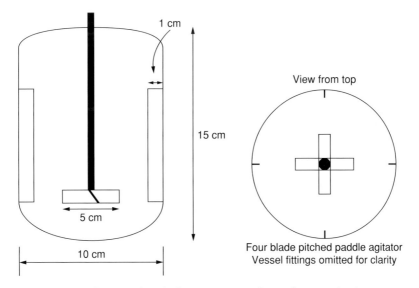

**Fig. 5.12**  Stirred reactor for use with multiphase reactions in chemical process development.

### 5.3.1.1   *Gas–liquid reactions*

For the measurement of the gas–liquid mass transfer rate, gas is injected adjacent to the bottom of the wall baffle. Water in the vessel is first deoxygenated with nitrogen and then the rate of approach to oxygen saturation is measured when air is sparged in. Finally, the oxygen desaturation rate can be measured as oxygen is removed using a nitrogen sparge; Fig. 5.13 shows a typical result for the oxygen saturation stage [8].

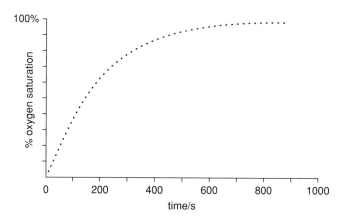

**Fig. 5.13** Plot of percentage saturation against time for the equipment of Fig. 5.12 with an agitator speed of 5 Hz and an air sparge rate of 1 L min$^{-1}$.

Analysis of these data, allowing for the time constant of the measuring probe, gave a value of $k_L a = 0.006\ \text{s}^{-1}$, corresponding to a half-life for saturation of nearly 2 minutes. At higher agitator speeds, $k_L a$ values of up to about 0.1 s$^{-1}$ are achievable in this equipment. Note that $k_L a$ has the same units as a first-order rate constant and is the rate constant for the transfer of gas to solvent. Hydrogen and oxygen are commonly used gases and have solubilities in water of around millimolar at atmospheric pressure and 25°C. So a $k_L a$ value of 0.1 s$^{-1}$ with a concentration 'driving force' of $10^{-3}$ M corresponds to a gas transfer rate of $10^{-4}$ M$^{-1}$ s$^{-1}$. It is common for gas–liquid reactions to show zero-order kinetics in this equipment simply because the rate-limiting step is the mass transfer of the reactive gas. Thus, it is not usually possible to extract true chemical reaction rate constants using this equipment.

If equivalent performance is required on the transfer of a gas–liquid reaction to a different item of equipment, it is necessary to maintain the $k_L a$ constant. Failure to do so is a frequent cause of difficulty in chemical process development.

### 5.3.1.2 Dispersed liquid–liquid systems

It is common to carry out reactions in dispersed liquid–liquid systems. The rates of such reactions show different dependences on the intensity of agitation depending on whether the reaction occurs in a bulk phase, within the diffusion film, or at the interface [9].

A reaction occurring in a bulk phase will show an increase in the rate with the area as shown in Fig. 5.3; for a reaction occurring in the film or at the interface, the rate will be linearly dependent on the interfacial area. The interfacial area in a dispersed two-phase liquid–liquid system can be estimated by measuring the rate of a suitable test reaction in a reactor with the known interfacial area (a flat interface, Section 5.3.2.1), and comparing it with the reaction rate in a dispersed system [6, 15]. A convenient reactive system for this purpose is a formate ester and 1–2 M aqueous NaOH. Formate esters are very reactive to hydroxide ion ($k_2$ typically around 25 M$^{-1}$ s$^{-1}$), so the reaction is complete inside the diffusion film, and the reaction rate is proportional to the interfacial area. A plot of the interfacial area per unit volume against the agitator speed obtained in this way in the author's laboratory for the equipment shown in Fig. 5.12 is shown in Fig. 5.14 [8].

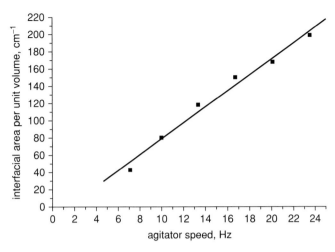

**Fig. 5.14** Plot of the interfacial area per unit volume against the agitator speed for the reactor shown in Fig. 5.12 for the reactive pair, *n*-hexyl formate and 1 M aqueous NaOH.

### 5.3.1.3 Liquid–solid reactions in a stirred reactor

Chemical kinetics deductions are, in some circumstances, possible from a reaction system using a dispersed solid. If the solid is entirely insoluble, for example a supported catalyst, true surface kinetics can be obtained provided (i) it can be shown that the chemical reaction on the surface is much slower than the associated mass transfer, and (ii) the surface area of the solid can be obtained. These conditions applied in the case of the oxidation of an aqueous solution of hydrazine using a dispersion of insoluble barium chromate [16]. Another case is where it can be shown that an increase in the amount of the solid component does not increase the reaction rate. In this case, exemplified by the formation of benzyl acetate from benzyl bromide and solid sodium acetate in toluene solvent, it is likely that the reaction occurs in the solution phase and that the reaction is proceeding at the saturation concentration of the solid reactant in the liquid phase [17].

In a study of the dissolution of potassium bicarbonate in dimethylformamide (DMF), Compton's group used ultrasound (25 kHz, 8 W cm$^{-2}$) to provide mixing [18]. The rate was monitored via the homogeneous deprotonation of 2-cyanophenol by the dissolved potassium bicarbonate (Fig. 5.15), using spectroscopic and electrochemical techniques; the

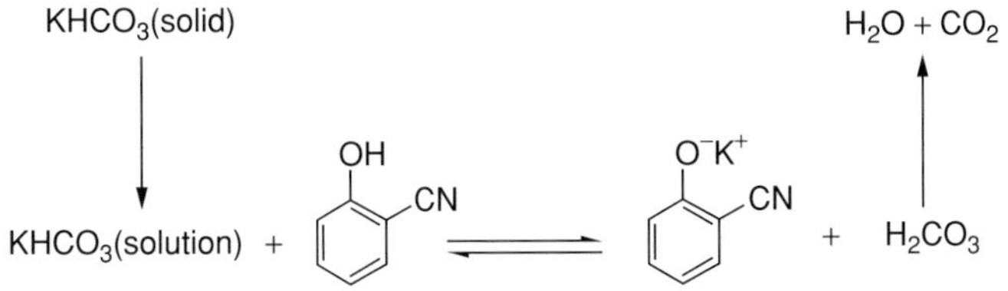

**Fig. 5.15** Reaction of potassium bicarbonate with 2-cyanophenol in DMF to monitor the rate of dissolution of the potassium bicarbonate.

process was shown to be a slow surface-controlled dissolution, rather than mass transfer limited in the solution phase (see also Section 5.3.2.2).

## 5.3.2    Techniques providing control of hydrodynamics

There have been several different approaches to the problem of controlling hydrodynamics adjacent to interfaces in order to be able to extract true chemical kinetics.

### 5.3.2.1    Techniques based on the Lewis cell

The simplest device providing rudimentary control of hydrodynamics is the Lewis cell [19], of which a variation by the present author is easier to construct. It consists of a reactor containing two contra-rotated agitators positioned on either side of a liquid–liquid interface, Fig. 5.16.

At rotation speeds up to about 2 Hz, the interface is stirred but not physically disrupted, so the interfacial area is calculable. This system has been used mainly to study liquid–liquid extraction kinetics, but can be used for gas–liquid reactions. The range of $k_L$ values obtainable is fairly small and Lewis in his original paper gives a range of only $1.7–4 \times 10^{-3}$ cm s$^{-1}$. The range of chemical reaction rates for which this device can be used to obtain reaction kinetics is limited. Consider a reaction which occurs in a liquid phase of depth 5 cm in a parallel-sided vessel. The area per unit volume is thus 0.2 cm$^{-1}$, so $k_L a$ will be around $2 \times 10^{-4}$ s$^{-1}$, giving a half-life for simple physical saturation of about 1 hour. From Equation 5.4, it is evident that, in order to study the kinetics of a reaction which occurs in the bulk liquid phase, the reaction will have to have a first-order rate constant significantly less than $k_L a$, which severely restricts the range of use of this technique. It can be of use in the study of fast 'in film' reactions, provided that the kinetics are first order or can be forced to be pseudo first order. Sharma has studied the hydrolysis of reactive esters by this technique [6, 20].

The present author has used the method to obtain an approximate rate constant for the oxidation of an arylhydrazine with oxygen, and to show that the reaction occurred 'in film'. A saturated solution of the hydrazine in an aqueous sodium chloride solution was reacted

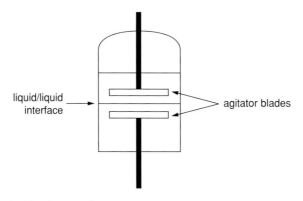

**Fig. 5.16**   Flat interface Lewis-type cell.

with oxygen in the reactor of Fig. 5.12 with an agitator speed of only 1 Hz. The interface was flat and the oxygen mass transfer coefficient $k_L$ was found to be 0.7 cm s$^{-1}$. This is sufficiently higher than the maximum expected $k_L$ value for purely physical transport, i.e. $\sim 5 \times 10^{-3}$ cm s$^{-1}$ [21], that it can be concluded that the reaction is occurring 'in film'.

This technique is not applicable in cases where the concentration of the reacting component in the receiving phase can be expected to be depleted in the diffusion film, since the range over which the mass transfer coefficient can be varied is too small to permit meaningful extrapolation. The necessary conditions for the acquisition of second-order rate constants in the case of a 'fast in-film' reaction have been discussed [15].

Interfacial processes can only be studied in the flat interface stirred cell in cases when there is no depletion within the diffusion film, and when there are no complicating surface adsorption effects.

### 5.3.2.2 The rotated disc reactor

A rotated disc in a fluid behaves as a pump which draws fluid towards the disc and throws it out radially, Fig. 5.17. Flow to a rotated disc has the surprising property that, under laminar flow conditions, the film thickness is uniform across the whole of the disc surface, excluding edge effects, and the film thickness ($\delta$) is rigorously calculable using fluid mechanics theory [22]. It is given by Equation 5.12 where $D$ is the diffusion coefficient in cm$^2$ s$^{-1}$, $v$ is the kinematic viscosity in cm$^2$ s$^{-1}$ and $\omega$ is the rotation speed in rad s$^{-1}$:

$$\delta = 1.805 \, D^{\frac{1}{3}} v^{\frac{1}{6}} \omega^{-\frac{1}{2}} [0.8934 + 0.316(D/v)^{0.36}]. \tag{5.12}$$

This device has been used for many years to measure diffusion coefficients and, more recently, to study the rates of chemical reactions involving solids. In the latter case, a circular pellet of reactant is embedded in the disc. For a dissolving solid which reacts with first-order kinetics (rate constant $k_1$) in the receiving phase (Fig. 5.18) the flux $j$ is given by Equation 5.13 [13]:

$$\frac{1}{j} = \frac{1}{C_{\text{sat}} - C_{\text{bulk}}} \left( \frac{1}{k_1} + \frac{\delta}{D} \right). \tag{5.13}$$

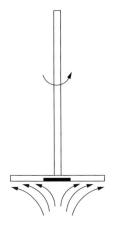

**Fig. 5.17** Flow pattern at a rotated disc.

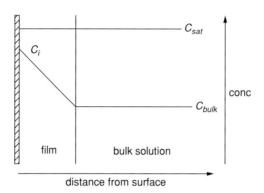

**Fig. 5.18**  Concentration profile for dissolution and first-order reaction for a dissolving solid on a rotated disc.

By combining Equations 5.12 and 5.13, we see that a plot of $j^{-1}$ against $\omega^{-1/2}$ gives a straight line which will pass through the origin when the process is diffusion controlled $(1/k_1 \ll \delta/D)$ or will show a positive intercept when there is a rate-limiting surface process. The majority of dissolution processes involving solids are purely transport controlled, i.e. they behave as if the solid is in equilibrium with its saturated solution at the point of contact with the fluid. This is not always the case, and it has been shown using this device that the rate-determining step for the dissolution of potassium fluoride in hot dimethylformamide occurs at the solid surface [13], cf. the dissolution of potassium bicarbonate in dimethylformamide, Section 5.3.1.3. The rotated disc has also been used to study nitration processes using solid nitronium tetrafluoroborate as the reagent [23].

### 5.3.2.3   Rotated diffusion cell

This was the earliest device to permit control of film thickness in the study of two-phase liquid/liquid reactions, Fig. 5.19, and the reaction takes place at the lower side of the porous

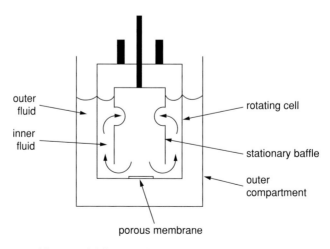

**Fig. 5.19**  Schematic of the rotated diffusion cell.

membrane. Film thicknesses can be varied by changing the rotation speed and, under appropriate conditions, double reciprocal plots, cf. Equation. 5.13, can be used to extrapolate to infinite rotation speed and hence to extract true surface reaction rates. The apparatus was used to elucidate the mechanism of the solvent extraction of cupric ions from water by oxime ligands, Fig. 5.11, Section 5.2.3 [4]. However, it suffers from the disadvantage that the resistance of the porous membrane to liquid transport is significant, and has been superseded by microelectrode techniques.

## 5.3.2.4    Channel flow techniques

The channel flow cell has been used to study a wide variety of liquid–solid reactions. It consists of a tube of rectangular cross section, typically 4.5 cm long by 0.6 cm wide by 0.1 cm deep (Fig. 5.20). The reactive solid is embedded in the base of the cell and a detector downstream of the reactive solid is used to monitor reactant and product concentrations.

substrate

detector
electrode

flow

**Fig. 5.20**   Schematic of the channel flow cell.

The mass transfer by convection and diffusion within the channel is precisely calculable. Numerical methods are used to fit measured concentrations at the detector electrode to the fluid flow in the cell, and to the reaction kinetics at the surface and in solution [24]. The technique was extensively developed during a study of the dissolution of calcite in dilute aqueous acid [25], and has latterly been applied to a number of organic reactions.

A recent example is the hydrolysis of chloranil with aqueous sodium hydroxide, which is a multistep process leading to the chloranilate dianion, Fig. 5.21 [26]. In a two-phase reaction system consisting of solid chloranil and aqueous sodium hydroxide, the rate-limiting step is the initial attack of hydroxide ion on chloranil. Hydrolysis rates at the 010, 100 and 001 cleavage planes of chloranil were separately measured [27], and different reactivities were found; this was rationalised by considering the exposure of the initially reactive functionality (assumed to be the carbonyl group) at the surface, as shown in Fig. 5.22. Hydrolysis at the 010 and 100 planes was shown to be a surface reaction driven by the hydroxide concentration adjacent to the interface; the 001 plane was shown to react more slowly by prior dissolution of chloranil before reaction.

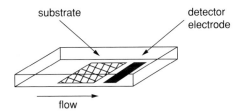

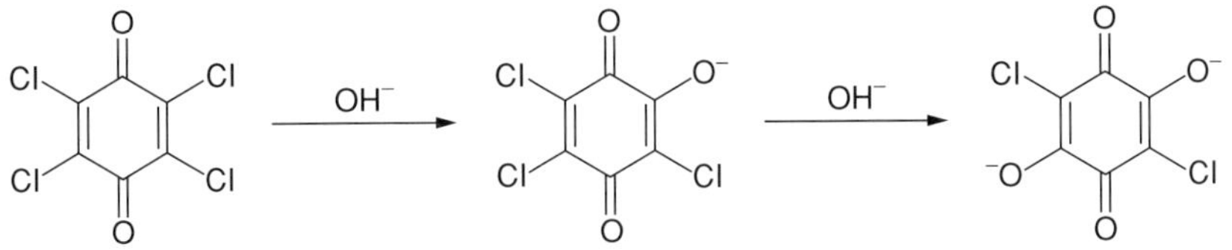

**Fig. 5.21**   Hydrolysis of chloranil.

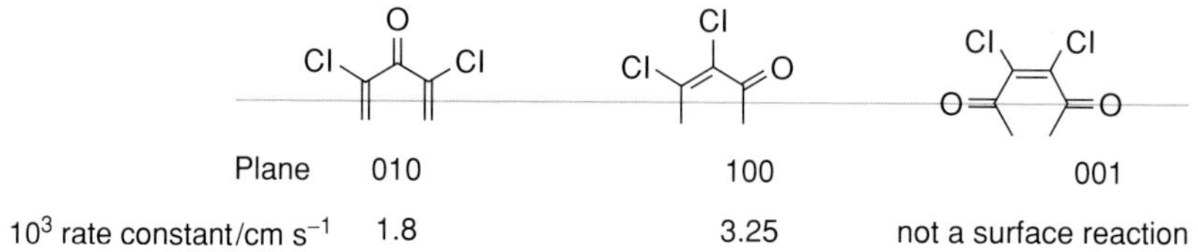

| Plane | 010 | 100 | 001 |
|---|---|---|---|
| $10^3$ rate constant/cm s$^{-1}$ | 1.8 | 3.25 | not a surface reaction |

**Fig. 5.22**   Different reactivities at the three cleavage planes of chloranil.

### 5.3.2.5   The jet reactor

Measurement of the kinetics of gas–liquid reactions is of great importance in the design of gas absorption equipment. The jet reactor provides a means of determining solubilities and reaction rates for gases which react rapidly with the liquid. A narrow jet of the liquid is passed through a reactive gas into a receiver, from which samples are taken to determine the amount of absorption, Fig. 5.23.

In a study of the absorption of phosgene in water and aqueous sodium hydroxide [28], the inlet jet diameter was 0.63 mm and the jet length in the range 0.75–7 cm. Contact times were

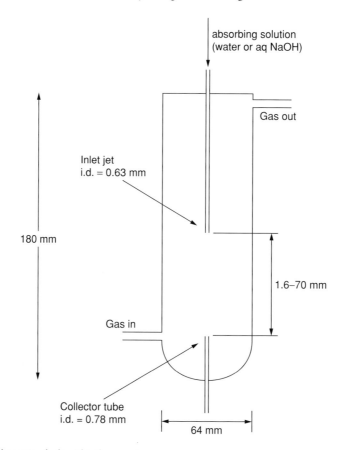

**Fig. 5.23**   Schematic of a liquid jet/gas reactor.

in the range 1.8–17 ms. At these short contact times, the gaseous phosgene penetrates only a small distance into the jet, and the kinetics simplify to those of diffusion into an infinite fluid. The authors showed that, for purely physical absorption, Equation 5.14 applies, where $q$ is the rate of absorption (in mol s$^{-1}$), $D$ is the diffusion coefficient (cm$^2$ s$^{-1}$), $v$ is the volumetric flow rate of the liquid (cm$^3$ s$^{-1}$) and $h$ is the height of the jet (cm):

$$q = 4C_{sat}\sqrt{Dvh}. \tag{5.14}$$

When the chemical reaction dominates the absorption rate, this expression becomes Equation 5.15 where $\bar{D}$ is the diameter of the jet:

$$q = C_{sat}\bar{D}h\pi\sqrt{Dk_1}. \tag{5.15}$$

This equipment was used to determine the solubility of phosgene in water (0.069 M atm$^{-1}$ at 25°C). It was also used to measure the first-order rate constant for the hydrolysis of phosgene in water ($k_1 = 6$ s$^{-1}$) and the second-order rate constant for the reaction with hydroxide ion ($k_2 = 1.6 \times 10^4$ M$^{-1}$ s$^{-1}$), both at 25°C.

Measurement of the solubility of a material with a half-life of about 100 ms is possible because it is deduced from the physical absorption rate at contact times sufficiently short ($\sim$2 ms) that no chemical reaction occurs, and Equation 5.14 is applicable.

### 5.3.2.6    Expanding drop methods

A recent development, termed by the inventors *microelectrochemical measurements at expanding droplets* (MEMED) [29], is a technique based on forming small droplets of a phase containing a reactant in a second immiscible liquid phase (Fig. 5.24). An ultramicroelectrode (UME, see Section 5.3.2.8 and Chapter 6) measures an electrochemical response as the droplet expands towards it, from which a concentration profile can be derived and, hence, the kinetics of related processes. Because the surface is continuously refreshed, it avoids

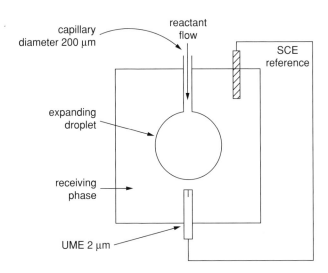

**Fig. 5.24**   The expanding droplet technique.

problems of surface contamination. Mass transfer coefficients of around $10^{-2}$ cm s$^{-1}$ are achieved.

The MEMED technique has been used to study the hydrolysis of aliphatic acid chlorides in a water/1,2-dichloroethane (DCE) solvent system [3]. It was shown unambiguously that the reaction proceeds via an interfacial process and shows saturation kinetics as the concentration of acid chloride in the DCE increases; the data were well fitted to a model based on a pre-equilibrium involving Langmuir adsorption at the interface. First-order rate constants for interfacial solvolysis of $CH_3(CH_2)_nCOCl$ were $300 \pm 150 (n = 2), 200 \pm 100 (n = 3)$ and $120 \pm 60$ s$^{-1}$ $(n = 8)$.

### 5.3.2.7    Confluence microreactor

In this recently developed device shown in Fig. 5.25 [30], two immiscible liquid streams converge in a micro-channel and then separate. A well-defined flow has been demonstrated, with no entrainment of the opposed flows, for flow rates from $10^{-4}$ to $1.0$ cm$^3$ s$^{-1}$. The highest flow rate corresponds to solution contact times around 1 ms, corresponding to a $k_L$ of the order of 0.1 cm s$^{-1}$. Thus, the confluence microreactor has the potential to measure liquid–liquid interfacial rate constants much higher than those accessible by current methods, and interesting applications are anticipated. The device itself has been patented [31].

Section through confluence reactor viewed from top. Not to scale

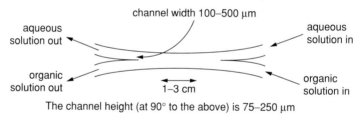

**Fig. 5.25**    The confluence microreactor.

### 5.3.2.8    Microelectrode techniques

Scanning electrochemical microscopy (SECM) using ultramicroelectrodes (UMEs) is a recently developed powerful tool for the study of rate processes occurring at liquid interfaces [32]. There are three essential components to the method: (i) the ultramicroelectrode (UME), (ii) equipment to permit precise positioning of the UME with respect to the interface and (iii) the mathematical modelling technique necessary to extract kinetic information from the data. The UME consists of a metal wire of diameter 0.6–25 μm sealed into a glass capillary, the end being polished flat. The diameter of the glass support is typically 10–20 times that of the metal electrode. Hemispherical diffusion to the electrode tip provides much greater mass transfer rates than to a macroscopic plane electrode (Fig. 5.26). As the electrode tip is brought closer to an inert surface, the current decreases due to hindered diffusion, and this change can be precisely modelled.

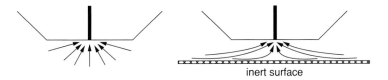

**Fig. 5.26** Diffusion to a microelectrode.

In a reacting system, the rate of reactant transfer is quantified by the electrode response at different distances from the interface, and the kinetics are fitted to these 'approach curves' in order to obtain the interfacial transfer rates. The electrode may be used in passive mode to monitor a transfer process, or may be used to perturb equilibrium conditions by removing one component electrochemically (see also Chapter 6). The technique has been used to study a range of processes including the kinetics of extraction of cupric ion from water [33], and the oxidation of inorganic complexes. A study involving a Langmuir trough permitted measurement of the reduction in the mass transfer rate of oxygen across a solvent–water interface as a 1-octadecanol monolayer was compressed [34].

### 5.3.3 Use of atomic force microscopy (AFM)

AFM has been used in real time to image reacting surfaces in contact with a liquid phase [35]. A study of the self-passivating reaction between *p*-chloranil and an aromatic amine included monitoring the surface topography of the *p*-chloranil in contact with the amine in aqueous solution at 25-second intervals, Fig. 5.27. Under the conditions employed, passivation by the reaction product was shown to be complete in 5 minutes; a concomitant hydroxide induced dissolution could also be monitored [7].

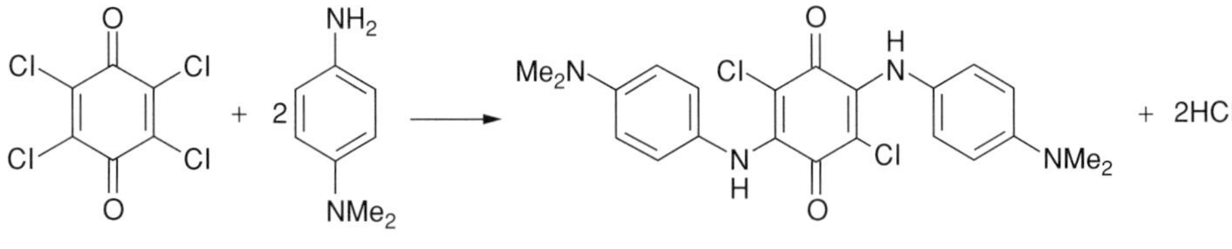

**Fig. 5.27** Reaction of chloranil with *N*,*N*-dimethyl-*p*-phenylenediamine.

## 5.4 Information requirements and experimental design

There are vastly different levels of knowledge requirement in the study of multiphase chemical reactions, depending on the objective of the study. Development of a chemical process rarely requires an in-depth understanding of the kinetics and mechanism, and industrial chemists work on a strict 'need to know' basis, since the acquisition of non-essential information is regarded as a waste of money. A key and difficult judgement is that of deciding what constitutes a minimum information requirement. For the industrial chemist, it will

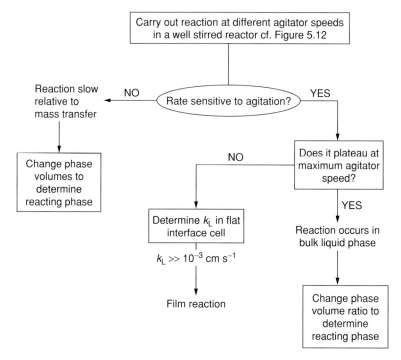

**Fig. 5.28**   Procedure to diagnose reacting phase and reaction regime.

usually be sufficient to gain an understanding of the locus of the reaction (in which phase it occurs, and whether it is a surface, film or bulk reaction), and to characterise the mass transfer requirements needed to provide adequate performance. The necessary information can usually be acquired in an appropriately configured stirred reaction vessel. For the purpose of acquiring data for process development, it is often useful to study the kinetics in the receiving phase under homogeneous conditions, and to obtain relevant data on solubilities or distribution coefficients. Multiphase reactions often occur at low concentrations in the reacting phase, so single-phase investigations will provide data under conditions similar to those in the real situation, leaving just the need to understand the mass transfer aspects.

   A procedure for diagnosing the reacting phase and kinetic regime (film or bulk) using readily available equipment is shown in Fig. 5.28. This level of investigation will rarely satisfy the requirements of an academic study, for which one or more of the techniques described in Section 5.3 will be required.


## 5.5   Summary

It is evident from the foregoing that the study of multiphase reactions requires a set of skills additional to those provided in the undergraduate education and training of most chemists and chemical engineers. The new concepts required are easy to grasp, as is their application to chemical process development in the laboratory, but the more sophisticated techniques for comprehensive dissection of mass transfer and chemical kinetics do require a considerable investment in time and equipment.

## Bibliography

Atherton, J.H. (1994) Mechanism in two-phase reaction systems: coupled mass transfer and chemical reaction. *Research in Chemical Kinetics*, **2**, 193.

Cooper, J.A. and Compton, R.G. (1998) Channel electrodes: a review. *Electroanalysis*, **10**, 141.

Cussler, E.L. (1984) *Diffusion: Mass Transfer in Fluid Systems*. Cambridge University Press, New York.

Dankwerts, P.V. (1970) *Gas-Liquid Reactions*. McGraw-Hill, New York.

Doraiswamy, L.K. and Sharma, M.M. (1984) *Heterogeneous Reactions: Analysis, Examples and Reactor Design*. Wiley, New York.

Hanna, G.J. and Noble, R.D. (1985) Measurement of liquid-liquid interfacial kinetics. *Chemical Reviews*, **85**, 583.

Starks, C.M., Liotta, C.L. and Halpern, M. (1994) *Phase-Transfer Catalysis*. Chapman and Hall, London.

Volkov, A.G. (Ed.) (2001) *Liquid Interfaces in Chemical, Biological and Pharmaceutical Applications*. Dekker, New York.

## References

1. Astarita, G. (1966) *Mass Transfer and Chemical Reaction*. Elsevier, Amsterdam.
2. Starks, C.M., Liotta, C.L. and Halpern, M. (1994) *Phase-Transfer Catalysis*. Chapman and Hall, London.
3. Zhang, J., Atherton, J.H. and Unwin, P.R. (2004) *Langmuir*, **20**, 1864.
4. Albery, W.J. and Choudhery, R.A. (1988) *Journal of Physical Chemistry*, **92**, 1142.
5. Long, F.A. and McDevit, W.F. (1952) *Chemical Reviews*, **52**, 119.
6. Nanda, A.K. and Sharma, M.M. (1966) *Chemical Engineering Science*, **21**, 707.
7. Booth, J., Compton, R.G. and Atherton, J.H. (1998) *Journal of Physical Chemistry*, **102**, 3980.
8. Atherton, J.H. unpublished work.
9. Doraiswamy, L.K. and Sharma, M.M. (1984) *Heterogeneous Reactions: Analysis, Examples and Reactor Design*. Wiley, New York.
10. Davies, H.S., Thomson, G. and Crandall, G.S. (1932) *Journal of the American Chemical Society*, **54**, 2340.
11. Naik, S.D. and Doraiswamy, L.K. (1998) *AIChE Journal*, **44**, 612.
12. Makosza, M. and Fedorynski, M. (2003) *Catalysis Reviews*, **45**, 321.
13. Atherton, J.H. (1994) *Research in Chemical Kinetics*, **2**, 193.
14. Atherton, J.H. and Carpenter, K.J. (1999) *Process Development: Physicochemical Concepts*. Oxford University Press, Oxford.
15. Sharma, M.M. and Dankwerts, P.V. (1970) *British Chemical Engineering*, **15**, 522.
16. Baumgartner, E., Blesa, M.A., Larotonda, R. and Maroto, A.J.G. (1985) *Journal of the Chemical Society Faraday Transactions I*, **81**, 1113.
17. Yadav, G.D. and Sharma, M.M. (1981) *Industrial and Engineering Chemistry, Process Design and Development*, **20**, 385.
18. Forryan, C.L., Klymenko, O.V., Brennan, C.M. and Compton, R.G. (2005) *Journal of Physical Chemistry B*, **109**, 2862.
19. Lewis, J.B. (1954) *Chemical Engineering Science*, **3**, 248.
20. Fernandes, J.B. and Sharma, M.M. (1967) *Chemical Engineering Science*, **22**, 1267.

21. Jhaveri, A.S. and Sharma, M.M. (1968) *Chemical Engineering Science*, **23**, 1.
22. Gregory, D.P. and Riddiford, A.C. (1956) *Journal of the Chemical Society*, 3756.
23. Guk, Yu.V., Golod, E.L. and Gidaspov, B.V. (1977) *Journal of Organic Chemistry of the USSR*, **13**, 14.
24. Unwin, P.R., Barwise, A.J. and Compton, R.G. (1989) *Journal of Colloid and Interface Science*, **128**, 208.
25. Compton, R.G. and Unwin, P.R. (1990) *Philosophical Transactions of the Royal Society of London A*, **330**, 1 and 47.
26. Booth, J., Saunders, G.H.W., Compton, R.G., Atherton, J.H. and Brennan, C.M. (1997) *Journal of Electroanalytical Chemistry*, **440**, 83.
27. Hill, E., Lloyd-Williams, R.R., Compton, R.G. and Atherton, J.H. (1999) *Journal of Solid State Electrochemistry*, **3**, 327.
28. Manogue, W.H. and Pigford, R.L. (1960) *AIChE Journal*, **6**, 494.
29. Slevin, C.J., Unwin, P.R. and Zhang, J. (2001) Hydrodynamic techniques for investigating reaction kinetics at liquid-liquid interfaces: historical overview and recent developments. In: A.G. Volkov (Ed.) *Liquid Interfaces in Chemical, Biological and Pharmaceutical Applications*. Dekker, New York, Chapter 13.
30. Yunus, K., Marks, C.B., Fisher, A.C., Allsopp, D.W.E., Ryan, T.J., Dryfe, R.A.W., Hill, S.S., Robert, E.P.L. and Brennan, C.M. (2002) *Electrochemistry Communications*, **4**, 579.
31. EP 0 790 849 B1 (application date 20.10.1995). Central Research Laboratories Ltd, Middlesex, UK.
32. Barker, A.L., Slevin, C.J., Unwin, P.R. and Zhang, J. (2001) Scanning electrochemical microscopy as a local probe of chemical processes at liquid interfaces. Chapter 12 in: A.G. Volkov (Ed.) *Liquid Interfaces in Chemical, Biological and Pharmaceutical Applications*. Dekker, New York. See also Chapter 6.
33. Slevin, C.J., Umbers, J.A., Atherton, J.H. and Unwin, P.R. (1996) *Journal of the Chemical Society Faraday Transactions*, **92**, 5177.
34. Slevin, C.J., Ryley, S., Walton, D.J. and Unwin, P.R. (1998) *Langmuir*, **14**, 5331.
35. Shakesheff, K.M., Davies, M.C., Domb, A., Jackson, D.E., Roberts, C.J., Tendler, S.J.B. and Williams, P.M. (1995) *Macromolecules*, **28**, 1108.

# Chapter 6

# Electrochemical Methods of Investigating Reaction Mechanisms

*Ole Hammerich*

## 6.1  What is organic electrochemistry?

Organic electrochemistry is concerned with the exchange of electrons between a substrate and an electrode and the chemical reactions that result from such electron transfer processes [1, 2]. Thus, organic electrochemical processes are conceptually related to other organic reactions that include one or more electron transfer steps [3] including, for instance, photoinduced electron transfer [4] and many organometallic [5, 6] and biological processes [7].

Equivalents to organic electrochemistry are also found in conventional organic redox chemistry, for instance in the oxidation and reduction of organic compounds with inorganic reagents [8–10] where, in the electrochemical process, the oxidation agent is replaced by the anode and the reduction agent by the cathode. Two reactions, the oxidation of a primary alcohol to a carboxylic acid (Equations 6.1a and 6.1b) and the reduction of a nitro compound to an amine (Equations 6.2a and 6.2b), serve to illustrate this difference. The notation used in the equations labelled with an 'a' refers to the conventional redox reaction, whereas the equations labelled with a 'b' refer to the corresponding electrochemical process:

$$3Ar\text{—}CH_2OH + 2Cr_2O_7{}^{2-} + 16H^+ \rightarrow 3Ar\text{—}CO_2H + 4Cr^{3+} + 11H_2O \tag{6.1a}$$

$$Ar\text{—}CH_2OH + H_2O \rightarrow Ar\text{—}CO_2H + 4e^- + 4H^+ \tag{6.1b}$$

$$Ar\text{—}NO_2 + 3Sn + 7H^+ \rightarrow Ar\text{—}NH_3{}^+ + 3Sn^{2+} + 2H_2O \tag{6.2a}$$

$$Ar\text{—}NO_2 + 6e^- + 7H^+ \rightarrow Ar\text{—}NH_3{}^+ + 2H_2O. \tag{6.2b}$$

However, the mechanisms of conventional redox reactions and electrochemical reactions may be quite different. Within the formalism of electron transfer theory, the electron transfer reactions at electrodes are usually of the outer-sphere type, whereas those that involve inorganic ions are often of the inner-sphere type [11].

Electrochemical reactions are attractive alternatives to conventional redox reactions for at least three reasons. First, the oxidising power of the anode and the reducing power of the cathode can be varied continuously through the electrode potential which is under the control of the experimentalist; this enhances the selectivity of the process. Second, the electron is a clean reagent and the removal of by-products, such as $Cr^{3+}$ or $Sn^{2+}$ in the examples given above, is avoided during work-up. For this reason, electrochemistry is often

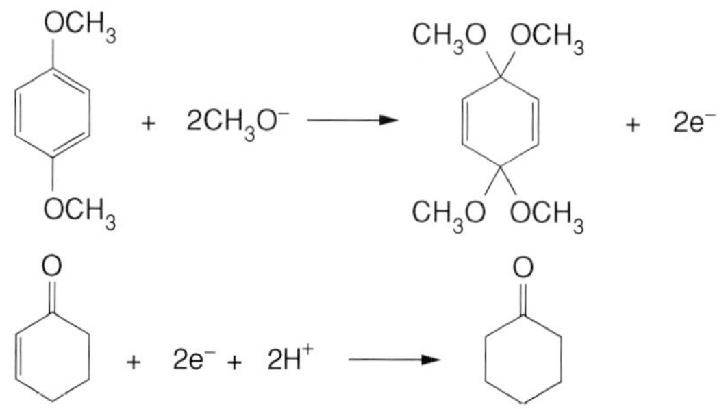

**Scheme 6.1**  Examples of electrochemical additions.

referred to as a green technology [12]. Third, electrochemistry offers many conversions that can only be achieved with difficulty by conventional organic synthesis. Some of these are given as examples below.

Organic electrochemical reactions are classified in the same way as other organic reactions [1, 2]. The most important prototypes include additions (Scheme 6.1) [13, 14], eliminations (Scheme 6.2) [15, 16], substitutions (Scheme 6.3) [17, 18], couplings and dimerisations (Scheme 6.4) [19–21], cleavages (Scheme 6.5) [22, 23], and catalytic reactions (Scheme 6.6) [24, 25]. Hundreds of other examples may be found in the literature [1, 2].

Electrochemical reactions may be carried out at any scale from the smallest to the largest and progress in nanotechnology has made it possible to address electron transfer at the single molecule level [26, 27]. Conversions at the laboratory scale are well established and have been addressed by numerous authors [1, 2] and, at the industrial scale, more than 50 electrochemical processes have reached a respectable level with the reductive hydrodimerisation of acrylonitrile to adiponitrile topping the list with an annual production of about 300 000 tons [28].

In addition to the direct conversions shown above, electrochemistry is often used in an indirect fashion, e.g. for the in situ generation of (harmful) reagents such as bromine or iodine by oxidation of bromide and iodide ion, respectively, or of $Ce^{4+}$ by oxidation of $Ce^{3+}$ [28]. Also, the regeneration of oxidation products such as dichromate, Equations 6.1a and 6.3, has been put to use [28]:

$$2Cr^{3+} + 7H_2O \rightarrow Cr_2O_7^{2-} + 6e^- + 14H^+. \tag{6.3}$$

Related to these indirect methods is the electrochemical generation of acids and bases [29].

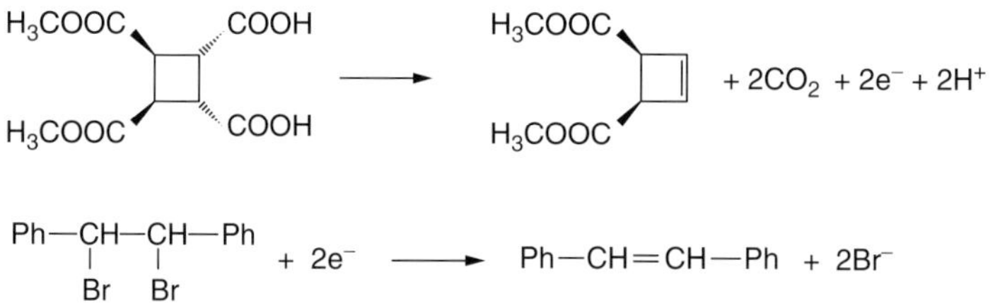

**Scheme 6.2**  Examples of electrochemical eliminations.

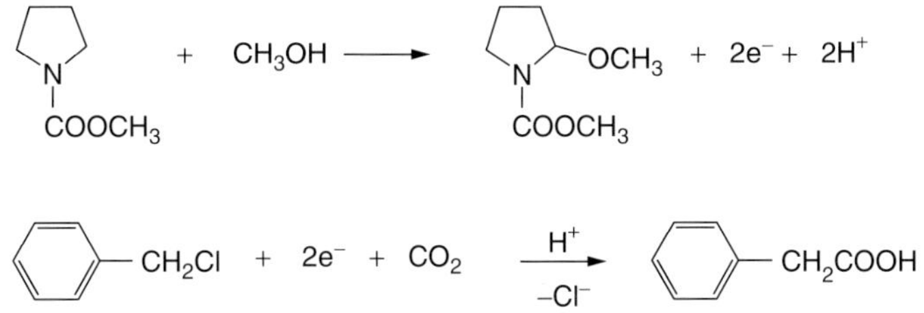

**Scheme 6.3**   Examples of electrochemical substitutions.

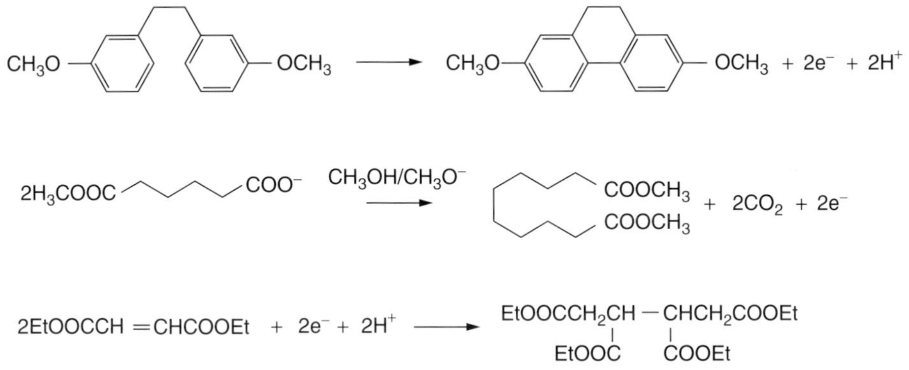

**Scheme 6.4**   Examples of electrochemical couplings and dimerisations.

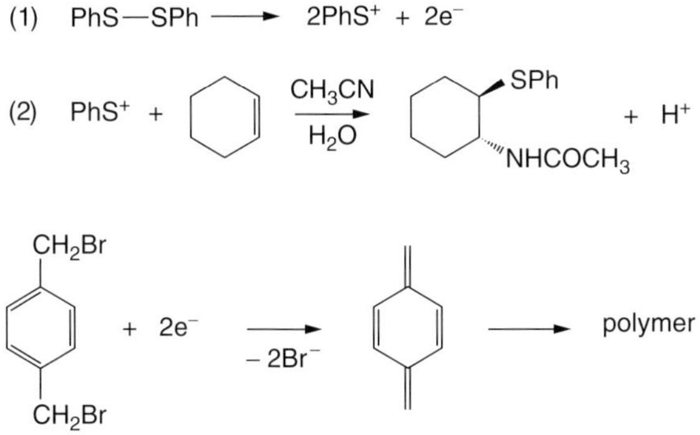

**Scheme 6.5**   Examples of electrochemical cleavages.

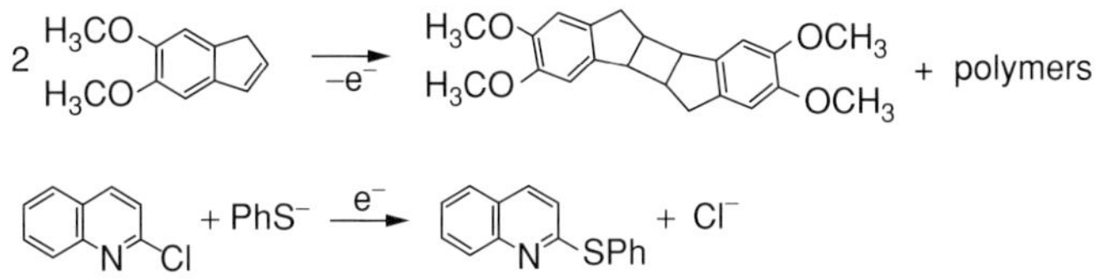

**Scheme 6.6**   Examples of electrocatalytic reactions.

## 6.2 The relationship between organic electrochemistry and the chemistry of radical ions and neutral radicals

The transformations shown in the previous section, except for the catalytic ones, all involve the exchange of electrons in multiples of 2. This, of course, is related to the fact that most organic compounds, in contrast to many inorganic ones, have an even number of electrons. However, electrons are transferred one by one and any organic multi-electron reaction may be broken down into a series of one-electron reactions separated by one or more chemical steps including structural and solvent reorganisation steps. Thus, the primary intermediate in organic electrochemistry is a radical ion or a neutral radical and, in addition, the reaction sequence leading to the final product(s) may also include one or more radical species (see Chapter 10 for further discussion of reactions involving radical intermediates). The reduction of anthracene, A in Scheme 6.7, in the presence of a proton donor such as phenol serves to illustrate this point [30]. Here, the primary intermediate is the anthracene radical anion, $(A^{\bullet-})$. Rate-determining protonation leads to the neutral radical, $AH^{\bullet}$, which

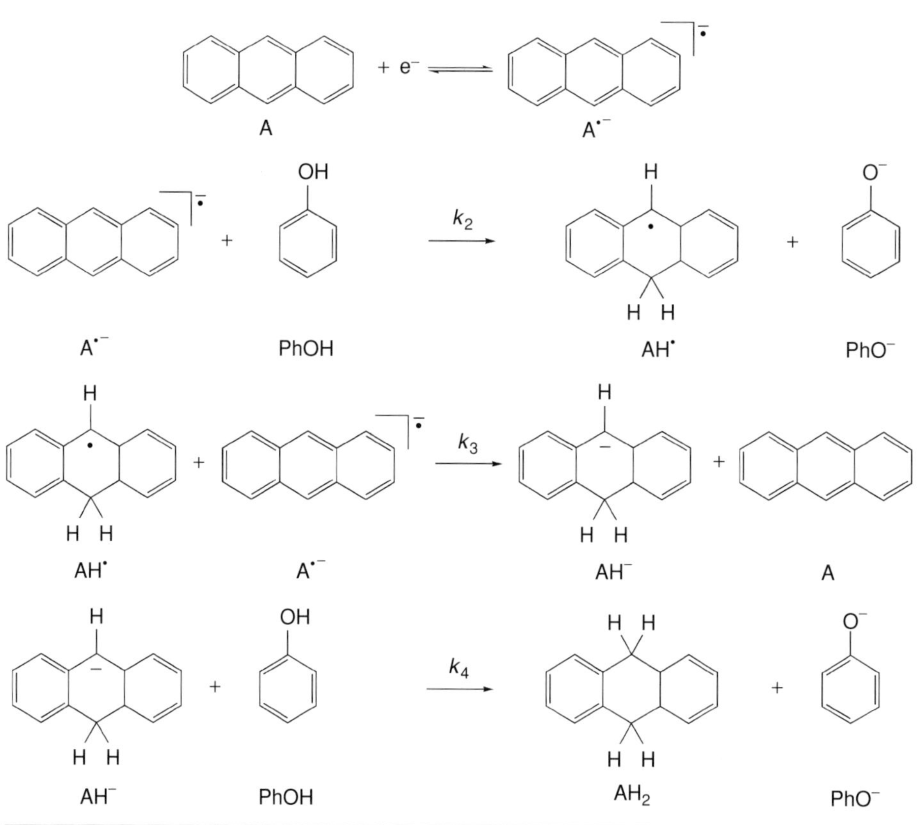

Stoichiometry:      $A + 2e^- + 2PhOH \longrightarrow AH_2 + 2PhO^-$

**Scheme 6.7**    The mechanism for the reduction of anthracene with phenol as proton donor.

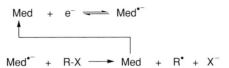

**Scheme 6.8**   The reduction of an alkyl halide facilitated by a mediator.

is further reduced by $A^{•-}$ to the anion, $AH^-$. Protonation of $AH^-$ completes the mechanism resulting in the formation of 9,10-dihydroanthracene ($AH_2$) and an overall stoichiometry including two electrons and two protons.

Radical ions are typically formed as the primary intermediate when the substrate is a neutral compound of the $\pi$-electron type. In contrast, the electron transfer to or from a saturated organic compound is usually associated with bond breaking. This has led to the term *dissociative electron transfer* and is observed, for example, during the reduction of alkyl halides (Equation 6.4) where it is seen that the primary organic intermediate in such cases is a neutral radical ($R^•$) [31]. Radicals are also formed by the oxidation of anions or by the reduction of cations:

$$R\text{—}X + e^- \rightarrow R^• + X^-. \tag{6.4}$$

The electrochemical formation of a radical ion from an aromatic compound or other highly conjugated species is, generally, fast and, therefore, the kinetics of the heterogeneous electron transfer process usually do not interfere with the kinetics of the follow-up reactions to be studied. For species with only one or two double bonds, the initial electron transfer process is often slow and may even be rate determining. In such cases, the kinetics of the follow-up reactions may be studied only with some difficulty. One method is to use a so-called mediator (Med) which serves to shuttle electrons between the substrate and the electrode. Thus, the slow electron transfer between the substrate and the electrode is replaced by two fast electron transfer processes, between the mediator and the electrode, and between the oxidised or reduced mediator and the substrate. In this event, the single reaction of Equation 6.4 is replaced by the two reactions in Scheme 6.8 [32]. It is seen that the mediator is recycled and consequently needs be present in only small, non-stoichiometric amounts.

## 6.3   The use of electrochemical methods for investigating kinetics and mechanisms

The fact that electrochemical processes are tied to electron transfer processes makes electrochemical methods generally less applicable for kinetics and mechanism studies than, for instance, spectroscopic methods. On the other hand, if the reaction under scrutiny involves a radical or radical-like species, electrochemical methods are invaluable tools that often provide a wealth of mechanistic detail. A major advantage of electrochemical methods for kinetics and mechanism studies is that intermediates (radical ions, radicals, etc.) may be formed and their chemical reactions studied at the same electrode in the same operation.

Many electrochemical methods have been developed to serve different purposes, but it is beyond the scope of this chapter to cover them all, so the presentation is purposefully highly selective. The focus will be on the methods that primarily would be helpful to the practising organic chemist and which do not require an advanced experimental set-up. Among these, linear sweep and cyclic voltammetry play a special role and will be treated in some detail. Instrumentation for these methods is now commonly available in most research laboratories. Methods that require highly specialised electrochemical skills, or equipment that is not generally available, such as rotating electrodes and equipment for spectroelectrochemistry, will not be discussed. More detailed information is available elsewhere [1, 33].

The application of any electrochemical method is usually initiated by preliminary investigations the aim of which is to provide information about the reactivity of the primary intermediate as well as other possible intermediates formed during the course of the reaction. Also, it is necessary initially to know the stoichiometry of the electrochemical reaction, including the number of electrons transferred; of course, it is necessary to know the product(s) of the reaction if this information is not available already. Preliminary experiments leading to this kind of information are briefly described in the appendix at the end of this chapter.

## 6.4 Experimental considerations

It is useful to distinguish here between *electroanalytical methods*, in which the material conversion is so low that the bulk concentrations of the substrate and reagents remain practically unchanged during the course of the experiment, and *exhaustive methods*, which are those that rely on the complete or nearly complete conversion of the substrate into products, e.g. coulometry and preparative electrolyses.

### 6.4.1 Two-electrode and three-electrode electrochemical cells

The simplest set-up for an electrochemical conversion consists of two electrodes which, inserted in an electrode holder, are immersed in the solution containing the substrate and an appropriate electrolyte (see Section 6.4.4). An example of such a two-electrode assembly is shown in Fig. 6.1. The electrodes are connected to an appropriate DC current source and the current is usually kept constant during the experiment. This simple set-up is adequate, for instance, for the oxidative 1,4-addition of methoxide to 1,4-dimethoxybenzene (Scheme 6.1) or the Kolbe-type conversion of the methyl hydrogen hexanedioate anion to dimethyl decanedioate (Scheme 6.4). In both of these cases, the cathode reaction is the reduction of hydroxylic protons of the solvent to hydrogen gas. In other cases, for instance where the product formed at one electrode would undergo a redox reaction at the other electrode, it is necessary to separate physically the anode and cathode. This is easily accomplished by using a two-compartment cell with the anode and the cathode compartments being separated by a fine-porosity sintered glass disc or an ion exchange membrane.

The voltage difference between the two electrodes, $\Delta V$, comprises the difference between the electrode potentials, $\Delta E$, and the so-called ohmic drop, $iR_s$, as shown in Equation 6.5, where $i$ is the cell current and $R_s$ is the solution resistance:

$$\Delta V = \Delta E + iR_s. \tag{6.5}$$

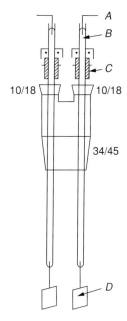

**Fig. 6.1** Two-electrode assembly: (A) platinum wire; (B) mercury-filled glass tube; (C) thermometer adapter, 10/18 standard taper joint, Teflon; (D) platinum electrodes 25 mm × 30 mm. Reprinted with permission [20].

Considering that the current in a typical laboratory cell is in the mA-to-A range and that the resistance in non-aqueous solvents may easily amount to several hundred ohms, the $iR_s$ drop can reach several volts; it follows that $\Delta V$ cannot be directly related to $\Delta E$. However, we are usually concerned only with the potential of the electrode at which the conversion of interest takes place, i.e. the anode in oxidations and the cathode in reductions. This electrode is referred to as the *working electrode* and the other as the *counter electrode*. The solution to the problem of measuring the potential of the working electrode is to introduce a third electrode, a *reference electrode*, and then measure the potential of the working electrode relative to that of the reference electrode in a separate measurement in which very little or no current flows (see Section 6.4.5).

## 6.4.2   Cells for electroanalytical studies

Potential control or potential measurements are fundamental to electroanalytical studies, so the cells used are usually of the three-electrode type. A typical cell for electroanalytical work, such as linear sweep and cyclic voltammetry, is shown in Fig. 6.2.

The cell consists typically of a 10–20 mL vessel and an electrode holder made of plastic or Teflon with holes for the working (W), reference (R) and counter (C) electrodes. In addition, there is an inlet for an inert gas, usually nitrogen or argon, by which the solution is purged before the measurements are made. Usually also, a gentle stream of the inert gas is maintained over the surface of the solution during the electroanalytical measurements.

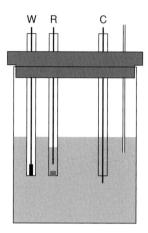

**Fig. 6.2** Schematic of a typical cell for electroanalytical studies. (W) Working electrode; (R) reference electrode; (C) counter electrode. An inlet for an inert gas is also shown.

There are two reasons for this precaution. First, molecular oxygen is easily reduced (to the superoxide ion), which may interfere with studies of other reduction processes; second, the radical ions or and/or radicals formed as a result of the electrode process may react with oxygen.

### 6.4.3   Electrodes for electroanalytical studies

#### 6.4.3.1   The working electrode (W)

The standard working electrode is a circular disc made of an inert conducting material such as platinum, gold or glassy carbon, and is held with the electrical connection in a non-conducting material such as glass or plastic. The electrode surface is polished to mirror quality. The diameter of the conducting disc is typically between 0.5 and 5 mm. Such electrodes are available commercially in a number of sizes and qualities. Ultramicroelectrodes with disc diameters in the μm range are also commercially available.

It is often recommended by manufacturers or others that working electrodes be polished prior to the measurements. We cannot recommend this as a general procedure. In our experience platinum electrodes, for example, improve in quality by being used and, after a measurement or a series of measurements, they should only be rinsed by an organic solvent and wiped to dryness by a soft paper tissue. Only when this gentle treatment is not sufficient to remove material from the surface, for instance a deposit, should the electrode be re-polished.

For reductions, hanging mercury drop electrodes or mercury film electrodes are frequently used owing to their microscopic smoothness and because of the large overpotential for hydrogen evolution characteristic for this electrode material. Mercury film electrodes are conveniently prepared by the electrochemical deposition of mercury on a platinum electrode from an acidic solution of an $Hg^{2+}$ salt, e.g. the nitrate. However, the oxidation of mercury metal to mercury salts or organomercurials at a low potential, 0.3–0.4 V versus the saturated calomel electrode (SCE), prevents the use of these electrodes for oxidations.

### 6.4.3.2    The counter electrode (C)

There are no special requirements to be met for the counter electrode. Platinum wires or platinum wires sealed into glass with approximately 1 cm of exposed wire as shown in Fig. 6.2 are adequate.

### 6.4.3.3    The reference electrode (R)

Reference electrodes are usually the SCE or the Ag/AgCl electrode, which are commercially available. However, the use of these types of electrodes may involve complications, such as the need to separate the reference and working compartments of the cell by a proper salt bridge in order to avoid exchange of dissolved species, or if the electrochemical process is to be studied under non-aqueous conditions. In the latter case, it is often more convenient to use non-aqueous reference electrodes, e.g. a silver wire immersed in the same solvent-supporting electrolyte solution as that in the working compartment but separated from it by a porous ceramic plug. A disadvantage of such a pseudo-reference electrode is that the potential cannot be predicted. Thus, it is necessary to calibrate the electrode if the measured potentials are to be used in comparison with published values. A commonly used calibration standard is the ferrocene/ferrocenium ($Fc/Fc^+$) redox couple.

## 6.4.4    The solvent-supporting electrolyte system

The ideal solvent for electrochemical studies should satisfy a number of requirements. In addition to the properties required for any good solvent for organic chemistry, such as a high solvating power and a low reactivity towards common intermediates, solvents for electrochemical use should be difficult to oxidise or reduce in the potential range of interest. Traditionally, the recommended potential limits are $+3$ V (versus the SCE) for oxidations and $-3$ V for reductions. Also, the solvent should have a dielectric constant higher than about 10 in order to ensure that the supporting electrolyte is well dissociated. Commonly used solvents are acetonitrile (MeCN) and dichloromethane for oxidations, and MeCN, $N,N$-dimethylformamide (DMF) and dimethyl sulfoxide (DMSO) for reductions.

Most electrochemical work requires that the electrical resistance of the solution containing the substrate to be investigated is low. This is achieved by using a solvent containing a well-dissociated supporting electrolyte which also serves to minimise migration effects (see Section 6.5.3). It is necessary that the supporting electrolyte, like the solvent, has a low reactivity and (in particular) is difficult to oxidise and/or reduce. For electroanalytical work, tetralkylammonium salts are frequently used, e.g. tetrabutylammonium hexafluorophosphate which is crystalline, non-hygroscopic and commercially available.

Solvents and supporting electrolytes of the highest commercial quality may often be used without further purification. However, it is recommended that DMF be distilled at reduced pressure prior to application.

## 6.4.5    The electronic instrumentation

In electroanalytical measurements, it is necessary to control the potential of the working electrode, which is usually accomplished by a so-called potentiostat. The potentiostat

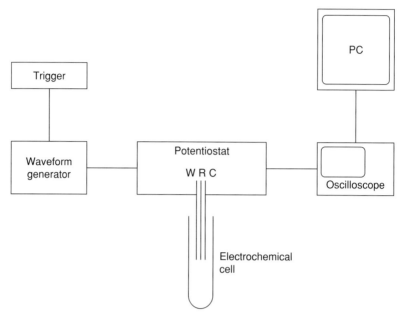

**Fig. 6.3** Schematic of an experimental set-up for electroanalytical measurements.

maintains the voltage across the working and counter electrodes at the value required to keep the potential between the working and the reference electrodes at the pre-set value. The compensation for the solvent resistance is normally accomplished by a positive feedback circuit incorporated into the potentiostat [33]. In addition to the potentiostat, the essential electronic units for electroanalytical work are a trigger, a waveform generator and a recording device (e.g. a digital oscilloscope or a transient recorder). The recording device is usually interfaced to a standard PC, and a schematic of a typical set-up is shown in Fig. 6.3.

Measurements are initiated by the trigger which activates the waveform generator. The required potential–time waveform produced by the waveform generator is fed into the potentiostat to which the three electrodes are connected. The resulting response curves are digitised by the digital oscilloscope or the transient recorder and are then submitted to the PC for data handling. Commercial equipment with the units combined into one instrument controlled by a PC, including software for instrument control and data handling, is available from a number of manufacturers. The reader is referred to the Internet for details.

## 6.5   Some basics

The application of electrochemical methods for the study of the kinetics and mechanisms of reactions of electrochemically generated intermediates is intimately related to the thermodynamics and kinetics of the heterogeneous electron transfer process and to the mode of transport of material to and from the working electrode. For that reason, we review below some basics, including the relationship between potential and current (Section 6.5.1), the electrochemical double layer and the double layer charging current (Section 6.5.2), and the relationship between current and mass transport (Section 6.5.3). Note that the abbreviations for concentrations are not uniform throughout this chapter. We use the notation that is most frequently met in the electrochemical literature; it includes brackets, [ ], for concentrations in kinetic equations, for example; $C_O^*$ and $C_X^*$ are used for the bulk concentrations of the species O and X, respectively, and $C_O(x, t)$ when we need to specify the concentration of O at distance $x$ from an electrode surface at time $t$.

### 6.5.1   Potential and current

An electron transfer reaction, Equation 6.6, is characterised thermodynamically by the standard potential, $E^\circ$, i.e. the value of the potential at which the activities of the oxidised form (O) and the reduced form (R) of the redox couple are equal. Thus, the second term in the Nernst equation, Equation 6.7, vanishes. Here and throughout this chapter $n$ is the number of electrons (for organic compounds, typically, $n = 1$), $R$ is the gas constant, $T$ is the absolute temperature and $F$ is the Faraday constant. Parentheses, ( ), are used for activities and brackets, [ ], for concentrations; $f_O$ and $f_R$ are the activity coefficients of O and R, respectively. However, what may be measured directly is the formal potential $E^{\circ\prime}$ defined in Equation 6.8, and it follows that the relationship between $E^\circ$ and $E^{\circ\prime}$ is given by Equation 6.9. Usually, it may be assumed that the activity coefficients are unity in dilute solution and, therefore, that $E^{\circ\prime} = E^\circ$.

$$O + ne^- \underset{k_s^{ox}}{\overset{k_s^{red}}{\rightleftarrows}} R\ (E^\circ) \tag{6.6}$$

$$E = E^\circ + \frac{RT}{nF} \ln \frac{(O)}{(R)} = E^\circ + \frac{RT}{nF} \ln \frac{f_O[O]}{f_R[R]} \tag{6.7}$$

$$E = E^{\circ\prime} + \frac{RT}{nF} \ln \frac{[O]}{[R]} \tag{6.8}$$

$$E^{\circ\prime} = E^\circ + \frac{RT}{nF} \ln \frac{f_O}{f_R} \tag{6.9}$$

The kinetics of an electron transfer reaction are described by the heterogeneous electron transfer rate constants, $k_s^{red}$ and $k_s^{ox}$, where the subscript 's' indicates that the process takes place at an electrode surface. The values of $k_s^{red}$ and $k_s^{ox}$ depend exponentially on $E$ as seen in Equations 6.10 and 6.11:

$$k_s^{red} = k^\circ \exp[-\alpha nF(E - E^\circ)/(RT)] \tag{6.10}$$

$$k_s^{ox} = k^\circ \exp[(1 - \alpha)nF(E - E^\circ)/(RT)]. \tag{6.11}$$

Here, $k^\circ$ is the standard heterogeneous electron transfer rate constant and $\alpha$ is the electro-chemical transfer coefficient [33], which corresponds in electrochemistry to the Brønsted coefficient in organic chemistry. It is seen from Equations 6.10 and 6.11 that $k_s^{red}$ and $k_s^{ox}$ are both equal to $k^\circ$ at $E = E^\circ$.

The relationship between $k_s^{red}$, $k_s^{ox}$ and the net current, $i$, is given by the Butler–Volmer equation (Equation 6.12) where $A$ is the electrode area, and $[O]_{x=0}$ and $[R]_{x=0}$ are the concentrations of O and R at the electrode surface ($x$ is the distance from the electrode surface).

In discussions of electrochemical phenomena, it is occasionally useful to use the current density, which is the current per unit of the electrode area, $i/A$, instead of the current:

$$i = nFA(k_s^{red}[O]_{x=0} - k_s^{ox}[R]_{x=0})$$
$$= nFAk^o\{[O]_{x=0} \exp[-\alpha nF(E - E^o)/(RT)]$$
$$-[R]_{x=0} \exp[(1 - \alpha)nF(E - E^o)/(RT)]\}. \qquad (6.12)$$

Electron transfer reactions are classified as reversible, quasi-reversible or irreversible depending on the ability of the reaction to respond to changes in $E$, which, of course, is related to the magnitude of $k^o$. The distinction is important, in particular, for the (correct) application of linear sweep and cyclic voltammetry, and for that reason further discussion of this classification will be postponed until after the introduction of these techniques in Section 6.7.2.

### 6.5.2 The electrochemical double layer and the charging current

From the above discussion, the reader might have got the impression that the application of a potential to an electrode immersed in a solution in the absence of an electroactive substrate would not result in a current response within the potential window of the solvent-supporting electrolyte system. This, however, is not the case. The interphase between the electrode and the solvent-supporting electrolyte behaves like an electric capacitor, i.e. a charged electrode attracts ions of the opposite charge resulting in a so-called double layer. This is illustrated in Fig. 6.4 for a negatively charged electrode. The layer of positive ions then attracts negative

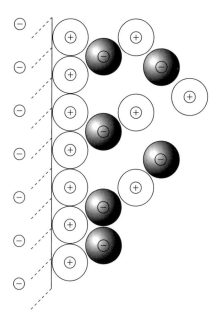

**Fig. 6.4** The so-called structured layer model of the electrochemical double layer close to a negatively charged electrode.

ions from the solution, but the electric field resulting from the layer of positive ions is partly compensated by the negatively charged electrode, so the second layer is less dense than the first layer, and so on. The charging current, $i_c$, observed when the potential of the electrode is suddenly shifted by the value $\Delta E$ is given by Equation 6.13, where $t$ is the time after the onset of the potential shift and $C_d$ is the double layer capacity:

$$i_c = (\Delta E / R_s) \exp(-t/(R_s C_d)). \tag{6.13}$$

When working electrodes with diameters between 0.5 and 1 mm are used, the charging current has usually decayed to a negligible value after 0.5 millisecond or less and may be neglected in experiments lasting 1 millisecond or more [33]. This decay time is reduced to the microsecond time regime when ultramicroelectrodes are being used.

## 6.5.3 Mass transport and current

It follows from Equation 6.12 that the current depends on the surface concentrations of O and R, i.e. on the potential of the working electrode, but the current is, for obvious reasons, also dependent on the transport of O and R to and from the electrode surface. It is intuitively understood that the transport of a substrate to the electrode surface, and of intermediates and products away from the electrode surface, has to be effective in order to achieve a high rate of conversion. In this sense, an electrochemical reaction is similar to any other chemical surface process. In a typical laboratory electrolysis cell, the necessary transport is accomplished by magnetic stirring. How exactly the fluid flow achieved by stirring and the diffusion in and out of the stationary layer close to the electrode surface may be described in mathematical terms is usually of no concern; the mass transport just has to be effective. The situation is quite different when an electrochemical method is to be used for kinetics and mechanism studies. Kinetics and mechanism studies are, as a rule, based on the comparison of experimental results with theoretical predictions based on a given set of rate laws and, for this reason, it is of the utmost importance that the mass transport is well defined and calculable. Since the intention here is simply to introduce the different contributions to mass transport in electrochemistry, rather than to present a full mathematical account of the transport phenomena met in various electrochemical methods, we shall consider transport in only one dimension, the $x$-coordinate, normal to a planar electrode surface (see also Chapter 5).

The flux of the species O, $J_O(x, t)$, is described in terms of the three components that constitute the Nernst–Planck equation (Equation 6.14). The parameters are defined below.

$$J_O(x, t) = -D_O \frac{\partial C_O(x, t)}{\partial x} - \frac{z_O F}{RT} D_O C_O(x, t) \left( \frac{\partial \Phi(x, t)}{\partial x} \right)$$
$$+ v(x, t) C_O(x, t) \tag{6.14}$$

The first two terms describe the contributions to the mass transport that are the result of forces that act on the species O. The first term is the *diffusion* component, which relates to the forces that act on O in the concentration gradient close to an electrode. ( $D_O$ is the diffusion coefficient of O, $C_O(x, t)$ is the concentration of the species O at the distance $x$ and the time $t$ and thus $\partial C_O(x, t)/\partial x$ is the concentration gradient.) The second term is the *migration* component, which relates to the forces that act on a charged species in the electrical potential gradient close to a charged electrode. ($z_O$ is the charge of O and $\partial\Phi(x, t)/\partial x$ is the electrical potential gradient along the $x$-axis.) Here, it should be remembered that at least one of the two species in an electron transfer reaction carries a charge. In the presence of a large excess of a well-dissociated electrolyte, this term can be neglected. Usually, an electrolyte concentration of 0.1 M is sufficient for substrate concentrations in the 0.5–5 mM range. The third and final term describes the *convection* component; this is related to the forced displacement of a small volume element of the solution in which O is dissolved. ($v(x, t)$ is the velocity of the fluid normal to the electrode surface.) It follows that, in the presence of a large excess of an electrolyte and in the absence of convective forces such as stirring, the last two terms in Equation 6.14 may be neglected, leaving only the diffusion component. Accordingly, the expression for the flux is reduced to Equation 6.15, known as Fick's first law. The flux at the electrode surface ($x = 0$) and the current are then given by Equations 6.16 and 6.17 for the case where there is no accumulation of material at the surface, which is usually the case in organic electrochemistry. It remains to specify how the concentration of O at a given distance from the electrode surface depends on the flux in and out of a small volume element at that distance. This is described by Fick's second law, Equation 6.18.

$$J_O(x, t) = -D_O \frac{\partial C_O(x, t)}{\partial x} \tag{6.15}$$

$$J_O(0, t) = -D_O \left( \frac{\partial C_O(x, t)}{\partial x} \right)_{x=0} \tag{6.16}$$

$$i(t) = -nFAJ_O(0, t) = nFAD_O \left( \frac{\partial C_O(x, t)}{\partial x} \right)_{x=0} \tag{6.17}$$

$$\frac{\partial C_O(x, t)}{\partial t} = D_O \frac{\partial^2 C_O(x, t)}{\partial x^2} \tag{6.18}$$

Fick's first and second laws (Equations 6.15 and 6.18), together with Equation 6.17, the Nernst equation (Equation 6.7) and the Butler–Volmer equation (Equation 6.12), constitute the basis for the mathematical description of a simple electron transfer process, such as that in Equation 6.6, under conditions where the mass transport is limited to linear semi-infinite diffusion, i.e. diffusion to and from a planar working electrode. The term *semi-infinite* indicates that the electrode is considered to be a non-permeable boundary and that the distance between the electrode surface and the wall of the cell is larger than the thickness, $\delta$, of the diffusion layer defined as Equation 6.19 [1, 33]:

$$\delta = \sqrt{2D_O t}. \tag{6.19}$$

By introduction of a typical value for $D_O$, $10^{-5}$ cm$^2$ s$^{-1}$, it is seen that the value of $\delta$ after, for example, 5 seconds amounts to 0.1 mm. At times larger than 10–20 seconds, natural convection begins to interfere and the assumption of linear diffusion as the only means of mass transport is no longer strictly valid. At times larger than approximately 1 minute, the deviations from pure diffusion are so serious and unpredictable that the current observed experimentally cannot be related to a practical theoretical model.

The mathematical treatment of hemi-spherical diffusion, as encountered in work with ultramicroelectrodes at long experiment times (Section 6.7.3), includes the convenient transposition of Fick's first and second law to polar coordinates. The mathematical details will be omitted here and the interested reader is referred to the specialist literature [34].

## 6.6    The kinetics and mechanisms of follow-up reactions

### 6.6.1    Nomenclature

The electrochemical literature abounds with symbols and abbreviations for mechanisms that are not always strictly logical and often unfamiliar to the non-electrochemist. Here, we have adopted the commonly used notation whereby the letter 'e' indicates an electron transfer process and the letter 'c' indicates a chemical reaction in solution. It is helpful to distinguish between reversible (fast) and rate-determining (slow) steps by using lower case letters for reversible and capital letters for rate-determining steps. Since electron transfer reactions can take place either at an electrode (heterogeneous electron transfer) or in solution (often referred to as homogeneous electron transfer), those taking place in solution are given by the subscript 'h'. Examples of the use of this terminology includes the eC mechanism, a reversible electron transfer followed by an irreversible chemical step, the eC(dim) mechanism, a reversible electron transfer followed by an irreversible dimerisation, and the $eCe_h$ mechanism, an example of which has been given already in Scheme 6.7.

### 6.6.2    Mechanisms and rate laws

Let us again consider the protonation of the anthracene radical anion by phenol (Scheme 6.7). The set of rate expressions corresponding to this reaction sequence is given by Equations 6.20–6.24:

$$d[A]/dt = k_3[AH^\bullet][A^{\bullet-}] \tag{6.20}$$

$$d[A^{\bullet-}]/dt = -k_2[A^{\bullet-}][PhOH] - k_3[AH^\bullet][A^{\bullet-}] \tag{6.21}$$

$$d[AH^\bullet]/dt = k_2[A^{\bullet-}][PhOH] - k_3[AH^\bullet][A^{\bullet-}] \tag{6.22}$$

$$d[AH^-]/dt = k_3[AH^\bullet][A^{\bullet-}] - k_4[AH^-][PhOH] \tag{6.23}$$

$$d[PhOH]/dt = -k_2[A^{\bullet-}][PhOH] - k_4[AH^-][PhOH]. \tag{6.24}$$

In this particular case, the protonation of the radical anion is rate determining [30]. Furthermore, all the experimental evidence indicates that the two intermediates, $AH^\bullet$ and $AH^-$, have only a transient existence and therefore it is justified to apply the steady-state approximation for these two species, i.e. $d[AH^\bullet]/dt = d[AH^-]/dt = 0$. Thus, the rather complex set of rate expressions given by Equations 6.20–6.24 may be reduced to the mathematically more tractable Equations 6.25–6.27:

$$d[A]/dt = k_2[A^{\bullet-}][PhOH] \tag{6.25}$$

$$d[A^{\bullet-}]/dt = -2k_2[A^{\bullet-}][PhOH] \tag{6.26}$$

$$d[PhOH]/dt = -2k_2[A^{\bullet-}][PhOH]. \tag{6.27}$$

The mathematical complexity may be further reduced if the reaction is studied under pseudo-first-order conditions, i.e. $C_{PhOH}^* \gg C_A^*$. Accordingly, $k_2[PhOH] \approx k_2 C_{PhOH}^* = k_2'$, in which case Equations 6.25 and 6.26 may be rewritten as Equations 6.28 and 6.29:

$$d[A]/dt = k_2'[A^{\bullet-}] \tag{6.28}$$

$$d[A^{\bullet-}]/dt = -2k_2'[A^{\bullet-}]. \tag{6.29}$$

The rate expressions 6.25–6.29 are all of the general form shown in Equation 6.30, where X is a reagent, here phenol. (Note that, in the electrochemical literature, the electron transfer reaction is sometimes written as $A + e^- \rightleftarrows B$, rather than $O + e^- \rightleftarrows R$, and the reaction orders as '*a*' and '*b*', rather than '*o*' and '*r*'; we are using this aspect of the original notation.) Thus, the first task in the analysis of the kinetics of a reaction is to determine the values of *a*, *b* and *x*, and then to measure $k_{obs}$. This is a major issue in Section 6.7.

$$\text{rate} = k_{obs}[O]^a[R]^b[X]^x \tag{6.30}$$

### 6.6.3    The theoretical response curve for a proposed mechanism

The basic strategy in the application of electroanalytical methods in studies of the kinetics and mechanisms of reactions of radicals and radical ions is the comparison of experimental results with predictions based on a mechanistic hypothesis. Thus, equations such as 6.28 and 6.29 have to be combined with the expressions describing the transport. Again, we restrict ourselves to considering transport governed only by linear semi-infinite diffusion, in which case the combination of Equations 6.28 and 6.29 with Fick's second law, Equation 6.18, leads to Equations 6.31 and 6.32 (note that we have now replaced the notation for concentration introduced in Equation 6.18 earlier by the more usual square brackets). Also, it is assumed here that the diffusion coefficients of A and $A^{\bullet-}$ are the same, i.e. $D_A = D_{A^{\bullet-}} = D$.

$$d[A]/dt = Dd^2[A]/dx^2 + k_2'[A^{\bullet-}] \tag{6.31}$$

$$d[A^{\bullet-}]/dt = Dd^2[A^{\bullet-}]/dx^2 - 2k_2'[A^{\bullet-}] \tag{6.32}$$

The task now at hand is to find solutions to these second-order differential equations under the boundary conditions defined by the electroanalytical method in question. Nowadays, this is most often accomplished by numerical integration, known in electroanalytical chemistry as *digital simulation*. It is beyond the scope of this chapter to go into the mathematical details, and the interested reader is referred to the specialist literature [33]. Commercial user-friendly software for linear sweep and cyclic voltammetry is available (DigiSim©); software for other methods has been developed and is available through the Internet.

## 6.7    The response curves for common electroanalytical methods

We will now discuss how the relationship between potential, mass transport and current manifests itself experimentally in some situations typically met in electroanalytical chemistry. In Sections 6.7.1–6.7.3, we discuss the response curves for two families of electroanalytical methods that are conducted under conditions where linear semi-infinite diffusion to/from planar working electrodes prevail. These are *chronoamperometry* and *double potential step chronoamperometry* (Section 6.7.1), and *linear sweep* and *cyclic voltammetry* (Section 6.7.2). In Section 6.7.3, we present the response curves observed at ultramicroelectrodes at long experiment times where the mass transport is governed by hemi-spherical diffusion.

The discussions will be based on a *reduction* process, Equation 6.6 reproduced below, studied under the conditions where the substrate O is the only electroactive species initially present in the solution and we will *initially* assume that the product of the electrode process, R, is stable under the conditions of the experiment. (The transposition to oxidation processes is straightforward. Be aware, however, that the minus and plus signs in some of the equations given below will then have to be interchanged.) This is followed by a discussion of how follow-up reactions involving R affect the response curves.

$$O + ne^- \underset{k_s^{ox}}{\overset{k_s^{red}}{\rightleftarrows}} R\,(E^\circ) \tag{6.6}$$

The response curves presented have, unless stated otherwise, been produced using DigiSim© (version 3.03). We will be using the 'classical' sign convention in plots of the current as a function of time or potential, i.e. reduction currents are considered positive and oxidation currents negative. The potential scale has positive values increasing to the left and negative values decreasing to the right.

### 6.7.1   Potential step experiments (chronoamperometry and double potential step chronoamperometry)

The term *amperometry* refers to measurements of the current and *chronoamperometry* (CA) to measurements of the current as a function of the time. Let us now examine the current–time behaviour of the reduction process in Equation 6.6 in a potential step experiment in which the potential, $E$, at time $t = 0$ is changed from a value, $E_{initial}$, when no current flows, to a potential, $E_{final}$, when all the molecules of O that reach the surface of the working electrode are reduced to R. (In the discussion to follow, the value of $E^\circ$ is arbitrarily set to 0 V.) The initial conditions correspond to $E_{initial} - E^\circ \gg 0$ and thus $[O]_{x=0} \approx C_O^*$ and the final conditions to $E_{final} - E^\circ \ll 0$ and thus $[O]_{x=0} \approx 0$. The potential–time program is shown graphically in Fig. 6.5 (top) for the situation where $E_{initial} = 0.5$ V and $E_{final} = -0.5$ V; the current–time response curve is shown in Fig. 6.5 (bottom). Semi-infinite linear diffusion is assumed, i.e. the diameter of the planar working electrode, $d$, is large relative to the thickness of the diffusion layer, $\delta$ (see Section 6.7.3 for a further discussion of this point). In this case, the current varies with time according to the Cottrell equation (Equation 6.33), i.e. its decay is proportional to $t^{-1/2}$ (the contribution from the double layer charging current, $i_c$, Equation 6.13, is omitted).

$$i = \frac{nFAC_O^*\sqrt{D_O}}{\sqrt{\pi t}} \tag{6.33}$$

That the current decays with time is easily understood by inspection of Fig. 6.6a, which illustrates the concentration profiles for this experiment – plots of how the concentrations

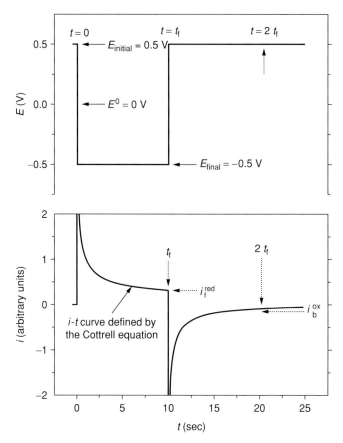

**Fig. 6.5** Potential–time program (top) and current–time response curves (bottom) for chronoamperometry ($t \leq t_f$) and double potential step chronoamperometry ($t \leq 2t_f$).

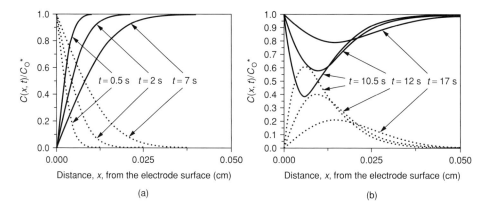

**Fig. 6.6** Relative concentration profiles for O (solid line) and R (dots) near a planar working electrode (a) at $t = 0.5$, 2 and 7 seconds, corresponding to a chronoamperometry experiment, and (b) at $t = 10.5$, 12 and 17 seconds, corresponding to the second half of a double potential step chronoamperometry experiment.

of O and R close to the electrode surface vary with the distance from the surface at a given time. It is seen that the concentration gradient for O at the electrode surface, $(\frac{\partial C_O(x,t)}{\partial x})_{x=0}$, and thus the current, continuously decreases with time. We also observe that pairs of profiles for O and R are symmetrically distributed around the line $C(x,t)/C_O^* = 0.5$, i.e. R is assumed to be stable in the solution and $[O] + [R] = C_O^*$, at all distances and all times. If the potential after the time $t = t_f$, where the reduction current has decayed to the value $i_f^{red}$ (Fig. 6.5), is shifted from $E_{final}$ back to the original value, $E_{initial}$, the molecules of R that reach the electrode surface will be oxidised back to O, that is, $[R]_{x=0} \approx 0$. This extension of the method is known as *double potential step chronoamperometry* (DPSCA). The key parameter in DPSCA is the ratio $-i_b^{ox}/i_f^{red}$, where $i_b^{ox}$ is the oxidation current measured at $t = 2t_f$. The value of this ratio, which is independent of $t_f$, is 0.2929. This means that less than 30% of R is being re-oxidised in spite of the fact that the potential is such that $[R]_{x=0} \approx 0$. The origin of this phenomenon is that most of R, even after the potential has been shifted back to $E_{initial}$, still diffuses away from the electrode surface. This can be appreciated by inspection of Fig. 6.6b, which shows the development of the concentration profiles during the second potential step. It is seen that the profile for R possesses a maximum and all molecules to the right of the maximum, of course, diffuse away from the working electrode.

The application of CA for monitoring the progress of a chemical reaction following the heterogeneous electron transfer is limited by the nature of the process. It is clear from the Cottrell equation, Equation 6.33, that the only parameter that may be affected by a follow-up reaction is the number of electrons, $n$. Thus, important mechanisms such as eC and eC(dim) cannot be distinguished by CA. This limits the application of CA for kinetic studies. In contrast, DPSCA is a most useful method.

It follows from the above discussion that the effect of a chemical follow-up reaction in which the primary intermediate R is consumed is that the current ratio, $-i_b^{ox}/i_f^{red}$, is smaller than the no-reaction-value, 0.2929. The faster the reaction of R, the smaller the ratio, and in the limit where R reacts so rapidly that the oxidation back to O cannot be observed, the ratio is, of course, zero. It is convenient to normalise the current ratio by the no-reaction-value by introduction of the parameter $R_I$ defined in Equation 6.34; clearly, $R_I$ may vary from 1 (no reaction) to 0 (complete reaction).

$$R_I = -i_b^{ox}/\left(0.2929 i_f^{red}\right) \tag{6.34}$$

It is convenient at this juncture to introduce a concept that, in electroanalytical chemistry, sometimes is referred to as the *reaction order approach*. Consider first the half-life-time, $t_{1/2}$, which in conventional homogeneous kinetics refers to the time for the conversion of half of the substrate into product(s). From basic kinetics, it is well known that $t_{1/2}$ is independent of the substrate concentration for a reaction that follows a first-order rate law and that $1/t_{1/2}$ is proportional to the initial concentration of the substrate for a reaction that follows a second-order rate law. Similarly, in electroanalytical chemistry it is convenient to introduce a parameter that reflects a certain constant conversion of the primary electrode intermediate. In DPSCA, it is customary to use $\tau_{1/2}$ (or $\tau_{0.5}$), which is the value of $t_f$ required to keep the value of $R_I$ equal to 0.5. The reaction orders (see Equation 6.30) are then given by Equations 6.35 and 6.36, where $R_{A/B} = a + b$, and $R_X = x$ (in reversal techniques such as DPSCA, in which O and R are in equilibrium at the electrode surface, it is not possible to separate the reactions orders $a$ and $b$); $C_X^*$ is the bulk concentration of X.

$$R_{A/B} = 1 + \frac{d \log(1/\tau_{1/2})}{d \log C_O^*} \qquad (6.35)$$

$$R_X = \frac{d \log(1/\tau_{1/2})}{d \log C_X^*} \qquad (6.36)$$

The determination of the reaction orders then includes recordings of $\tau_{1/2}$ at a series of concentrations of O and, if appropriate, the reagent X. For reactions following a simple rate law of the type in Equation 6.30, plots of $\log(1/\tau_{1/2})$ versus $\log C_O^*$ or $\log C_X^*$ would then result in straight lines with slopes that are equal to $R_{A/B} - 1$ and $R_X$, respectively.

In principle, the value of $k_{obs}$ for a given reaction mechanism may be calculated directly from the value of $\tau_{1/2}$ [1]. However, unless the reaction is already well characterised, we cannot recommend this strategy since the value of $k_{obs}$ obtained by this approach relies on data corresponding to only the conversion that is defined by $\tau_{1/2}$. Thus, for a complicated reaction scheme where, for example, the rate-determining step changes with the concentration of O (or X), this procedure might lead to erroneous results. Instead, it is recommended that the analysis of the kinetics in general be based on recordings of the full $R_I$ versus $\log(1/\tau_{1/2})$ curve for each concentration of O (and X). Curves of this type should then be compared with the theoretical curve, a *working curve*, calculated for the proposed mechanism. A good match of the experimental curve and the working curve may be taken as support of the proposed mechanism. The match of the scales for the two curves gives access to the rate constant, $k_{obs}$. An example related to the protonation of the anthracene radical anion by phenol [30, 35] (Scheme 6.7) is shown in Fig. 6.7.

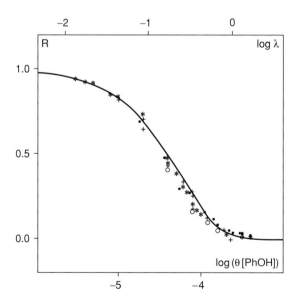

**Fig. 6.7** The double potential step chronoamperometry working curve for the eCe$_h$ mechanism (full line) and experimental data for the protonation of the anthracene radical anion by phenol (points). The scale at the top corresponds to the working curve and the scale at the bottom to the experimental data. (The parameter $\theta$ in the figure corresponds to $t_f$ in the text.) Note that the data for the variation of $t_f$ and [PhOH] have been plotted on the same working curve. Reprinted with permission [35].

The approximate kinetic window of DPSCA at electrodes with a diameter in the 0.5–1 mm range is given by the expression $0.3 \text{ s}^{-1} < k < 300 \text{ s}^{-1}$, where $k$ is a first-order or pseudo-first-order rate constant. With the application of microelectrodes, the upper limit can be extended considerably.

### 6.7.2  Potential sweep experiments (linear sweep voltammetry and cyclic voltammetry)

The term *voltammetry* refers to measurements of the current as a function of the potential. In linear sweep and cyclic voltammetry, the potential steps used in CA and DPSCA are replaced by linear potential sweeps between the potential values. A triangular potential–time waveform with equal positive and negative slopes is most often used (Fig. 6.8). If only the first half-cycle of the potential–time program is used, the method is referred to as *linear sweep voltammetry* (LSV); when both half-cycles are used, it is *cyclic voltammetry* (CV). The rate by which the potential varies with time is called the *voltage sweep* (or *scan*) *rate*, $v$, and the potential at which the direction of the voltage sweep is reversed is usually referred to as $E_{switch}$.

In voltammetry, the relevant kinetic parameters for the electrode process (in addition to $v$) are $k^{\text{o}}$ and $\alpha$ (see Equations 6.10–11). The mutual dependence of $k^{\text{o}}$ and $v$ is conveniently expressed through the magnitude of a dimensionless parameter, $\Lambda$, defined in Equation 6.37, in which the diffusion coefficients for O and R are assumed to be identical ($D_O = D_R = D$):

$$\Lambda = \frac{k^{\text{o}}}{\left(\dfrac{DvnF}{RT}\right)^{1/2}}. \tag{6.37}$$

A reversible electron transfer reaction is the limiting case where O and R are in thermodynamic equilibrium at the electrode surface, i.e. the electron transfer reaction responds instantaneously to a change in $E$. Thus, the ratio between $[O]_{x=0}$ and $[R]_{x=0}$ is given by the Nernst equation, Equation 6.7. In principle, the equilibrium condition implies an infinitely

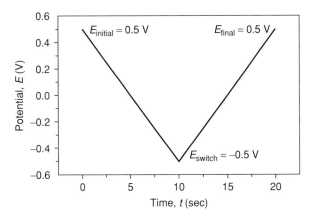

**Fig. 6.8**  Typical cyclic voltammetry waveform corresponding to a voltage sweep rate ($v$) of 0.1 V s$^{-1}$; $E_{\text{initial}} = E_{\text{final}} = 0.5$ V, and $E_{\text{switch}} = -0.5$ V.

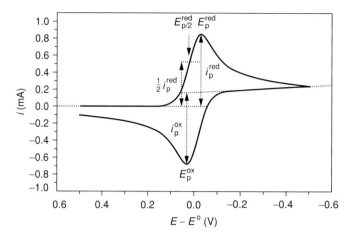

**Fig. 6.9** Cyclic voltammogram for a reversible one-electron reduction; $v = 1$ V s$^{-1}$ (DigiSim©).

large value of $\Lambda$, which may result from an infinitely large value of $k^{\circ}$ and/or an infinitely small value of $v$. In practical work, an electron transfer process is called reversible if the deviations from this limiting case are so small that they cannot be detected experimentally. This, of course, depends on the measurement precision, which for routine experiments is typically $\pm 2$ mV. In such cases, deviations from reversible behaviour cannot be detected for $\Lambda$ larger than approximately 12 [36].

The response curve, a *voltammogram*, for a reversible one-electron reduction is shown in Fig. 6.9; its characteristic feature is the presence of two peaks. The position, the height and the broadness of each peak are given by the peak potentials, $E_{\text{p}}^{\text{red}}$ and $E_{\text{p}}^{\text{ox}}$, the peak currents, $i_{\text{p}}^{\text{red}}$ and $i_{\text{p}}^{\text{ox}}$, and the half-peak widths, $E_{\text{p}}^{\text{red}} - E_{\text{p/2}}^{\text{red}}$ and $E_{\text{p}}^{\text{ox}} - E_{\text{p/2}}^{\text{ox}}$, where $E_{\text{p/2}}^{\text{red}}$ and $E_{\text{p/2}}^{\text{ox}}$ are the values of the potential at $i = \frac{1}{2} i_{\text{p}}^{\text{red}}$ and $i = \frac{1}{2} i_{\text{p}}^{\text{ox}}$, respectively. The presence of peaks in the voltammogram is understood by inspection of the concentration profiles shown in Fig. 6.10. In comparison with CA experiments discussed above, it is seen that the gradient during the early part of the CV experiment is not very steep, which is because $[O]_{x=0}$ in the

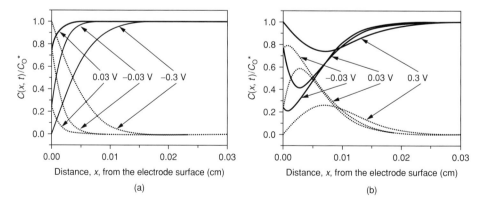

**Fig. 6.10** Concentration profiles corresponding to (a) the forward sweep, and (b) the reverse sweep in cyclic voltammetry. O: full line; R: dots. The values of $E - E^{\circ}$ corresponding to each set of profiles are given in the figure (DigiSim©).

early part of the experiment does not differ much from that in the bulk, which (in turn) is because $E - E^\circ$ is relatively large. When $E$ and $[O]_{x=0}$ decrease, the gradient, and thus the current, increases until (after a time) the gradient again begins to decrease. As a result, the current passes through a maximum and a peak is observed.

Under equilibrium conditions, the values of $E_p^{red}$, $E_p^{red} - E_{p/2}^{red}$ and $i_p^{red}$ are independent of the kinetic parameters $k^\circ$, $\alpha$ and $v$. Ideally, the part of the voltammogram recorded during the forward sweep (here the reduction wave) satisfies the following three criteria, where the numerical values given in units of mV refer to $T = 298$ K.

1. The position of the peak relative to $E^\circ$ is given by Equation 6.38:

$$E_p^{red} - E^\circ = -1.11\frac{RT}{nF}V = -\frac{28.5}{n}mV. \qquad (6.38)$$

2. The height of the peak, $i_p^{red}$, is given by Equation 6.39: note that the peak height increases linearly with $v^{1/2}$:

$$i_p^{red} = 0.4463\, nFAC_O^* D_O^{1/2} v^{1/2} \left(\frac{nF}{RT}\right)^{1/2}. \qquad (6.39)$$

3. The broadness of the peak, $E_p^{red} - E_{p/2}^{red}$, is given by Equation 6.40:

$$E_p^{red} - E_{p/2}^{red} = -2.20\frac{RT}{nF}V = -\frac{56.5}{n}mV. \qquad (6.40)$$

Additionally, if the second half of the voltammogram (here, the re-oxidation wave) is recorded, we have the following.

4. The peak separation, $\Delta E_p = E_p^{ox} - E_p^{red}$, is approximately 57 mV, the exact value depending on the potential $E_{switch}$, i.e. when the voltage sweep is reversed. Equation 6.41 relates to the limiting case when $E^\circ - E_{switch} = \infty$.

$$\Delta E_p = E_p^{ox} - E_p^{red} = 2 \times 1.11\frac{RT}{nF}V = \frac{57.0}{n}mV \qquad (6.41)$$

5. The peak current ratio $-i_p^{ox}/i_p^{red}$ is unity and independent of $v$.

In practice, the value of $E^\circ$ may be taken to be the average of $E_p^{ox}$ and $E_p^{red}$, Equation 6.42, often referred to as the *midpoint potential*.

$$E^\circ \approx \left(E_p^{ox} + E_p^{red}\right)/2 \qquad (6.42)$$

Reversible electron transfers are typically observed for aromatic compounds or other compounds with extended electronic $\pi$-systems, i.e. compounds for which the electron transfer is associated with only little geometrical change.

In the general case, referred to as *quasi-reversible*, the electron transfer reaction in Equation 6.6 does not respond instantaneously to changes in $E$. In other words, $[R]_{x=0}$ and $[O]_{x=0}$ are determined not only by the value of $E - E^\circ$, but also by the magnitudes of $k^\circ$ and $\alpha$ through Equations 6.10 and 6.11. Typical voltammograms for quasi-reversible electron transfers are shown in Fig. 6.11. There are no simple analytical expressions for $E_p^{red} - E^\circ$, $i_p^{red}$, $E_p^{red} - E_{p/2}^{red}$, $\Delta E_p$ and $-i_p^{ox}/i_p^{red}$ for quasi-reversible electron transfers. Values for a given set of $v$, $k^\circ$ and $\alpha$ are, when needed, most conveniently obtained by digital simulation.

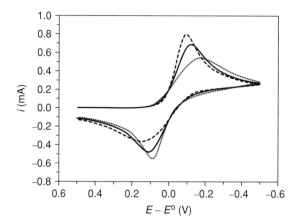

**Fig. 6.11**  Cyclic voltammograms for a quasi-reversible one-electron reduction at $v = 1$ V s$^{-1}$ and $k^o = 3 \times 10^{-3}$ cm s$^{-1}$; $\alpha = 0.3$ (dots), 0.5 (solid), 0.7 (dashes) (DigiSim©).

Quasi-reversible electron reactions are observed for many aromatic compounds at high voltage sweep rates and, at low voltage sweep rates, for compounds that have only small electronic $\pi$-systems composed of one or two double bonds.

When $\Lambda$ becomes progressively smaller, the shape of the voltammogram continues to change. Experimentally, a constant shape is reached when $\Lambda$ is smaller than about 0.2 [36]. The value of $E$ required to obtain an appreciable rate of reduction of O is now so far on the negative side of $E^o$ that the second term in Equation 6.12 may be neglected. In other words, the electron transfer has become *irreversible*. By the same type of argument, it is clear that the oxidation of R back to O during the reverse sweep proceeds irreversibly as well. Typical voltammograms for irreversible electron transfers are shown in Fig. 6.12.

Irreversible electron transfers such as those shown in Fig. 6.12 are rarely observed in organic electrochemistry. The radical ions resulting from slow electron transfer reactions

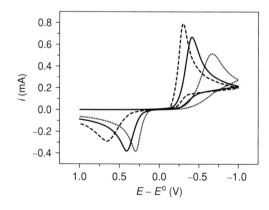

**Fig. 6.12**  Cyclic voltammograms for an irreversible one-electron reduction at $v = 1$ V s$^{-1}$ and $k^o = 10^{-5}$ cm s$^{-1}$; $\alpha = 0.3$ (dots), 0.5 (solid), 0.7 (dashes) (DigiSim©).

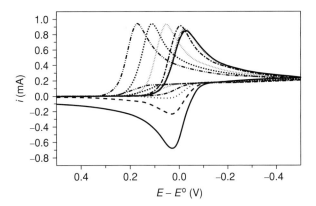

**Fig. 6.13** Cyclic voltammograms for an eC mechanism at $v = 1$ V s$^{-1}$ and $k = 2$ s$^{-1}$ (dash), 10 s$^{-1}$ (dot), $10^2$ s$^{-1}$ (dash-dot), $10^4$ s$^{-1}$ (short dot), $10^6$ s$^{-1}$ (short dash) and $10^8$ s$^{-1}$ (dash-dot-dot), respectively. The solid line corresponds to the no-reaction-case, $k = 0$ (DigiSim©).

are usually highly reactive, which prevents the observation of the oxidation peak on the return sweep.

It follows from the above discussion that an electron transfer reaction that appears reversible at one (low) sweep rate may change to a quasi-reversible or even an irreversible process at higher sweep rates. This should be kept in mind since the application of LSV and CV in kinetics studies usually includes the recording of voltammograms at a number of different sweep rates (see below), and the analysis of the data is usually based on the assumption that the electron transfer is reversible.

We shall now discuss the response for a reversible one-electron reduction followed by an irreversible chemical reaction, for example an eC mechanism with a first-order rate constant $k$. The voltammograms resulting from different values of $k$ at $v = 1$ V s$^{-1}$ are shown in Fig. 6.13. Given this value of $v$, it is seen that both the shape and the position of the voltammogram depend on the magnitude of $k$. On the other hand, for a given value of $k$, it is intuitively understood that the effect of the chemical reaction will gradually diminish if the sweep rate is allowed to increase. In the limit, the experiment time is so short that the chemical reaction does not have the time to manifest itself and, consequently, the voltammogram observed is just that for the electron transfer reaction, Equation 6.6.

The effect of the relative magnitudes of $k$ and $v$ on the position and shape of a voltammogram is conveniently discussed in terms of the dimensionless parameter $\lambda$ in Equation 6.43, given for both first- and second-order kinetics:

$$\lambda = \frac{kRT}{vnF} \text{ (first order)} \quad \text{or} \quad \lambda = \frac{kC_O^* RT}{vnF} \text{ (second order).} \tag{6.43}$$

### 6.7.2.1   CV conditions

Beginning with $\lambda = 0$, we see from Fig. 6.13 that the effect of increasing values of $\lambda$ is initially that the current associated with the oxidation of R back to O during the reverse sweep gradually disappears. The position of the peak is, so far, only little affected. The region of $\lambda$ values, from the point where the effect of the chemical reaction becomes experimentally detectable to the point where the current for the oxidation of R has totally disappeared, is given by the expression $0.11 < \lambda < 1.89$ [36]. In this region the system is under mixed diffusion and kinetic control, conveniently referred to as CV conditions. Introducing $n = 1$ and $T = 298$ K, and assuming that the range of applicable $v$ values is $0.1$–$500$ V s$^{-1}$, translates into a kinetic window of $0.4$ s$^{-1} < k < 4 \times 10^4$ s$^{-1}$. The time scale of the experiment in this region is of the same order as the half-life of R.

As mentioned above, the characteristic feature of processes in this kinetic region is that the peak current ratio $-i_p^{ox}/i_p^{red}$ varies from about unity to zero. Thus, a procedure for studying the kinetics would be to record values of $-i_p^{ox}/i_p^{red}$ at different sweep rates and compare these with a working curve for the proposed mechanism in a way analogous to that discussed for DPSCA above. However, a problem with this approach is the difficulty of defining a baseline for the reverse sweep (see below) and, for that reason, CV suffers from some limitations when used in quantitative work. This has led to the development of *derivative cyclic voltammetry* (DCV) [37].

The voltammogram shown in Fig. 6.9 is redrawn in Fig. 6.14 (left) showing $i$ as a function of $t$. The problem of defining a baseline for the measurement of $i_p^{ox}$ is illustrated by the two broken lines. The differentiated curve, $di/dt$ versus $t$, is shown in Fig. 6.14 (right). The peaks labelled $i_f'$ and $i_b'$, corresponding to the maximum steepness of the voltammogram during the forward and backward sweeps, respectively, reflect the CV peak heights. Both $i_f'$ and $i_b'$ are measured relative to the zero line. Since the slope of the baseline for $i_p^{ox}$, i.e. the extension of the reduction wave, is not far from zero where the measurement is made, it follows that the magnitude of $i_b'$ is essentially baseline independent.

The ratio $R_I' = -i_b'/i_f'$ plays the same role in DCV as the ratio $-i_p^{ox}/i_p^{red}$ in CV, or the ratio $-i_b^{ox}/i_f^{red}$ in DPSCA. It should be noted, however, that $-i_b'/i_f'$, in contrast to $-i_p^{ox}/i_p^{red}$ and $-i_b^{ox}/i_f^{red}$, does not approach zero for increasing values of $\lambda$, but a value close to 0.1, the exact value being dependent on $E_{switch}$. This is because the derivative curve during the reverse sweep is not zero, even when the peak due to R has completely vanished.

The rate law necessary for making a mechanistic proposal is conveniently determined by DCV using the reaction order approach introduced in Section 6.7.1. Usually, the value of $v$ required to keep $R_I'$ equal to 0.5 is used and referred to as $v_{1/2}$ (or $v_{0.5}$). The relationships between $v_{1/2}$ and the reaction orders of Equation 6.30 are given by Equations 6.44 and 6.45

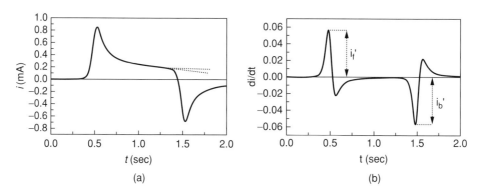

**Fig. 6.14** The voltammogram in Fig. 6.9 plotted as the current–time curve (left) and the differentiated current–time curve (right).

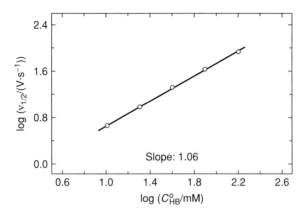

**Fig. 6.15**  Data for log $v_{1/2}$ as a function of log $C^*_{PhOH}$ for the protonation of the anthracene radical anion with phenol. Reprinted with permission [30b].

in a fashion similar to that for DPSCA:

$$R_{A/B} = 1 + \frac{d \log v_{1/2}}{d \log C^*_O} \tag{6.44}$$

$$R_X = \frac{d \log v_{1/2}}{d \log C^*_X}. \tag{6.45}$$

An example of the reaction order analysis using Equation 6.45 is shown Fig. 6.15 for data obtained for the protonation of anthracene radical anion by phenol in DMSO [30b]. The slope of the regression line in this case is 1.06 indicating that the rate law is first order in the concentration of phenol.

An example of the working curve approach, resulting from a study of the stereochemistry of the electrohydrodimerisation of cinnamic acid esters [38], is shown in Fig. 6.16.

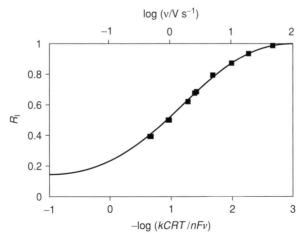

**Fig. 6.16**  Data from derivative cyclic voltammetry for the dimerisation of the radical anions of (−)-bornyl cinnamate. The rate constant, $k_{dim}$, obtained by matching the two scales is $5.6 \times 10^2$ M$^{-1}$ s$^{-1}$. Reprinted with permission [38].

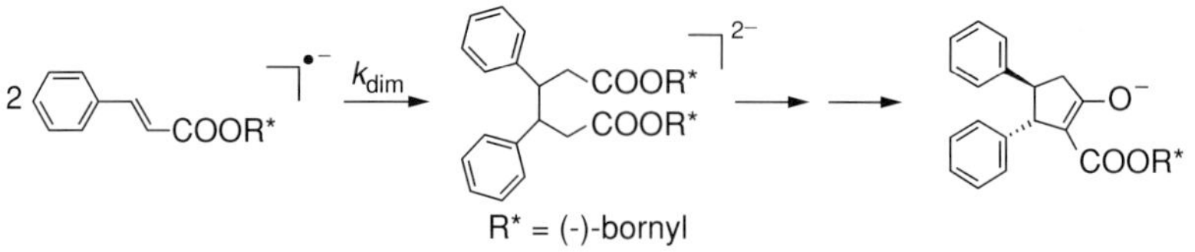

**Scheme 6.9**  The dimerisation of a chiral cinnamic ester radical anion.

Here the experimental data for the dimerisation of the (−)-bornyl cinnamate radical anion (Scheme 6.9) are fitted to the working curve for the eC(dim) mechanism.

### 6.7.2.2   LSV conditions

At values of $\lambda$ higher than approximately 1.9, the shape of the voltammogram is essentially independent of $\lambda$ and an increase in $\lambda$ only results in a displacement of the voltammogram in the positive direction. This is illustrated by the voltammograms corresponding to $k = 10^4$, $10^6$ and $10^8 s^{-1}$ shown earlier in Fig. 6.13. A stationary state has now been established in solution by mutual compensation of the chemical reaction of R and the diffusion process, and the system is said to be under purely kinetic control or under LSV conditions. The characteristic feature of the voltammogram in this region of $\lambda$ values is the absence of a peak during the reverse sweep. The position of the voltammogram, as measured by $E_p$, is displaced in the positive direction relative to that for the no-reaction-case, Equation 6.6, and depends markedly on the value of $\lambda$.

The dependences of $E_p$ on changes in $\log v$, $\log C_O^*$ and $\log C_X^*$ are linear with the slopes given by Equations 6.46–6.48, where, again, the reaction orders are defined by the rate law in Equation 6.30:

$$\frac{dE_p}{d\log v} = -\frac{1}{b+1}\frac{RT}{nF}\ln 10 \tag{6.46}$$

$$\frac{dE_p}{d\log C_O^*} = \frac{a+b-1}{b+1}\frac{RT}{nF}\ln 10 \tag{6.47}$$

$$\frac{dE_p}{d\log C_X^*} = \frac{x}{b+1}\frac{RT}{nF}\ln 10. \tag{6.48}$$

Introduction of the reaction orders, e.g. for the eC mechanism under pseudo-first-order conditions ($a = 0$, $b = 1$ and $x = 1$), results in $dE_p/d\log v = -29.6\,\mathrm{mV}$, $dE_p/d\log C_O^* = 0$ and $dE_p/d\log C_X^* = 29.6\,\mathrm{mV}$ at $n = 1$ and $T = 298$ K.

When $E^\circ$ for the initial electron transfer reaction is known, the measurements of $E_p$ give direct access to the rate constant, $k$. An example of the relationship between $E_p - E^\circ$ and $k$ for the eC-mechanism is given by Equation 6.49 [36]:

$$E_p - E^\circ = \left(-0.783 + \frac{1}{2}\ln\frac{kRT}{vnF}\right) \times \left(\frac{RT}{nF}\right)\mathrm{V}. \tag{6.49}$$

### 6.7.2.3   Fitting simulated voltammograms to experimental voltammograms

Advances in computer technology have stimulated the development of data analysis methods that use more than just a few experimental points, such as $E_p - E^\circ$, $E_p - E_{p/2}$ or $-i_p^{ox}/i_p^{red}$,

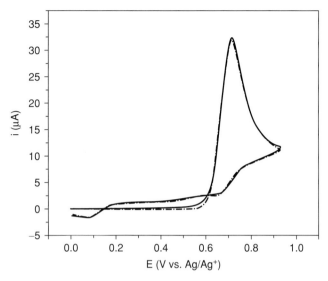

**Fig. 6.17** Cyclic voltammetry data (solid line) for the oxidation of 2,4-dimethyl-3-ethylpyrrole in MeCN at $v = 0.2$ V s$^{-1}$ and the simulated curve (dash-dot, DigiSim$^©$) corresponding to the reaction sequence in Scheme 6.10. Reprinted with permission [39].

along the voltammetric wave. The strategy is to fit a simulated voltammogram for the proposed mechanism to the experimental one by an iterative procedure in which the kinetic and thermodynamic parameters in the simulation are varied systematically. This procedure is included in some of the commercial software for instrument control and data handling that accompany modern electroanalytical instrumentation.

This approach would at first glance seem to be a major advantage. However, it should be borne in mind that the problems originating from, for example, background currents and non-proper adjustments of the electronic equipment remain; it still requires some skill, and experience, to incorporate correctly parameters for the uncompensated electrical solution resistance and the double layer capacity, for example, in the data treatment. Another problem is that voltammograms, even for very different mechanisms, are usually very similar and, consequently, this method should be used with care. In our opinion, the method works best when the voltammogram presents structural features that are characteristic of the mechanism in question. An example, the oxidation of 2,4-dimethyl-3-ethylpyrrole, originating from our recent work on the oxidation of pyrroles [39] is shown in Fig. 6.17. The additional structural feature in this case is the trace crossings, observed at 0.6 and 0.15 V, caused by the slow proton transfer from the dimer dication and monocation to the unoxidised substrate (steps 2 and 3 in Scheme 6.10).

### 6.7.3 Potential sweep experiments with ultramicroelectrodes

In the above discussion, semi-infinite linear diffusion has been assumed. In practical work, this means that the diameter of the working electrode, $d$, is much larger than the thickness of the diffusion layer, $\delta$. The question that then arises is: what happens when the diameter of the electrode is allowed to decrease and finally reaches the thickness of the diffusion layer?

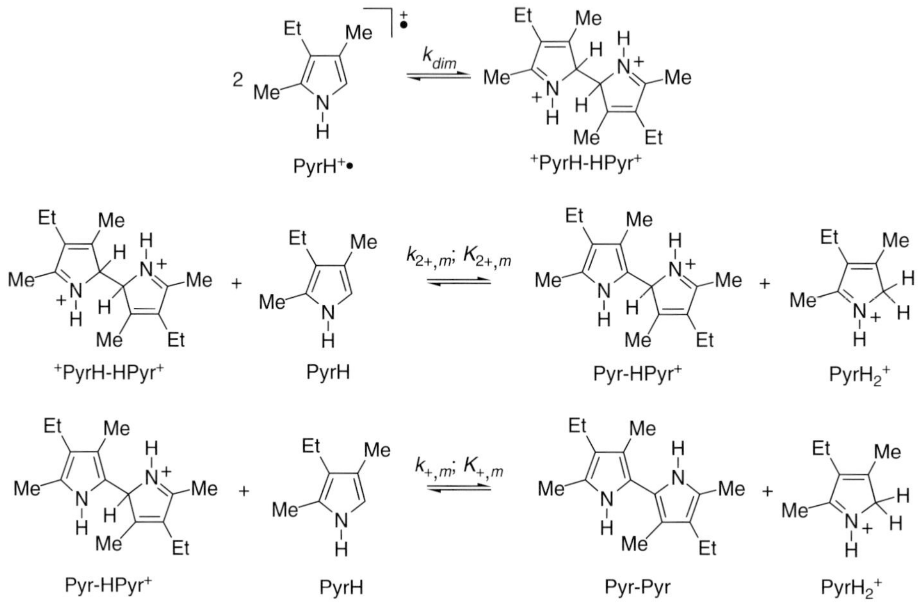

**Scheme 6.10**   The first three steps of the mechanism for the dimerisation of the 2,4-dimethyl-3-ethylpyrrole radical cation.

Obviously, as illustrated in Fig. 6.18, the deviations from linear diffusion always observed at the edges of the working electrode become more and more significant until, in the limit, the diffusion is hemi-spherical as shown in Fig. 6.18c.

The effect of this gradual change of the diffusion pattern on the voltammogram obtained by CV for a reversible electron transfer reaction is shown in Fig. 6.19. It is seen that the change in the diffusion pattern has a profound influence on the magnitude of the current density – it increases with decreasing values of $d$. This is because the transport by diffusion is more effective at the edges than at the centre of the electrode as illustrated in Fig. 6.18. Also, the appearance of the voltammogram is strongly dependent on the diffusion pattern. We see in Fig. 6.19 that the voltammograms gradually change from the characteristic peak

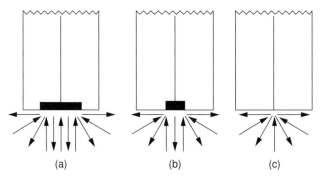

(a)                      (b)                      (c)

**Fig. 6.18**   Diffusion pattern for planar electrodes with decreasing diameters.

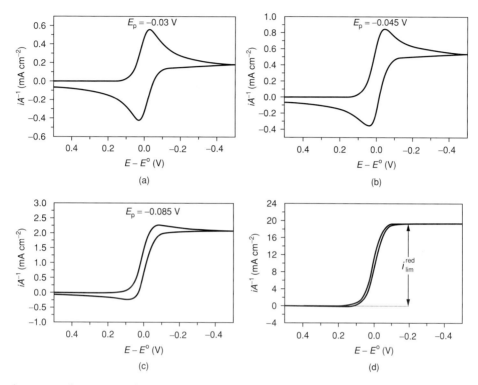

**Fig. 6.19** Voltammograms for a reversible one-electron reduction at circular disc electrodes with the diameters (a) 1.4 mm, (b) 0.1 mm, (c) 0.02 mm and (d) 0.002 mm; $v = 1$ V s$^{-1}$ (DigiSim©).

shaped curve obtained when linear diffusion prevails, to an S-shaped curve characteristic of hemi-spherical diffusion.

An important feature of the voltammogram shown in Fig. 6.19d is the absence of a peak. Instead, the current reaches a plateau indicating that a steady state has been obtained. In this case, the diffusion of R away from the electrode is so effective that reverse current for the re-oxidation of R to O cannot be observed, even when R is non-reactive.

The limiting current, $i_{lim}^{red}$, observed at the plateau (Fig. 6.19d) is given by Equation 6.50, where $r_o$ is the radius of the *conducting part* of the electrode:

$$i_{lim}^{red} = 4nFD_OC_O^*r_o. \tag{6.50}$$

It is important to stress at this point that the hemi-spherical diffusion pattern that results in the S-shaped response curve is the result of a diffusion layer thickness that is *large* relative to the diameter of the electrode as observed, for instance, at ultramicroelectrodes at long experiment times. When the experiment time becomes shorter and shorter, by increasing the voltage sweep rate, the thickness of the diffusion layer becomes smaller and smaller and the diffusion pattern gradually changes from hemi-spherical to the linear semi-infinite diffusion pattern observed when the diffusion layer thickness is *small* relative to the diameter of the electrode. The two extremes are nicely illustrated by the experimental voltammograms shown in Fig. 6.20 for the oxidation of ferrocene at an ultramicroelectrode ($d = 13\mu$m) at

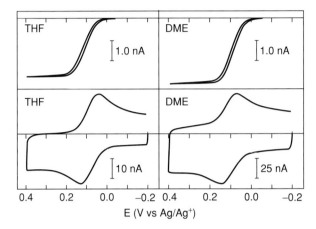

**Fig. 6.20**  Cyclic voltammograms obtained at an Au ultramicroelectrode ($d = 13\,\mu\text{m}$) for the oxidation of ferrocene in THF and DME at $v = 10\,\text{mV s}^{-1}$ (top), $500\,\text{V s}^{-1}$ (bottom, left) and $1000\,\text{V s}^{-1}$ (bottom, right). Reprinted with permission [40].

$v = 10\,\text{mV s}^{-1}$ (top) and $1000\,\text{V s}^{-1}$ (bottom) [40]. Note also that the contribution from the charging current (see Equation 6.13) is significant at the high voltage sweep rates necessary for the observation of peak-shaped voltammograms at ultramicroelectrodes.

Most electroanalytical methods, and in particular LSV and CV, usually involve electrodes with diameters in the mm range and substrate concentrations between 1 and 10 mM. Concentrations much larger than 10 mM lead to distortions caused by the ohmic drop associated with the large currents resulting from high concentrations; the higher the concentration, the more serious the distortions. Consequently, conclusions regarding kinetics and mechanism become increasingly unreliable as the substrate concentration increases. The development of the ultramicroelectrode technology has changed this situation. With electrode areas in the $10$–$100\,\mu\text{m}^2$ range, the currents observed even in highly concentrated solutions are small and so are the distortions resulting from inadequate $iR_s$ compensation.

Using ultramicroelectrodes, it is possible to study reactions under the conditions of synthesis, including electrosynthesis. An example is the electrohydrodimerisation of acrylonitrile to adiponitrile (Scheme 6.11, top) mentioned in the introduction; in industry this is typically carried out with an emulsion of acrylonitrile in an aqueous phosphate buffer as electrolyte. At substrate concentrations in the mM level, the reduction of acrylonitrile takes another route leading to saturation of the C—C double bond (Scheme 6.11, bottom). This precludes studies of the dimerisation using substrate concentrations at the mM level and thereby working electrodes of conventional sizes. The transition between the two mechanisms could be studied conveniently using an ultramicroelectrode as the working electrode

$$2\text{CH}_2{=}\text{CHCN} \;+\; 2e^- \;+\; 2\text{H}_2\text{O} \;\longrightarrow\; \begin{matrix}\text{CH}_2\text{CH}_2\text{CN}\\ |\\ \text{CH}_2\text{CH}_2\text{CN}\end{matrix} \;+\; 2\text{OH}^-$$

$$\text{CH}_2{=}\text{CHCN} \;+\; 2e^- \;+\; 2\text{H}_2\text{O} \;\longrightarrow\; \text{CH}_3\text{CH}_2\text{CN} \;+\; 2\text{OH}^-$$

**Scheme 6.11**  The electrohydrodimerisation (top) and hydrogenation (bottom) of acrylonitrile.

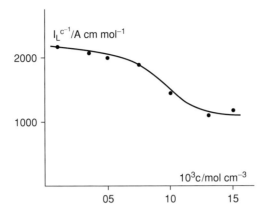

**Fig. 6.21** A plot of $i_{lim}^{red}/C_O^*$ as a function of $C_O^*$ for the reduction of acrylonitrile in an aqueous phosphate buffer. Reprinted with permission [41].

under conditions of hemi-spherical diffusion, i.e. at low voltage sweep rates [41]. A plot of $i_{lim}^{red}/C_O^*$, which is proportional to $n$ (Equation 6.50), versus $C_O^*$ is shown in Fig. 6.21 and the transition from a one-electron process at high substrate concentrations to a two-electron process at low concentrations is seen to take place in the range 0.7 M $< C_O^* <$ 1.2 M.

### 6.7.4 Concluding remarks

It is evident from the examples given above that the electroanalytical response curves are highly dependent on the type of mass transport that prevails during the experiment. The potential sweep methods, LSV and CV, result in the more structured curves including the presence of one or more peaks. Thus, deviations from the behaviour of a simple one-electron process, for example caused by a chemical follow-up reaction, are more easily detected during LSV and CV than in other methods. In addition, as we shall see (Appendix), electroactive intermediates resulting from follow-up reactions may sometimes be detected. This is a major reason for the success of these methods. Voltammetry at ultramicroelectrodes offers the advantage that the currents at moderate to low voltage sweep rates are small and thus the proper electronic compensation of the solution resistance is less important. Highly resistive media may be employed, and importantly, substrate concentrations at the synthetic level. The effective transport resulting from hemi-spherical diffusion results in an electrode system that is relatively insensitive to natural convection.

## Appendix

### A.1 The preliminary experiments

In studies of the kinetics and mechanisms of electrochemical reactions it is important to know, at least approximately, the oxidation or reduction potential of the substrate and also to have some prior knowledge about the reactivity of the electroactive intermediate(s) formed as a result of the initial electron transfer process. Nowadays, the method of choice for obtaining this type of information is CV. Below, we show by examples how to use CV in such preliminary work and also discuss briefly the application of coulometry and product studies.

## A.2 Preliminary studies by cyclic voltammetry

For preliminary work, a circular disc electrode with a diameter in the 1–5 mm range is usually used. The substrate concentrations are typically between 1 and 10 mM and the voltage sweep rates are usually between 50 mV s$^{-1}$ and 500 V s$^{-1}$. Sometimes the voltage sweep is continued to include several $E - t$ half-cycles.

Examples of experimental voltammograms are shown in Figs 6.22–6.24. These are all from our current research on the oxidation of substituted pyrroles [39, 42]. Numerous other examples may be found in the literature [1, 2]. The voltammograms shown are recorded with a potentiostat without electronic compensation of the solution resistance in order to illustrate better some of the problems caused by the solution resistance. The peak potentials below are given with a precision of ±10 mV, which is typical for values reported in the literature. These values originate from the data files and cannot be determined with this precision directly from the figures reproduced below.

The first example (Fig. 6.22) shows the voltammogram resulting from the oxidation of 2,3,4,5-tetraphenylpyrrole (2345tpp) at $v = 0.2$ V s$^{-1}$. An oxidation peak, O1, with $E_p^{ox} = 0.95$ V is observed during the potential sweep in the positive direction, and a reduction peak, R1, with $E_p^{red} = 0.88$ V is observed during the potential sweep in the negative direction.

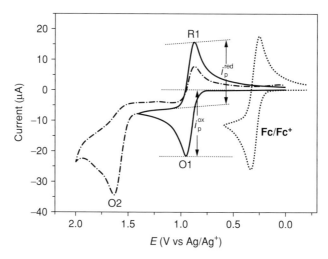

**Fig. 6.22** Cyclic voltammogram for the oxidation of 2,3,4,5-tetraphenylpyrrole (2 mM) in MeCN at $v = 0.2$ V s$^{-1}$ with the direction of the sweep being reversed at 1.4 V (full line) and 2.0 V (dash-dot). The voltammogram for the ferrocene/ferrocenium one-electron redox couple (dot) is shown for comparison [42].

These data result in a peak separation, $\Delta E_p$, of approximately 70 mV. It is seen also that the peak current ratio, $-i_p^{ox}/i_p^{red}$, is close to 1. For comparison, the voltammogram recorded for the oxidation of ferrocene, Fc, with $E_p^{ox} = 0.33$ V and $E_p^{red} = 0.26$ V is included in Fig. 6.22. Two important points should be noticed. First, the peak separation for the reversible one-electron oxidation of Fc to $Fc^+$ is similar to that for 2345tpp, i.e. close to 70 mV, which illustrates, as mentioned above, that the mere observation of peak separations larger than approximately 60 mV cannot be taken as evidence that the electron transfer process is quasi-reversible. In fact, in this case, the 'too large' peak separation is caused by the lack of proper $iR_s$-compensation. Second, the peak heights observed for 2345tpp are slightly smaller than those for Fc. This is caused by the smaller diffusion coefficient for 2345tpp relative to that for Fc. Thus, the data indicate that the oxidation of 2345tpp corresponds to a reversible one-electron process, i.e. the formation of a radical cation, and that the radical cation is stable on the time scale of the experiment.

When the sweep in the positive direction is extended to higher potentials, the further oxidation of the radical cation to the dication is observed at $E_p = 1.63$ V (O2). In this case, the current corresponding to the reduction of the dication back to the radical cation is not observed, indicating that the dication undergoes a (fast) chemical reaction. The voltammogram provides no information of the fate of the dication.

The second example, Fig. 6.23, shows the voltammogram for 1-methyl-2,3,5-triphenylpyrrole (1m235tpp) at two different voltage sweep rates, $\nu = 0.05$ V s$^{-1}$ (left) and $\nu = 5$ V s$^{-1}$ (right). From the voltammogram in Fig. 6.23a, it is seen that the peak corresponding to the reduction of the radical cation, R1, is barely visible at $\nu = 0.05$ V s$^{-1}$ and that a peak system, O2/R2, resulting from one or more products is observed in the range 1.1–1.3 V. Thus, it is clear that the radical cation of this compound undergoes a follow-up reaction which is almost complete on the time scale of the experiment. An increase of $\nu$ to 5 V s$^{-1}$ (Fig. 6.23b) has two effects. First, the peak corresponding to the reduction of the radical cation is now fully developed and, second, the peak system resulting from the product has essentially disappeared. Thus, on this time scale, the reaction of the radical cation has progressed very little. The voltammograms provide no information about the

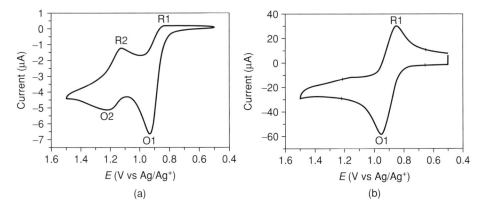

**Fig. 6.23** Cyclic voltammogram for the oxidation of 1-methyl-2,3,5-triphenylpyrrole (2 mM) in MeCN at two different voltage sweep rates: (a) $\nu = 0.05$ V s$^{-1}$, and (b) $\nu = 5$ V s$^{-1}$ [42].

origin of the redox activity observed at O2/R2, only that this peak system results from one or more products formed by reaction of the 1m235tpp radical cation.

The third example, Fig. 6.24, shows the oxidation of 2,4-dimethylpyrrole (24dmp) at $v = 0.2$ V s$^{-1}$ in the absence (left) and the presence (right) of a non-nucleophilic base, in this case 2,6-di-*tert*-butylpyridine [42]. For this compound, there is no indication of a reduction current corresponding to the radical cation, not even at $v = 500$ V s$^{-1}$ (not shown), indicating that the radical cation undergoes a very fast follow-up reaction, here a dimerisation. An intermediate is observed at O2. Addition of a non-nucleophilic base to the voltammetry solution increases the height of the oxidation peak, O1, and the oxidation peak at O2 disappears. The increase of O1 may have two different origins. The presence of the base could increase the nucleophilicity of the solvent system with the result that the radical cation (partly) reacts with a nucleophile resulting in a radical intermediate that is further oxidised at the potential of O1. This further oxidation would result in an increase of the current. Product studies of the type described in Section A.4 show, however, that this is not the explanation. The other possibility is that the substrate, in the absence of the non-nucleophilic base, (partly) serves as a base accepting the protons liberated during the reaction of the radical cation as shown in Scheme 6.10 earlier for the oxidation of 2,4-dimethyl-3-ethylpyrrole. Since the protonated 24 dmp is not oxidised at O1, the height of O1 is smaller than the peak that would have been observed. The fact that O2 disappears upon addition of a base indicates that this peak is caused by the oxidation of a protonated intermediate.

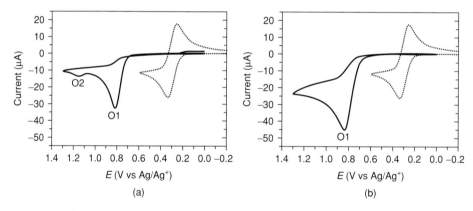

**Fig. 6.24** Cyclic voltammograms for the oxidation of 2,4-dimethylpyrrole (2 mM) in (a) MeCN and (b) MeCN in the presence of an excess of 2,6-di-*tert*-butylpyridine; $v = 0.2$ V s$^{-1}$ [42].

In addition to investigating the effect of changing the voltage sweep rate and of addition of nucleophiles/bases (oxidation) or electrophiles/acids (reductions), the preliminary work often also includes investigation of how the voltammograms are affected by changes in the substrate concentration.

## A.3 Determination of the number of electrons, n (coulometry)

The number of electrons, $n$, transferred during an electrochemical process is an important stoichiometric parameter, and may depend on the experiment time. For example, when the

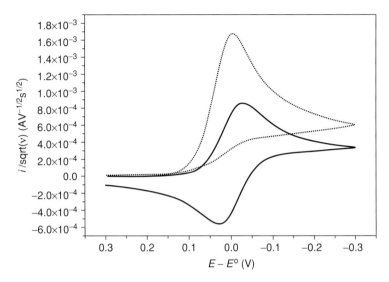

**Fig. 6.25** Cyclic voltammograms for an eCe$_h$ mechanism with $k = 20$ s$^{-1}$ at $v = 100$ V s$^{-1}$ (solid) and $v = 0.1$ V s$^{-1}$ (dots). Note that the $y$-axis shows values of $i/\sqrt{v}$ (DigiSim©).

protonation of the anthracene radical anion by phenol (Scheme 6.7, section 6.2) is studied by cyclic voltammetry, it is observed that the data obtained at high voltage sweep rates, e.g. at $v = 100$ V s$^{-1}$, correspond to $n = 1$. This indicates that the proton transfer reaction on this time scale has not had time to manifest itself kinetically. However, the data obtained at low sweep rates, e.g. $v = 0.1$ V s$^{-1}$, correspond to $n = 2$, i.e. to the overall stoichiometry of the reaction. This effect of the sweep rate is illustrated in Fig. 6.25 and it is seen that the value of $i/\sqrt{v}$ at $v = 100$ V s$^{-1}$ is approximately half of that observed at $v = 0.1$ V s$^{-1}$.

The number of electrons exchanged on a time scale similar to that of a preparative electrolysis is determined by coulometry. A coulometry experiment involves the complete conversion of the substrate to product(s) and, accordingly, $C_O^*$ decreases with time, in principle to zero. This is in contrast to the electroanalytical methods where $C_O^*$ stays essentially constant during the experiments. Coulometry is carried out at either constant potential or constant current and, usually, the solution is stirred magnetically.

During coulometry at constant potential, the current decays exponentially in the manner of a first-order reaction. The amount of charge passed after the time $t$ is obtained by integration of the current–time curve by, for example, a digital integrator, often referred to as a coulometer. The improved selectivity obtained by potential control during the coulometry experiment is paid for by a rather long electrolysis time. In principle, the end point ($i = 0$) is not reached until infinite time has elapsed. In practice, however, the electrolysis is stopped when the current has decayed to a few per cent of the initial current. The error introduced in this way may be neglected for all practical purposes.

Coulometry at constant current is often considered as being less attractive than coulometry at constant potential. However, when the current density is low, the potential of the working electrode stays almost constant until approximately 90% of the substrate is consumed. Control of the current rather than the potential has, however, a number of advantages. First, the charge consumed during the reaction is directly proportional to the electrolysis time,

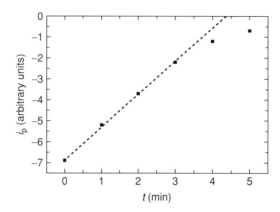

**Fig. 6.26**  Values of $i_p^{ox}$ recorded by voltammetry at 1-minute intervals during constant current coulometry (25 mA) of 2,4-dimethylpyrrole (0.1 mmol) in MeCN/Bu$_4$NPF$_6$ (0.1 M) [42].

and second, simple electronic equipment, such a DC current source, may be used instead of a potentiostat. Considering that 1 F equals 96 486 C, it is easily seen that the total conversion of, for example, 0.1 mmol of a substrate at $i = 25$ mA requires an electrolysis time of 6.43 minutes for $n = 1$. The decay in the substrate concentration during the electrolysis is conveniently followed by measurements of the LSV peak height of the substrate at regular time intervals. Thus, the coulometry experiment is typically carried out in the following way. First a voltammogram is recorded, without stirring, before electrolysis is started. Then, stirring is initiated and the constant current is switched on. After, say, 1 minute, the current and the stirring are switched off and a new voltammogram is recorded, this time corresponding to $t = 1$; this is continued until the peak height of the voltammogram shows that the substrate is essentially consumed. Extrapolation of the plot of $i_p$ versus $t$ to $i_p = 0$ gives the value of $t$ corresponding to the amount of charge for the total consumption of the substrate, and thus the value of $n$ is easily calculated. The results shown in Fig. 6.26 are for the oxidation of 0.1 mmol of 24 dmp at $i = 25$ mA. It is seen that the current decays in an essentially linear fashion for the first 3 or 4 minutes, and extrapolation to $i_p = 0$ results in $t = 4.3$ minutes and $n = 0.67$. This result illustrates that non-integer $n$-values may be observed, which is not unusual in studies of reactions that proceed through complex reaction mechanisms.

## A.4 Preparative or semi-preparative electrolysis, identification of products

It may seem unnecessary to stress that it is important to know the product or the product composition of any electron transfer reaction before it is meaningful to discuss the kinetics and mechanism (see Chapter 2). Nevertheless, this simple rule is often violated. One reason probably is that it is not always a trivial task to isolate and identify the products from an electrochemical reaction using the usual arsenal of methods available to the organic chemist. A major problem is that often it is not an easy task to separate the product from the supporting electrolyte. In such cases, direct analysis of the product mixture without the need of a work-up, e.g. by LC-UV/vis-MS, is desirable. Analysis by this method of the product mixture resulting from the oxidation of 1,2,5-trimethylpyrrole is shown in Fig. 6.27. It is

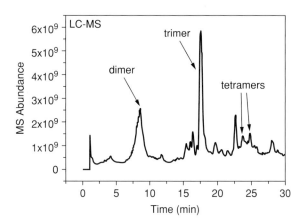

**Fig. 6.27**  LC-ESI/MS chromatograms of a solution resulting from the electrochemical oxidation of 1,2,5-trimethylpyrrole in MeCN/LiCF$_3$SO$_3$ (0.025 M) [42].

seen that the oxidation of this particular pyrrole leads to a mixture of a dimer and a trimer as the major products, but that several tetramers plus numerous other minor products are also formed.

# References

1. Lund, H. and Hammerich, O. (Eds) (2001) *Organic Electrochemistry* (4th edn). Dekker, New York.
2. Schäfer, H. (Ed.) (2005) *Organic Electrochemistry* (Volume 8 of *Encyclopedia of Electrochemistry*, A.J. Bard and M. Stratmann (Eds)). Wiley, New York.
3. Balzani, V. (Ed.) (2001–2002) *Electron Transfer in Chemistry* (vols 1–5). Wiley-VCH, Weinheim.
4. Horspool, W. and Lenci, F. (Eds) (2003) *Handbook of Organic Photochemistry and Photobiology* (2nd edn). CRC Press, New York.
5. Kochi, J.K. (1994) *Advances in Physical Organic Chemistry*, **29**, 185.
6. Evans, D.H. (1990) *Chemical Reviews*, **90**, 739.
7. Gray, H.B. and Winkler, J.R. (1996) *Annual Review of Biochemistry*, **65**, 537.
8. Birch, A.J. (1996) *Pure and Applied Chemistry*, **68**, 553.
9. Lund, H. (2001) Amalgam and related reductions. Chapter 28 in: H. Lund and O. Hammerich (Eds) *Organic Electrochemistry* (4th edn). Dekker, New York.
10. Sheldon, R.A. and Kochi, J.K. (1981) *Metal-Catalyzed Oxidations of Organic Compounds*. Academic Press, New York.
11. Eberson, L. (1987) *Electron Transfer Reaction in Organic Chemistry*. Springer, Berlin.
12. Scott, K. (2002) Electrochemistry and sustainability. In: D.J.H. Clark and D. Macquarrie (Eds) *Handbook of Green Chemistry and Technology*. Blackwell, Oxford, pp. 433–65.
13. Margaretha, P. and Tissot, P. (1977) *Organic Syntheses*, **57**, 92.
14. Chapuzet, J.M., Lasia, A. and Lessard, J. (1998) Electrocatalytic hydrogenation of organic compounds. In: J. Lipkowski and P.N. Ross (Eds) *Electrocatalysis*. Wiley-VCH, New York, pp. 155–96.
15. Leftin, J.H., Redpath, D., Pines, A. and Gil-Av, E. (1973) *Israel Journal of Chemistry*, **11**, 75.
16. Lund, H. and Hobolth, E. (1976) *Acta Chemica Scandinavica*, **B30**, 895.
17. Shono, T., Matsumura, Y. and Tsubata, K. (1985) *Organic Syntheses*, **63**, 206.
18. Baizer, M.M. and Chruma, J.L. (1972) *Journal of Organic Chemistry*, **37**, 1951.

19. Ronlan, A., Hammerich, O. and Parker, V.D. (1973) *Journal of the American Chemical Society*, **95**, 7132.
20. White, D.A. (1981) *Organic Syntheses*, **60**, 1.
21. White, D.A. (1981) *Organic Syntheses*, **60**, 58.
22. Bewick, A., Coe, D.E., Mellor, J.M. and Owton, W.M. (1985) *Journal of the Chemical Society, Perkin Transactions 1*, 1033.
23. Utley, J.H.P. and Gruber, J. (2002) *Journal of Materials Chemistry*, **12**, 1613.
24. Cedheim, L. and Eberson, L. (1976) *Acta Chemica Scandinavica*, **B30**, 527.
25. Amatore, C., Chaussard, J., Pinson, J., Saveant, J.M. and Thiebault, A. (1979) *Journal of the American Chemical Society*, **101**, 6012.
26. Bard, A.J. and Fan, F.R. (1996) *Accounts of Chemical Research*, **29**, 572.
27. Hodes, G. (Ed.) (2001) *Electrochemistry of Nanomaterials*. Wiley-VCH, Weinheim.
28. Pütter, H. (2001) Industrial electroorganic chemistry. Chapter 31 in: H. Lund and O. Hammerich (Eds) *Organic Electrochemistry* (4th edn). Dekker, New York.
29. Nielsen, M.F. (2005) Electrogenerated acids and bases. Chapter 14 in: H. Schäfer (Ed.) *Organic Electrochemistry* (Volume 8 of *Encyclopedia of Electrochemistry*, A.J. Bard and M. Stratmann (Eds)). Wiley, New York.
30. (a) Nielsen, M.F. and Hammerich, O. (1992) *Acta Chemica Scandinavica*, **46**, 883 and references cited therein; (b) Nielsen, M.F. and Hammerich, O. (1989) *Acta Chemica Scandinavica*, **43**, 269.
31. Maran, F., Wayner, D.D.M. and Workentin, M.S. (2001) *Advances in Physical Organic Chemistry*, **36**, 85.
32. Lund, H., Daasbjerg, K., Lund, T., Occhailini, D. and Pedersen, S.U. (1997) *Acta Chemica Scandinavica*, **51**, 135.
33. Bard, A.J. and Faulkner, L.R. (2001) *Electrochemical Methods: Fundamentals and Applications* (2nd edn). Wiley, New York.
34. Amatore, C. (1995) Electrochemistry at ultramicroelectrodes. Chapter 4 in: I. Rubinstein (Ed.) *Physical Electrochemistry*. Dekker, New York.
35. Amatore, C., Gareil, M. and Savéant, J.-M. (1983) *Journal of Electroanalytical Chemistry*, **147**, 1.
36. Nadjo, L. and Savéant, J.-M. (1973) *Journal of Electroanalytical Chemistry*, **48**, 113.
37. Parker, V.D. (1986) Electrochemical applications in organic chemistry. Chapter 2 in: A.J. Fry and W.E. Britton (Eds) *Topics in Organic Electrochemistry*. Plenum, New York.
38. Fussing, I., Güllü, M., Hammerich, O., Hussain, A., Nielsen, M.F. and Utley, J.H.P. (1996) *Journal of the Chemical Society, Perkin Transactions 2*, 649.
39. Hansen, G.H., Henriksen, R.M., Kamounah, F.S., Lund, T. and Hammerich, O. (2005) *Electrochimica Acta*, **50**, 4936.
40. Howell, J.O. and Wightman, R.M. (1984) *Journal of Physical Chemistry*, **88**, 3915.
41. Montenegro, M.I. and Pletcher, D. (1988) *Journal of Electroanalytical Chemistry*, **248**, 229.
42. Hansen, G.H., Lund, T. and Hammerich, O., unpublished work.

# Chapter 7
# Computational Chemistry and the Elucidation of Mechanism

*Peter R. Schreiner*

## 7.1 How can computational chemistry help in the elucidation of reaction mechanisms?

### 7.1.1 General remarks

The quote by Schleyer that "computational chemistry is to model all aspects of chemistry by calculation rather than experiment" tells us that practically every mechanistic question can be tackled by computational methods. This is true in principle, but it says nothing about the *quality* of the computed numbers, and Coulson said, "Give us insights, not numbers", which emphasises this point and relates to the fact that it is *easy* – fortune or curse – to compute numbers. This chapter presents some guidelines regarding the value and the interpretation of the 'numbers' when it comes to elucidating reaction mechanisms with computational chemistry approaches.

"Why theory?" some hard-core mechanistically orientated experimentalists will ask. "Everything is a theory" might be a theoretician's trite answer. The truth is somewhere in between, and we must acknowledge that there is no such thing as an experiment without an underlying theory for its interpretation. For instance, the chemist's much loved crystal structures are the results of theoretically solving (nowadays computationally) the Bragg relationships for determining electron densities from diffraction measurements, and matching these with electron density distributions. The $R_w$ values then tell us how well theory and experiment for a structural proposal agree. A low value lends high credibility to the proposed result and this is often taken to indicate that the pleasing static picture tells us just about everything we wish to know about the structure of a particular molecule. We all know that this is far from being true: all molecules vibrate, even at the absolute zero. That is, the X-ray structure (and that by every other elucidation method) is an *averaged* picture – perhaps an excellent one – of the molecular structure as it was determined by matching experiment and theory. The term *structure* itself is in that sense somewhat ill defined because it rests on the theoretical assumption that the nuclear and electronic motions can be cleanly separated. Invoking this so-called Born–Oppenheimer approximation is the dividing line between chemistry and physics, and a healthy theoretical approximation to understand chemistry at the structural level.

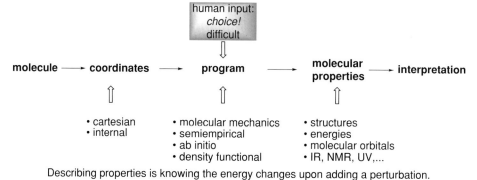

Chemistry is knowing the energy as a function of the nuclear coordinates.

• cartesian
• internal

• molecular mechanics
• semiempirical
• ab initio
• density functional

• structures
• energies
• molecular orbitals
• IR, NMR, UV,...

Describing properties is knowing the energy changes upon adding a perturbation.

**Fig. 7.1**  Typical workflow for using computational chemistry.

Computational methods typically employ the Born–Oppenheimer approximation in most electronic structure programs to separate the nuclear and electronic parts of the Schrödinger equation that is still hard enough to solve approximately. There would be no potential energy (hyper)surface (PES) without the Born–Oppenheimer approximation – how difficult mechanistic organic chemistry would be without it!

Explicitly optimising molecular structures at rest and adding the effects of internal energy distributions such as vibrations and rotations later is a huge advantage because of the separability of interactions. Computational approaches put a handle on such imprecise but important concepts as steric hindrance, electrostatics, partial charges, strain, aromaticity and others. The convenience of such descriptors can be related to proper physical phenomena that are accessible by mathematically and physically sound computations. The conceptual approach for using computations to elucidate reaction mechanisms is outlined in Fig. 7.1. The changes in the energy while changing the nuclear coordinates reveal the potential energy surface of molecules of a particular composition. The chemical properties then follow as a result of perturbations to the structures and energies of the *stationary points* (zero gradient, i.e. minima or maxima on the PES) along the *reaction path*.

Computational methods have already become an indispensable tool for the elucidation of reaction mechanisms. In the following, rather than offering an extensive introduction to the growing and ever changing arsenal of computational chemistry methods, we introduce the basic concepts of the underlying theories and make some general remarks regarding typical strengths and weaknesses. There are excellent books on computational methodology that give details on the most commonly used methods and their underlying theory [1, 2]. Molecular dynamics (MD) approaches, important as they are for an even better understanding of reaction mechanisms, will not be covered in the present chapter because ab initio MD studies are currently restricted to rather small molecular systems.

### 7.1.2    Potential energy surfaces, reaction coordinates and transition structures

Potential energy surfaces are at the heart of all mechanistic chemical interpretations. Nevertheless, it is important to distinguish between pictorial presentations of mechanistic

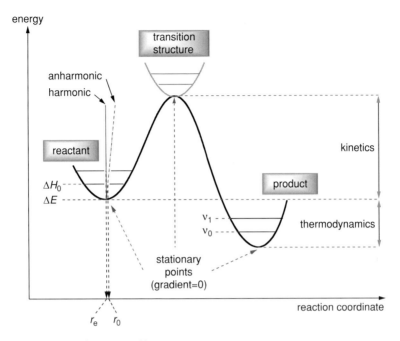

**Fig. 7.2** Generic potential energy profile.

proposals that agree with experimental measurements, and what is computed. Virtually all single-molecule, non-dynamic computations, as they are typically employed for examining reactive intermediates along a reaction path, silently *but incorrectly* imply that it is straightforward to convert atomic and molecular properties of single, isolated molecules to a property of an entire mole (e.g. energy in kcal mol$^{-1}$). Computing thermodynamic properties relies on being able to make this transition but we should not forget that it is a huge leap of faith of partition functions and statistical mechanics.

The computational construction of a potential energy surface is similar to the construction of a reaction profile from kinetics and thermodynamics experiments (Fig. 7.2). Several details should be pointed out, however, because they often lead to misunderstandings or even disputes between experimentalists and computational chemists. As noted above, experimental structures are averaged structures because the molecules *anharmonically* vibrate. As a consequence, an experimental bond length ($r_0$) by various experimental approaches is *longer* than the computed bond length corresponding to the bottom of the energy minimum (equilibrium structure, bond length $r_e$) for the molecule at rest because $r_0$ values are determined for the zeroth vibrational energy levels. Furthermore, most computations accompanying mechanistic studies employ the *harmonic* approximation, and a correction to $r_0$, although straightforward in principle (but computationally demanding), is not carried out. Hence, computed and experimental structures *cannot* agree perfectly, unless much care is taken in correcting for vibrational and other (e.g. relativistic) effects. At the same time, one should be very sceptical if computed bond lengths are considerably *longer* than those of the corresponding experimental ones.

Molecular species in the *transition state* are usually approximated by a single molecular *transition structure* (TS) although we know that a *state* consists of a large number of structures with Gaussian energy distributions (in classical transition state theory) under specified conditions of temperature, pressure, etc. Again, this is a huge leap of faith, and it is remarkable that many energy barriers associated with computed transition structures agree so well with measured barriers. Although this may be due to fortuitous error cancellation in many instances, it is comforting to see that it is the rule rather than the exception.

Transition structures are characterised by one and only one imaginary vibrational frequency; highly symmetric molecules may display a degenerate set of several imaginary frequencies. Higher order saddle points with more than one imaginary frequency are usually chemically irrelevant but exceptions are known for highly symmetric cases when, for instance, the downhill path from a transition structure leads onto a ridge with a bifurcation into several degenerate minima [3]. Finding a transition structure computationally is often considered an art because it is not easy to minimise all coordinates but one. This also has to do with the notion of the *reaction coordinate* which is the lowest energy pathway between the various stationary points. Sometimes it is trivial to identify a single vibration as being an extremely good description of the reaction coordinate (e.g. in bond-breaking processes) but usually the true reaction coordinate is a combination of several vibrations whose relative contributions to the developing imaginary vector might even change along the reaction path (e.g. in rearrangements). In terms of finding transition structures, there are many strategies and the best is perhaps the one relying on chemical intuition and following some simple empirical rules. Automated transition structure generators are available but usually operate by geometrically averaging the coordinates of starting materials and products. This leaves out important information; for example, the overall energy change will tell knowledgeable chemists whether to expect an early or late transition structure according to the Hammond postulate. Second, symmetry may be an excellent help in finding transition structures. Consider the degenerate $S_N2$ reaction between a fluoride ion and methyl fluoride. While the latter has $C_{3v}$ symmetry, we immediately recognise that the TS is of even higher symmetry ($D_{3h}$) as it must lie along the path between the identical minima. As all bending coordinates for the TS are thus completely fixed, the optimisation proceeds quickly to the expected transition structure. Symmetry should always be employed for seeking transition structures because even if a highly symmetric TS turns out to have more than one imaginary frequency, it is easy to inspect the lowest of several imaginary modes, follow it to the lower symmetry and thus find the correct TS faster. Another useful device is to go backwards from the TS, instead of thinking in the forward direction from a starting material to the TS; this is especially helpful for reactions highly endergonic in the forward direction. Finally, the choice of coordinates will also affect the expeditious locating of transition structures. Internal coordinates are advantageous over Cartesian ones because they not only imply molecular structure but also make the interpretation of vibrational modes much easier (one of them will be a major contributor to the reaction coordinate).

## 7.1.3   Absolute and relative energies; isodesmic and homodesmotic equations

As outlined below, molecular mechanics and semiempirical computations directly report heats of formation, a convenience that is not directly available in ab initio computations, which give *absolute electronic energies* (in atomic units, au). These have little meaning to

$$C_3H_8 \quad + \quad CH_4 \quad \longrightarrow \quad 2\,C_2H_6$$

| | | | |
|---|---|---|---|
| $\Delta_f H^0$ (kcal mol$^{-1}$) | ? | $-17.9$ | $-20.0$ |
| E[B3LYP/6-31G(d)]($-$au) | 119.14425 | 40.51839 | 79.83042 |

$$\Delta H_R = 2\Delta_f H^0\,(C_2H_6) - \Delta_f H^0\,(CH_4) - \Delta_f H^0\,(C_3H_8) =$$

$$2\,E\,(C_2H_6) - E\,(CH_4) - E\,(C_3H_8) = +1.13 \text{ kcal mol}^{-1}$$

$$\Delta_f H^0\,(C_3H_8) = \Delta H_R + \Delta_f H^0\,(CH_4) - 2\,\Delta_f H^0\,(C_2H_6) = 24.1 \text{ kcal mol}^{-1}$$

**Scheme 7.1**   Isodesmic equation for the conversion of methane and propane into ethane (the abbreviation for the theoretical method, B3LYP/6-31G(d), indicates the use of Becke's three-parameter hybrid exchange functional with the Lee–Yang–Parr correlation functional and a double-$\zeta$ Pople-type split valence basis set with one set of d-type polarization functions).

chemists directly but *differences* lead to meaningful quantitative energy changes. Heats of formation can also be computed indirectly very accurately by ab initio methods through various approaches; two are particularly straightforward: (i) from atomisation energies and (ii) using *isodesmic* or *homodesmotic* equations.

Atomisation energies rely on the ability to compute very accurate energies for atoms and molecules, something that is not easily achieved. An isodesmic equation is a hypothetical reaction where the number and type of bonds on the reactant side equal the number and type on the product side. As the number of bonds of each given type on both sides of the reaction is the same, the energies of isodesmic reactions can be computed quite accurately even with relatively simple theoretical models because of systematic error cancellation. Predicting the heat of formation of unknown organic compounds is, therefore, straightforward. As the heats of formation of all but one component, the compound of interest, are known, one may solve for the heat of formation of this unknown component. Consider the following simple example: you know the heats of formation of methane and ethane and would like the $\Delta_f H^0$ of propane (Scheme 7.1).

The agreement with the experimental result for propane ($\Delta_f H^0 = 25.0$ kcal mol$^{-1}$) is excellent and almost within error bars even with this relatively fast density functional theory approach, B3LYP/6-31G(d). This example works so well because it is formally a *homodesmotic* equation; this is a special type of isodesmic equation where not only are all bonds the same, but the hybridisation of the atoms that make up the bonds is also the same (here, sp$^3$). Homodesmotic equations give more accurate results when compared with other isodesmic equations since the structural similarity between products and reactants is greater leading to better error cancellation. Isodesmic and homodesmotic equations (there are also other types of error-cancelling equations) generally not only yield more accurate heats of formation than atomisation energies but can also be used for computing a large variety of other thermodynamic data, even for large systems. For instance, if one wishes to compute aromatic stabilisation energies, the equation in Scheme 7.2, which does not even need computer-generated numbers (all heats of formation were taken from the NIST database), might be quite suitable. The result of 33.6 kcal mol$^{-1}$ for the aromatic stabilisation energy of benzene calculated simply from heats of formation is quite acceptable and agrees reasonably well with values computed from the three different levels of theory shown.

| | | | | | $\Delta H_R$ |
|---|---|---|---|---|---|
| $\Delta_f H^0$ (kcal mol$^{-1}$) | +25.00 | −1.03 | +19.82 | −29.43 | **−33.6** |
| HF/6-31G(d) (−au) | 231.83190 | 233.01965 | 230.70314 | 234.20801 | **−37.4** |
| B3LYP/6-31G(d) (−au) | 233.41894 | 234.64829 | 232.24866 | 235.88044 | **−38.8** |
| MP2/6-31G(d) (−au) | 232.59362 | 233.79141 | 231.45773 | 234.99243 | **−40.9** |

**Scheme 7.2**  Homodesmotic equation for conversion of cyclohexa-1,3-diene and cyclohexene into benzene and cyclohexane.

## 7.2   Basic computational considerations

Applying computational techniques to chemical problems first requires a careful choice of the theoretical method. Basic knowledge of the capabilities and the drawbacks of the various methods is an absolute necessity. However, as no practical chemist can be expected to be well versed in the language and fine details of computational theory, we approach the subject by briefly (and by no means completely) reminding the reader of the underlying concepts of particular methods and of their often less well-documented limitations.

### 7.2.1   Molecular mechanics

Molecular mechanics (MM) [4] force fields (MMFF), originating in the 1930s [5], typically use empirical formulae to approximate the interatomic interactions in an average fashion with a variable set of corrective terms to account for electronic interactions that are not included by design; force fields combining molecular mechanics and electronic structure theory have also been developed [6]. Empirical corrections are usually derived from comparisons with experimental data or (these days) high-level computations. Owing to the simplicity of the model, these computations are very fast and allow very large systems to be investigated. Because of the enormous effort that goes into the parameterisation, practically all force fields are tailor-made for particular classes of compounds (hydrocarbons, carbohydrates, peptides, etc.). While the first force fields were explicitly developed for determining the heats of formation of hydrocarbons, these $\Delta_f H^0$ values as well as the corresponding structures can be computed very accurately nowadays with ab initio methods so that MMFF approaches are mostly used currently for very large structures (and their large numbers of isomers) for the evaluation of their macroscopic properties. Every force field is as good as its parameterisation set and vice versa, implying that it is always possible to develop a new and highly accurate force field for a particular set of compounds. This force field should then *only* be applied to compounds that are comparable in their structures and properties. MMFF computations are attractive for the quick determination of molecular structures and heats of formation that are typically in excellent agreement with results of higher level computations, but at very much lower computational costs. The computed structures can also be characterised as minima by computing the vibrational frequencies through second derivatives of the energy with respect to all coordinates; maxima, i.e. transition structures, are typically not part of the parameterisation step and should therefore *not* be computed by common force field methods. These harmonic vibrational frequencies, as well as the computed heats of formation, can be used to judge the quality of the force field employed and to aid in the experimental structure elucidation.

*In a nutshell (MM2 force field)*: No electrons, purely mechanical model • Structural accuracy: bond lengths, 0.01 Å; bond angles, 1° ; torsion angles, a few degrees • Conformational energies: accurate to 1 kcal mol$^{-1}$ • Vibrational frequencies: accurate to 20–30 cm$^{-1}$.

## 7.2.2 Wave function theory

The description of electron motion and electronic states that is at the heart of all of chemistry is included in wave function theory, which is also referred to as self-consistent-field (SCF) or, by honouring its originators, Hartree–Fock (HF) theory [7]. In principle, this theory also includes density functional theory (DFT) approaches *if* one uses densities derived from SCF densities, which is common but not a precondition [2]; therefore, we treat density functional theory in a separate section. Many approaches based on wave function theory date back to when desktop supercomputers were not available and scientists had to reduce the computational effort by approximating the underlying equations with data from experiment. This approach and its application to the elucidation of reaction mechanisms are outlined in Section 7.2.3.

An important point is also that the wave function cannot be interpreted directly because it simply contains too much information and is not in human-readable format. Only the *electron density* has meaning in the sense that the normalised square of the total wave function represents the electron density. There are many ways to interpret wave functions, and molecular orbitals, which are used to construct the total wave function, are the most obvious. It should not be forgotten, however, that *canonical* (delocalised) orbitals are typically used for computations while *localised* orbitals are much preferable for structural interpretation. This requires a localisation step that introduces some degree of arbitrariness; orbitals themselves are only small parts of the entire wave function information, so they should be interpreted with great care. A similar situation arises for many very useful properties such as atomic or group charges, polarisability, electrostatic potentials, none of which are the direct result (expectation value) of a quantum mechanical operator. That is, the wave function does not yield these quantities *directly*; they have to be derived by an arbitrary scheme for partitioning the wave function. Different approaches yield different results; the nowadays old-fashioned Mulliken population analysis, which also gives Mulliken charges, is very sensitive to the choice of basis set and fails for highly delocalised structures. Natural bond orbital (NBO) analysis is much preferred but still not perfect. This is demonstrated with the rotational barrier of ethane, a case study below.

## 7.2.3 Semiempirical methods

In contrast to molecular mechanics force fields, modern semiempirical methods are classified as an SCF electron-structure theory (wave function-based) method [8]. Older (pre-HF) semiempirical approaches, such as extended Hückel theory, involve drastic approximations and rely on the researcher's intuition and ability to extrapolate from simple computations to meaningful chemistry.

In line with chemists' valence electron models of chemical bonding, only valence electrons are considered in the computational procedures. To avoid the computation of the interaction of every electron with every other electron, multicentre integrals are neglected and empirical corrections based on experimental results are added. The integrals are either determined directly from experimental data or are computed from the corresponding analytical formulae or suitable parametric expressions. These simplifications speed up the computations considerably and still include most electronic effects; in particular, electron correlation, which is not introduced explicitly but 'through the back door' by means of fitting the parameters to experimental data, is included. The interpretation of semiempirical computations is straightforward as all typically relevant properties can be computed using the underlying physical principles (eigenvalues of operators). There are many flavours of semiempirical theory and they continue to be developed because many very large structures simply cannot be computed with ab initio or DFT methods [9]. The common abbreviations stand either for the place of their development (e.g. the popular AM1, Austin Model 1) [10] or for their level of approximation (e.g. MNDO, modified neglect of differential diatomic overlap) [11]. As semiempirical molecular orbital methods are parameterised to reproduce experimental reference data or high-level theoretical predictions, reference properties directly related to the intended applications should be selected. The quality of semiempirical results depends strongly on the effort put into the parameterisation [12].

In its widely used form, the MNDO approximation and its variants (AM1 and PM3) are suitable for the computation of properties of compounds of first row elements. The computational effort (often referred to by computational chemists as 'cost') scales approximately with $N^2$ ($N$ = number of particles – nuclei and electrons). These methods serve as efficient tools for searching large conformational spaces, e.g. as a preliminary to subsequent higher level computations; however, errors in the computations are less systematic than in ab initio methods. This is particularly evident when an error cannot be related to a physically measurable quantity.

A key advantage of semiempirical methods is that they give heats of formation directly. Small cyclic hydrocarbons are typically computed to be too stable, and sterically crowded structures are predicted to be too unstable. This is because semiempirical methods do not describe weak interactions well, e.g. those arising from London dispersion forces; thus, they would not be suitable to describe, for instance, molecular structures that rely heavily on hydrogen bonding interactions.

*In a nutshell (AM1)*: Valence electrons only; neglect of multicentre integrals; empirical corrections • Faster than ab initio or DFT methods: $N^2$ scaling for energies, $N^3$ for gradients • Mean absolute deviations: 6.3 kcal mol$^{-1}$ for heats of formation; 0.014 Å for bond lengths; 2.8° for bond angles; 0.48 eV for first ionisation potentials • Sterically crowded molecules appear too unstable • Four-membered rings appear too stable • Hypercoordinate compounds appear too unstable • Nonclassical structures normally appear too unstable compared with corresponding classical structures • Rotational barriers are often underestimated • Hydrogen bonds appear too weak and too long • Calculated activation barriers are typically somewhat too high • Pericyclic reactions: biradicaloid mechanisms favoured.

## 7.2.4  Hartree–Fock theory

Self-consistent field computations in which every electron is treated as though it experiences all the other electrons as an average field are referred to as Hartree–Fock (HF) methods. HF theory minimises the interactions of all electrons with respect to the atomic coordinates, and it adheres to the variational theorem, i.e. any arbitrary function used in the Schrödinger equation will give an energy ($E_{HF}$) that is *higher* than the 'true' energy ($E$) of the real system. Hence, the energies computed with HF theory present an upper limit of the true energy.

The energy difference between the true energy of a system and $E_{HF}$ is the electron correlation energy ($E_{corr}$), i.e. $E = E_{HF} + E_{corr}$, and it arises from the spontaneous mutual repulsion of electrons that goes beyond the average field assumption; all so-called post-HF methods attempt to recover this small but important energy term. Part of the idea of semiempirical theories was to parameterise the semiempirical equations in such a way that most of the correlation energy can be recovered very simply. Naturally, this led to a division of theory with one side trying to solve the Schrödinger equation as rigorously as possible while the other tries to devise efficient and practicable ways to utilise the theory to determine molecular properties.

The omission of electron correlation has serious consequences. As the spontaneous repulsion of electrons occupying the same space at the same time is only averaged in HF theory, geometries at this level usually have bond lengths that are too short. Highly delocalised systems such as carbocations, polycyclic arenes and strained rings are generally not described well. Bond-breaking/bond-making processes are also described poorly because the effects of multideterminant descriptions are not included in the basic HF formalism, so dissociation energies have larger errors.

Despite these deficiencies, HF theory is extremely useful. First of all, virtually all methods with explicit electron correlation rely on a HF reference wave function that is amended with elaborate schemes to evaluate the correlation energy. Many DFT approaches also determine the HF electron density first and then use it as the operand in the DFT treatment. Moreover, systematic errors such as constant amounts of electron correlation energy that are omitted in, for instance, comparisons of relative energies of isomers, often cancel and HF energies allow good estimates of the relative energies of structurally related species. The use of balanced but hypothetical isodesmic equations makes use of this advantageous error cancellation. This approach is particularly useful for similar structures for which the errors in the bond energies, etc., are assumed to be transferable and therefore likely to cancel when isodesmic equations are used.

In contrast to semiempirical theory, HF deals reasonably well with triple bonds and strained rings (Table 7.1). While the error for the isomer energy differences between hydrogen cyanide and hydrogen isocyanide (no. 1), between their methylated analogues (no. 2) and between propene and cyclopropane (no. 3) are just about acceptable, the energy difference between two seemingly simple molecules such as acetaldehyde and ethylene oxide (no. 4) is poorly described at this level of theory. For an open-shell species such as hydroxymethylene with a singlet ground state that requires multideterminant treatment (the singlet state must be described by several distributions of the two paired electrons into two non-equivalent orbitals), the errors can be enormous (no. 5).

**Table 7.1** Energy differences (kcal mol$^{-1}$) between isomers computed at the HF/6-31G(d) level compared with experimental values.

| No. | Isomers | HF/6-31G(d) | Experiment | Error |
|-----|---------|-------------|------------|-------|
| 1 | H—C≡N and $^-$C≡N$^+$—H | 12.3 | 14.5 | 2.1 |
| 2 | CH$_3$C≡N and $^-$C≡N$^+$—CH$_3$ | 20.8 | 20.9 | 0.1 |
| 3 | H$_2$C═CH—CH$_3$ and △ | 8.2 | 6.9 | 1.3 |
| 4 | CH$_3$CHO and △ | 33.4 | 26.2 | 7.2 |
| 5 | H$_2$CO and H—C—OH | 52.6 | 4.9 | 47.7 |

*In a nutshell (HF/6-31G(d))*: Bond lengths and angles of 'normal' organic molecules quite accurate (within 2%) • Conformational energies accurate to 1–2 kcal mol$^{-1}$ • Vibrational stretching frequencies for most covalent bonds systematically 10–12% too high • Errors in zero point vibrational energies (ZPVEs): ∼1–2 kcal mol$^{-1}$ • Isodesmic reaction energies accurate to 2–5 kcal mol$^{-1}$ • Entropies accurate to 0.5 cal K$^{-1}$ mol$^{-1}$ • Protonation/deprotonation energies (gas phase): errors ∼10 kcal mol$^{-1}$ • Atomisation and homolytic bond-breaking reactions: errors 25–40 kcal mol$^{-1}$.

### 7.2.5    *Electron-correlation methods*

Although HF theory is quite remarkable in describing molecular structures and energies, it is clear that the simplified assumption of one electron moving in the average field of all the others is not sufficient to take into account important electronic effects that make all the difference when describing phenomena such as delocalisation, polarisation, aromaticity and so forth. *Relative* energy comparisons alleviate many of these inadequacies *if* the computational errors are transferable between the different species that are being compared. This is often the case for energy differences between isomers but certainly not for complex mechanisms that involve several bond-breaking and bond-making steps.

The simplest way of including electron correlation is *Møller–Plesset* (MP) *perturbation theory* (also referred to as MBPT = many-body perturbation theory), which introduces electron correlation as a small perturbation to the HF Hamiltonian [13]. The energy and wave function are expressed in infinite additive terms ($k$) which correspond to the order of the MP perturbation; the sum of the zero- and first-order energy terms represents the unperturbed HF energy. This series is often truncated with the second term (MP2, $k = 2$) with the serious consequence that the variational theorem is *not* fulfilled (the computed energy is *not* an upper limit of the true energy of a given system). Second, including higher levels of MP theory (MP3, MP4 and so on) leads to a non-convergent series at infinite order that does not converge to the true energy. One of the problems of MP theory is the underlying assumption that the perturbation is small; hence, if HF is grossly wrong for a particular problem, MP methods cannot be expected to be able to correct for all the deficiencies of the HF reference.

Bond dissociation energies (BDEs) are at the heart of organic thermochemistry and many mechanistic considerations, so it is imperative to be able to compute and predict these quantities accurately. The problem is that, upon lengthening a bond in the early stages of cleaving

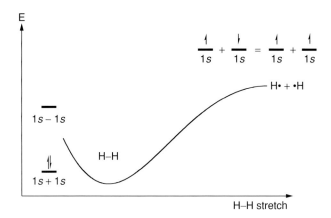

**Fig. 7.3**  Bond dissociation of molecular hydrogen and electronic changes.

it, the electronic structure undergoes extreme changes. Consider the trivial example of $H_2$ in Fig. 7.3: in its molecular state, the two electrons are paired and occupy, with opposite spins, one low-lying molecular orbital that is derived from the constructive (binding) overlap of the two 1s orbitals in equal parts; another antibonding MO is considerably higher in energy but unoccupied. The energy difference between this singlet state and its corresponding triplet (at the same geometry) is large. When the molecule is fully dissociated, we have two orbitals, degenerate in energy and shape, with one electron in each (Hund's rule); the energy difference between singlet and triplet states is now zero. Even the overall endothermicity of the reaction will be difficult without electron correlation because a closed-shell singlet state ($H_2$) will have a *different* amount of electron correlation energy from two individual open-shell doublet radicals (H). MP theory does an excellent job in recovering major parts of the electron correlation energy and the improvement in going to just MP2 for BDEs (Table 7.2) is spectacular *although* the HF reference wave function is *not* good for determining corrections by small perturbations. While MP2 typically leads to an over-correction in one direction, MP3 often does the opposite and is not necessarily better; MP4 is more reliable (and considerably more elaborate) and brings the BDE values back closer to experiment.

**Table 7.2**  Bond dissociation energies (kcal mol$^{-1}$) of some important bond types at various levels of theory using a 6–31G(d,p) basis set and comparison with experiment.

| Reaction | HF | MP2 | MP3 | MP4 | Experiment |
|---|---|---|---|---|---|
| $H_2 \rightarrow H^\bullet + H^\bullet$ | 85 | 101 | 105 | 106 | 109 |
| $LiH \rightarrow Li^\bullet + H^\bullet$ | 45 | 48 | 49 | 58 | 32 |
| $BeH \rightarrow Be^\bullet + H^\bullet$ | 52 | 52 | 49 | 47 | 50,56 |
| $FH \rightarrow F^\bullet + H^\bullet$ | 93 | 131 | 127 | 128 | 141 |
| $^3CH_2 \rightarrow CH^\bullet + H^\bullet$ | 101 | 109 | 108 | 107 | 105,106 |
| $^1CH_2 \rightarrow CH^\bullet + H^\bullet$ | 70 | 89 | 90 | 91 | 98 |
| $H_2O \rightarrow OH^\bullet + H^\bullet$ | 86 | 119 | 115 | 116 | 126 |
| $BH_3 \rightarrow BH_2^\bullet + H^\bullet$ | 90 | 106 | 108 | 109 | 57–107 |
| $NH_3 \rightarrow NH_2^\bullet + H^\bullet$ | 83 | 110 | 108 | 109 | 116 |
| $CH_4 \rightarrow CH_3^\bullet + H^\bullet$ | 87 | 109 | 110 | 110 | 113 |

*Configuration interaction* (CI) is the next higher and theoretically logical development to include electron correlation. In CI theory, selected electron excitations are generated from the HF reference wave function in terms of configuration state functions (CSFs); most typical are single and double excitations (termed singles and doubles) leading to the acronym CISD. The inclusion of all excitations, which is only practicable for very small molecules [14], is then referred to as *full configuration interaction* (FCI). While CISD gives excellent results for small systems, it is generally not a suitable method for increasingly large structures because of the enormous computational effort (CISD scales with about $N^6$) and, more severely, because it is not size-extensive due to its truncation at the SD excitation level. That is, the amount of electron correlation energy recovered does not scale proportionally with molecular size. CI methods are, therefore, excellent benchmark methods for small molecules but they are impracticable for large structures.

Configuration interaction treatments can also be easily extended to multideterminant problems but these require even more care and an exorbitant computational effort [15]. Multi-reference CI approaches are ideally suited to determining, for example, electronic spectra, as they can deal with excited states in a straightforward fashion [16].

*Coupled cluster* theory (CC) [17], often referred to as the 'gold standard' of computational quantum chemistry, includes electron correlation by expanding the HF reference wave function through a cluster operator that is weighted by the cluster amplitude [18]. As with other electron correlation methods, the infinite series expansion is truncated at a certain level with single and double substitutions (CCSD) being currently the most practicable. The effects of triple excitations are often estimated by perturbation theory and this is expressed by placing 'T' in parentheses as in CCSD(T). This level of theory has proven to be extremely accurate provided that large well-balanced basis sets are used, e.g. DZP, TZP, or preferably those designed for high levels of electron correlation such as the cc-pVXZ series [19]. Although coupled cluster wave functions are non-variational, the beneficial effect of a finite basis set is typically much larger than the potential error introduced by not using a variational method; coupled cluster methods are size-extensive. The use of CC methods is hampered only by the exorbitant computational effort as CCSD(T), for instance, scales with $N^7$; so-called local or even linear scaling methods are being developed and show great promise for the application of this mathematically elegant and accurate approach to large systems [20]. A huge advantage of coupled cluster methods, compared with MP and CI treatments, is their insensitivity to the quality of the reference wave function. That is, even structures with pronounced multi-reference character can often be described with the formally single-determinant coupled cluster approach. Sometimes, problems arise from unrealistically large singles amplitudes when near-degeneracies of orbitals lead to state problems and CCSD(T) fails to give accurate answers; these problems are usually diagnosed by inspection of the so-called $T_1$ diagnostic terms that should be below 0.02 [21].

The corrections in going from HF to MP2 are large (Table 7.3) and the computational effort increases dramatically for small improvements in energies and geometries. With a limited basis set such as DZP, it is clear that CCSD(T), which should be superior to MP4, cannot live up to its promise because the basis set is too small. A major conclusion is that there must be a good balance between attempted amount of electron correlation recovery and basis set size. It is advisable to use at least triple-$\zeta$ quality type basis sets for highly correlated methods such as coupled cluster theory.

**Table 7.3**  Differences between bond lengths (Å) computed by various methods (with a double-$\zeta$ basis set) and experimental values; a minus sign indicates that the bond length is underestimated. The amount of electron correlation energy recovered (i.e. the quality of the energy calculation) improves from left to right.

| Molecule | Coordinate | HF | MP2 | MP4 | CCSD(T) |
|---|---|---|---|---|---|
| $H_2O$ | O—H | −0.006 | 0.005 | 0.005 | 0.005 |
| $NH_3$ | N—H | −0.011 | 0.002 | 0.004 | 0.004 |
| $CH_4$ | C—H | −0.001 | 0.003 | 0.006 | 0.006 |
| $C_2H_4$ | C—H | −0.002 | 0.006 | 0.008 | 0.008 |
|  | C=C | −0.012 | 0.024 | 0.025 | 0.023 |
| $H_2CO$ | C—H | −0.005 | 0.003 | 0.008 | 0.006 |
|  | C=O | −0.015 | 0.022 | 0.023 | 0.021 |
| HCN | C—H | −0.003 | 0.004 | 0.006 | 0.006 |
|  | C≡N | −0.017 | 0.032 | 0.030 | 0.024 |
| $CO_2$ | C=O | −0.015 | 0.022 | 0.028 | 0.018 |

*In a nutshell (electron correlated methods)*: Complete electronic description, straightforward interpretation • Highly accurate methods that can rival experiment for small organic molecules • Very time consuming • CI methods are not size-extensive; CC methods are non-variational • Slow convergence • Strongly basis set dependent • Systematic improvement straightforward.

## 7.2.6  *Density functional theory*

While Hartree–Fock theory is based on the proper construction of a complex wave function, the main objective of density functional theory (DFT) is to replace the many-body electronic wave function with the electron density as a simpler quantity to deal with. Although DFT has its conceptual roots in the uniform electron gas model and had been around in the physics community for a long time, it is only due to the work of Hohenberg and Kohn (HK theorem) who proved the direct relationship between the ground state electron density and the ground state electronic wave function of a many-particle system, that DFT is used for chemistry as well. According to the HK theorem, the ground state density minimises the total electronic energy of the system. The HK theorem is only an existence theorem, stating that a direct relationship exists, but it does not provide a way to establish it; hence, approximations for describing the density and its functional dependence on the energy have to be made. One of the oldest approaches is the local-density approximation (LDA), which is exact for the uniform electron gas. In practice, the HK theorem is rarely used for computational chemistry. Instead, the most common implementation of DFT is the Kohn–Sham method which reduces the intractable many-body problem of interacting electrons in a static external potential to a tractable problem of non-interacting electrons moving in an effective potential. The effective potential includes the external potential and the effects of Coulombic interactions between the electrons. As a consequence, DFT requires considerably less computer time and disk space than ab initio computations making it feasible to deal with much larger atom and molecular systems. Although it is in principle an exact quantum mechanical method (if the true functional is known), we treat DFT separately from traditional

**Table 7.4** Performance of B3LYP/6-31G(d) compared with standard ab initio methods (with a DZP basis) for two very difficult cases.

| | $F^\bullet + H_2 \rightarrow FH + H^\bullet$ | | $Cr_2$ properties | |
|---|---|---|---|---|
| | Barrier (kcal mol$^{-1}$) | $R_e$ (TS) (Å) | $D_e$ (kcal mol$^{-1}$) | Vibration (cm$^{-1}$) |
| HF | 14.2 | 1.465 | −19.4 | 1151 |
| CCSD | 3.3 | 1.560 | −2.9 | 880 |
| CCSD(T) | 2.3 | 1.621 | 0.5 | 705 |
| DFT | 3.6 | 1.590 | 1.5 | 597 |
| Experiment | 2.0 | 1.679 | 1.4 | 470 |

ab initio methods because the arbitrariness in choosing the functional combinations makes the logical and systematic improvement of DFT currently rather difficult [2]. And although this is not the place to elaborate on the various functionals, we should add a few words about B3LYP which is currently the most commonly applied DFT method for finite systems. For B3LYP, the weights of the exchange and correlation functionals (VWN and LYP) are determined by minimising the root mean square error over a molecular test set [22]; B3LYP has proven to be robust and generally applicable to a large variety of chemical problems. Also, while classical ab initio methods systematically overestimate barrier heights, B3LYP typically (but not necessarily systematically) underestimates them [23]. Some of these problems are evident when partial open-shell character is present in certain molecular structures for which the self-interaction part in the B3LYP functional is energetically overemphasised; while chemical structures are still acceptable, the energies can err significantly.

The robust and convincing performance of a 'cheap' DFT level such as B3LYP/6-31G(d) is emphasised by the data in Table 7.4, which presents some notoriously problematical cases. Considering the time and effort (i.e. 'cost', cf. Table 7.5), DFT does unbelievably well! This, however, does *not* imply that this level of theory treats every problem well; careful validation is always required, and this is particularly true of DFT methods.

**Table 7.5** Approximate average errors and timings of typical quantum chemical methods with a double-$\zeta$ basis set; $O$ = number of occupied orbitals; $N$ = number of particles (electrons/nuclei).

| Method | Approximate average error (kcal mol$^{-1}$)* | Approximate time factor |
|---|---|---|
| HF | 5–30 | $ON^{2-3}$ |
| DFT | 2–10 | $ON^3$ |
| MP2 | 17 | $ON^4$ |
| MP3 | 14 | $ON^{4-5}$ |
| MP4 | 4 | $ON^5$ |
| MP5 | 3 | $ON^6$ |
| CISD | 14 | $ON^6$ |
| CCSD | 4 | $O^2N^6$ |
| CCSD(T) | 1 | $O^2N^7$ |
| CCSDT | 0.5 | $O^2N^8$ |
| CCSDTQ | 0.0 | $O^2N^{>8}$ |

*Compared with full configuration interaction (FCI).

*In a nutshell (DFT methods)*: Electron density description of chemical phenomena, straightforward interpretation • Large molecular systems can be treated • Variable accuracy, validation necessary • Little basis set dependence • No systematic improvements.

## 7.2.7 Symmetry

As many small molecules are highly symmetric or are built up from repetitive symmetric subunits, symmetry should also be employed in the computations, not only to satisfy the experimental observations but also, more importantly, to speed up the computations by means of reducing the dimensionality of the coordinate space to be optimised with respect to the total energy. Consider cyclopropane which formally has a 21-dimensional potential energy surface that would be rather difficult to manage. In reality, we know that there are only three different internal degrees of freedom (assuming an equilateral triangle of carbons and overall $D_{3h}$ symmetry), namely the C—H and C—C bond distances, and the H—C—H bond angle. It is obvious that the proper consideration of symmetry very much reduces the computational complexity and should, therefore, always be employed. It often makes searches for transition structures much easier as the flexibility of a particular structure is reduced. A stationary point should in any event be characterised by computing the second derivative, i.e. the Hessian matrix in its full dimensionality, to make sure that it is a minimum (no imaginary frequencies) or a transition structure (just one imaginary frequency); very often, the abbreviation NIMAG (number of imaginary frequencies) is used.

One should be cautious, however, of some difficult and (for closed-shell systems) rare cases when the molecular symmetry and the wave function symmetry are *not* the same. A trivial example would be the two possible closed-shell singlet states of methylene (:CH$_2$) that are both $C_{2v}$ symmetric. The state with the doubly occupied methylene $3a_1$ orbital is the lowest lying singlet state of methylene, labelled $\tilde{a}\ ^1A_1$, with the open-shell $\tilde{b}\ ^1B_1$ and the closed-shell $\tilde{c}\ ^1A_1$ states being higher lying singlet states [24]. If such a situation occurs, it is imperative to use symmetry to assign the computed states properly by their symmetry labels (they are all $A$ in $C_1$).

## 7.2.8 Basis sets

All wave function based methods use mathematical functions to describe the hydrogen-type orbitals; each molecular orbital is constructed as a linear combination of the atomic orbitals (LCAO approach) with coefficients that minimise the iterative HF procedure. This is usually done either with Slater-type orbitals (STOs, used in semiempirical and some DFT methods) or with products of primitive Gaussian-type orbitals (GTOs, used in virtually all other approaches) that approximate the STO behaviour of the hydrogen orbitals. STOs are normally used in computationally inexpensive semiempirical methods where the integral computation is not the computational bottleneck. Owing to their quadratic exponent, GTOs are easier to integrate; many GTOs are needed to describe properly the regions in space far away from the atomic centres because STOs have long delocalisation 'tails'. This is particularly important for highly delocalised structures and weak interactions. Hence, one has to try to use as large a basis set as is computationally feasible. Diffuse ('+') functions, which are often necessary for the computation of carbanions [25], often lead to convergence problems because they have a large radial extent that can span several atoms.

The choice of a basis set can be crucial and is particularly important for post-HF methods. HF itself and DFT methods are not so basis set dependent, but this should not be taken as an excuse for not checking a few basis sets even for these levels. The popular Pople-type basis sets [26], e.g. 6-31G(d), have proven their value for HF and DFT, and are well behaved for structures involving first row and many heavier elements. The cc-pVXZ basis sets [19], which are also sometimes used in conjunction with DFT computations, were explicitly designed to recover as much correlation energy as possible for highly correlated methods and are, therefore, not necessarily guaranteed to perform well with DFT methods.

### 7.2.9   Validation

Computational methods are typically employed to rationalise experimental findings such as molecular structures and spectroscopic data *after the fact*. It is clear, however, that the true strength of computational chemistry is the *prediction* of new, hitherto unknown structures and reactions. In an ideal situation, this would greatly reduce experimental effort and, consequently, time and resource allocations. The key problem is that although many computational methods are well tested on a limited set of structures, weaknesses of particular methods sometimes unexpectedly surface; this is especially true when fleeting intermediates are examined in mechanistic studies. It is imperative to validate every computational approach properly with respect to the problem under consideration; this is particularly important for semiempirical and DFT methods that are geared towards stable ground state structures. It is clear that reliable predictions are difficult, and they are only possible when the computational approach has been properly validated by comparing test output against *high-quality* experimental results or other reliable computational data.

Ultimately, every computational investigation is a compromise between computational effort and accuracy. Ideally, one would use huge basis sets in conjunction with highly correlated methods. This, however, is impracticable as the scaling for such approaches is huge (Table 7.5). For instance, if a 10-electron problem (e.g. methane) with a small basis set such as 6-31G(d) takes 1 minute to optimise at the HF level with a good starting geometry (four to five optimisation steps), it will take about 2–3 hours at CCSD(T) with the same basis set. Doubling the size of the system will increase the time by a factor of 128, which then roughly corresponds to a CPU time of 2 weeks! It is, therefore, up to the researcher to pick an appropriate method, validate the approach and be aware of its limitations.

## 7.3   Case studies

### 7.3.1   The ethane rotational barrier and wave function analysis

Simple yet important 'mechanistic' cases concern the computation of rotational barriers around single bonds. A timely case in point is the rotational barrier of ethane, an old yet much debated subject [27, 28]. While the notion of hindered rotation in ethane is often credited to Kemp and Pitzer (1936) [29], Ebert (1929) [30] and then Wagner (1931) [31] were the first to propose it. Teller and Weigert established the presence of a hindered rotation by quantum theoretical computations on the temperature dependence of the molar heat capacity of ethane [32]. Needless to say, even then, these propositions were discussed vigorously, and the Kemp and Pitzer paper (which does not cite the earlier work of Ebert or Wagner) seems to settle the dispute between the various groups [29]. Independently, Wilson (incidentally, not citing the 1936 Kemp and Pitzer publication) emphasised the presence of an appreciable rotational barrier of about 3 kcal mol$^{-1}$ [33]. This early work demonstrated that the internal rotational potential energy of ethane has three degenerate minima corresponding to a single stable conformation, the staggered (S) conformer. These conformers are separated from each other by degenerate eclipsed (E) conformations lying about 3 kcal mol$^{-1}$ higher [34]. But what is the *origin* of the rotational barrier? Steric repulsion between bonds due to the overlap of two filled C—H-bond orbitals (and hence violating the Pauli exclusion principle) [35] in the eclipsed conformation seems to be an attractive explanation. Computationally 'switching off' this interaction shows that the staggered conformation is preferred *regardless* of the presence of exchange (Pauli) repulsion. Another, less obvious, quantum-mechanical interaction is clearly responsible for the observed conformational preference, namely hyperconjugation [36]. This interaction arises from partial electron transfer from the filled $\sigma_{C-H}$ bonding orbital to the unfilled $\sigma^*_{C-H}$ antibonding orbitals (Fig. 7.4). Despite the fact that a lower lying occupied orbital ($\sigma_{C-H}$) interacts with a higher lying empty ($\sigma^*_{C-H}$) orbital, quantum mechanics tells us that this interaction is stabilising [37]. If the hyperconjugative interactions between the methyl groups are turned off in model computations, the preference for the staggered conformation of ethane disappears.

Natural bond orbital (NBO) analysis emphasises the importance of hyperconjugative interactions [38] (Fig. 7.5A), despite the overwhelming number of textbook presentations of the seductively transparent steric repulsion explanation. The latter is favoured by a so-called *energy decomposition analysis* (EDA, Fig. 7.5B) which is mostly concerned with the fate of the Pauli repulsion when considering relaxed (optimised) versus unoptimised methyl group fragments [28]. When the bond lengths of the methyl groups and the C—C bond are computationally not allowed to relax, the eclipsed form displays a maximum in the Pauli repulsion curve, irrespective of hyperconjugative effects. Finally, a valence bond (VB) theory study attempts to reconcile the dispute about the relative importance of Pauli repulsion and hyperconjugation, and suggests that *both* are important (Fig. 7.5C). For mechanistic

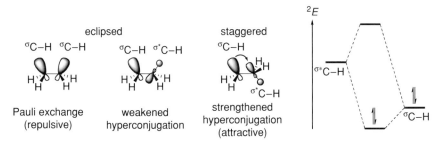

**Fig. 7.4** Molecular oribital interaction in ethane, and their interaction diagram emphasising the stabilisation by C—H hyperconjugation.

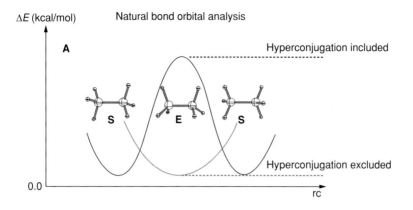

$\Delta E$ (kcal/mol)          Natural bond orbital analysis

**A**

Hyperconjugation included

S          E          S

Hyperconjugation excluded

0.0

rc

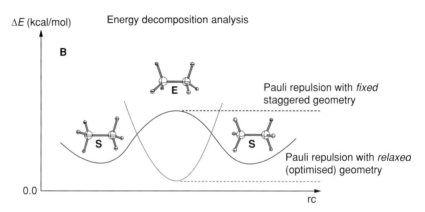

$\Delta E$ (kcal/mol)          Energy decomposition analysis

**B**

E

Pauli repulsion with *fixed* staggered geometry

S          S

Pauli repulsion with *relaxed* (optimised) geometry

0.0

rc

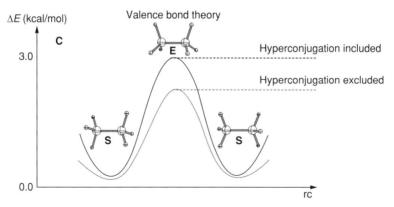

$\Delta E$ (kcal/mol)          Valence bond theory

**C**

E

Hyperconjugation included

3.0

Hyperconjugation excluded

S          S

0.0

rc

**Fig. 7.5**  Rotational energy profiles of ethane according to different ways of analysing the wave function (energies not drawn to scale, exaggerated for clarity).

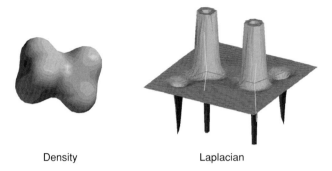

Density                    Laplacian

**Fig. 7.6** Electron density (left) and Laplacian of the electron density of ethane (very large peaks are truncated)

considerations, it is important to ask why three different ways of analysing the same ab initio wave function result in two entirely opposing views from NBO and EDA approaches.

As noted earlier, the wave function cannot be interpreted directly and the ways to distil chemically meaningful information are arbitrary and incomplete by *choice*. There are only a few approaches that work directly with the *electron density* which does have a measurable physical meaning. Apart from the spatial extension of a molecule and a detailed shape analysis, the electron density itself tells us little because most larger molecules look as if they are wrapped in a thick foil that does not allow one to discern the various atoms and bonds (see ethane depicted on the left of Fig. 7.6). A much better approach is to consider how the electron density *changes*, i.e. by taking the gradients of the density; this valuable approach is commonly referred to as 'atoms in molecules' (AIM) [39]. With this method, it is straightforward to identify the atoms and bonds in a molecule; these are characterised by 'critical points' where the first gradient of the density is zero while higher derivatives are non-zero to identify maxima and minima in the density gradients. The latter gives an $n \times n$ matrix of second derivatives of the electron density that is referred to as the *Hessian matrix* of the charge density; the *Laplacian* of the electron density, $\nabla^2 \rho$ is also invariant to the coordinate system (Fig. 7.6, right). The critical points determined this way reveal much about bonding, structure, electronegativity and other properties of molecules.

In the ethane case, however, the AIM analysis helps in understanding the overlap of the bonds and the location of the electrons as derived from the density picture, but it does not tell us anything about the origin of the rotational barrier. For that, we need methods that quantitatively give us energies that can be associated with the effects of donor–acceptor bonding (hyperconjugation) and electron–electron repulsion (Pauli repulsion) as noted above.

A simple and robust quantitative MO-type approach (as opposed to density approaches) is the ubiquitous *Mulliken population analysis* [40]. The key concept of this easily programmed and fast method is the distribution of electrons based on occupations of atomic orbitals. The atomic populations do *not*, however, include electrons from the overlap populations, which are divided exactly in the middle of the bonds, regardless of the bonding type and the electronegativity. As a consequence, differences of atom types are not properly accommodated and the populations per orbital can be larger than 2, which is a violation of the Pauli principle; a simple remedy for this error is a *Löwdin* population analysis that

**Table 7.6** Comparison of Mulliken and natural bond orbital (NBO)/natural population analysis (NPA) charges in LiH and LiF using different basis sets and computational approaches.

| Level | Li (in LiH) | Li (in LiF) |
| --- | --- | --- |
| *Mulliken* | | |
| HF/STO-3G | −0.016 | +0.229 |
| HF/6-31G(d) | +0.170 | +0.666 |
| HF/6-311+G(d,p) | +0.358 | +0.718 |
| B3LYP/6-31G(d) | +0.105 | +0.518 |
| B3LYP/6-311+G(d,p) | +0.295 | +0.616 |
| MP2/6-31G(d) | +0.178 | +0.662 |
| MP2/6-311+G(d,p) | +0.359 | +0.703 |
| *NBO/NPA* | | |
| HF/6-31G(d) | +0.731 | +0.917 |
| HF/6-311+G(d,p) | +0.815 | +0.977 |
| B3LYP/6-31G(d) | +0.685 | +0.826 |
| B3LYP/6-311+G(d,p) | +0.778 | +0.960 |
| MP2/6-31G(d) | +0.727 | +0.922 |
| MP2/6-311+G(d,p) | +0.817 | +0.975 |

symmetrically transforms the Mulliken analysis into an orthogonal coordinate system so that the Pauli exclusion principle is obeyed. Furthermore, a strong basis set dependence for Mulliken (and Löwdin) population analysis is observed (Table 7.6), and larger basis sets sometimes produce charges that are not credible [41].

A much more balanced way is offered by a *natural population analysis* (NPA) which is a part of *natural bond orbital (NBO)* analysis methods; these use orbitals which diagonalise the two-centre subdeterminants of the one-particle density matrix (*natural orbitals*, NO) [37, 38]. The key advantage is the possibility for very detailed analyses of the various interactions (electron transfer, donor–acceptor bonds, conjugation, hyperconjugation) with inclusion of all atomic properties. Also, NBO methods do not suffer from strong basis set dependence (Table 7.6). It is comforting to see that three different theoretical approaches, HF, DFT and MP2, largely give the same NBO charges. One critical point often raised with NBO methods is that they do not leave the occupied orbital space untouched, but mix the occupied and unoccupied spaces when transforming to bonding and antibonding orbitals. The reason is the aim of NBO analyses to maximise occupation of the localised bonding orbitals so that the interpretation is as straightforward as possible. Since MO theory does not provide a unique way for localising the occupied orbital space, NBO methods derive the localised orbitals from a delocalised wave function.

*Electron decomposition analysis* (EDA, also known as Kitaura–Morokuma (KM) analysis) [42] attempts to dissect the HF energy contributions of an interaction such as those of two methyl groups connected by a single bond in ethane into Pauli repulsion, electrostatic and orbital relaxation terms. This is achieved through dissecting the respective terms from a *nonorthogonal* molecular orbital basis, which may lead to basis-related numerical instabilities, and a physically unsound description of charge transfer and cooperative resonance interactions. This goes as far as to be able to make the barrier in ethane disappear by varying the treatment of the nonorthogonality. Of course, this is not to say that this type of analysis is intrinsically flawed but it is an expert method that requires careful attention.

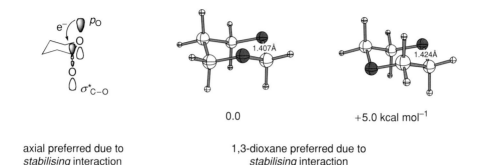

0.0            +5.0 kcal mol$^{-1}$

axial preferred due to
*stabilising* interaction

1,3-dioxane preferred due to
*stabilising* interaction

**Fig. 7.7** Anomeric effect in sugars (left) preferentially stabilising the axial position of the 'anomeric' C—O bond, and isomer energy difference between 1,3- and 1,4-dioxane at B3LYP/6-31G(d) as clear evidence against steric (Pauli) repulsion arguments for the anomeric effect.

The results of a valence bond treatment of the rotational barrier in ethane lie between the extremes of the NBO and EDA analyses and seem to reconcile this dispute by suggesting that *both* Pauli repulsion and hyperconjugation are important. This is probably closest to the truth (remember that Pauli repulsion dominates in the higher alkanes) but the VB approach is still imperfect and also is mostly a very powerful expert method [43]. VB methods construct the total wave function from linear combinations of covalent resonance and an array of ionic structures; as the covalent structure is typically much lower in energy, the ionic contributions are included by using highly delocalised (and polarisable) so-called Coulson–Fischer orbitals. Needless to say, this is not error free and the brief description of this rather old but valuable approach indicates the expert nature of this type of analysis.

The bottom line is that analysing wave functions is not an easy task. From the methods briefly introduced above, it seems that the NBO approach is far superior in terms of ease of use, numerical stability and interpretability.

The hyperconjugative model also explains related conformational preferences such as the anomeric effect. This preferred axial orientation of the anomeric hydroxyl in sugars, e.g. glucose, is a result of the electronic preference of an OCOC unit for the *gauche* rather than *trans* relationship, Fig. 7.7 [44]. Very often, the anomeric effect is incorrectly ascribed to a dipole–dipole repulsive interaction [45] ('rabbit ears' on oxygens in sugars, for instance), rather than an *attractive* one. A simple comparison of the heats of formation of 1,3- and 1,4-dioxane, for example, emphasises the inherently stabilising nature of this effect: 1,3-dioxane is 5.0 kcal mol$^{-1}$ more stable [46]. This conclusion is also supported by the significantly different C—O bond lengths in the two dioxanes. Were the effect destabilising, these bonds would be *longer* in 1,3-dioxane.

### 7.3.2 The nonclassical carbocation problem and the inclusion of solvent effects

The controversy regarding the classical versus nonclassical nature of the 2-norbornyl cation was one of the central topics of physical organic chemistry [47, 48]. It also has become amenable to computations but it is clear that this is all but easy because the solvolysis

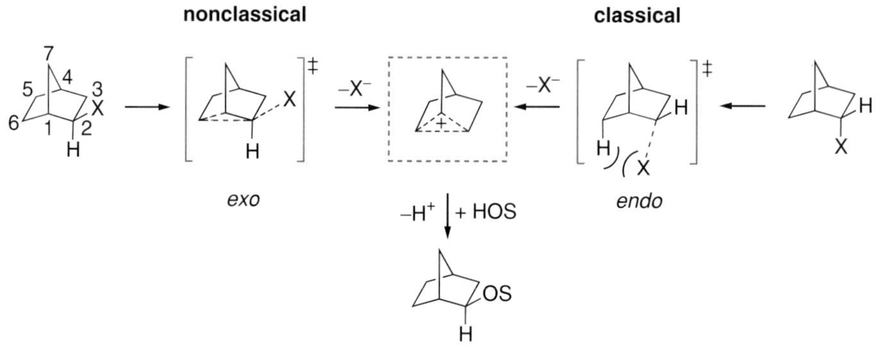

**Fig. 7.8** Pictorial representation of the celebrated 2-norbornyl solvolysis problem where both enantiomerically pure 2-*endo* and 2-*exo* substrates lead through a common symmetric intermediate to the more stable racemic 2-*exo* substitution product. Much mechanistic emphasis was placed on the central 2-norbornyl cation (cf. Fig. 7.11).

transition structures, the nature of the intermediate carbocation(s) and solvent effects have to be considered.

As early as 1949, Winstein and Trifan reported the distinctively different solvolytic behaviour of 2-*exo*- and 2-*endo*-norbornyl derivatives, and noted that the titrimetric rate constant for solvolysis of 2-*exo*-norbornyl brosylate in acetic acid is 350 times faster than that of the 2-*endo* isomer; both gave 2-*exo*-norbornyl acetate (Fig. 7.8) [49]. The *exo/endo* rate ratio is even larger when the rates were measured polarimetrically using optically active starting materials. Starting with optically active 2-*exo*-norbornyl brosylate, the polarimetric rate constant exceeded the titrimetric one for acetolysis by a factor of about 100, while the polarimetric and titrimetric rate constants were *equal* for the acetolysis of optically active 2-*endo*-norbornyl brosylate. According to Winstein, the *exo* reaction proceeds faster through a symmetrical ion pair intermediate which can undergo internal return in competition with product formation, and both routes give only racemic *exo* products.

The alternative possibility of two rapidly equilibrating enantiomeric norbornyl cations or ion pairs did not, in Winstein's view, account for the large *exo/endo* rate ratio; rather, this was attributed to what we now call *neighbouring group participation* (NGP) by the well-aligned $C^2$—$C^6$ bond during the *exo* but not the *endo* solvolysis. Winstein argued that NGP in the reaction of *exo* substrates leads to a delocalised 'nonclassical' transition structure and then to the symmetrical nonclassical norbornyl cation. In contrast, solvolysis of *endo* substrates was suggested to proceed by a 'classical' transition structure in the rate-limiting step, which then collapses to the same bridged ion, from which the products are formed.

Brown argued that the *exo/endo* ratios of some tertiary norbornyl systems were also large, despite the expected classical character of tertiary cation intermediates. Large *exo/endo* rate ratios, in his view, are more likely *steric* in origin, and he argued that *endo* derivatives suffer from steric hindrance to ionisation, i.e. the $C^6$-*endo* hydrogen hinders departure of the leaving group in the solvolytic transition state. Brown considered the *exo* solvolysis rate 'normal', while the *endo* solvolysis is low due to this steric hindrance. A comparison of the solvolysis rates of various secondary substrates in non-nucleophilic solvents showed, however, that the 2-*endo*-norbornyl solvolysis rate is quite 'normal' while that of the *exo* isomer is clearly exceptional (the rate enhancement for the *exo* isomer was originally termed *anchimeric*

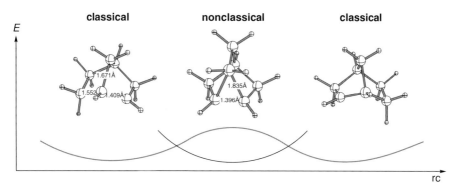

**Fig. 7.9** The classical/nonclassical 2-norbornyl cation problem. Grey: a pair of rapidly equilibrating classical cations with a nonclassical, bridged transition structure; black: the nonclassical cation as the minimum cation. The fully MP2/6-31G(d) optimised 2-norbornyl cations are depicted; the nonclassical ion is 13.6 kcal mol$^{-1}$ more stable at this and comparable levels of theory.

*assistance*, to be distinguished from NGP which is the mechanistic cause of the phenomenon). The discussion then turned to the nature of the intermediate 2-norbornyl cations on the assumption that they would resemble very closely the respective *exo* and *endo* transition structures. This focusing on the intermediate was triggered by Olah's discovery of superacid media and instrumental developments which subsequently enabled the direct spectroscopic observation of the norbornyl cation by Schleyer and Olah [50]. Although the geometry of the 2-norbornyl cation is still not known experimentally, ingenious experimental methods in combination with high-level ab initio computations finally established its symmetrically bridged structure (Fig. 7.9) [51].

While there is now solid agreement between theory and experiment for the nonclassical, symmetrically bridged minimum energy structure of the 2-norbornyl cation, it is not always straightforward to compare experimental data obtained mostly in solution with computed gas phase values. One could argue that the positive charge in the classical 2-norbornyl cation is more localised and therefore much more stabilised in highly polar solvents. Could this make the computed energy difference of ca. 14 kcal mol$^{-1}$ between classical and nonclassical structures shrink to 3–7 kcal mol$^{-1}$ between the solvolysis transition structures? The classical 2-norbornyl cation is a transition structure that can be structurally characterised *only* by computations (or through indirect experimental evidence, e.g. secondary deuterium kinetic isotope effects [52]). The theoretical task ahead, therefore, was to estimate the solvation energy *difference* for the two cations.

There are various ways to include solvation, e.g.

(a) by self-consistent reaction field (SCRF) models [53];
(b) by statistical Monte Carlo (MC) solution simulations [54]; and
(c) by explicitly including a limited number of solvent molecules in the ab initio computations.

SCRF methods use cavities for the solute which interact with the solvent as a field; the induced dipoles lead to an energy stabilisation that is summed up as the solvation energy. MC simulations utilise a combination of force fields for the solvent molecules that statistically (large sampling) interact with solutes; usually, differences for solutes in the same solvent

**Table 7.7** Difference in free energies of solvation (kcal mol$^{-1}$) for the classical and nonclassical 2-norbornyl cations by three very different computational approaches.

| Model | SCRF | MC | Explicit solvent |
|---|---|---|---|
| $\Delta\Delta G_{solv}$ | 1.2 | 0.7 | 0.5 |

volume are computed and the sum of all energy changes, while perturbing from one solute to another, gives the changes in free energy of solvation ($\Delta\Delta G_{solv}$).

For the 2-norbornyl cation problem, the effects of solvation on the classical and non-classical ion were studied by methods (a)–(c) to compare the performance of these largely different models for the same problem (*validation*) [55]. Water was chosen as the solvent since one expects the difference in solvation energies to be large in highly polar media. First, the SCRF computations at HF/6-311+G(d,p) for the two cations in water only require the dielectric constant of water as additional input. Second, by the MC method, the differences in free energies of hydration ($\Delta\Delta G_{solv}$) were computed by a gradual geometric transition between the two MP2/6-31G(d) fully optimised cation structures using statistical perturbation theory. Third, one water molecule was included as an explicit solvent molecule in the MP2/6-31G(d) optimisation procedure to keep the cations fixed (water would immediately react with a cation and this sort of equilibrium process cannot yet be studied computationally). The key results are that the solvation energy difference between the classical and nonclassical 2-norbornyl cations is only about 1 kcal mol$^{-1}$, *regardless* of the solvation model employed (Table 7.7)! That is, the positive charge in the classical cation is *not* substantially more localised than in the symmetrical nonclassical cation! In less polar solvents, the solvation energy difference is expected to be even smaller. Consequently, the differences in rates of solvolysis of 2-*exo* and 2-*endo* norbornyl derivatives can *only* be explained in terms of the differences between the *exo* and *endo* transition structures.

But how can one model the obviously difficult transition structures? First, a suitable compromise between theoretical rigour and practicability of the computations had to be found. At the time when this case study was originally published (1997) [48], and even at the time of printing this book chapter, it was exceedingly difficult to localise transition structures on flat potential energy hypersurfaces for large molecules without symmetry. It is imperative for such difficult transition structures to compute second derivatives at several points along the optimisation path. As DFT is often quite close to MP2, the B3LYP/6-31G(d) level was chosen because it offered the best compromise between practicability and desired accuracy. Higher level energies were determined at B3LYP/6-311+G(d); the diffuse functions should take into account the large charge dispersion expected for these transition structures and they should also significantly reduce *basis set superposition errors* (BSSE) which always need to be considered with weakly interacting fragments [56]. As a validation, the final level B3LYP/6-311+G(d)//B3LYP/6-31G(d) reproduces cation–water complexation energies satisfactorily. As heterolytic bond cleavages are difficult to model computationally in the absence of solvent stabilisation, positively charged substrates were employed where the leaving groups are neutral molecules (Fig. 7.10). Water was eventually chosen as the preferred leaving group to model not only the heterolytic bond cleavage of, for example, a protonated alcohol or an ester, but also the reaction of a solvent leading to a product.

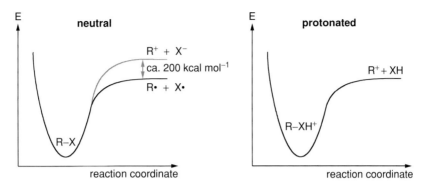

**Fig. 7.10** Dissociation curves for neutral and protonated $S_N1$ substrates in the gas phase.

The computed activation energies for the model water leaving group are, as expected, much smaller (*exo*, 2.8 kcal mol$^{-1}$; *endo*, 5.3 kcal mol$^{-1}$) than the experimentally found activation barriers for the solvolysis of 2-*exo*- and 2-*endo*-norbornyl tosylates (18–26 kcal mol$^{-1}$) since the latter involve the dissociation of neutral species where charge separation and ion solvation take place. The computed energy difference between TS$_{exo}$ and TS$_{endo}$ is 3.7 kcal mol$^{-1}$ (Fig. 7.11) and the ground state energy difference between the *exo* and *endo* protonated alcohols is 1.2 kcal mol$^{-1}$, the more stable *exo* isomer being the more reactive. These theoretical results lead to a difference in activation energies from the respective starting materials, $\Delta\Delta H^{\ddagger}$, for water as the leaving group, of 2.5 kcal mol$^{-1}$ which, in turn, leads to a computed value of about 70 for the *exo/endo* solvolysis rate ratio at 298 K. As these results

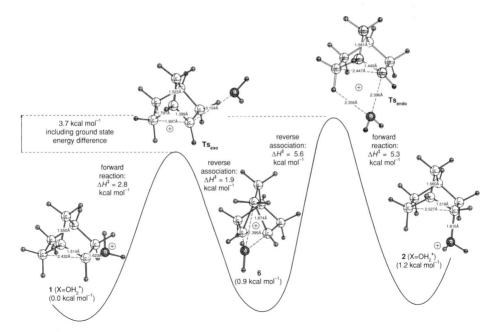

**Fig. 7.11** B3LYP/6-31G(d) computed norbornyl solvolysis from the protonated *endo*- and *exo*-2-alcohols.

are in general accord with experiment, one can now address the central question: Why is $\Delta\Delta H^{\ddagger}$ much smaller than the ca. 14 kcal mol$^{-1}$ energy difference between the classical and nonclassical 2-norbornyl cations, i.e. what is the origin of the compensating stabilisation of the *endo*- over the *exo*-transition structure?

Although the C$\cdots$O distances in the transition structures are large, the interaction energies of the cationic moieties with $H_2O$ are quite substantial. The C$\cdots$O and the two closest H$\cdots$O distances for TS$_{endo}$ (C$\cdots$O = 2.396 Å; H$\cdots$O = 2.356 and 2.319 Å) are much smaller than those for TS$_{exo}$ (C$\cdots$O = 3.104 Å; H$\cdots$O = 2.463 and 2.603 Å). Since the O$\cdots$H distances are within the usual hydrogen-bonding range, TS$_{endo}$ is *stabilised* more through the interaction with the leaving group than is TS$_{exo}$. Although this stabilisation may be attenuated in solution, the usual leaving groups (from neutral solvolysis substrates) are negatively charged and CH$\cdots$X-hydrogen bonding will be more important. In marked contrast to Brown's suggestion, the computational analysis reveals that there is a *stabilising* rather than a repulsive steric interaction with the *endo*-hydrogen in the 6-position in TS$_{endo}$ (Fig. 7.11)!

## 7.4  Matching computed and experimental data

One of the most powerful ways to use computational chemistry methods for the elucidation of reaction mechanisms is the direct matching of computed and spectroscopic data. This involves comparison of optimised structures with experimental structural data keeping in mind that different experimental and theoretical methods give different averaged values for the geometrical parameters of molecules. That is, a perfect match is not expected unless the theoretical treatment is complete and includes often neglected effects such as vibrational anharmonicities, exhaustive inclusion of electron correlation, relativistics, spin–orbit as well as Coriolis couplings, and more. These sizeable effects are intentionally neglected in most routine theoretical applications when only approximate structures and relative energies matter. When spectroscopic accuracy is desired, the effort is magnified and the theoretical treatment can become very elaborate.

Without going into concrete examples, Table 7.8 presents the author's current view on the availability of highly accurate methods depending on the property or quantity under consideration. Needless to say, this listing is subjective and incomplete but nevertheless serves as a guideline for what is possible to date, and it is disappointing to see that reaction rates cannot routinely be predicted. This is because of the importance of *reaction dynamics*, which are much harder to include in high-level computations than static molecules. This is both a question of theory, which is currently progressing rapidly, and the availability of increasingly powerful computers to map potential energy surfaces very accurately. Still, we are quite far from modelling reactions along accurate trajectories, and approximate methods help to estimate reaction rates.

On the other hand, it is pleasing to see that the organic chemist's standard evidence for structural identification such as NMR and IR spectral data can be computed quite accurately; UV/vis spectra can also be computed with *time-dependent* methods, which, however, cannot yet be used to optimise the excited states; as the Franck–Condon principle is a common assumption in electron spectroscopy, this is no serious drawback for computing

**Table 7.8**  Listing of different properties of molecules and molecular systems with the current degree of difficulty; this list is subjective and incomplete.

| Ready-to-use methods | Expert methods (continuing development) | Methods in progress (to varying degrees) |
|---|---|---|
| Geometries | Molecular dynamics | Full reaction dynamics |
| Energies | Full relativistics | Reaction rates |
| IR | Density matrix methods | Excited states (adiabatic) |
| NMR | Linear scaling | Improvement of DFT |
| MW | Local correlation methods | Melting/boiling points |
| UV/vis, ECD, VCD | $r_{12}$-methods | Crystal structure predictions |
| Photoelectron spectra | Spin–orbit couplings | Band structure methods |
| Excited states (vertical) | ESR | Solid state dynamics |
| Solvent effects | | Infinite systems |
| | | Geminal functional theory |

UV/vis spectra. The accurate characterisation of excited states, however, poses formidable conceptual and theoretical challenges.

Linear scaling methods that attempt to reduce the computation of all possible multicentre integrals in the HF formalism are well underway and will allow ever larger molecular systems to be examined computationally; local correlation and $r_{12}$-methods (including a term that accounts explicitly in the wave function for the interelectronic distances) go hand-in-hand with this development. This will also improve the way 'infinite' systems can be treated, to examine, for example, the transition between MO and band structures. Crystal structure predictions by statistically sampling and optimising large ensembles of molecules will also improve with linear scaling. Macroscopic properties such as boiling and melting points are very difficult to compute ab initio because large numbers of molecules are needed.


## 7.5   Conclusions and outlook

It is obvious that computational chemistry methods are already an extremely valuable tool for examining reaction mechanisms. As computations *always* yield numbers it is up to the chemist to judge critically their value. Hence, validation (either through literature precedents or by one's own work) separates good from not-so-good computations. In general, absolute values are still rather difficult to compute accurately; comparisons, for example differences between the energies of related molecular species, are much more reliable and can, if computed properly, rival experimentally derived relative energies. Many spectroscopic data are now routinely and reliably computed, and compare well with experimental results.

Much is left to be desired regarding reaction rates which currently cannot be computed well owing to theoretical and computational limitations. This is clearly a hot area with much current development and forthcoming studies look extremely impressive. Before too long, it should be possible to map very accurately entire potential energy surfaces on a par with experiment but *much* faster (and, of course, cheaper).

## 7.6 List of abbreviations

| | |
|---|---|
| AM1 | Austin model 1 (a semiempirical theory) |
| B3LYP | Becke's three-parameter hybrid exchange functional with the Lee–Yang–Parr correlation functional |
| cc-pVDZ | Correlation-consistent polarised valence $x$–$\zeta$ basis set ($x = 2$–$n$) |
| CCSD | Coupled cluster theory including singe and double excitations |
| CCSD(T) | CCSD including perturbatively determined triple excitations |
| CI | Configuration interaction |
| CSF | Configuration state function |
| DFT | Density functional theory |
| DZP | Double-$\zeta$ plus polarisation basis set |
| $E_{corr}$ | Correlation energy |
| $E_{hf}$ | HF energy |
| FCI | Full configuration interaction |
| FF | Force field |
| $Gn$ | Gaussian-$n$ set of reference molecules |
| GTO | Gaussian-type orbital |
| HF | Hartree–Fock |
| MBPT | Many-body perturbation theory |
| MM | Molecular mechanics |
| MNDO | Modified neglect of differential diatomic overlap |
| $MPn$ | $n$th-order Møller–Plesset perturbation theory |
| MR | Multi-reference |
| PM3 | Parameterised model 3 |
| RHF | Restricted HF |
| RMS | Root mean square |
| $S$ | Spin operator |
| SCF | Self-consistent field (analogous to HF theory) |
| STO | Slater-type orbital |
| TZP | Triple-$\zeta$ plus polarisation basis set |
| UHF | Unrestricted HF |

## References

1. Cramer, C.J. (2004) *Introduction to Computational Chemistry.* Wiley-Interscience, Hoboken, NJ; Jensen, F. (1999) *Introduction to Computational Chemistry.* Wiley, Chichester; Schleyer, P.v.R., Allinger, N.L., Clark, T., Gasteiger, J., Kollman, P.A., Schaefer III, H.F. and Schreiner, P.R. (1998) *The Encyclopedia of Computational Chemistry.* Wiley, Chichester.
2. Koch, W. and Holthausen, M.C. (2001) *A Chemist's Guide to Density Functional Theory* (2nd edn). Wiley-VCH, Weinheim.
3. Kapp, J., Schreiner, P.R. and Schleyer, P.v.R. (1996) *Journal of the American Chemical Society*, **118**, 12154.
4. Burkert, U. and Allinger, N.L. (1982) *Molecular Mechanics.* American Chemical Society, Washington, D.C.

5. Kettering, C.F., Shutts, L.W. and Andrews, D.H. (1930) *Physical Review*, **36**, 531; Hendrickson, J.B. (1961) *Journal of the American Chemical Society*, **83**, 4537.
6. Mapelli, C., Castiglioni, C., Zerbi, G. and Mullen, K. (1999) *Physical Review B*, **60**, 12710; Ohno, K. and Takahashi, R. (2002) *Chemical Physics Letters*, **356**, 409.
7. Gill, P.M.W. (1998) In: P.v.R. Schleyer, N.L. Allinger, T. Clark, J. Gasteiger, P.A. Kollman, H.F. Schaefer III and P.R. Schreiner (Eds) *The Encyclopedia of Computational Chemistry* (vol 1). Wiley, Chichester, p. 678.
8. Dewar, M.J.S. (1975) *The PMO Theory of Organic Chemistry*. Plenum, New York; Stewart, J.J.P. (1990) In: K.B. Lipkowitz and D.B. Boyd (Eds) *Reviews in Computational Chemistry* (vol 1). Wiley-VCH, New York, p. 45.
9. Clark, T. (2000) *Journal of Molecular Structure (Theochem)*, **530**, 1.
10. Dewar, M.J.S., Zoebisch, E.G., Healy, E.F. and Stewart, J.J.P. (1985) *Journal of the American Chemical Society*, **107**, 3902.
11. Dewar, M.J.S. and Thiel, W. (1977) *Journal of the American Chemical Society*, **99**, 4899.
12. Thiel, W. (1998) In: K.K. Irikura and D.J. Frurip (Eds) *Thermochemistry from Semiempirical Molecular Orbital Theory, ACS Symposium Series* (vol 677). American Chemical Society, Washington, D.C., p. 142.
13. Cremer, D. (1998) In: P.v.R. Schleyer, N.L. Allinger, T. Clark, J. Gasteiger, P.A. Kollman, H.F. Schaefer III and P.R. Schreiner (Eds) *The Encyclopedia of Computational Chemistry* (vol 3). Wiley, Chichester, p. 1706; Møller, C. and Plesset, M.S. (1934) *Physical Review*, **46**, 618.
14. Sherrill, C.D., Van Huis, T.J., Yamaguchi, Y. and Schaefer, H.F. (1997) *Journal of Molecular Structure (Theochem)*, **400**, 139.
15. Hanrath, M. and Engels, B. (1997) *Chemical Physics*, **225**, 197; Engels, B., Hanrath, M. and Lennartz, C. (2001) *Computers & Chemistry*, **25**, 15.
16. Lischka, H., Shepard, R., Brown, F.B. and Shavitt, I. (1981) *International Journal of Quantum Chemistry*, **515**, 91; Werner, H.J. and Knowles, P.J. (1991) *Journal of Chemical Physics*, **94**, 1264.
17. Cizek, J. (1966) *Journal of Chemical Physics*, **45**, 4256; Purvis, G.D. and Bartlett, R.J. (1981) *Journal of Chemical Physics*, **75**, 1284; Urban, M., Noga, J., Cole, S.J. and Bartlett, R.J. (1985) *Journal of Chemical Physics*, **83**, 4041; Bartlett, R.J. and Noga, J. (1988) *Chemical Physics Letters*, **150**, 29; Bartlett, R.J., Watts, J.D., Kucharski, S.A. and Noga, J. (1990) *Chemical Physics Letters*, **165**, 513.
18. Gauss, J. (1998) In: P.v.R. Schleyer, N.L. Allinger, T. Clark, J. Gasteiger, P.A. Kollman, H.F. Schaefer III and P.R. Schreiner (Eds) *The Encyclopedia of Computational Chemistry* (vol 1). Wiley, Chichester, p. 615.
19. Dunning Jr., T.H., Peterson, K.A. and Woon, D.E. (1998) In: P.v.R. Schleyer, N.L. Allinger, T. Clark, J. Gasteiger, P.A. Kollman, H.F. Schaefer and P.R. Schreiner (Eds) *The Encyclopedia of Computational Chemistry* (vol 1). Wiley, Chichester, p. 88.
20. Ozment, J.L., Schmiedekamp, A.M., Schultz-Merkel, L.A., Smith, R.H. and Michejda, C.J. (1991) *Journal of the American Chemical Society*, **113**, 397; Bates, K.R., Daniels, A.D. and Scuseria, G.E. (1998) *Journal of Chemical Physics*, **109**, 3308; Flocke, N. and Bartlett, R.J. (2004) *Journal of Chemical Physics*, **121**, 10935; Korona, T., Pfluger, K. and Werner, H.J. (2004) *Physical Chemistry Chemical Physics*, **6**, 2059; Schutz, M. and Manby, F.R. (2003) *Physical Chemistry Chemical Physics*, **5**, 3349; Werner, H.J., Manby, F.R. and Knowles, P.J. (2003) *Journal of Chemical Physics*, **118**, 8149; Schutz, M. (2002) *Physical Chemistry Chemical Physics*, **4**, 3941; Liang, W.Z., Shao, Y.H., Ochsenfeld, C., Bell, A.T. and Head-Gordon, M. (2002) *Chemical Physics Letters*, **358**, 43; Schutz, M. (2002) *Journal of Chemical Physics*, **116**, 8772; Li, S., Ma, J. and Jiang, Y. (2002) *Journal of Computational Chemistry*, **23**, 237; Mazziotti, D.A. (2001) *Journal of Chemical Physics*, **115**, 8305; Schutz, M. and Werner, H.J. (2001) *Journal of Chemical Physics*, **114**, 661; Schutz, M. (2000) *Journal of Chemical Physics*, **113**, 9986; Kudin, K.N. and Scuseria, G.E. (2000) *Physical Review B*, **61**, 16440; Scuseria, G.E. and Ayala, P.Y. (1999) *Journal of Chemical Physics*, **111**, 8330.

21. Lee, T.J., Rendell, A.P. and Taylor, P.R. (1990) *Journal of Physical Chemistry*, **94**, 5463; Lee, T.J., Rice, J.E., Scuseria, G.E. and Schaefer, H.F. (1989) *Theoretica Chimica Acta*, **75**, 81.
22. Becke, A.D. (1993) *Journal of Chemical Physics*, **98**, 1372; Becke, A.D. (1988) *Physical Review A*, **38**, 3098; Becke, A.D. (1993) *Journal of Chemical Physics*, **98**, 5648.
23. Lynch, B.J. and Truhlar, D.G. (2001) *Journal of Physical Chemistry*, **105**, 2936.
24. Bettinger, H.F., Schleyer, P.v.R., Schreiner, P.R. and Schaefer III, H.F. (1997) In: E.R. Davidson (Ed.) *Modern Electronic Structure Theory and Applications in Organic Chemistry*. World Scientific, River Edge, NJ, p. 89; Bettinger, H.F., Schreiner, P.R., Schleyer, P.v.R. and Schaefer, H.F. (1998) In: P.v.R. Schleyer, N.L. Allinger, T. Clark, J. Gasteiger, P.A. Kollman, H.F. Schaefer and P.R. Schreiner (Eds) *The Encyclopedia of Computational Chemistry* (vol 1). Wiley, Chichester, p. 183.
25. Clark, T., Chandrasekhar, J., Spitznagel, G.W. and Schleyer, P.v.R. (1983) *Journal of Computational Chemistry*, **4**, 294; Spitznagel, G.W., Clark, T., Chandrasekhar, J. and Schleyer, P.v.R. (1982) *Journal of Computational Chemistry*, **3**, 363; Treitel, N., Shenhar, R., Aprahamian, I., Sheradsky, T. and Rabinovitz, M. (2004) *Physical Chemistry Chemical Physics*, **6**, 1113.
26. Binkley, J.S., Pople, J.A. and Hehre, W.J. (1980) *Journal of the American Chemical Society*, **102**, 939; Hariharan, P.C. and Pople, J.A. (1973) *Theoretica Chimica Acta*, **28**, 213; Hehre, W.J., Stewart, R.F. and Pople, J.A. (1969) *Journal of Chemical Physics*, **51**, 2657.
27. Pitzer, R.M. (1983) *Accounts of Chemical Research*, **16**, 207; Schreiner, P.R. (2002) *Angewandte Chemie (International Edition)*, **41**, 3579; Weinhold, F. (2003) *Angewandte Chemie (International Edition)*, **42**, 4188; Mo, Y.R., Wu, W., Song, L.C., Lin, M.H., Zhang, Q. and Gao, J.L. (2004) *Angewandte Chemie (International Edition)*, **43**, 1986.
28. Bickelhaupt, F.M. and Baerends, E.J. (2003) *Angewandte Chemie (International Edition)*, **42**, 4183.
29. Kemp, J.D. and Pitzer, K.S. (1936) *Journal of Chemical Physics*, **4**, 749.
30. Ebert, L. (1929) *Leipziger Vorträge*, 74.
31. Wagner, C. (1931) *Zeitschrift fur Physikalische Chemie*, **B14**, 166.
32. Teller, E. and Weigert, K. (1933) *Nachrichten der Göttinger Gesellschaft der Wissenschaften*, 218.
33. Wilson Jr., E.B. (1938) *Journal of Chemical Physics*, **6**, 740.
34. Csaszar, A.G., Allen, W.D. and Schaefer, H.F. (1998) *Journal of Chemical Physics*, **108**, 9751.
35. Badenhoop, J.K. and Weinhold, F. (1997) *Journal of Chemical Physics*, **107**, 5406.
36. Mulliken, R.S. (1939) *Journal of Chemical Physics*, **7**, 339.
37. Weinhold, F. (1999) *Journal of Chemical Education*, **76**, 1141.
38. Pophristic, V. and Goodman, L. (2001) *Nature*, **411**, 565; Carpenter, J.E. and Weinhold, F. (1988) *Journal of Molecular Structure (Theochem)*, **169**, 41; Brunck, T.K. and Weinhold, F. (1979) *Journal of the American Chemical Society*, **101**, 1700.
39. Bader, R.F.W. (1994) *Atoms in Molecules – A Quantum Theory*. Clarendon, Oxford.
40. Mulliken, R.S. (1962) *Journal of Chemical Physics*, **36**, 3428; Mulliken, R.S. (1955) *Journal of Chemical Physics*, **23**, 1833.
41. Martin, F. and Zipse, H. (2005) *Journal of Computational Chemistry*, **26**, 97.
42. Kitaura, K. and Morokuma, K. (1976) *International Journal of Quantum Chemistry*, **10**, 325.
43. Shaik, S. and Shurki, A. (1999) *Angewandte Chemie (International Edition)*, **38**, 587; Shaik, S. (1998) In: P.v.R. Schleyer, N.L. Allinger, T. Clark, J. Gasteiger, H.F. Schaefer and P.R. Schreiner (Eds) *Encyclopedia of Computational Chemistry* (vol 5). Wiley, Chichester, p. 3143; Shaik, S. and Hiberty, P.C. (2003) *Helvetica Chimica Acta*, **86**, 1063.
44. Petillo, P. and Lerner, L. (1993) *The Anomeric Effect and Associated Stereoelectronic Effects*. American Chemical Society, New York; Cramer, C.J., Kelterer, A.M. and French, A.D. (2001) *Journal of Computational Chemistry*, **22**, 1194.
45. Collins, P.M. and Ferrier, R.J. (1995) *Monosaccharides: Their Chemistry and Their Roles in Natural Products*. Wiley-Interscience, New York.
46. Bystrom, K. and Mansson, M. (1982) *Journal of Chemical Society, Perkin Transactions 2*, 565.

47. Brown, H.C. and Schleyer, P.v.R. (1977) *The Nonclassical Ion Problem*. Plenum, New York; Lenoir, D., Apeloig, Y., Arad, D. and Schleyer, P.v.R. (1988) *Journal of Organic Chemistry*, **53**, 661; Olah, G.A. (1995) *Angewandte Chemie*, **107**, 1519; Olah, G.A. and Molnár, A. (2003) *Hydrocarbon Chemistry* (2nd edn). Wiley, Hoboken, NJ.
48. Schreiner, P.R., Schleyer, P.v.R. and Schaefer, H.F. (1997) *Journal of Organic Chemistry*, **62**, 4216.
49. Winstein, S. and Trifan, D.S. (1952) *Journal of the American Chemical Society*, **74**, 1154; Winstein, S. and Trifan, D.S. (1949) *Journal of the American Chemical Society*, **71**, 2953; Winstein, S. and Trifan, D.S. (1952) *Journal of the American Chemical Society*, **74**, 1147.
50. Schleyer, P.v.R., Watts, W.E., Fort, R.C., Comisarow, M.B. and Olah, G.A. (1964) *Journal of the American Chemical Society*, **86**, 4195.
51. Schleyer, P.v.R. and Sieber, S. (1993) *Angewandte Chemie (International Edition)*, **32**, 1606; Sieber, S., Buzek, P., Schleyer, P.v.R., Koch, W. and Carneiro, J.W.d.M. (1993) *Journal of the American Chemical Society*, **115**, 259.
52. Maskill, H. (1976) *Journal of the American Chemical Society*, **98**, 8482; Maskill, H. (1975) *Journal of Chemical Society, Perkin Transactions 2*, 1850; Maskill, H. (1976) *Journal of Chemical Society, Perkin Transactions 2*, 1889.
53. Miertus, S., Scrocco, E. and Tomasi, J. (1981) *Chemical Physics*, **55**, 117.
54. Allen, M.P. and Tildesley, D.J. (1987) *Computer Simulation of Liquids*. Clarendon, Oxford; Beveridge, D.L. and DiCapua, F.M. (1989) *Annual Review of Biophysics and Biophysical Chemistry*, **18**, 431; Zwanzig, R.W. (1954) *Journal of Chemical Physics*, **22**, 1420.
55. Schreiner, P.R., Severance, D.L., Jorgensen, W.L., Schleyer, P.v.R. and Schaefer, H.F. (1995) *Journal of the American Chemical Society*, **117**, 2663.
56. Boys, S.F. and Bernardi, F. (1970) *Molecular Physics*, **19**, 553.

# Chapter 8
# Calorimetric Methods of Investigating Organic Reactions

## U. Fischer and K. Hungerbühler

## 8.1   Introduction

In the pharmaceutical and fine chemical industries, process development and optimisation start when the target chemical structure and a possible synthetic path have been identified by chemical research. Chemical process development ends when the production has been successfully implemented in the final production facility.

At the core of every chemical process there is an intended reaction generally accompanied by unwanted side reactions. The intended reaction may proceed in one single step or, more often, takes place in several chemical transformations. Also, side reactions may proceed in multiple steps thus leading to complex reaction schemes. Process development and process control aim at choosing operating conditions favouring the synthesis of the main product and minimising unwanted by-products. A high yield signifies not only a higher economic profit from product sales but also an efficient use of raw materials, energy (e.g. less energy required for separations), and utilities, as well as the generation of less waste and lower emissions. These last mentioned advantages also improve the profitability of a process because smaller amounts of raw materials and less energy have to be paid for, and less waste has to be treated and disposed of.

Many tasks of process development and optimisation can be carried out, or are significantly supported, only if a reaction model and the corresponding parameters are available. However, the reliability and usefulness of the data calculated strongly depend on the chosen reaction model and the quality of the reaction parameters used. A fundamental understanding of the thermokinetics is also a prerequisite for an investigation of process safety.

Of course, most of the chemical reactions employed in the production of fine chemicals and pharmaceuticals are rather complex from a mechanistic point of view. However, it should be possible to propose reasonable empirical models for most of the reactions from basic chemical knowledge. An empirical reaction model has to fulfil the needs of the early process development, but does not have to represent deep insight into the actual reaction mechanism. Thus, an empirical reaction model need only describe the most important main and side reactions with as few reaction parameters as possible. This will minimise the effort needed to quantify the proposed parameters and increase the robustness of the model in the later application.

As mentioned, all reaction models will include initially unknown reaction parameters such as reaction orders, rate constants, activation energies, phase change rate constants, diffusion coefficients and reaction enthalpies. Unfortunately, it is a fact that there is hardly any knowledge about these kinetic and thermodynamic parameters for a large majority of reactions in the production of fine chemicals and pharmaceuticals; this impedes the use of model-based optimisation tools for individual reaction steps, so the identification of optimal and safe reaction conditions, for example, can be difficult.

Although many different analytical techniques have been developed during the past decades, and various mathematical algorithms exist to extract the desired information from experimental data, these methods nevertheless suffer from some fundamental drawbacks. For example, many analytical techniques requiring calibration and sampling still take too long; with regard to sampling, therefore, online analytical techniques offer an important advantage. Furthermore, not all of the desired reaction parameters can be measured directly and some can only be obtained by complex processing of the basic measurement data. Such determinations are often time consuming or require sophisticated mathematical techniques. In the early stages of process development, there might also be insufficient quantities of the essential test compounds available to carry out the required analyses. These facts call for a further development of the available analytical techniques, or the invention of new ones. Otherwise, considerable potential for the improvement of many chemical processes, which in fact needs to be achieved, might remain elusive.

## 8.2 Investigation of reaction kinetics and mechanisms using calorimetry and infrared spectroscopy

The kinetic and thermodynamic characterisation of chemical reactions is a crucial task in the context of thermal process safety as well as process development, and involves considering objectives as diverse as profit and environmental impact. As most chemical and physical processes are accompanied by heat effects, calorimetry represents a unique technique to gather information about both aspects, thermodynamics and kinetics. As the heat-flow rate during a chemical reaction is proportional to the rate of conversion (expressed in mol s$^{-1}$), calorimetry represents a differential kinetic analysis method [1]. For a simple reaction, this can be expressed in terms of the mathematical relationship in Equation 8.1:

$$q_{react}(t) \sim r(t) \, V_r, \tag{8.1}$$

where $q_{react}$ is the reaction heat-flow rate (with units W) measured by a calorimeter, $r$ is the rate of reaction (mol m$^{-3}$ s$^{-1}$) and $V_r$ is the reaction volume (m$^3$). All three variables ($q_{react}$, $r$ and $V_r$) are functions of time and the progress of the chemical reaction, and thus change during the investigation of the reaction. For complex reactions, the reaction heat-flow rate is influenced by the different reaction steps, and its allocation to individual steps might be difficult.

In contrast to calorimetry, most of the analytical techniques that are applied to the study of kinetics, such as concentration measurements or online measurement of reaction spectra (e.g. UV–vis, near infrared, mid infrared and Raman), can be related to integral kinetic analysis methods [1]. This can be expressed in terms of the proportionality in Equation 8.2:

$$s_i(t) \sim c_i(t), \tag{8.2}$$

where $s_i$ represents the value measured by one of the analytical sensors mentioned above, which corresponds to the $i$th component in the reaction system with the concentration–time profile $c_i(t)$ (expressed in mol m$^{-3}$). From this, it becomes clear that any combination of a differential analysis method, such as calorimetry, with an integral analysis method could lead to a significant improvement in the kinetic analysis. Here, infrared spectroscopy, in particular *attenuated total reflectance infrared spectroscopy* (IR-ATR), will be discussed in more detail. Compared with calorimetry, this analytical method provides more information about individual reaction steps and possible intermediates.

## 8.2.1  Fundamentals of reaction calorimetry

For the determination of reaction parameters, as well as for the assessment of thermal safety, several thermokinetic methods have been developed such as differential scanning calorimetry (DSC), differential thermal analysis (DTA), accelerating rate calorimetry (ARC) and reaction calorimetry. Here, the discussion will be restricted to reaction calorimeters which resemble the later production-scale reactors of the corresponding industrial processes (batch or semi-batch reactors). We shall not discuss thermal analysis devices such as DSC or other micro-calorimetric devices which differ significantly from the production-scale reactor.

Calorimetric applications can also be differentiated by the way in which the reaction temperature is controlled, i.e. isothermal, adiabatic, temperature programmed and isoperibole (constant coolant temperature) modes exist. For the purpose of scale-up, as well as for kinetic and thermodynamic analysis of a desired synthetic reaction, isothermal reaction measurements are mostly preferred. This mode is supposed to be the easiest in application because no heat accumulation by the reactor content has to be considered, so no heat capacities as a function of temperature are required. Therefore, we will focus on isothermal reaction calorimetric measurements. However, it should be mentioned that mainly non-isothermal measurements are carried out, especially in the field of safety analysis, in order to investigate undesired decomposition reactions. Since non-isothermal experiments provide information about the temperature dependence of the chemical reaction system under investigation, their information content is obviously larger, compared with isothermal measurements. This may be an advantage when sophisticated evaluation methods are available, but (especially for complex reaction systems) the information density of non-isothermal reaction measurements is often too large for the common analysis methods. In addition, isothermal conditions have the advantage that the temperature dependences of any signals obtained from additional integral analytical sensors, which may be combined with the calorimetric measurements, do not need to be considered.

## 8.2.2  Types of reaction calorimeters

Most of the existing reaction calorimeters consist of a reaction vessel and a surrounding jacket with a circulating fluid that transports the heat away from the reactor (see Fig. 8.1)

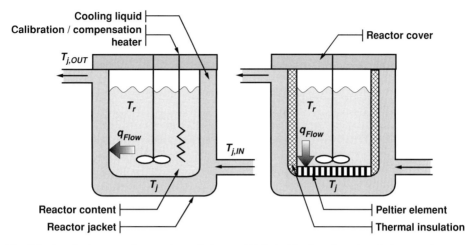

**Fig. 8.1**　Standard set-up of a reaction calorimeter [4]. Left side: heat-flow, heat-balance and power-compensation calorimeters. Right side: Peltier calorimeters.

[2–4]. Such devices can be classified according to their measurement and control principles as follows.

### 8.2.2.1　Heat-flow calorimeters

The temperature of the reactor content ($T_r$, see Fig. 8.1) is controlled by varying the temperature of the cooling liquid ($T_j$). The heat-flow rate from the reactor content through the wall into the cooling liquid ($q_{Flow}$) is determined by measuring the temperature difference between the reactor content and the cooling liquid. In order to convert this temperature signal into a heat-flow signal (usually expressed in W), a heat-transfer coefficient has to be determined using a calibration heater. To allow a fast control of $T_r$, the flow rate of the cooling liquid through the jacket should be high. The heat-flow principle was developed by Regenass and co-workers [3], and most of the commercially available reaction calorimeters, such as the RC1 from Mettler Toledo, the SysCalo devices from Systag and the Simular from HEL, are based on this principle.

### 8.2.2.2　Power-compensation calorimeters

The temperature of the reactor content ($T_r$) is controlled by varying the power of a compensation heater inserted directly into the reactor content. As with an electrical heater, cooling is not possible, so the compensation heater always maintains a constant temperature difference between the reactor jacket and the reactor content. Thus 'cooling' is achieved by reducing the power of the compensation heater. The heat-flow rate from the reactor content through the wall into the cooling liquid ($q_{Flow}$) is typically not determined because the heat-flow rate of the reaction is directly visible in the power consumption of the compensation heater. The temperature of the cooling liquid ($T_j$) is maintained at a constant temperature by an external cryostat. The power-compensation principle was first implemented by Andersen and was further developed by several researchers [3]. Recently, small-scale power-compensation calorimeters have been developed [5, 6]. Commercial power-compensation calorimeters are AutoMate and Simular (combined with heat flow) from HEL.

### 8.2.2.3    Heat-balance calorimeters

The temperature of the reactor content ($T_r$) is controlled by varying the temperature of the cooling liquid ($T_j$). The heat-flow rate from the reactor content through the wall into the cooling liquid ($q_{Flow}$) is determined by measuring the difference between the jacket inlet ($T_{j,IN}$) and outlet ($T_{j,OUT}$) temperatures and the mass flow of the cooling liquid. Together with the heat capacity of the cooling liquid, the heat-flow signal is directly determined without calibration. The heat-balance principle was first implemented by Meeks [3]; commercial versions are the RM200 from Chemisens, the SysCalo 2000 Series from Systag and the ZM-1 from Zeton Altamira (developed in collaboration with Moritz and co-workers [3]).

### 8.2.2.4    Peltier calorimeters

The temperature of the reactor content ($T_r$) is controlled by varying the power of the Peltier elements. In contrast to the heaters in power-compensation calorimeters, Peltier elements can be used for cooling and heating. The heat-flow rate from the reactor content through the Peltier element into the cooling liquid ($q_{Flow}$) is calculated on the basis of the required electrical power and the measured temperature gradient over the Peltier elements. The temperature of the cooling liquid ($T_j$) is maintained at a constant temperature by an external cryostat. Becker designed the first calorimeter using Peltier elements [3], and a similar one was described by Nilsson and co-workers [3]. The latter is similar to the one shown in Fig. 8.1 (right-hand side), but the whole reactor is immersed in a thermostat bath that replaces the reactor jacket. The reactor base consists of Peltier elements and the rest of the reactor wall is insulated; consequently, the main heat flow out of the reactor is through the Peltier elements. Nilsson's design was the basis for the commercially available CPA200 from Chemisens. However, in the CPA200, the heat flow through the reactor base is not calculated from the power consumption of the Peltier elements but by a heat-flow sensor which is incorporated between the reactor base and the Peltier elements.

## 8.2.3    Steady-state isothermal heat-flow balance of a general type of reaction calorimeter

The only heat-flow rate discussed so far has been the heat flow through the reactor jacket ($q_{Flow}$ in Fig. 8.1). For the general case of an isothermal reaction, the main heat-flow rates that have to be considered in a reaction calorimeter are shown in Fig. 8.2 and will be discussed next. In this discussion, ideal isothermal control of the reaction temperature, $T_r$, will be assumed [4]. Consequently, no heat accumulation terms of the reaction mixture and the reactor inserts are shown in Fig. 8.2. However, this underlying assumption does not hold for all applications and apparatuses.

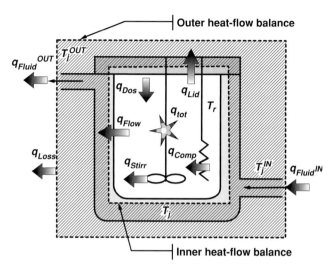

**Fig. 8.2** Main heat-flow rates that have to be considered in heat-flow, heat-balance and power-compensation reaction calorimeters running under strictly isothermal conditions [4]. The heat-flow rates inside a Peltier calorimeter are analogous (compare with Fig. 8.1). The direction of the heat-flow arrows corresponds to a positive heat-flow rate. For explanation of the different heat-flow rates, see the text.

The task of the calorimeter is to determine the total heat-flow rate, $q_{tot}$ (the units being W), during a chemical reaction. Generally, any kind of chemical or physical process in which heat is released or absorbed is included. Therefore, $q_{tot}$ can be expressed by Equation 8.3:

$$q_{tot} = q_{React} + q_{Mix} + q_{Phase},\qquad(8.3)$$

where $q_{React}$ is the reaction heat-flow rate, $q_{Mix}$ is the heat-flow rate due to mixing enthalpies when different fluids are mixed and $q_{Phase}$ is the heat-flow rate due to phase changes (all expressed in W). For a reaction at constant pressure, the reaction heat-flow rate component can be expressed by Equation 8.4:

$$q_{React} = - \sum_{j=1,\dots,N_R} \Delta_r H_j\, r_j\, V_r,\qquad(8.4)$$

where $\Delta_r H_j$ is the enthalpy of the $j$th reaction (in J mol$^{-1}$), $V_r$ is the volume of the reaction mixture (in m$^3$), $r_j$ is the $j$th rate of reaction (in mol m$^{-3}$ s$^{-1}$ with a positive sign) and $N_R$ is the number of reactions. Note that, in the field of reaction calorimetry, the total heat-flow rate $q_{tot}$ is generally defined as positive when heat is released by the chemical reaction. Therefore, a negative sign is introduced into Equation 8.4 to ensure that $q_{React}$ is positive for an exothermic reaction (negative $\Delta_r H$).

The heat evolved by the stirrer, $q_{Stirr}$ (in W), can be described by Equation 8.5:

$$q_{Stirr} = Ne\,\rho_r n_S^3 d_R^5,\qquad(8.5)$$

where $Ne$ is the dimensionless Newton number, $\rho_r$ is the density of the reaction mixture (in kg m$^{-3}$), $n_S$ is the stirrer frequency (Hz) and $d_R$ is the diameter of the stirrer (m). The heat-flow rate caused by the addition of reactants, $q_{Dos}$ (W), is given by Equation 8.6:

$$q_{Dos} = f\,c_{p,Dos}(T_{Dos} - T_r),\qquad(8.6)$$

where $f$ is the reactant flow rate (in mol s$^{-1}$), $c_{p,\text{Dos}}$ is the specific heat capacity of the added liquid (in J mol$^{-1}$ K$^{-1}$) and $T_{\text{Dos}}$ (K) is the temperature of the added liquid. The crucial heat-flow rate, $q_{\text{Flow}}$, shown in Figs 8.1 and 8.2 is generally expressed by the following steady-state equation, Equation 8.7,

$$q_{\text{Flow}} = UA\,(T_r - T_j), \tag{8.7}$$

where $A$ is the total heat-transfer area (m$^2$) and $U$ is the overall heat-transfer coefficient (W m$^{-2}$ K$^{-1}$).

The parameter $U$ consists of the two main coefficients of heat transfer shown in Equation 8.8:

$$\frac{1}{U} = \frac{1}{h_r} + \frac{1}{\varphi}, \tag{8.8}$$

where $h_r$ is the solution-to-wall coefficient for the steady-state heat transfer (in W m$^{-2}$ K$^{-1}$) and $\varphi$ is a device-specific heat-transfer coefficient (in W m$^{-2}$ K$^{-1}$). For a standard reaction calorimeter with a cooling jacket (see Fig. 8.1, left-hand side), $\varphi$ can be resolved further as in Equation 8.9:

$$\frac{1}{\varphi} = \frac{L}{\lambda_W} + \frac{1}{h_j}, \tag{8.9}$$

where $h_j$ is the wall-to-jacket coefficient for the steady-state heat transfer (in W m$^{-2}$ K$^{-1}$), $\lambda_W$ is the heat conductivity of the reactor wall (in W m$^{-1}$ K$^{-1}$) and $L$ is the thickness of the reactor wall (in m). If the reactor wall contains a Peltier element, a more sophisticated description for the device-specific heat transfer is required, but the following discussion is still valid.

The two steady-state heat-transfer coefficients, $h_r$ and $h_j$, could be further described in terms of the physical properties of the system. The solution-to-wall coefficient for heat transfer, $h_r$ in Equation 8.8, is strongly dependent on the physical properties of the reaction mixture (heat capacity, density, viscosity and thermal conductivity) as well as on the fluid dynamics inside the reactor. Similarly, the wall-to-jacket coefficient for heat transfer, $h_j$, depends on the properties and on the fluid dynamics of the chosen cooling liquid. Thus, $U$ generally varies during measurements on a chemical reaction mainly for the following two reasons.

(1) $h_r$ varies because the physical properties of the reaction mixture change during a reaction (e.g. viscosity increases during a polymerisation reaction).
(2) Depending on the calorimetric system chosen, $h_j$ may vary because the jacket temperature ($T_j$) changes during measurements on the reaction and, consequently, the physical properties of the cooling liquid that determine $h_j$ change as well (this only applies for reaction calorimeters with a cooling jacket, see Fig. 8.1, left-hand side).

Not only $U$ but also the heat-transfer area, $A$ in Equation 8.7, can change during a reaction because of volume changes caused by density changes or addition of reactants. For a Peltier calorimeter designed according to Fig. 8.1 (right-hand side), however, $A$ does not change.

Equation 8.8 is only valid under steady-state conditions when the heat-flow rate through the reactor wall is constant. However, if a reaction is taking place, the heat-flow rate through the reactor wall might vary depending on the calorimetric principle being applied. Therefore, heat accumulation occurs inside the reactor wall, or inside the Peltier element, as well as the reactor- and jacket-sided film layers. Recently, heat-flow models for the reactor wall have been proposed [7, 8], but neither the dynamic heat transfer within the reaction mixture nor that within the cooling liquid was considered. Due to the complexity of an exact physical consideration, the dynamically changing aspect of $q_{Flow}$ is generally neglected completely, and the steady-state equation 8.8 is used for the heat-transfer components.

## 8.2.4  Infrared and IR-ATR spectroscopy

Vibrations and rotations of molecules generally absorb electromagnetic irradiation in the infrared range (400 to 4000 $cm^{-1}$) as long as the dipole moment changes during the vibration or rotation. In the liquid phase, the rotations of the molecules are strongly influenced by intermolecular interactions. A consequence is that rotational fine structure cannot be resolved, so vibrational absorption bands appear broad in the infrared spectrum. However, the vibrational bands remain characteristic of specific functional groups of the molecules. Consequently, infrared spectroscopy is often used to characterise substances and, in combination with reaction calorimetry, is used to record changes in a reaction mixture as a function of time. These changes originate in the chemical transformation of the compounds in the reaction and, therefore, give important information about their concentration–time profiles.

Conventionally, infrared spectroscopy is carried out in the transmission mode, where the light passes through a sample cell with a defined thickness. There are two main disadvantages of this technique for the purpose of reaction analysis.

(1)  Either samples have to be withdrawn, or a flow-through cell has to be constructed.
(2)  Some solvents, such as water, absorb strongly over a wide range of the infrared spectrum; the sample thickness, therefore, has to be very small, otherwise quantitative analysis will be very inaccurate. In practice, however, small sample thicknesses are difficult to achieve.

A general solution to both problems is the application of attenuated total reflectance (ATR) in combination with infrared spectroscopy. The theory of ATR spectroscopy is well described in several books and articles which also demonstrate the applicability of the Beer–Lambert law to ATR spectroscopy [9]. The combination of reaction calorimetry and ATR spectroscopy is now rather common [10–13] typically using commercially available calorimeters.

The technique of IR-ATR spectroscopy is easy to apply in reaction analysis as no sampling or flow-through cells are required. As most organic compounds are infrared-active, the technique is useful for many reaction types. However, there are some matters that should always be kept in mind when the reaction's IR-ATR spectrum is interpreted.

- The penetration depth of the IR beam is in the μm range from the ATR surface. Consequently, the assumption that the reaction observed in this surface layer equals the reaction in the bulk medium must be verified. If slurries are involved, only the liquid phase can be analysed.
- Infrared spectra are generally temperature dependent as the observed vibrations of the molecules in the liquid phase depend on the temperature. Additionally, the effective thickness of the ATR 'sample' depends on physical properties that may vary with temperature. Consequently, IR-ATR measurements should be carried out at constant temperature.

- The effective thickness of the ATR 'sample' depends on the absorption of the reaction mixture. During reaction measurements, this absorption is changing, so the Beer–Lambert law may not always be obeyed. Generally, these disturbances can be neglected but care should be taken when strong absorption bands are observed.
- The technique might fail to identify components at low concentrations because their contributions to the total measured absorbance may not be detected above the background noise.
- The ATR crystal also absorbs light over a certain interval of the infrared range which, therefore, will not be available for measurements.
- In order to increase the signal-to-noise ratio for a measured spectrum, several spectra are typically recorded and averaged. The duration of a standard measurement, therefore, is in the range of 10 to 60 seconds.

### 8.2.5 Experimental methods for isothermal calorimetric and infrared reaction data

#### 8.2.5.1 Experimental methods for isothermal calorimetric reaction data

The task of any reaction calorimeter is to determine the total heat-flow rate, $q_{tot}$ (in W), during a reaction. In the following, a summary of different methods for the isothermal determination of $q_{tot}$ will be given. The aim of all methods described below is to determine the enthalpy and the kinetic model parameters (such as reaction orders and associated rate constants) of the reaction under investigation [4]. If the temperature dependence of the reaction has to be studied, isothermal measurements at several different temperatures have to be carried out. The results of the individual investigations can then be plotted, e.g. in an Arrhenius plot, to determine the activation energy. Some of the proposed techniques also allow a simultaneous determination of several isothermal measurements at different temperatures by replacing rate constants using Equation 8.10:

$$k = k(T_{ref}) \exp\left(-\frac{E_A}{R}\left(\frac{1}{T_r} - \frac{1}{T_{ref}}\right)\right),  \tag{8.10}$$

where $T_{ref}$ (K) is a reference temperature, $E_A$ (J mol$^{-1}$) is the Arrhenius activation energy and $R$ (J mol$^{-1}$ K$^{-1}$) is the ideal gas constant.

As indicated in Equation 8.3, $q_{tot}$ is not generally simply equal to the reaction heat-flow rate $q_{React}$ (see Equation 8.4) but is affected by other physical or chemical processes which have heat changes, e.g. mixing or phase changes. As will be shown in Section 8.3, even for a simple reaction such as the hydrolysis of acetic anhydride, a significant heat of mixing occurs which must be taken into account. Furthermore, it should always be kept in mind that the $q_{tot}$ values determined by a reaction calorimeter also contain measurement errors such as base line drifts, time distortions or ambient temperature influences.

First we will discuss measurement methods that do not require postulation of a reaction model such as the determination of the reaction enthalpy by integration of $q_{tot}$, which is the simplest of the model-free methods that still leads to a physically meaningful result. The integration of the measured $q_{tot}$ signal leads to Equation 8.11:

$$Q_{tot} = \int_{t=0}^{t=t_f} q_{tot} dt = \sum_{i=1,...,N_R} (-\Delta_r H_i) n_{M,i} + Q_{mix} + Q_{Phase} + Q_{Error}, \qquad (8.11)$$

where $t_f$ is the time integration limit (in s), $Q_{tot}$ is the integral of the total heat-flow rate (in J), $n_{M,i}$ is the number of moles of the $i$th reaction component (mol), $Q_{mix}$ is the heat of mixing (J), $Q_{Phase}$ is the heat released or absorbed by phase change processes (J) and $Q_{Error}$ is the sum of all measurement errors (J). Note that the appropriate selection of the integration time limit has a significant influence on the result for $Q_{tot}$.

For demonstration purposes, we shall assume here that the values of $Q_{mix}$, $Q_{Phase}$ and $Q_{Error}$ are negligible, meaning that $q_{tot} = q_{React}$. Furthermore, the reaction is assumed to take place in one single step. If these assumptions are valid, the reaction enthalpy $\Delta_r H$ can be calculated directly by Equation 8.12:

$$-\Delta_r H \approx -\Delta H = \frac{Q_{tot}}{n_M}, \qquad (8.12)$$

where $\Delta H$ is the total enthalpy change (J mol$^{-1}$) and $n_M$ is the number of moles transformed in the reaction investigated. However, in real applications, $Q_{mix}$, $Q_{Phase}$ and $Q_{Error}$ are generally not zero and will, therefore, be fully integrated into the reaction enthalpy. Once the reaction enthalpy has been determined, it is possible to calculate the thermal conversion or fractional heat evolution of the reaction by Equation 8.13:

$$X_{thermal}(t) = \frac{\int_{\tau=0}^{\tau=t} q_{tot} d\tau}{Q_{tot}}. \qquad (8.13)$$

This thermal conversion can be compared to the chemical conversion of the investigated reaction as long as the assumptions mentioned above are valid, or the correspondence between chemical and thermal conversions has been verified by another analytical technique.

If the assumptions made above are not valid, and/or information about the rate constants of the investigated reactions is required, model-based approaches have to be used. Most of the model-based measurements of the calorimetric signal are based on the assumption that the reaction occurs in one single step of $n$th order with only one rate-limiting component concentration; in the simplest case, this would be pseudo-first-order kinetics with all components except one in excess. The reaction must be carried out in batch mode ($V_r$ = constant) in order to simplify the determination, and the general reaction model can, therefore, be written as Equation 8.14 with component $A$ being rate limiting:

$$A + \cdots \rightarrow Prod \qquad r_A(t) = -kC_A(t)^n \qquad q_{React}(t) = \Delta_r H \, r_A(t) V_r. \qquad (8.14)$$

In this equation, $C_A$ is the concentration (in mol m$^{-3}$) of the rate-limiting component $A$, $k$ is the $n$th-order rate constant (with units m$^{3(n-1)}$ mol$^{1-n}$ s$^{-1}$), $n$ is the order of the reaction and $r_A$ is the rate of reaction (units, mol m$^{-3}$ s$^{-1}$). As already mentioned, in the field of reaction calorimetry, $q_{React}$ is generally defined as positive for an exothermic reaction (negative $\Delta_r H$). The aim of the determination is to calculate the kinetic parameters $k$ and (possibly) $n$. Some methods also determine the thermodynamic parameter $\Delta_r H$ on the basis of this reaction model.

Another option in calorimetric experiments is the determination of the kinetic parameters based on the reaction enthalpy determined by prior integration of $q_{tot}$ according to Equation 8.11. Assuming that $Q_{mix}$, $Q_{Phase}$ and $Q_{Error}$ in Equation 8.11 are negligible, $\Delta_r H$ as well as the thermal conversion curve can be calculated according to Equations 8.11–8.13. The rate of reaction, $r_A$, can then be expressed as Equation 8.15 where $C_{A,0}$ is the initial concentration of component $A$:

$$r_A(t) = \frac{q_{tot}(t)}{V_r \Delta_r H} = -kC_{A,0}^n \left(1 - X_{thermal}(t)\right)^n.$$
(8.15)

A plot of $\log(r_A)$ versus $\log(1 - X_{thermal})$ should be linear, and the reaction order ($n$) with respect to component $A$ can be determined from the slope, and the rate constant ($k$) from the intercept.

If the reaction order ($n$) with respect to component $A$ is known in advance, the reaction model in Equation 8.14 can be integrated. Assuming the reaction is first order in component $A$ ($n = 1$), the rate constant, $k$, can be determined by the non-linear least-squares optimisation indicated in Equation 8.16:

$$\min_k \sum_{i=1}^{N_t} \left[X_{thermal}(t_i) - \left(1 - e^{-kt_i}\right)\right]^2,$$
(8.16)

where $t_i$ is the $i$th calorimetric measurement and $N_t$ is the total number of measurements.

The last possibility discussed here for a simple model-based determination is the separate determination of the rate constant, $k$, as well as the reaction enthalpy, $\Delta_r H$, based on the postulated reaction model. For first-order reactions ($n = 1$ in Equation 8.14), the reaction model can be integrated and $q_{tot}$ can be expressed by Equation 8.17 assuming that $q_{tot} = q_{React}$:

$$q_{tot}(t) = V_r(-\Delta_r H)kC_{A,0}e^{-kt}.$$
(8.17)

A plot of $\log(q_{tot})$ versus time should result in a straight line with gradient $= -k$. Any period of the reaction when the assumption $q_{tot} = q_{React}$ is not valid will be evident in the plot as a deviation from the straight line. Such periods can then be excluded from the determination of $k$.

The methods for calorimetric measurements discussed above can only be applied for single-step reactions in batch mode with one single rate-limiting component concentration. If the evaluation of the calorimetric signal is to be extended to the general case of the semi-batch operation mode ($V_r = V_r(t)$), or to multiple reaction systems including eventual mass-transfer processes, these methods will fail. More general evaluation methods have been developed for such circumstances. The basis of these more general methods is a reaction model represented by a system of ordinary differential equations. The reaction model can now include more than one chemical reaction as well as mass-transfer or dosing processes. In general, analytical solutions for these reaction models do not exist, so integration is carried out by numerical methods.

The task is the determination of the parameters of the reaction model. These reaction model parameters can be rate constants, activation energies, reaction orders or mass-transfer parameters. Additionally, the reaction enthalpies of the different reaction steps have to be determined because the integration approach represented by Equations 8.11 and 8.12 is no longer feasible. The parameters to be determined are obtained by fitting the postulated model to the calorimetric measurements, i.e. the difference between measured $q_{tot}$ and calculated $q_{tot}$ (each as a function of time) is minimised using (for example) the non-linear least-squares optimisation method indicated in Equation 8.18:

$$\min_{\theta_{1,\dots,N_P},\,\Delta_r H_{1,\dots,N_R}} \sum_{i=1}^{N_t} \left[ q_{tot}(t_i) - \sum_{j=1}^{N_R} V_r(t_i)(-\Delta_r H_j) r_j(t_i, \theta_{1,\dots,N_P}) \right]^2, \tag{8.18}$$

where $\theta_{1,\dots,N_P}$ are the unknown reaction model parameters, $N_P$ is the number of these model parameters, $\Delta_r H_j$ is the $j$th reaction enthalpy (J mol$^{-1}$), $N_R$ is the number of reactions, $N_t$ is the number of time samples and $r_j$ is the $j$th rate of reaction (in mol m$^{-3}$ s$^{-1}$ with a positive sign). The application of this approach involves all the possible pitfalls inherent in non-linear optimisation, such as numerous local minima which distract the search for the desired global minimum; however, for complex reaction systems, this is the only viable way.

### 8.2.5.2 Experimental methods for isothermal infrared reaction data

The basis of all determinations under this heading is the Beer–Lambert law [14–16]. All reaction spectra obtained by any spectroscopic sensors can be used as long as the Beer–Lambert law is obeyed. For spectra of a reaction containing several components and absorbances measured a number of times at several wavelengths, the matrix form of Equation 8.19 can be used:

$$A(N_t \times N_{\tilde{v}}) = C(N_t \times N_C) \times E(N_C \times N_{\tilde{v}}) \equiv$$

$$\begin{bmatrix} a_{1,1} & \cdots & \cdots & \cdots & a_{1,N_{\tilde{v}}} \\ \cdots & \cdots & A & \cdots & \cdots \\ a_{N_t,1} & \cdots & \cdots & \cdots & a_{N_t,N_{\tilde{v}}} \end{bmatrix} = \begin{bmatrix} c_{1,1} & \vdots & c_{1,N_C} \\ \vdots & \vdots & \vdots \\ \vdots & C & \vdots \\ \vdots & \vdots & \vdots \\ c_{N_t,1} & \vdots & c_{N_t,N_C} \end{bmatrix} \tag{8.19}$$

$$\times \begin{bmatrix} e_{1,1} & \cdots & \cdots & \cdots & e_{1,N_{\tilde{v}}} \\ \cdots & \cdots & E & \cdots & \cdots \\ e_{N_C,1} & \cdots & \cdots & \cdots & e_{N_C,N_{\tilde{v}}} \end{bmatrix}.$$

Here, $A$ is the reaction's measured IR spectral absorbance, $N_t$ is the number of measurements at different times, $N_{\tilde{v}}$ is the number of wavelengths, $C$ is the concentration matrix with the concentration–time profiles of each absorbing component in the columns, $N_C$ is the number of chemical components and $E$ is the 'pure spectra matrix' with the spectral absorption at each wave number of each pure absorbing component in the rows. If a chemical component does not absorb, the corresponding spectrum of the pure chemical will be a vector of zeros.

It should be noted that, by measuring reaction spectra for the purpose of estimating reaction-model parameters (such as rate constants or activation energies), a new set of unknown parameters is introduced, i.e. the spectral absorbances of the pure chemical components involved in the reaction (matrix $E$).

A special evaluation method that requires the postulation of a reaction model is *single peak evaluation*. Generally, a single column of matrix $A$ (see Equation 8.19), corresponding to the absorbance at a selected wavelength as a function of time, $a$, is evaluated in order to determine kinetic parameters of the reaction system. It is assumed that only one chemical component of the reaction system is absorbing at this wavelength. The corresponding evaluation methods are the same as when the concentration profile of this component is known.

Often, the univariate evaluation technique of a single peak will fail because spectral regions where only one single pure chemical component is absorbing cannot be assumed. Consequently, multivariate techniques which allow peak overlapping were developed. Most of them were developed for the recording of NIR reaction spectra where peak overlapping is a serious problem. The same techniques can also be applied to evaluate mid-IR reaction spectra; however, here the spectra are more specific and peak overlapping less severe.

The aim of the multivariate evaluation methods is to fit a reaction model to the measured reaction spectrum on the basis of the Beer–Lambert law and thus identify the kinetic parameters of the model. The general task can be described by the non-linear least-squares optimisation described in Equation 8.20:

$$\min_{\theta_{1,...,N_P},\,\hat{E}} \left\| A - C_{calc}(\theta_{1,...,N_P}) \times \hat{E} \right\|_2^2 \quad \text{with} \quad A(N_t \times N_{\tilde{v}}),\, C_{calc}(N_t \times N_C),\, \hat{E}(N_C \times N_{\tilde{v}}),$$

(8.20)

where $\theta_{1,...,N_P}$ are the unknown model parameters, $N_P$ is the number of model parameters and $A$ is the measured reaction spectrum. Matrix $C_{calc}$ corresponds to $C$ in Equation 8.19 and contains the calculated concentration profiles of the chemical components in the columns. The concentration profiles are simulated using a given reaction model and are, therefore, only dependent on the model parameters, $\theta_{1,...,N_P}$. Simple kinetic reaction models can be integrated analytically and $C_{calc}$ is a direct function of $\theta_{1,...,N_P}$. If the reaction model chosen is too complex for analytical integration, the ordinary differential equations of the reaction model have to be integrated numerically at each iteration step. Matrix $\hat{E}$ corresponds to $E$ in Equation 8.17 and contains the estimated spectral absorbance of each pure chemical component.

The unknown reaction model parameters $\theta_{1,...,N_P}$, as well as the unknown 'pure spectra matrix' $\hat{E}$, have to be identified by solving the optimisation problem expressed in Equation 8.20. The direct solution of this non-linear optimisation is not feasible because far too many unknown parameters would have to be identified at one time. However, the matrix elements of $\hat{E}$ are linear whereas $\theta_{1,...,N_P}$ are non-linear parameters. Thus, Equation 8.20 can be solved by separating the overall optimisation into a linear optimisation for $\hat{E}$ and a non-linear optimisation for $\theta_{1,...,N_P}$. Several solutions for this separation of linear and non-linear parameters have been reported in the literature. The easiest and most straightforward is to replace $\hat{E}$ by its linear least-squares estimate. The linear least-squares problem can be carried out with physical constraints for $\hat{E}$ such as non-negativity. For the solution of the non-linear optimisation, standard mathematical procedures are applied. Another option is the application of *principal component analysis* (PCA) or similar techniques [14–16].

### 8.2.5.3 Methods for combined determination of isothermal calorimetric and infrared reaction data

Equation 8.18 presented the general, non-linear optimisation method for calorimetric measurements based on a reaction model, and Equation 8.20 presented the general, non-linear optimisation method for spectroscopic measurements based on a reaction model. A comparison of these two equations reveals that the non-linear parameters $\theta_{1,...,N_P}$, which are defined by the reaction model, are common to both equations. Only the linear parameters in these equations ($\hat{E}$ and $\Delta_r H_{1,...,N_R}$) are different. As the linear parameters in both equations can be replaced by linear least-squares estimates, the combination of the two non-linear optimisations is relatively straightforward and can be carried out as described by Equation 8.21:

$$\min_{\theta_{1,...,N_P}} \left\{ \begin{array}{l} W_Q \min_{\Delta_r H_{1,...,N_R}} \left( \sum_{i=1}^{N_t} \left[ q_{tot}(t_i) - \sum_{j=1}^{N_R} V_r(t_i)(-\Delta_r H_j) r_j(t_i, \theta_{1,...,N_P}) \right]^2 \right) \\ + W_{IR} \min_{\hat{E}} \left( \left\| A - C_{calc}(\theta_{1,...,N_P}) \times \hat{E} \right\|_2^2 \right) \end{array} \right\}. \quad (8.21)$$

In this equation, $W_Q$ and $W_{IR}$ are weighting factors that express the importance of the residuals obtained in the calorimetric and infrared determinations, respectively. The definition of these weightings is crucial for the results that are obtained, but is not at all straightforward. Recently, an approach to this problem based on an automated sensitivity analysis has been reported [17]. Besides tackling this problem of mathematically combining the evaluation of two different signals measured for the same experiment, we shall demonstrate in Section 8.3 that the application of both measurement techniques in parallel has synergistic effects for the clarification of the physical and chemical processes that are involved in the one experiment.

## 8.3 Investigation of reaction kinetics using calorimetry and IR-ATR spectroscopy – examples of application

The aim of this section is to demonstrate how reaction calorimetry in combination with IR-ATR spectroscopy can be used for the determination of kinetic and thermodynamic parameters. Several examples of chemical reactions will be discussed, each highlighting a different aspect in the application of reaction calorimetry. The reactions considered are the hydrolysis of acetic anhydride, the sequential epoxidation of 2,5-di-*tert*-butyl-1,4-benzoquinone and the hydrogenation of nitrobenzene. The results discussed in this section were obtained using a new calorimetric principle presented below.

### 8.3.1 Calorimetric device used in combination with IR-ATR spectroscopy

The calorimeter that has been used to obtain the results presented in this section basically combines the power-compensation and heat-balance principles (see Sections 8.2.2.2 and 8.2.2.3). The heat-balance principle is implemented by Peltier elements [18]. This new

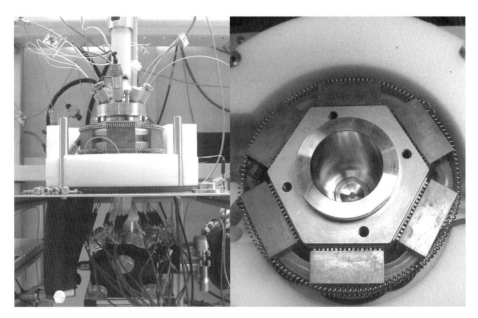

**Fig. 8.3**   Left: front-side view of the new calorimeter/FTIR system. Right: top view of the open reactor with the IR-ATR window at the bottom.

combination of power-compensation and heat-balance calorimetry has been patented [19], and different devices using this principle have been developed. The latest development (see Fig. 8.3) also allows the measurement of gas consumption in hydrogenation reactions.

The Hastelloy® interchangeable vessel has a diameter of 40 mm and a height of approximately 50 mm, and the stand-alone IR-ATR probe by ASI Applied System is fixed in the bottom. The reaction mixture is stirred by a magnetically coupled stirrer, while the pressures of the reactor and reservoirs (to determine the gas consumption) are measured online. The six Peltier elements are located on the outer surface of the hexagonal prism jacket. The coolers are connected to a cryostat (Huber CC150) using a mixture of water and ethanol which allows a cooling temperature of $-30°C$ ($T_{cry}$) to be reached. The temperature of the symmetrical jacket is monitored by 18 thermocouples. Since special high-temperature Peltier elements are installed, the maximum jacket temperature is about 200°C; the minimum is about $-20°C$, which is limited by the cryostat used. The reactor is supplied with eight inserts, two of which are for feed streams and one is to allow samples of the reaction mixture to be withdrawn for external analysis (e.g. by GC or GCMS). In addition, it is optionally possible to introduce an endoscope to enable visual observations. The ATR probe is connected to a Mettler Toledo FTIR spectrophotometer (IR4000). The addition of the feed is by a Jasco pump and the feed temperature ($T_{Dos}$) is measured by an additional thermocouple placed inside the feed tube. All connection tubes are made of PEEK™. The reactor and all the peripherals are controlled by the LabVIEW® programme which also controls the acquisition of the raw calorimetric and IR-ATR data. More details on the experimental set-up can be found elsewhere [18]. The calorimeter has been calibrated using different test reactions for which parameters obtained were in good agreement with literature values [17, 18, 20].

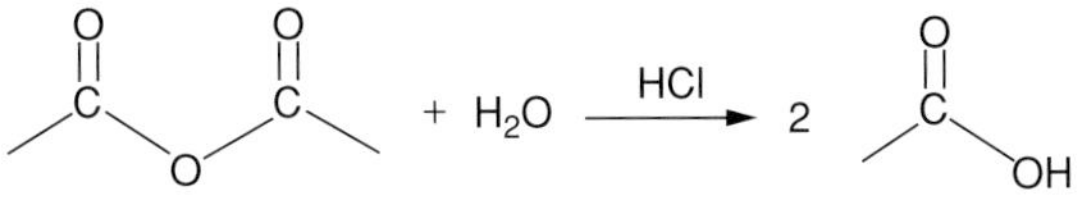

**Scheme 8.1** The hydrolysis of acetic anhydride.

### 8.3.2  Example 1: Hydrolysis of acetic anhydride

As pointed out in Section 8.2, most physical and chemical processes, not just the chemical transformation of reactants into products, are accompanied by heat effects. Thus, if calorimetry is used as an analytical tool and such additional processes take place before, during, or after a chemical reaction, it is necessary to separate their effects from that of the chemical reaction in the measured heat-flow signals. In the following, we illustrate the basic principles involved in applying calorimetry combined with IR-ATR spectroscopy to the determination of kinetic and thermodynamic parameters of chemical reactions. We shall show how the combination of the two techniques provides extra information that helps in identifying processes additional to the chemical reaction which is the primary focus of the investigation. The hydrolysis of acetic anhydride is shown in Scheme 8.1, and the postulated pseudo-first-order kinetic model for the reaction carried out in 0.1 M aqueous hydrochloric acid is shown in Equation 8.22:

$$-\frac{dc_{AcOAc}}{dt} = kc_{AcOAc}. \tag{8.22}$$

#### 8.3.2.1  Materials and methods

First, a reference background spectrum for the IR spectrophotometer was obtained; then the reactor was charged with 35 mL of 0.1 M hydrochloric acid. The stirrer speed was set to 600 rpm and the reaction temperature, $T_r$, was set. Next, 2 g of a mixture of 10.7 mmol of acetic anhydride and 15.1 mmol of acetic acid was added at a constant dosing rate of 5 mL min$^{-1}$. Three experiments were carried out at each of three reaction temperatures, $T_r = 25$, 40 and 55°C. For the determination of $q_{Dos}$, the heat capacity of the feed mixture (1.83 kJ kg$^{-1}$ K$^{-1}$) was calculated using the mass fraction and the heat capacities of the pure components (acetic anhydride: $c_p = 1.65$ kJ kg$^{-1}$ K$^{-1}$ and acetic acid: $c_p = 2.05$ kJ kg$^{-1}$ K$^{-1}$).

#### 8.3.2.2  Results and discussion

The $q_{tot}$ values for the hydrolysis of acetic anhydride at three different temperatures (25, 40 and 55°C) are shown as functions of time in Fig. 8.4. The $q_{tot}$ curve at 25°C shows a significant peak at the beginning of the reaction; this corresponds to the heat of mixing during the dosing phase.

By means of backward extrapolation towards time zero, we were able to separate the heat of mixing at the beginning of the experiment from the heat of reaction, as shown in Fig. 8.5a. In this way, and using Equation 8.18, results at 25°C of $\Delta_r H = -61 \pm 2$ kJ mol$^{-1}$ and $Q_{mix} = -6$ kJ mol$^{-1}$ were obtained. At the same time, the pseudo-first-order rate constant, $k$, was determined to be $2.8 \pm 0.1 \times 10^{-3}$ s$^{-1}$. The measurements of the enthalpies and the rate constant were repeated at 40 and 55°C. The activation energy was then determined

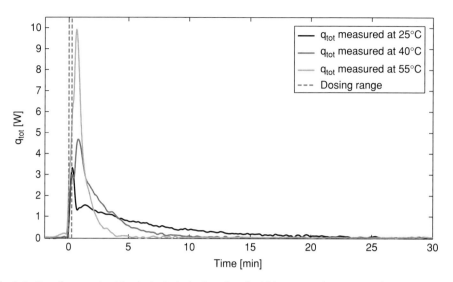

**Fig. 8.4** Heat flow rate ($q_{tot}$) for the hydrolysis of acetic anhydride measured at 25, 40 and 55°C. Reprinted in modified form with permission [18].

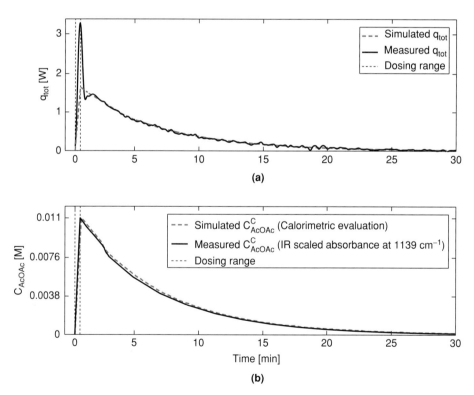

**Fig. 8.5** Hydrolysis of acetic anhydride investigated separately at $T_r = 25°C$ (a) by calorimetry and (b) by infrared spectroscopy. Graph (a) shows measured and simulated reaction power; graph (b) shows measured and simulated concentration–time curves of acetic anhydride. The simulated curve is from the kinetic parameters obtained from the calorimetric measurements, and is compared with the one determined by the IR measurements at 1139 cm$^{-1}$. Reprinted in modified form with permission [18].

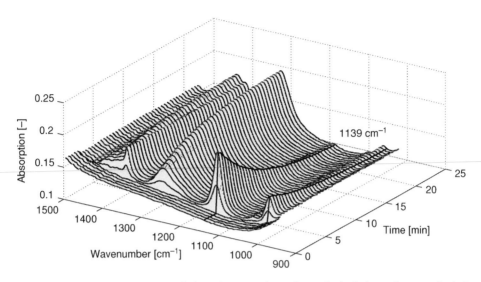

**Fig. 8.6** Part of the IR spectrum recorded as a function of time during the hydrolysis of acetic anhydride at 25°C. The peak indicated at 1139 cm$^{-1}$ was used to monitor the decreasing concentration of acetic anhydride during its hydrolysis (see Fig. 8.5b).

from the rate constants at the three temperatures using the Arrhenius equation ($E_a = 56$ kJ mol$^{-1}$).

The hydrolysis reaction was also investigated with the integrated IR-ATR probe. To determine the first-order rate constant from the IR measurements, the easiest and most efficient way is to find a wavelength where only one component is absorbing; the peak at 1139 cm$^{-1}$ was chosen (see Fig. 8.6), which corresponds to the C—O—C stretching vibration of acetic anhydride. The absorbance data, $A_{AcOAc}$, were converted into corresponding concentrations, $C_{AcOAc}^A$, using Equation 8.23:

$$C_{AcOAc}^A = \left(A_{AcOAc} - A_{AcOAc}^{final}\right) \frac{\max\left(C_{AcOAc}^C\right)}{\max\left(A_{AcOAc} - A_{AcOAc}^{final}\right)}. \qquad (8.23)$$

As can be seen from Fig. 8.5b, the concentrations of acetic anhydride determined from the IR absorbance data ($C_{AcOAc}^A$) and the corresponding concentrations determined from calorimetric measurements excluding the heat of mixing ($C_{AcOAc}^C$) are in good agreement.

The reaction rate constant, $k$, can be determined from the $C_{AcOAc}^A$ data by linear regression as well as from the thermal data. The regression was done over the same time period that was used for the regression of the thermal measurements and the result ($k = 2.6 \pm 0.1 \times 10^{-3}$ s$^{-1}$ at 25°C) is in good agreement with the value obtained from the calorimetric measurements ($2.8 \pm 0.1 \times 10^{-3}$ s$^{-1}$). The determination of the rate constant from spectroscopic measurements was also repeated at 40 and 55°C and, again, the activation energy was calculated; the result ($E_a = 55$ kJ mol$^{-1}$) is in excellent agreement with the value determined from the calorimetric measurements (56 kJ mol$^{-1}$).

When the measured heat-flow rate ($q_{tot}$) curve at 25°C is compared with the results obtained from the IR signal (see Fig. 8.5), it again becomes clear that the initial peak of the $q_{tot}$ curve, which is not visible in the IR signal, is not related to the chemical reaction but to the mixing. In this simple case, we have an excellent example of how the *simultaneous* measurement of a signal other than the heat-flow rate is able to distinguish between chemical and physical heat effects.

### 8.3.3 Example 2: sequential epoxidation of 2,5-di-tert-butyl-1,4-benzoquinone

A fundamental problem of reaction simulation is the choice of an appropriate reaction model. No standard procedure for this problem can be found in the literature. It is essential, therefore, that model-based measurements of reaction data support the task of model selection. Generally, the residuals in the comparison of the data from the modelled reaction with the experimental measurements are taken as an indication of the quality of the reaction model. However, the robustness of the model fit generally decreases with increasing number of reaction parameters (such as rate constants, activation energies, reaction enthalpies or spectral absorbances) that have to be determined. In this example, we demonstrate how different reaction models can be postulated and then tested on the basis of calorimetric and IR-ATR measurements.

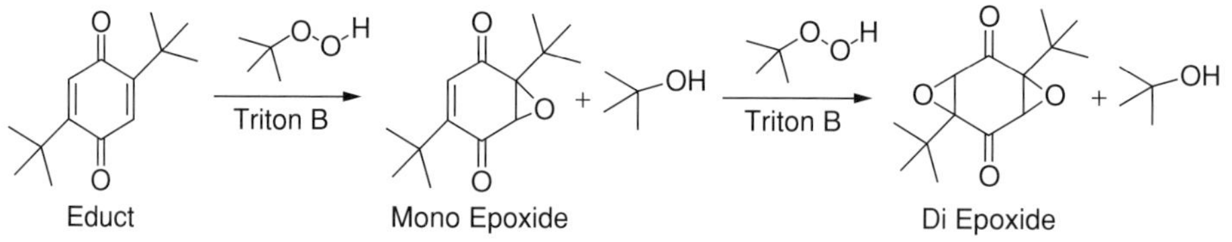

**Scheme 8.2** The sequential epoxidation of 2,5-di-*tert*-butyl-1,4-benzoquinone with *tert*-butyl hydroperoxide.

The sequential epoxidation of 2,5-di-*tert*-butyl-1,4-benzoquinone with *tert*-butyl hydroperoxide is shown in Scheme 8.2. In the experiments discussed below, Triton-B was added to the mixture as a catalyst, and the basic reaction model is written as Equations 8.24:

$$\frac{dn_{\text{Educt}}}{dt} = -r_1(t, k_1) V_r(t) \qquad \frac{dn_{\text{Hydroperoxide}}}{dt} = \{-r_1(t, k_1) - r_2(t, k_2)\} V_r(t)$$

$$\frac{dn_{\text{Mono Epoxide}}}{dt} = \{r_1(t, k_1) - r_2(t, k_2)\} V_r(t) \qquad \frac{dn_{\text{Di Epoxide}}}{dt} = r_2(t, k_2) V_r(t)$$

$$\frac{dn_{\text{tButanol}}}{dt} = \{r_1(t, k_1) + r_2(t, k_2)\} V_r(t)$$

$$\frac{dn_{\text{Solvent}}}{dt} = 0 \qquad \frac{dn_{\text{Methanol}}}{dt} = V_{\text{dos}} C_{\text{dos, Methanol}} \qquad \frac{dn_{\text{TritonB}}}{dt} = V_{\text{dos}} C_{\text{dos, TritonB}}$$

$$\frac{dV_r}{dt} = V_{\text{dos}}$$

$$r_1(t, k_1) = k_1 \frac{n_{\text{Educt}}(t)}{V_r(t)} \frac{n_{\text{TritonB}}(t)}{V_r(t)} C_{\text{Hydroperoxide},0}$$

$$r_2(t, k_2) = k_2 \frac{n_{\text{Mono Epoxide}}(t)}{V_r(t)} \frac{n_{\text{TritonB}}(t)}{V_r(t)} C_{\text{Hydroperoxide},0},$$

$$(8.24)$$

where $n_j$ is the number of moles of component $j$ (mol), $V_r$ is the volume of the reaction mixture (dm$^3$), $r_i$ is the $i$th reaction rate (mol dm$^{-3}$ s$^{-1}$), $k_i$ is the $i$th rate constant

$(dm^6 \, mol^{-2} \, s^{-1})$, $v_{dos}$ is the dosing rate $(s^{-1})$, $c_{dos,Methanol}$ is the concentration of methanol in the feed and $c_{dos,TritonB}$ is the concentration of the Triton B. Although the concentration of the Triton B is included in $r_1$ and $r_2$, it varies only during the short addition phase (24 seconds) and remains constant during the rest of the experiment. As results from several isothermal experiments at different temperatures are evaluated together, the temperature dependence of the rate constants is expressed using Equation 8.10. Instead of just the rate constants $k_1$ and $k_2$, two activation energies ($E_{A,1}$ and $E_{A,2}$), as well as the two rate constants $k_1(T_{ref})$ and $k_2(T_{ref})$ at the reference temperature, have to be quantified. Furthermore, two reaction enthalpies, $\Delta_r H_1$ and $\Delta_r H_2$, as well as the spectral absorbances of the eight pure chemical components at each temperature, are assumed to be unknown.

### 8.3.3.1 Materials and methods

After the reactor had been cleaned, evacuated and purged with $N_2$, 1.29 g of 2,5-di-*tert*-butyl-1,4-benzoquinone (5.85 mmol) were added, taking care that the benzoquinone did not touch the ATR sensor. After the desired jacket temperature had been reached, a reference background infrared spectrum was recorded. Then, 19.2 mL dioxan, 8 mL EtOH and 8 mL *tert*-butyl hydroperoxide (70% solution in water, 58.5 mmol) were added, the stirrer was turned on to 400 rpm and the desired reaction temperature was set. After degassing the solution for 3 minutes with $N_2$, Triton B (0.8 mL of a 40% solution in methanol, 1.78 mmol) was added within 24 seconds into the closed reactor to start the reaction.

This experiment was carried out four times at 17°C and three times at each of 24, 30 and 36°C. The spectral absorbances of all experiments were then concatenated into a single $A_{Data}$ matrix. Similarly, the calorimetric data were concatenated into a single $q_{Data}$ vector.

Three additional experiments were carried out at 30°C using only 5 mL of *tert*-butyl hydroperoxide (70% solution in water, 36.4 mmol) instead of 8 mL. The reaction data collected from all six experiments at 30°C were then concatenated to a second $A_{Data}$ matrix and a second $q_{Data}$ vector. The experimental procedure is described in further detail elsewhere [20].

### 8.3.3.2 Results and discussion

The results for this reaction were obtained by applying three different protocols.

(1) The approach by Zogg and co-workers, which was mentioned in Section 8.2.5.3 and which uses an automated sensitivity analysis to obtain weighting factors for the combined manipulation of calorimetric and spectroscopic data according to Equation 8.21.
(2) A separate calorimetric determination.
(3) A separate infrared determination.

Furthermore, three different case studies have been investigated for this reaction; in the following, the different case studies and protocols are referred to as, for example, A1 for the combined evaluation (method 1 above) of case study A.

*Case study A:* $\Delta_r H_1$ *is allowed to differ from* $\Delta_r H_2$. In addition to the measurement data (sets of 4, 3, 3 and 3 at 17, 24, 30 and 36°C, respectively, at identical concentrations) and the reaction model described by Equations 8.24, limits for the unknown reaction parameters are required in order to apply the combined method 1. For the unknown reaction parameters, $k_1$ and $k_2$ (at $T_{ref} = 25°C$), a range of 0 to 38 $dm^6 \, mol^{-2} \, min^{-1}$ was chosen, and for the unknown activation energies, $E_{A,1}$ and $E_{A,2}$, a range of 10 to 150 kJ $mol^{-1}$. Additionally, the two reaction enthalpies, $\Delta_r H_1$ and $\Delta_r H_2$, were restricted to the range of 0 to $-1000$ kJ mol$^{-1}$ for all temperatures. Finally, feasible limits for the eight-component spectral absorbances (reactant, hydroperoxide, mono-epoxide, di-epoxide, alcohol, methanol, solvent and Triton B) had to be specified. As no measured spectra of pure compounds were used, all absorbance values at all wave numbers for all pure components were constrained to the range of 0 to 5. For each temperature, a new set of pure component spectral absorbances (matrix $E$) was specified; otherwise it would not have been possible to describe the measured experimental data accurately. Based on these input data, determinations were carried out by the three protocols.

For the combined method, the quality of the fit to the model is shown in Fig. 8.7 for the measurements at 17 and 30°C. For illustration purposes, absorbance in the two lower

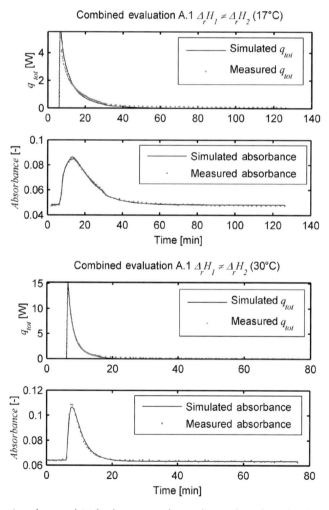

**Fig. 8.7** Application of protocol A1 for the sequential epoxidation of 2,5-di-*tert*-butyl-1,4-benzoquinone at 17 and 30°C using the combined evaluation algorithm [20]. Mean values from all experiments at each temperature are shown. Absorbance in the lower plots corresponds to a single wave number (1687 cm$^{-1}$) from the reaction spectrum.

**Table 8.1** Reaction parameters $k_1$, $k_2$, $E_{A,1}$ and $E_{A,2}$ for the sequential epoxidation of 2,5-di-*tert*-butyl-1,4-benzoquinone determined by the protocols A1, A2, A3 and B1, B2, B3* [20].

| | $k_1$, $k_2$ (dm$^6$ mol$^{-2}$ min$^{-1}$) | | | | |
| --- | --- | --- | --- | --- | --- |
| | 17°C | 24°C | 30°C | 36°C | $E_{A,1}$, $E_{A,2}$ (kJ mol$^{-1}$) |
| Combined (A.1) | 4.9 | 8.8 | 14.2 | 22.5 | 60 |
| | 1.2 | 2.4 | 4.2 | 7.3 | 70 |
| Separate calorimetric (A.2) | 3.2 | 5.8 | 9.4 | 15.1 | 61 |
| | 0.6 | 1.2 | 2.1 | 3.8 | 76 |
| Separate infrared (A.3) | 5.0 | 8.3 | 12.6 | 18.9 | 53 |
| | 1.3 | 2.6 | 4.7 | 8.2 | 73 |
| Combined (B.1) | 4.5 | 8.1 | 13.0 | 20.7 | 60 |
| | 1.3 | 2.5 | 4.4 | 7.3 | 69 |
| Separate calorimetric (B.2) | 4.2 | 7.6 | 12.3 | 19.6 | 60 |
| | 0.8 | 1.8 | 3.3 | 6.0 | 73 |
| Separate infrared (B.3) | 5.0 | 8.3 | 12.6 | 18.9 | 53 |
| | 1.3 | 2.6 | 4.7 | 8.2 | 73 |

*The rate constants $k_1$ and $k_2$ were determined at 30°C; then, using the activation energies $E_{A,1}$ and $E_{A,2}$, rate constants at 17, 24 and 36°C were calculated using Equation 8.10. The reaction model is described by Equations 8.24.

plots corresponds to a single wave number (1687 cm$^{-1}$); however, all wave numbers were actually used and peak overlapping was allowed. We conclude that the calorimetric as well as the infrared data were successfully accommodated by the specified reaction model. The error values corresponding to the two parts of Equation 8.21 are 201 W$^2$ for the calorimetric component (referred to as $\Delta q$ in the following) and 2.529 for the infrared (referred to as $\Delta A$ in the following). The reaction parameters derived from the model ($k_1$, $k_2$, $E_{A,1}$ and $E_{A,2}$) are listed in Table 8.1. In Table 8.2, the derived reaction enthalpies ($\Delta_r H_1$ and $\Delta_r H_2$) are listed and compared with the sum of $\Delta_r H_1$ and $\Delta_r H_2$ ($= \sum \Delta_r H_i$) determined by integration of the calorimetric signal.

For the separate calorimetric investigation, the quality of the fit to the model was similar to that for the combined investigation discussed above ($\Delta q = 173$ W$^2$, $\Delta A = 2.685$). In this determination, the error for the infrared measurements, $\Delta A$, was obtained by using the kinetic parameters obtained from fitting the calorimetric data to calculate concentration–time

**Table 8.2** Reaction enthalpies* $\Delta_r H_1$, $\Delta_r H_2$ and $\sum \Delta_r H_i = \Delta_r H_1 + \Delta_r H_2$ for the sequential epoxidation of 2,5-di-*tert*-butyl-1,4-benzoquinone determined using the protocols A1, A2, A3 and B1, B2, B3 [20]. The reaction model is described by Equations 8.24.

| Protocol | $\Delta_r H_1$ | $\Delta_r H_2$ | $\sum \Delta_r H_i$ |
| --- | --- | --- | --- |
| Combined (A1) | −160 | −200 | −360 |
| Separate calorimetric (A2) | −240 | −150 | −390 |
| Separate infrared (A3) | −180 | −170 | −350 |
| Combined (B1) | −180 | −180 | −360 |
| Separate calorimetric (B2) | −190 | −190 | −380 |
| Separate infrared (B3) | −180 | −180 | −360 |

*All values in kJ mol$^{-1}$.

curves which were then compared with the curves obtained from the infrared measurements. For the separate infrared determination, the results are also given in Tables 8.1 and 8.2. The quality of the fit to the model was similar to that for the combined determination ($\Delta q = 294 \, W^2$, $\Delta A = 2.528$). In this case, the error for the calorimetric measurements, $\Delta q$, was obtained by using the kinetic parameters obtained from fitting the infrared data to calculate concentration–time curves, which were then compared with the corresponding curves obtained from the calorimetric measurements. It should be noted that the reaction enthalpies by the 'separate infrared determination' are obtained by using the kinetic parameters determined with this protocol for fitting the reaction enthalpies to the calorimetric data.

By comparing the three sets of results obtained by the different protocols (Tables 8.1 and 8.2), we conclude that the separate determinations using calorimetric and infrared data do not result in the same reaction parameters. It is, therefore, essential to apply the combined protocol in order to obtain a single set of reaction parameters that represents an optimal solution for all measured data. Note also that some of the reaction parameters ($\Delta_r H_1$, $\Delta_r H_2$, $k_1$ and $E_{A,2}$) determined by the combined protocol are outside the range defined by the separate ones. It would be unreasonable, therefore, simply to average the reaction parameters of the separate evaluations in order to obtain a single set.

A comparison of the different reaction enthalpies reveals that the calorimetric determination shows a large difference between $\Delta_r H_1$ and $\Delta_r H_2$ (90 kJ mol$^{-1}$), whereas $\Delta_r H_1$ and $\Delta_r H_2$ obtained by the infrared method differ by only 10 kJ mol$^{-1}$. The combined protocol (difference $= 40$ k mol$^{-1}$) lies in between. In particular, the results by the calorimetric method appear to be unreasonable.

All three protocols showed $E_{A,1}$ to be smaller than $E_{A,2}$. The difference between $E_{A,1}$ and $E_{A,2}$ suggested by the infrared protocol is rather large (20 kJ mol$^{-1}$) whereas the results of the combined evaluation are the most reasonable from mechanistic considerations.

*Case study B: additional constraint, $\Delta_r H_1 = \Delta_r H_2$.* Based on the discussion of the values for $\Delta_r H_1$ and $\Delta_r H_2$ obtained in case study A, further determinations were carried out using the additional constraint of $\Delta_r H_1 = \Delta_r H_2$. Again, three protocols were used. For the combined protocol, the quality of the fit to the model is slightly worse than that in case study A but is still satisfactory and, when expressed graphically, appears virtually identical with Fig. 8.7 ($\Delta q = 210 \, W^2$, $\Delta A = 2.529$). The reaction parameters determined are included in Tables 8.1 and 8.2. For the separate calorimetric protocol, the quality of the fit to the model was similar to that from the combined protocol ($\Delta q = 183 \, W^2$, $\Delta A = 2.571$). The reaction parameters obtained by the separate infrared protocol are given in Table 8.1. They are identical with those from the separate infrared protocol obtained in case study A as the additional constraint has no influence here. The reaction enthalpies obtained ($\Delta_r H_1$ and $\Delta_r H_2$) are listed in Table 8.2. They are influenced by the additional constraint and thus are no longer identical with those of the separate infrared protocol obtained in case study A. The quality of fit to the model was similar to that for the combined protocol ($\Delta q = 305 \, W^2$, $\Delta A = 2.528$).

By comparing the results of case study B with those of case study A (see Tables 8.1 and 8.2), we conclude that the separate calorimetric protocols give significantly different results for most reaction parameters ($\Delta_r H_1$, $\Delta_r H_2$, $k_1$, $k_2$) whereas the separate infrared determinations give only slightly different reaction enthalpies; with the combined protocol, significantly different results are obtained only for the reaction enthalpies. The activation energies $E_{A,1}$ and $E_{A,2}$ in case study B are closely similar to those in case study A.

Compared to case study A, the separate determination of calorimetric and infrared data in case study B differ less (similar $\Delta_r H_1$, $\Delta_r H_2$, $k_1$, $E_{A,1}$, $E_{A,2}$, different $k_2$), but it is still essential to apply the combined protocol in order to obtain a mutually consistent set of reaction parameters. But as in case study A, some of the reaction parameters ($k_1$, $E_{A,2}$) determined by the combined protocol are outside of the range defined by the separate evaluations.

*Case study C: modified reaction model.* As mentioned above, three additional experiments at 30°C were carried out using less hydroperoxide. They were considered together with the three standard measurements at 30°C. Thus, the dependence of the reaction kinetics on the hydroperoxide concentration could be analysed, and the definitions of the reaction rates $r_1$ and $r_2$ in Equations 8.24 were replaced by those for a modified empirical reaction model shown in Equations 8.25:

$$r_1(t, k_1', ord_{TB}, ord_{HP}) = k_1' \frac{n_{Educt}(t)}{V_r(t)} \left[ \frac{n_{TritonB}(t)}{V_r(t)} \right]^{ord_{TB}} \left[ \frac{n_{Hydroperoxide}(t)}{V_r(t)} \right]^{ord_{HP}}$$

$$r_2(t, k_2', ord_{TB}, ord_{HP}) = k_2' \frac{n_{Mono\,Epoxide}(t)}{V_r(t)} \left[ \frac{n_{TritonB}(t)}{V_r(t)} \right]^{ord_{TB}} \left[ \frac{n_{Hydroperoxide}(t)}{V_r(t)} \right]^{ord_{HP}} \qquad (8.25)$$

In Equations 8.25, $k_1'$ and $k_2'$ are the rate constants of the two epoxidation steps and were limited to the range 0 to 1 [$(dm^3\,mol^{-1})^{(ord_{TB} + ord_{HP})}\,s^{-1}$] and, as experiments were carried out only at 30°C, no activation energies were determined. The reaction orders, $ord_{TB}$ and $ord_{HP}$, in Triton-B and hydroperoxide concentrations were limited to the ranges 0 to 3 and $-3$ to 3, respectively. As in case study A, the two reaction enthalpies $\Delta_r H_1$ and $\Delta_r H_2$ were restricted to the range 0 to $-1000$ kJ mol$^{-1}$ and all absorbance values at all wave numbers for all components were constrained to the range of 0 to 5. Based on these input limitations, determinations were carried out again employing the three different approaches.

Using the combined protocol, the calorimetric as well as the infrared data of all six experiments were successfully modelled by Equation 8.25 ($\Delta q = 58$ W$^2$, $\Delta A = 0.284$) and the associated reaction model parameters. The kinetic parameters determined ($k_1'$, $k_2'$, $ord_{TB}$, $ord_{HP}$) as well as the reaction enthalpies ($\Delta_r H_1$ and $\Delta_r H_2$) are listed in Table 8.3. The qualities of the fits to the model for the separate calorimetric ($\Delta q = 45$ W$^2$, $\Delta A = 0.326$) and infrared ($\Delta q = 104$ W$^2$, $\Delta A = 0.282$) protocols were similar to that for the combined protocol.

**Table 8.3** Kinetic ($k_1'$, $k_2'$, $ord_{TB}$, $ord_{HP}$) and thermodynamic ($\Delta_r H_1$, $\Delta_r H_2$) parameters for the sequential epoxidation of 2,5-di-*tert*-butyl-1,4-benzoquinone based on six measurements at 30°C at different hydroperoxide concentrations using the protocols C1, C2, C3 [20]. The reaction model in Equations 8.25 was applied.

| | Combined (C1) | Separate calorimetric (C2) | Separate infrared (C3) |
|---|---|---|---|
| $\Delta_r H_1$; $\Delta_r H_2$ (kJ mol$^{-1}$) | $-180$; $-180$ | $-240$; $-140$ | $-200$; $-140$ |
| $k_1'$, $k_2'$ [(dm$^3$ mol$^{-1}$)$^{(ord_{TB}+ord_{HP})}$ min$^{-1}$] | 36.6; 11.1 | 25.2; 5.8 | 39.6; 15.6 |
| $ord_{TB}$ | 1.13 | 1.12 | 1.20 |
| $ord_{HP}$ | $-0.16$ | $-0.15$ | $-0.18$ |

Table 8.3 shows that, as for case studies A and B, separate calorimetric and infrared protocols for case study C did not result in the same reaction parameters; the combined protocol is thus required. Again, some of the reaction parameters ($\Delta_r H_1$, $\Delta_r H_2$) determined by the combined protocol are outside the range defined by the separate protocols. Of all the reaction parameters determined, only the reaction enthalpies are similar to those obtained from case studies A and B. As with case study A, the difference between $\Delta_r H_1$ and $\Delta_r H_2$ by the calorimetric protocol (100 kJ mol$^{-1}$) is large and unreasonable. On the other hand, $\Delta_r H_1$ and $\Delta_r H_2$ determined by the combined protocol are reasonably found to be equal and close to the results from case studies A and B. The error values obtained by all three protocols (combined, separate calorimetric and separate infrared) are significantly lower for the modified reaction model, Equations 8.25, than for the original one, Equations 8.24. By this criterion, therefore, the modified model accommodates the experimental data better than the original one and can thus be considered more reasonable.

This example demonstrates how reaction calorimetry in combination with IR-ATR spectroscopy can be used to discriminate between different postulated reaction models, and to determine the kinetic and thermodynamic parameters for the selected model. In practical applications, when different (semi-) empirical models can be postulated, model discrimination is crucial.

### 8.3.4   Example 3: Hydrogenation of nitrobenzene

Catalytic hydrogenation of aromatic nitro compounds is an industrially important process for the introduction of amino functionality into pharmaceutical and agrochemical intermediates, and in the polyurethane industry. Aromatic nitro compounds are very easily hydrogenated, and hydrogenations have been carried out under a wide range of conditions, including the gas phase. The reaction example presented here is the hydrogenation of nitrobenzene to give aniline, Scheme 8.3.

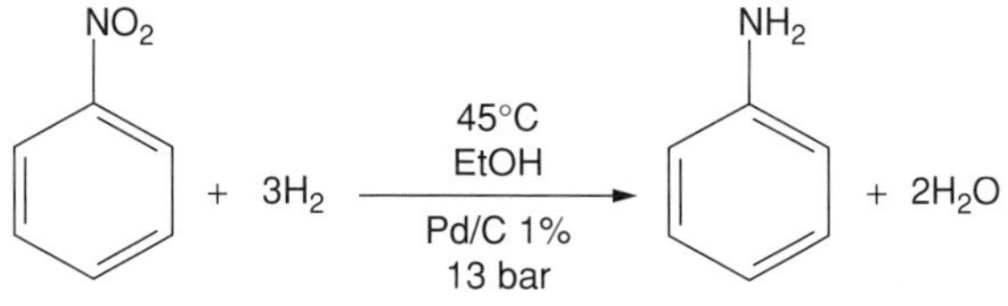

**Scheme 8.3**   Hydrogenation of nitrobenzene to give aniline.

This reaction is an example of a heterogeneous reaction with a solid catalyst with one reactant principally in solution and another in the gas phase; the gas–liquid–solid mixture has to be mixed thoroughly to promote conversion (see Chapter 5 for more detailed consideration of multiphase reactions). Compared with the examples above, the measurement of the hydrogen uptake delivers an additional signal, which can also be used for the determination of reaction parameters.

### 8.3.4.1   Materials and methods

The reactor was charged with 35 mL of absolute ethanol and 0.08 mg of 1% Pd/C, the stirrer was then turned on at 400–500 rpm, and the desired reaction temperature was set.

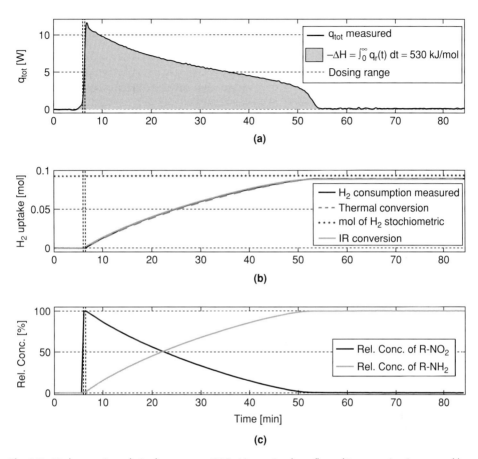

**Fig. 8.8** Hydrogenation of nitrobenzene at 45°C: (a) reaction heat flow; (b) conversion in terms of hydrogen consumed as obtained by three independent approaches; (c) concentrations of nitrobenzene and aniline obtained by PCA from the spectroscopic data [14–16].

The reactor was next flushed with nitrogen (1 atm) to remove air from the reactor which was then pressurised with hydrogen (13 bar). The reaction proceeded at a constant partial pressure of hydrogen which was taken to be the total pressure. The stirrer speed was increased to 1200 rpm and nitrobenzene (3.7 g, 0.0333 mol) was added at a constant rate of 8 mL min$^{-1}$.

### 8.3.4.2 *Results and discussion*

To obtain a simple kinetic model for this heterogeneous reaction, it has to be assumed that the concentration of hydrogen as a reactant, in terms of partial pressure, is constant, and that the hydrogen is thoroughly mixed with the reaction mass to avoid external mass-transfer limitations. Consequently, experiments were carried out in which the speed of the stirrer was varied from 600 to 1800 rpm. For stirrer speeds up to 1200 rpm, a strong dependence of the rate of reaction on the stirrer speed was found. For stirrer speeds above 1200 rpm, no significant increase in the reaction rate was found; therefore, a speed of 1200 rpm was used in determinations to avoid external mass-transfer limitations. Internal mass transfer was neglected because the catalyst particle sizes were below 10 μm.

In addition, the deactivation of the catalyst was studied qualitatively by repeated additions of nitrobenzene to the reaction mixture containing catalyst, solvent and product. In a double-logarithmic plot, the resulting reaction rate constants decrease linearly with time following sequential additions of nitrobenzene. This behaviour, which is typical of catalyst deactivation, was found to be independent of temperature. For modelling purposes, the deactivation relationship could be considered, or a rate constant corresponding to the average activity of the catalyst during each experiment could be used.

The reaction heat-flow rate measured at 45°C for the hydrogenation of nitrobenzene is shown in Fig. 8.8a. The integration of this heat-flow rate gave a reaction enthalpy of approximately $530 \pm 2$ kJ mol$^{-1}$. The measured consumption of hydrogen is shown in Fig. 8.8b. It is compared with the hydrogen consumption (in mol) calculated according to the thermal conversion (obtained from Fig. 8.8a) as well as according to the conversion indicated by IR spectroscopy as determined by PCA [14–16] (Fig. 8.8c). The three hydrogen consumption curves agree with each other well and demonstrate the reliability of the three sensors, all of which provide a sound basis for the determination of a kinetic model. Only minor traces of hydroxylamine (as a possible intermediate) were detected in the IR spectrum, a result confirmed by off-line analysis. From the perspective of practical application, therefore, a one-step reaction model can be postulated rather than a mechanistically orientated two-step model. In this example also, the synergy between different analytical sensors is evident.

## 8.4   Conclusions and outlook

The kinetic and thermodynamic characterisation of chemical reactions is a crucial task in the context of thermal process safety as well as process development and optimisation. As most chemical and physical processes are accompanied by heat effects, calorimetry represents a unique technique to gather information about both aspects, thermodynamics and kinetics. As the heat-flow rate during a chemical reaction is proportional to the rate of conversion, calorimetry represents a differential kinetic analysis technique. The combination of calorimetry with an integral kinetic analysis method, e.g. UV–vis, near infrared, mid infrared or Raman spectroscopy, enables an improved kinetic analysis of chemical reactions.

The examples presented in Section 8.3 demonstrate this synergy in an approach using calorimetry and IR-ATR spectroscopy. For the hydrolysis of acetic anhydride, the combination of the two analytical techniques enabled a differentiation between the heat effect due to the chemical reaction and that due to a physical phenomenon – in this case, mixing. Due to this separation of the physical heat effect, a more reliable value for the chemical heat effect was obtained. For the sequential epoxidation of 2,5-di-*tert*-butyl-1,4-benzoquinone, the importance of selection of an appropriate kinetic model has been demonstrated. For complex reaction systems, several models can be postulated. The appropriateness of these models can then be tested on the basis of experimental data. Combined analytical techniques provide an enriched data set for this purpose as has been demonstrated for this example. After the selection of the most appropriate model, the corresponding parameters can be used for further analysis. The example of the hydrogenation of an aromatic nitro-compound showed that these techniques can also be applied in heterogeneous systems involving a solid catalyst and a reactant in the gas phase, which have to be mixed thoroughly with the liquid phase to facilitate conversion. The measurement of the hydrogen uptake provided an additional signal that confirmed the results obtained from the other two. For other more complex reaction systems, this third signal might be required to obtain a unique outcome, i.e. to identify an appropriate reaction model and the corresponding reaction parameters. In summary, for the determination of kinetic and thermodynamic parameters, it is advantageous to use a combined approach, e.g. using reaction calorimetry in combination with IR-ATR spectroscopy. The examples presented in Section 8.3 demonstrate the benefits and the wide range of applicability of this approach.

Because of the importance of reaction kinetics in the context of chemical process safety and optimisation, a further development of tools is needed that enables the easy and quick determination of thermodynamic and kinetic parameters. Particular emphasis has to be put on calorimetric devices that correspond to the conditions in chemical production as far as possible but nevertheless have only a small volume. As already discussed in detail, the combination with additional analytical tools is essential. Furthermore, the devices have to have a wide range of applicability with regard to temperature, pressure, chemical regime, number and types of phases involved and so on. Finally, computer tools are needed that allow a quick and easy determination of kinetic and thermodynamic parameters from the measurements. The systematic application of such improved methods could result in a number of significant improvements in chemical processes in industry.

# References

1. Levenspiel, O. (1999) *Chemical Reaction Engineering* (3rd edn). Wiley, New York.
2. Landau, R.N. (1996) *Thermochimica Acta*, **289**, 101.
3. Regenass, W. (1997) *Journal of Thermal Analysis*, **49**, 1661.
4. Zogg, A., Stoessel, F., Fischer, U. and Hungerbühler, K. (2004) *Thermochimica Acta*, **419**, 1.
5. Pollard, M. (2001). *Organic Process Research Development*, **5**, 273.
6. Pastré, J., Zogg, A., Fischer, U. and Hungerbühler, K. (2001) *Organic Process Research and Development*, **5**, 158.
7. Karlsen, L.G. and Villadsen, J. (1987) *Data Treatment, Chemical Engineering Science*, **42**, 1165.
8. Zaldivar, J.M., Hernandez, H. and Barcons, C. (1996) *Thermochimica Acta*, **289**, 267.
9. Mirabella, F.M. (1993) *Internal Reflection Spectroscopy*. Dekker, New York.
10. LeBlond, C., Wang, J., Larsen, R., Orella, C. and Sun, Y.-K. (1998) *Topics in Catalysis*, **5**, 149.
11. am Ende, D.J., Clifford, P.J., DeAntonis, D.M., SantaMaria, C. and Brenek, S.J. (1999) *Organic Process Research and Development*, **3**, 319.
12. Ubrich, O., Srinivasan, B., Lerena, P., Bonvin, D. and Stoessel, F. (1999) *Journal of Loss Prevention in the Process Industries*, **12**, 485.
13. Nomen, R., Sempere, J. and Avilés, K. (2001) *Chemical Engineering Science*, **56**, 6577.
14. de Juan, A., Maeder, M., Martinez, M. and Tauler, R. (2000) *Chemometrics and Intelligent Laboratory Systems*, **54**, 123.
15. Dyson, R., Maeder, M., Neuhold, Y.-M. and Puxty, G. (2003) *Analytica Chimica Acta*, **490**, 99.
16. Jiang, J.-H., Liang, Y. and Ozaki, Y. (2004) *Chemometrics and Intelligent Laboratory Systems*, **71**, 1.

17. Zogg, A., Fischer, U. and Hungerbühler, K. (2004) *Chemometrics and Intelligent Laboratory Systems*, **71**, 165.

18. Visentin, F., Gianoli, S.I., Zogg, A., Kut, O.M. and Hungerbühler, K. (2004) *Organic Process Research and Development*, **8**, 725.

19. Zogg, A., Wohlwend, M., Hungerbühler, K. and Fischer, U. (2000) Patent No. EP 1184649, Application No. 00810797.1.

20. Zogg, A., Fischer, U. and Hungerbühler, K. (2004) *Chemical Engineering Science*, **59**, 5795.

# Chapter 9

# The Detection and Characterisation of Intermediates in Chemical Reactions

## C. I. F. Watt

## 9.1   Introduction: What is an intermediate?

Chemical reactions are induced by subjecting stable compounds to conditions under which bonding changes occur permitting reorganisations to new compounds, sufficiently stable to these reactions conditions for isolation and characterisation. In almost all cases, multiple bond changes are required, and these are rarely concerted, so that progression from reactants to stable products occurs in a stepwise fashion via less stable intermediate species of varying lifetimes. This chapter focuses on the detection and characterisation of these intermediates. Before considering the practicalities, however, we set the context in which experiments are to be designed and interpreted. Central to this are the meanings of terms such as 'concerted' and 'stepwise', already introduced without definition in this preamble.

### 9.1.1   Potential energy surfaces and profiles

In principle, and increasingly in practice also, modern computational methods permit the accurate calculation of the energy of a particular number of electrons moving in the field of a specified array of $N$ nuclei. Nuclear coordinates may be adjusted and the process repeated systematically to construct a potential energy (PE) surface in the $(3N - 6)$ independent coordinates required to define the geometry of any nuclear array larger than a diatomic, and well-established computational methods can then identify stationary points (maxima, minima and saddles) on this surface (see Chapter 7). For interconverting compounds, R and P, two minima on the PE surface correspond to the nuclear arrangements of stable molecules of R and P, and they will be connected by a saddle point associated with a third nuclear arrangement which will have partial bonds corresponding to those which must be made and broken in the conversion. This array of electrons and nuclei is the *transition structure* or *activated complex* ($\ddagger$) for the reaction, and the paths of steepest descent, in mass-scaled coordinates, from this structure connecting to the energy minima corresponding to R and P is the (*intrinsic*) *reaction coordinate* for their interconversion. This coordinate will contain contributions from motion of all atoms in the array, but will be dominated by those involved in the bonding changes corresponding to the interconversion.

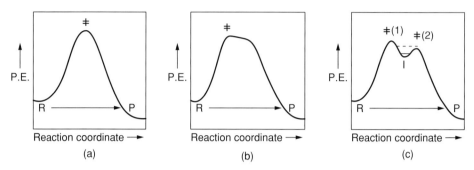

**Fig. 9.1**  PE profiles (a) for a concerted reaction and (c) for a stepwise process.

Importantly, motion along the reaction coordinate at the saddle point has no restoring force, so the transition structure has only $3N - 7$ real vibrations. Indeed, passage through the saddle corresponds to conversion of the vibration along the reaction coordinate into a translation, so that $k_B T = h\nu^{\ddagger}$, where $\nu^{\ddagger}$ is frequency of this hypothetical vibration. At 25°C, $\nu^{\ddagger} = 6 \times 10^{12}$ s$^{-1}$; equivalently, the lifetime of a transition structure is $1.7 \times 10^{-13}$ s and methods capable of observing labile species on this time scale ('transition state spectroscopy') have been developed, permitting many of the assumptions about their behaviour to be tested directly [1].

The plot of PE against progress along the reaction coordinate provides the familiar reaction profile of the mechanistic chemist, with the PE maximum in the profile corresponding to the saddle point on the PE surface; Fig. 9.1a shows the simplest case. Reactions where a single transition structure ($\ddagger$) separates reactants and products are said to be single-step or *concerted* elementary processes.

Barriers separating reactants and products may be broad or narrow (compare profiles **a** and **b** in Fig. 9.1) and continuation of the trend from **a** to **b** may eventually produce a local minimum (I) flanked by two barriers, ($\ddagger$1) and ($\ddagger$2) as in Fig. 9.1c. Provided these flanking barriers corresponding to the saddle points on the PE surface are high enough for the minimum to accommodate at least one vibrational energy level, this minimum represents a bound state with a full complement of $3N - 6$ vibrations, and describes a *reaction intermediate* in the conversion of R into P, and the complete process can be regarded as the stepwise sequence of two conversions, first R to I, and then I to P. The lifetime of the intermediate, I, is governed by the height of the lower of the two flanking barriers, ($\ddagger$2) in the case illustrated, but may not be less than the lifetime of a transition structure ($1.7 \times 10^{-13}$ s), and sufficient for several molecular vibrations (for C—H and C—C stretches, $\nu = 1.25 \times 10^{15}$ and $3.3 \times 10^{14}$ s$^{-1}$, respectively). Short-lived intermediates are often referred to as *reactive intermediates*. At the other extreme, intermediates may have substantial lifetimes and accumulate to readily observable concentrations.

These descriptions apply to the bonding changes occurring in a unimolecular chemical conversion where the energy for a reactant to achieve the transition structure is provided by collision with surrounding molecules, which in solution would be mainly solvent. For bimolecular reactions, a similar treatment applies, but the *chemical* steps proceed from an *encounter complex* formed by collision of the component species. In solution, reacting molecules must diffuse together through the solvent and, even when an encounter occurs, a chemical change may not occur until adjustments in the solvent encapsulating the complex are complete. The time scales for these physical processes are short, but exceed those for the changes occurring within a transition structure, so that rates of formation or decomposition of transition structures may be limited by the physical processes when barriers to chemical changes (i.e. bonding) are low.

### 9.1.2  From molecular potential energy to rates of reaction

Potential energy surfaces or profiles are descriptions of reactions at the molecular level. In practice, experimental observations are usually of the behaviour of very large numbers of molecules in solid, liquid, gas or solution phases. The link between molecular descriptions and macroscopic measurements is provided by *transition state theory*, whose premise is that activated complexes which form from reactants are in equilibrium with the reactants, both in quantity and in distribution of internal energies, so that the conventional relationships of thermodynamics can be applied to the *hypothetical* assembly of transition structures.

Thus, for the general reaction given in Equation 9.1, the concentration of the transition structures, $TS^{\ddagger}$, is governed by the equilibrium constant $K^{\ddagger}$, with no implication as to *how* $TS^{\ddagger}$ is formed from A and B:

$$a\text{A} + b\text{B} \underset{}{\overset{K^{\ddagger}}{\rightleftharpoons}} TS^{\ddagger} \overset{k^{\ddagger}}{\longrightarrow} \text{P}. \tag{9.1}$$

The rate of reaction is the concentration of $TS^{\ddagger}$ multiplied by the frequency, $\nu^{\ddagger}$, of their decomposition into product, usually given by $\nu^{\ddagger} = k_{B}T/h$; Equation 9.2 then applies:

$$d[\text{P}]/dt = (k_{B}T/h)K^{\ddagger}[\text{A}]^{a}[\text{B}]^{b}. \tag{9.2}$$

Under this treatment, the reaction order of an elementary unimolecular or bimolecular reaction must identify with molecularity, and $K^{\ddagger}$ is related in Equation 9.3 to the standard molar free energy difference, $\Delta G^{\ddagger}$, between reactants and *transition state* (a hypothetical construct comprising one mole of transition structures, see below):

$$\Delta G^{\ddagger} = -\text{RT} \ln K^{\ddagger}. \tag{9.3}$$

In turn, $\Delta G^{\ddagger}$ is composed of contributions from changes in the enthalpy and entropy between reactants and transition state, as shown in Equation 9.4:

$$\Delta G^{\ddagger} = \Delta H^{\ddagger} - T\Delta S^{\ddagger}. \tag{9.4}$$

For stable molecules, thermochemical and spectroscopic measurements are available, and free energies relative to specified standard states may be obtained directly from experiment, or from well-established additivity schemes [2] based on experimental data, at least for gas phase reactions. In principle, these are also calculable ab initio for both reactants and transition structures (see Chapter 7). Enthalpy differences correspond closely to potential energy differences between the stationary points, with allowance for zero-point energies. For entropic contributions, molecular weights, concentrations and structure permit calculation of translational and rotational contributions. The vibrational contribution, however, depends on the distribution of vibrational energy levels, governed by the shape of the PE surface at the stationary points. For transition structures, there are, of course, only $3N - 7$ vibrational modes to consider, and sophisticated theoretical treatments recognise that exact coincidence between the maxima in PE and in free energy along the reaction coordinate is likely to be exceptional, with entropic demand reflecting the section of the PE surface

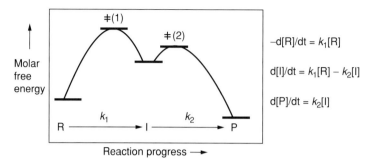

**Fig. 9.2** Treatment of a stepwise reaction via an intermediate.

orthogonal to the reaction coordinate [3]. Furthermore, the balance between enthalpic and entropic contributions to free energy depends on temperature, so the position of the free energy maximum will be temperature dependent. Because of these complications, reaction PE profiles are often replaced by those in which the energy axis is standard molar free energy, and the reaction coordinate then corresponds to progression through a series of hypothetical thermodynamic states, converting one mole of reactant to the product with all molecules reacting simultaneously. The *transition structure* and *transition state* are often used interconvertibly but should be recognised as distinct – the former is a molecular species; the latter is a hypothetical thermodynamic state.

Transition state theory thus allows the writing of a rate equation for any elementary reaction, and a transformation in which an intermediate is postulated can be treated as a sequence of elementary steps. For any particular sequence, a set of differential equations may be written. For the simplest of these, the sequence of two irreversible unimolecular reactions shown in Fig. 9.2, the exact integrated forms are available permitting calculation and plotting of the time course of anticipated concentration changes for a comparison with experimental data; see Chapters 3 and 4.

For complex reaction sequences, analytical integration becomes more challenging, but with the computational power available in the modern personal computer, the brute force methods of numerical integration are accessible to any user of spreadsheet packages such as *Excel* [4]. These integrations are easily set up for almost any sequence or combination of elementary reactions and even crude models can give useful indications of behaviour.[*] Fitting of models to experimental data is, however, much more difficult and better left to expertly developed specialist packages, a number of which are commercially available. Possibly the most sophisticated of these is SPECFIT/32™, a multivariate data analysis package for modelling and fitting chemical kinetics, and a variety of equilibrium titration 3D data sets. Typically, these comprise simultaneous measurements of absorbance and wavelength as a function of an independent variable such as time, pH or [Titrant]. Its globalised fitting treatment [5] can be adapted to a wide range of experimental observations that respond linearly to sample concentration or mole fraction.

---

[*] A useful Chemical Kinetics Simulator is available as freeware from: http://www.almaden.ibm.com/st/computational_science/ck/msim/

## 9.2    A systematic approach to the description of mechanism

The usual description of changes occurring in a chemical transformation exploits structural representations in which bonds between atoms are viewed as localised shared electron pairs. Identification of just which of these have been broken and made often requires no more than an intelligent examination of the structural and stoichiometric relationships between reactants and products (see Chapter 2). Remaining ambiguities may require resolution by experiments with isotopically labelled reactants and, in the case of molecular rearrangements, crossover experiments to distinguish between *inter-* and *intra*-molecular possibilities (see Chapters 2 and 12). The differentiating structural features of participating crossover reactants should be minimal and, indeed, such experiments are most reliable when reactants are distinguished by isotopic substitution, preferably at sites remote from those involved in bonding changes. When these essential preliminaries are in place, mechanisms and credible intermediates may be proposed. Modern mechanistic nomenclature affords a systematic approach which ensures that *all* possibilities are considered [6].

### 9.2.1    Reaction classification

Stable covalent compounds contain atoms whose bonding satisfies the rules of valence and stereochemistry. In conversion to a new stable compound, bonds must be made or broken (Associative or Dissociative processes). Bond cleavage may be heterolytic (generating electrophile/nucleophile pairs) or homolytic (generating radicals, see Chapter 10), and bonds may be formed by the reverse of these processes. Electron transfer (sometimes called *single electron transfer*, SET) to a stable molecule may also induce bond cleavage to generate a radical and an anion, see Chapters 6 and 10. Any one of the bond cleavages occurring within a stable species leads to one containing an atom with an unusual valence or electronic configuration. At least two processes (either of which may be a bond cleavage or formation) must occur in the conversion of one stable molecule into another, and this can be seen in the basic reactions – *elimination*, *substitution* and *addition* – and their intramolecular variants. At their simplest, these basic reactions are distinguished by the change in the number of ligands attached to a defined core atom or core set of atoms. In *eliminations*, two dissociative bonding changes decrease the number of ligands by two, with the formation of a ring or an increase in bond order between a core pair. *Substitutions* involve one dissociative and one associative change so that the number of ligands does not change. In *additions*, two associative changes occur, increasing the number of ligands by two. Some possibilities are outlined in Fig. 9.3, and molecular rearrangements can always be viewed as intramolecular variants of these. Note that the symbol C in these refers to atoms defined as belonging to the reacting *core*, and may be other than carbon. An assumption of tetravalence is made in the figure, but is not required by the logic. Elimination or addition at a single atom site necessarily leads to a species with an atom exhibiting *hypo-* or *hyper*-valence ($C^*$ in the diagram); this might reasonably be expected to be a high-energy species, and likely itself to engage in further reaction. For the conservation of valence in eliminations or additions, they must occur at multiple-atom cores, with elimination introducing unsaturation or ring closure in the core set, and addition effecting the reverse.

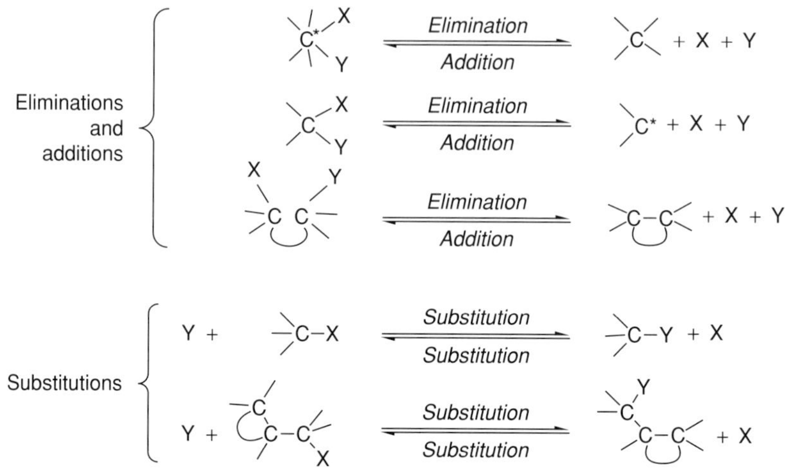

**Fig. 9.3** Bonding changes in eliminations, additions and substitutions (C* in the diagram indicates a *hypo-* or *hyper*-valence state of the Core atom, C).

## 9.2.2  *Consequences of uncoupled bonding changes*

All cases outlined above involve two bonding changes, and these may be concerted, i.e. a single activated complex separates reactants and products. Alternatively, however, the bonding changes may be sufficiently *uncoupled* that a species with real existence, containing a *hypo-* or *hyper*-valent atom, is involved. This species would correspond to the high-energy minimum encountered in our earlier discussion of PE surfaces and reaction profiles, and be flanked on that surface by two maxima corresponding to activated complexes in which the bonding changes are taking place sequentially, i.e. we now have a stepwise process. Barriers to reactions in which two bonding changes occur concertedly have been approximated as the sum of barriers to the two separate steps, with adjustment for benefit of coupling between them [7]. This benefit is believed to be small enough to ensure that concerted bonding changes compete with stepwise processes only when the barrier for conversion of the intermediate to the product would be small. Pericyclic processes provide a notable exception to this generalisation, and reactions involving cyclic conjugated arrays also have to be examined for the effects of aromaticity or anti-aromaticity, which are not taken into account in simple Lewis structures.

Since the ligands involved may be radicals, nucleophiles or electrophiles, and there are two bonding changes at the core atoms, there are no fewer than six possible intermediate structures for each of the basic processes, reflecting the permutations in nature and timing of the bonding changes. The possibilities are generated in Fig. 9.4 for substitutions at a single atom site (A + D or D + A reactions). Again, tetravalence has been assumed at the core atoms, and charged nucleophiles and electrophiles are shown, but are not essential to the argument. Cationic, anionic and radical intermediates are generated for consideration and testing against accumulating experimental evidence. In this case, the possibilities include atoms showing expansion of their normal valencies. Such possibilities might be rejected as non-viable at an early stage in any consideration when the core atom is carbon, nitrogen

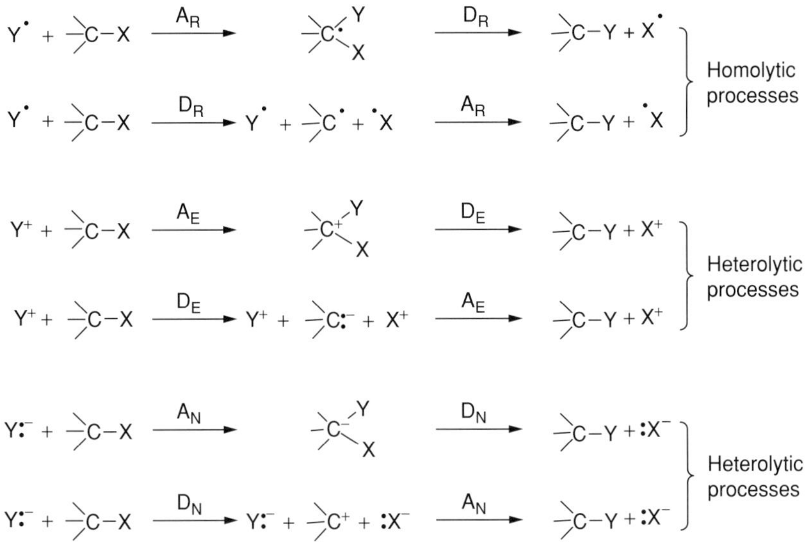

$$Y^\bullet + {>}C{-}X \xrightarrow{A_R} {>}C{\overset{Y}{\underset{X}{\cdot}}} \xrightarrow{D_R} {>}C{-}Y + X^\bullet$$

$$Y^\bullet + {>}C{-}X \xrightarrow{D_R} Y^\bullet + {>}C^\bullet + {}^\bullet X \xrightarrow{A_R} {>}C{-}Y + {}^\bullet X$$

Homolytic processes

$$Y^+ + {>}C{-}X \xrightarrow{A_E} {>}C{\overset{Y}{\underset{X}{+}}} \xrightarrow{D_E} {>}C{-}Y + X^+$$

$$Y^+ + {>}C{-}X \xrightarrow{D_E} Y^+ + {>}C{:}^- + X^+ \xrightarrow{A_E} {>}C{-}Y + X^+$$

Heterolytic processes

$$Y{:}^- + {>}C{-}X \xrightarrow{A_N} {>}C{\overset{Y}{\underset{X}{-}}} \xrightarrow{D_N} {>}C{-}Y + {:}X^-$$

$$Y{:}^- + {>}C{-}X \xrightarrow{D_N} Y{:}^- + {>}C^+ + {:}X^- \xrightarrow{A_N} {>}C{-}Y + {:}X^-$$

Heterolytic processes

**Fig. 9.4** Possible intermediates in substitutions at a single atom reaction site.

or oxygen, but considered seriously when dealing with substitutions at atoms of the second row elements.

Similar sets of possibilities may be generated for eliminations (D + D reactions) and the corresponding additions (A + A reactions). Again, a set of six cationic, anionic and radical intermediates is generated for consideration in each case, and a reader may care to explore these combinations.

### 9.2.3  Sequences of basic reactions

Real chemical reactions may comprise linked sequences of the trio of basic transformations, with the link between a pair of basic reactions being a transient molecular species which, although formally satisfying the valence rules for its constituent atoms, shows high reactivity under the conditions of the overall chemical transformation. This reactivity may reflect high energy associated with unfavourable bond dipole interactions (e.g. in the adducts of nucleophiles to carbonyl groups), repulsions between electron pairs on bonded atoms, or with unusual molecular geometries enforced by the incorporation of reaction sites into cyclic arrays (e.g. benzyne or bridgehead alkenes) or other steric constraints. The relationship between energy and stability, however, is not a simple one, as can be seen in the behaviour of Dewar benzenes. These are less stable than their benzene isomers by ca. 250 kJ mol$^{-1}$, but barriers to these profoundly exothermic rearrangements are high enough (ca. 90 kJ mol$^{-1}$) to permit isolation and storage of the Dewar benzenes under relatively conventional conditions [8]. At the other extreme, barriers to proton transfers within hydrogen-bonded acid–base pairs are usually so low that rates of exothermic dissociations of protic acids are limited not by the barrier to the proton transfer, but by diffusive separation of the new acid–base pair [9].

## 9.3    Evidence and tests for the existence of intermediates

As noted earlier, the identification of products from known reactants and the determination of reaction stoichiometry are the starting point of a quest for a reaction intermediate. Additionally, species necessary for reaction may be found, but in substoichiometric quantities. Such species include radical initiators in chain processes, see Chapter 10. In the extreme, necessary species may not appear at all in the stoichiometric equation, and therefore are true catalysts. Mechanisms of catalysis are dealt with in Chapters 11 and 12, but it is appropriate to note here that most function by opening new reaction pathways, replacing a single high-energy step involving concerted bonding changes by lower energy steps, so that the intervention of intermediates involving the catalytic species must be considered amongst the possibilities. Once the possibilities have been identified, experiments may be designed, first to test for the existence of a transient intermediate in a reaction and then to assign structure, energy and lifetime. Of course, an overall chemical transformation may involve more than a single intermediate, each with its own characteristics.

Prior knowledge of the behaviour of a proposed intermediate under a particular set of reaction conditions is often available and facilitates experimental design. For example, species which are transient under one set of conditions (solvent, temperature) may be stable under others, and then observable by conventional methods. Similar considerations apply to structural variation, which may stabilise charge or unusual valence states. Systematic studies of the effects of variation of conditions, or of structural variation on reactivity, often permit useful extrapolation to behaviour of a proposed intermediate under the conditions in question. Importantly, if extrapolations of this kind indicate that a proposed intermediate would have a lifetime of less than $10^{-13}$ s under a particular set of reaction conditions, then that proposal must be re-evaluated. Either the mechanism involving the proposed intermediate is fundamentally flawed, or the bonding changes involved in its formation and destruction are actually concerted.

More directly, a proposed intermediate may be generated from alternative precursors, in non-equilibrium concentrations, under conditions quite close to those of the reaction in question. Fast kinetic methods may then yield rates for decay of the intermediate, and products may be determined for comparison with observations on the reaction in question. The reader is directed to the recent review edited by Moss, Platz and Jones Jr. [10], for a modern comprehensive survey of progress in reactive intermediate chemistry. The chapters deal in some detail with the methods touched on as outlined above and the results of their application to carbocations, carbanions, carbenes, radicals and strained species.

In the rest of this chapter, we attempt to extract and illustrate some general principles in the design of experiments to test for the existence of intermediates in particular reactions.

### 9.3.1    Direct observation

The ability to monitor changes in the concentration of species over the course of a reaction is central to any mechanistic investigation, and time spent in selecting or developing an appropriate method is inevitably repaid both in terms of mechanistic understanding and in yield optimisation (see Chapters 2 and 3). Ideally, an analytical method permits continuous monitoring, and most exploit the direct measurement of a physical property of a reaction mixture simply related to the concentration of reactants or products in the reacting system. These include intensities of absorbed or emitted light, rotation of the plane of polarised light, conductance, amplitude of signals in NMR spectra, variation of pressure, volume, refractive index and (more recently – see Chapter 8) heat flow [11]. All of these respond relatively rapidly to variation of concentration, and applicability to faster reactions is usually limited by mixing times rather than the physics of the method. Reactions on time scales as short as $10^{-3}$ s may be followed when detectors are incorporated into stopped-flow reactors with high-efficiency mixers. Magnetic resonance spectroscopy as a tool for observing concentration changes has presented the greatest challenge to devising ways of minimising delays between mixing and first observation; but even here, rapid injection techniques are available which permit observation of reacting mixtures within milliseconds of reagent mixing [12]. (This use of NMR spectroscopy is distinct from dynamic NMR methods which can be used to investigate rates of conformational changes or some fast reactions when they are comparable in rate with nuclear relaxation processes.) All these methods are usually well adapted to observation of *change* of concentration, but often require careful calibration before they can be applied to determination of absolute concentrations. Electrochemical methods are dealt with separately (see Chapter 6), but we note here that ion selective electrodes offer a sensitive method of monitoring concentration changes in the species to which the electrode responds, although the response is often much slower than in the methods listed above.

Test reactions may be run in UV–vis cuvettes or NMR tubes, or other specialised vessels designed to facilitate observation by the particular physical method adopted, and reagent concentrations, choice of solvents, temperatures may also be adjusted to optimise observations. Extension of measurements obtained under these conditions to the behaviour of reactions under synthetic or process conditions is rarely straightforward. Observations under the 'real' conditions are desirable if results are to be applied in development of a synthetic process. Many of the methods listed above are easily adaptable for the construction of robust probes for insertion into reaction vessels. Notably, probes exploiting light absorption from mid-range IR to the UV, using attenuated total reflectance, light-pipe and fibre optic technology, are commercially available for insertion into reaction vessels.

Methods involving sampling and separate analysis are applicable to almost any reaction scale or concentration regime, provided time scales permit and quenches are available to stop the reaction at the time of sampling (see Chapter 2). Quenches may be chemical (often adjustment of the acidity of the medium) or physical (dilution or chilling of the reaction mixture), and analysis of samples by chromatography is powerfully applicable to many common reactions where the continuous monitoring methods mentioned above are not. Sophisticated and expensive methods may not be required for preliminary observations. Application of a sample to a pre-treated thin layer chromatography plate, for example, often provides an effective and instant acid or base quench. Similarly, simple dilution of the mixture into a UV cuvette may stop a reaction. Monitoring of a reaction by periodic TLC analysis or examination of a UV spectrum of a mixture has often provided the first indication of species transiently formed over the course of the reaction. With an appropriate quench, the analytical methods can be tailored to the compound of interest to give an absolute measure of its concentration.

Although the method of sampling and analysis is often regarded as applicable only to relatively slow reactions, modern instrumentation is available in the form of rapid quench-flow

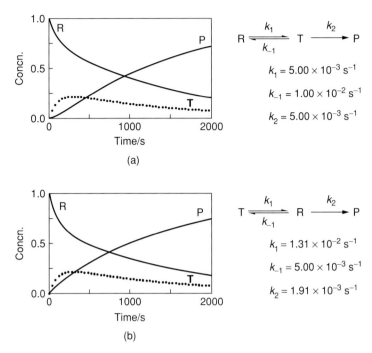

**Fig. 9.5**  Time courses of reactions where the transient T is (a) an intermediate and (b) a cul de sac.

systems exploiting stopped-flow technology, which reduce the time scales for sampling and quench to milliseconds*. In such systems, small volumes of reagent solutions are driven through a high-efficiency mixer and, rather than being delivered to a cell for immediate observation, flow into an ageing loop for measured times before addition of a chemical quench solution. The process can be repeated for different ageing times and collected quenched samples then be analysed by any appropriate method.

The detection of a transient species does not, however, establish it as a true intermediate in the conversion of reactants to products. The possibility that the transient is a mechanistic cul de sac must also be considered. Fig. 9.5 compares the possible behaviour of the simplest cases where the reactant (R), product (P) and transient (T) are linked by first-order reactions to show that observation of a single species does not discriminate between the possibilities. In both cases, T and P might reasonably be regarded as the kinetic and thermodynamic products of a reaction. Concentration changes as a function of time for conversion of R into P with T being a true intermediate are shown in Fig. 9.5a, calculated using the rate constants displayed; Fig. 9.5b shows a reaction sequence with the transient T being a mechanistic cul de sac. The set of rate constants shown produces a profile of concentration changes for T which is closely similar to that in Fig. 9.5a. The profiles for R and P in the two cases, however, do differ. Most obviously, the profile for the growth of P where the transient is an intermediate displays an initial delay while the concentration of the intermediate builds up.

---

* Rapid quench-flow systems are available from HI-TECH Scientific, Brunel Road, Salisbury, SP7 7PU, UK.

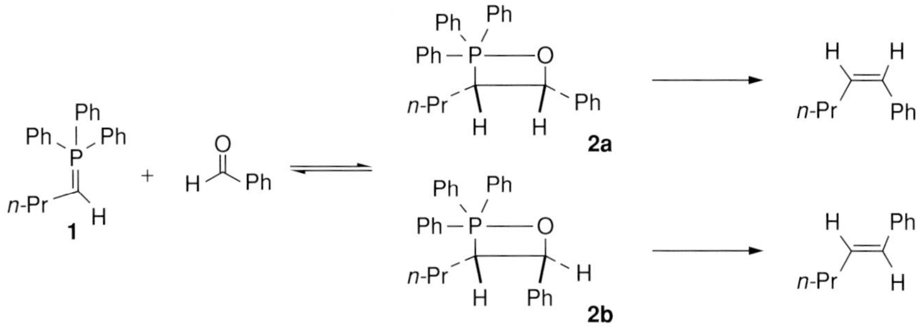

**Scheme 9.1**  Stereoisomeric oxaphosphetanes as NMR-detectable intermediates in the Wittig synthesis of alkenes.

Other combinations of rate constants in the two cases might produce identical profiles for *either* loss of reactant (R) *or* formation of product (P); in both cases, profiles for the *other* two components will differ. It is worth repeating that monitoring the concentration of a single component cannot distinguish between intermediate and cul de sac transient species.

In cases of the types shown in Fig. 9.5, where the transient rises to a relatively high maximum (ca. 22% of total), good quality data across the time course for any two of the three species might permit distinction. Even better, a timely application of an appropriate quench might permit isolation of the transient species for separate characterisation and kinetic examination. Subjection of the putative intermediate, after isolation or separate synthesis, to the same reaction conditions must yield the anticipated products, and at a rate not lower than that of the interrupted reaction.

Often, small variation in reaction conditions (reagent, concentration, order of addition, solvent, acidity, temperature) can transform a transient with only fleeting existence into the one that accumulates to readily observable concentrations. Reactions with transient species of both types are exemplified below.

In Scheme 9.1, stereoisomeric oxaphosphetanes are shown as observable intermediates in the Wittig olefin synthesis. Recognised very early in the development of the Wittig reaction as possible intermediates, they do not accumulate under the usual synthetic conditions which involve addition of ketone or aldehyde to a preformed ylid. However, these turned out to be directly observable by $^{31}P$ and $^{13}C$ NMR spectroscopy when reagents are mixed at low temperature. Thus, benzaldehyde and the ylid **1** were mixed in an NMR tube at $-30°C$. Under these conditions, the oxaphosphetanes **2a** and **2b** accumulated together to over 90% of the mixture after a few minutes. Their competing equilibration and decay to stereoisomeric alkenes could then be observed over a period of hours [13]. Oxaphosphetanes can now be synthesised by alternative routes and modern Wittig chemistry recognises them as versatile starting materials in olefin synthesis [14].

A transient also occurs in the reaction of 2,4-dinitrophenyl chloride (**3**) with sodium hydroxide in aqueous DMSO, Scheme 9.2 [15]. Isomeric $\sigma$-adducts, **4** and **5**, formed relatively rapidly from **3**, are cul de sac species and detectable by UV–vis spectroscopy. A slower separate reaction of **3** yields the phenolate, presumably via intermediate **6**, which did not accumulate to UV detectable concentrations.

In the sequences shown in Fig. 9.5, the maximum concentration of the transient decreases as the ratios $(k_{-1} + k_2)/k_1$ (case a) or $(k_1 + k_2)/k_{-1}$ (case b) increase. The most sensitive

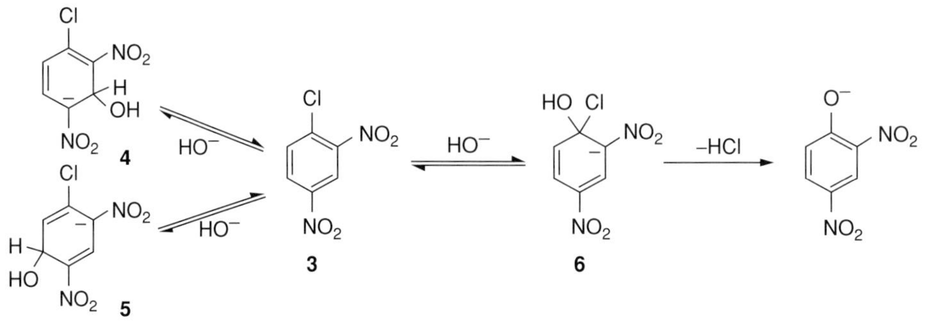

**Scheme 9.2**   UV-detectable cul de sac transient species, **4** and **5**, in the alkaline hydrolysis of 2,4-dinitrophenyl chloride, **3**.

methods (ESR, UV–vis spectroscopy, fluorimetry) may permit observation of very low concentration intermediates but, as the maximum concentration of the transient decreases, deviations of the profiles of R and P from simple first-order behaviour become increasingly difficult to detect. Under these circumstances, separate estimation of the concentrations of the transient and measurement of $k_2$ are essential to establish the nature of T. A corollary to this is that failure to observe directly a transient species does not exclude its intermediacy unless it can be separately shown that the sensitivity and time constant of the analytical methods would have permitted its detection at its anticipated concentration under the conditions of the reaction.

### 9.3.2   Deductions from kinetic behaviour

The Bodenstein approximation recognises that, after a short initial period in the reaction, the rate of destruction of a low concentration intermediate approximates its rate of formation, with the approximation improving as the maximum concentration of intermediate decreases (see Chapters 3 and 4). Equating rates of formation and destruction of a non-accumulating intermediate allows its concentration to be written in terms of concentrations of observable species and rate constants for the elementary steps involved in its production and destruction. This simplifies the kinetic expressions for mechanisms involving them, and Scheme 9.3 shows the situation for sequential first-order reactions. The set of differential equations

$$R \underset{k_{-1}}{\overset{k_1}{\rightleftharpoons}} I \xrightarrow{k_2} P \Longrightarrow \begin{array}{l} d[R]/dt = k_1[R] - k_{-1}[I] \\ d[I]/dt = k_1[R] - (k_{-1} + k_2)[I] = 0 \\ d[P]/dt = k_2[I] \end{array} \Longleftarrow$$

$$d[P]/dt = k_{obs}[R] \text{ where } k_{obs} = k_1 k_2/(k_{-1} + k_2) \left\{ \begin{array}{l} \text{If } k_2 \gg k_{-1}, k_{obs} \Rightarrow k_1 \\ \text{If } k_{-1} \gg k_2, k_{obs} \Rightarrow k_1 k_2/k_{-1} \end{array} \right.$$

**Scheme 9.3**   Application of the Bodenstein approximation to an intermediate in sequential first-order reactions.

governing the behaviour of the components may be written directly, and the relationship containing the Bodenstein approximation is arrowed.

When this is incorporated into expressions for rates of disappearance of reactant or formation of product, it is clear that the reaction is expected to display first-order behaviour so that, kinetically, the reaction is indistinguishable from a single-step conversion of R into P. The rate constant, $k_{obs}$, is composite, containing contributions from all the elementary steps, with simplification becoming possible when $k_2 \gg k_{-1}$ or $k_{-1} \gg k_2$. Kinetics alone cannot provide evidence for the existence of a transient intermediate species if all reactions in the sequence are first order.

The above basic scheme is readily adapted to situations where the elementary reactions are of higher molecularity, with the first-order rate constants then being replaced where appropriate by products of second-order rate constants and concentrations of species involved. This leads to rate laws of higher complexity and kinetic behaviour which now may signal the existence of a transient intermediate. It is not practicable to treat all possibilities here [16], but consideration of the simplest of such situations reveals useful patterns. Scheme 9.4 presents a reaction of known stoichiometry, and four possible alternative kinetic schemes involving a reversibly formed intermediate (I) consistent with that stoichiometry.

For each kinetic scheme in Scheme 9.4, the rate law obtained by applying the Bodenstein approximation to the intermediate (I) is presented and, for this discussion, we consider that the reactant R is the component whose concentration can be easily monitored. The reactions are all expected to be first order in [R], but the first-order rate constants show complex dependences on [X] and, in two cases, also on [Y]. All the rate laws contain sums of terms in the denominator, and the compositions of the transition structures for formation and destruction of the intermediate are signalled by the form of the rate law when each term of the denominator is separately considered. This pattern is general and can be usefully applied in devising mechanisms compatible with experimentally determined rate laws even for much more complex situations.

In case (a), R and I interconvert by unimolecular processes, and then I is converted into product in a bimolecular reaction with reagent X. Such reactions are expected to be first order in [R], but order with respect to [X] will vary between 0 and 1 depending on the relative magnitudes of $k_2[X]$ and $k_{-1}$. Since X is consumed during the reaction, the variation may be detectable as downward drift in the rate constant when monitoring disappearance of R.

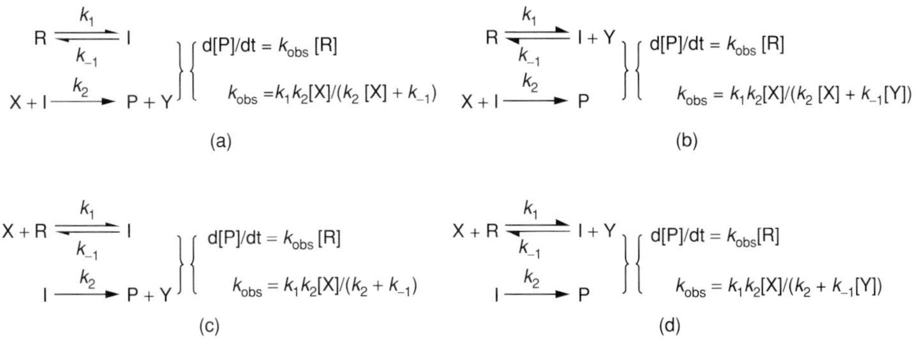

(a)

$R \underset{k_{-1}}{\overset{k_1}{\rightleftharpoons}} I$

$X + I \overset{k_2}{\longrightarrow} P + Y$

$d[P]/dt = k_{obs}[R]$

$k_{obs} = k_1 k_2 [X]/(k_2 [X] + k_{-1})$

(b)

$R \underset{k_{-1}}{\overset{k_1}{\rightleftharpoons}} I + Y$

$X + I \overset{k_2}{\longrightarrow} P$

$d[P]/dt = k_{obs}[R]$

$k_{obs} = k_1 k_2 [X]/(k_2 [X] + k_{-1}[Y])$

(c)

$X + R \underset{k_{-1}}{\overset{k_1}{\rightleftharpoons}} I$

$I \overset{k_2}{\longrightarrow} P + Y$

$d[P]/dt = k_{obs}[R]$

$k_{obs} = k_1 k_2 [X]/(k_2 + k_{-1})$

(d)

$X + R \underset{k_{-1}}{\overset{k_1}{\rightleftharpoons}} I + Y$

$I \overset{k_2}{\longrightarrow} P$

$d[P]/dt = k_{obs}[R]$

$k_{obs} = k_1 k_2 [X]/(k_2 + k_{-1}[Y])$

**Scheme 9.4** Possible mechanisms involving intermediates and second-order kinetic terms for the same overall transformation, $X + R \rightarrow P + Y$.

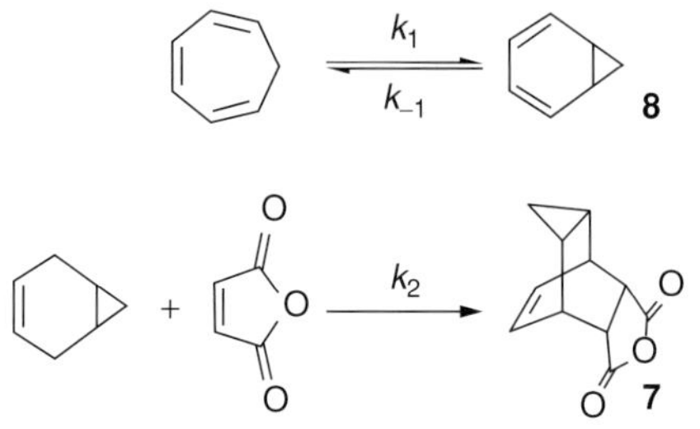

**Scheme 9.5**   Reaction of cycloheptatriene and maleic anhydride via bicyclo[4.1.0]hepta-2,4-diene.

Alternatively, and more controllably, disappearance of R in reactions run with substantial excesses of X ($[X]/[R] > 10$, preferably) should show good first-order behaviour, with pseudo-first-order rate constants increasing as the concentrations of X are increased, and approaching a limiting value of $k_1$. With sufficient data, curve fitting of a plot of $k_{obs}$ against [X] to the expression given above would yield values of $k_1$ and $k_{-1}/k_2$. Alternatively, a plot of $1/k_{obs}$ against $1/[X]$ should be linear, with intercept and slope yielding values of $1/k_1$ and $k_{-1}/(k_1 k_2)$, respectively.

More often, the extreme situations occur with $k_{-1} \gg k_2[X]$ or $k_{-1} \ll k_2[X]$. In the former case, second-order behaviour is anticipated and the reaction is kinetically indistinguishable from a single-step bimolecular reaction between R and X. Such a situation arises in the reaction of cycloheptatriene and maleic anhydride to yield the adduct 7, Scheme 9.5. The bicyclic valence isomer (8) of cycloheptatriene is reasonably postulated as a non-accumulating intermediate in the reaction, but the reaction kinetics are persistently first order in maleic anhydride and in cycloheptatriene [17], and the intermediacy of 8 is supported by evidence from other sources [18].

The alternative situation where $k_{-1} \ll k_2[X]$ is also common. Reactions are then zero order in [X], and the clear disparity between stoichiometric coefficients and reaction orders demands the existence of an intermediate. The situation occurs in acid- and base-catalysed reactions of ketones with reactive electrophiles (e.g. $X = Cl_2$, $Br_2$ or $I_2$), which are usually zero order in the electrophilic reagent, unless concentrations of the electrophiles are extremely low [19]. The intermediate reacting with the electrophile may be the enol tautomer of the ketone or its enolate anion, formed catalytically from the ketone.

In case (b) of Scheme 9.4, the initial formation of the intermediate is dissociative. When X is in large excess, for example as solvent, near first-order behaviour may be expected, but with the value of $k_{obs}$ showing a downward drift as Y accumulates over the course of the reaction, the extent and observability of the drift reflecting the ratio of $k_{-1}/k_2$. Such deviations are often not readily detectable with reactions run on dilute solutions and, even if obvious, they are difficult to quantify. The more controllable experiment again involves running a series of reactions in the presence of known and relatively large amounts of Y (preferably with $[Y]/[R] \geq 10$). For any particular reaction, disappearance of R should then be first order, but the values of $k_{obs}$ should decrease with the increase in [Y]. With sufficient data, curve fitting of a plot of $k_{obs}$ against [Y] to the expression given above would yield values of $k_1$ and $k_{-1}/k_2[X]$. Alternatively, a plot of $1/k_{obs}$ against [Y] should be linear, with intercept and slope yielding values of $1/k_1$ and $k_{-1}/(k_1 k_2[X])$, respectively, from a known constant value of [X].

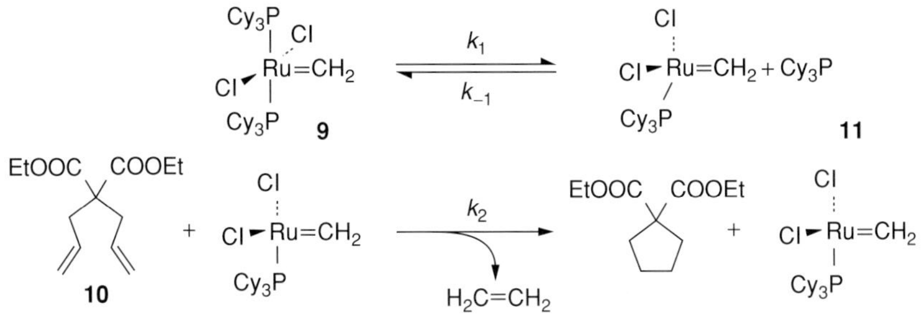

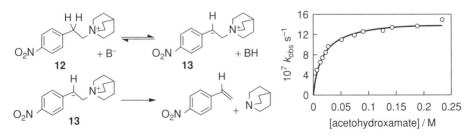

**Scheme 9.6**  Dissociative formation of active species in olefin metathesis (Cy = cyclohexyl).

Such cases are not uncommon, but full quantitative treatments are rare, since often relatively large amounts of Y must be added to obtain measurable effects. Complications may then arise from the effects of the added Y on the nature of the medium (see Chapters 2 and 3). These are particularly notable when Y and I are charged, as is often the case. Under those circumstances, maintenance of the constant ionic strength of the medium with a known non-participating ionic species is essential. The classic case of common ion depression in solvolysis of benzhydryl chloride is dealt with in Chapter 2. A more recent example of this kind of treatment with neutral reactants occurs in the elucidation of the mechanism of olefin metathesis [20], catalysed by the ruthenium methylidene **9**, Scheme 9.6. With ca. 5% of **9**, disappearance of diene **10** was clearly not first order. However, reactions run in the presence of large excesses of phosphine **11** were much slower and showed first-order kinetics. The plot of $k_{obs}$ against $1/[11]$ was linear, consistent with dissociation of **9** to yield an active catalytic species prior to engagement with the diene, with $k_{-1}[11] \gg k_2[\text{diene}]$. Because first-order kinetics were observed under these conditions, determination of order with respect to the catalytic species (as well as the diene) was simplified, and an outline for the mechanism could be constructed (see also Chapter 12 for more detailed consideration of catalysed olefin metathesis).

In cases (c) and (d) of Scheme 9.4, the intermediate is formed by associative processes. In the former, the kinetics will again be indistinguishable from those of a simple bimolecular process. In case (d), however, saturation behaviour similar to that in case (b) is anticipated. A good example occurs in the elimination of quinuclidine from *N*-(*p*-nitrophenylethyl)quinuclidinium ion (**12**) by buffered acetohydroxamate (Scheme 9.7).

**Scheme 9.7**  Saturation kinetics in the buffered base-induced elimination from *N*-(*p*-nitrophenylethyl)quinuclidinium.

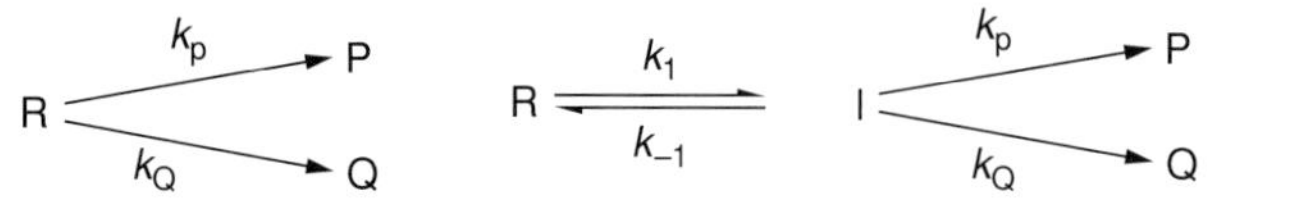

**Scheme 9.8**  Multiple products (a) by parallel routes from reactant and (b) via a single intermediate.

With the buffer ratio held constant ($[B^-]/[BH] = 0.5$), and the ionic strength maintained at 1 mol dm$^{-3}$ with potassium chloride, initial rates show saturation kinetics over the range $0.01 < [B^-] < 0.25$ mol dm$^{-3}$. This behaviour is compatible with the intermediacy of the carbanion **13**, and the rate-determining step changing from its rate-limiting formation at low buffer concentrations to rate-limiting expulsion of quinuclidine from the carbanion, formed rapidly and reversibly, at high buffer concentrations [21].

   In the reactions discussed and exemplified above, reactants, transient species and products are related by linear sequences of elementary reactions. The transient species can be regarded as a kinetic product and, if isolable, subject to the usual tests for stability to the reaction conditions. Multiple products, however, may also occur by a mechanism involving branching. Indeed, the case shown earlier in Fig. 9.5b, where the transient is a cul de sac species, is the one in which the branching to the thermodynamic product P and kinetic product T occurs directly from the reactant. In the absence of reversibility, the scheme becomes as that shown in Scheme 9.8a, where the stable products P and Q are formed as, for example, in the stereoselective reduction of a ketone to give diastereoisomeric alcohols. The reduction of 2-norbornanone to a mixture of *exo*- and *endo*-2-norbornanols by sodium borohydride is a classic case. The product ratio is constant over the course of the reaction and reflects directly the ratio of rate constants for the competing reactions. The pseudo-first-order rate constant for disappearance of R is the sum of the component rate constants.

   Branching may also occur from an intermediate and Scheme 9.8b shows the simplest scheme where P and Q are formed irreversibly from a Bodenstein intermediate (I). Kinetically, this mechanism is not distinguishable from that where branching occurs at the reactant. The product ratio [P]/[Q] is the ratio of the rate constants of the forward processes at the branch point. The reaction again shows first-order behaviour with respect to the reactant R, with the overall experimental pseudo-first-order rate constant dependent on all the microscopic rate constants.

## 9.3.3   Trapping of intermediates

Whether or not branching occurs at an intermediate, its existence may be demonstrated by running the reaction in the presence of a reagent designed to intercept that intermediate to yield a new and characteristic product. Observations at a qualitative level can be extremely informative (e.g. formation of cyclo-adducts when aryl halides are treated with a strong base in the presence of conjugated dienes to trap a benzyne intermediate) but even more information may be obtained from quantitative experiments, especially when product analyses are coupled with rate measurements.

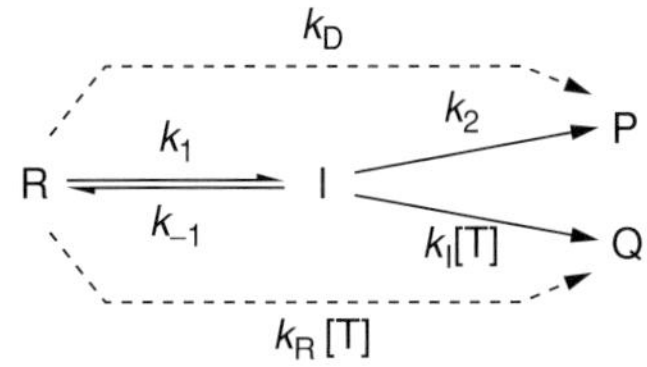

$$d[P]/dt = k_1 k_2 [R]/(k_{-1} + (k_2 + k_I[T])) \qquad (+ k_D[R])^*$$

$$d[Q]/dt = k_1 k_I[T][R]/(k_{-1} + (k_2 + k_I[T])) \qquad (+ k_R[R][T])^*$$

$$-d[R]/dt = k_1(k_2 + k_I[T])[R]/(k_{-1} + (k_2 + k_I[T])) (+ k_D[R] + k_R[R][T])^*$$

**Scheme 9.9** A general scheme for trapping an intermediate, **I**.

Scheme 9.9 adapts our basic reaction scheme to take account of the effect of a substantial excess of a trapping reagent (T) on the conversion of the reactant (R) to the product (P) via the intermediate, I. The new and characteristic product (Q) may be formed by the reaction of I with T and, to be effective, the trapping reaction must compete efficiently with the main reaction channel, i.e. $k_I[T] \approx k_2$ in Scheme 9.9. Competing processes are included as dashed lines in Scheme 9.9 whereby the product P is also formed by a pathway not involving the intermediate with rate constant $k_D$, and the trapping product Q is formed by the direct bimolecular reaction of R with the trap (T) with second-order rate constant $k_R$. These considerably complicate interpretations of experimental data, but may be recognised and their contributions quantified under some circumstances by analysis of rate–product correlations (see Chapter 2).

Included in Scheme 9.9 are expressions for the rates of formation of both P and Q, and for disappearance of reactant, derived by applying the Bodenstein approximation to the intermediate. Contributions to reactivity from the pathways not involving the intermediate (dashed lines) are gathered in the starred term of each of these relationships. A number of cases may be recognised.

In the extreme case that there is no reaction via the intermediate, i.e. $k_1 = 0$, the unstarred terms in the rate equations all vanish, and a little further manipulation establishes Equations 9.5 and 9.6 as the simple relationships between product distribution (either as the ratio Q:P or as the inverse of the mole fraction of Q, respectively) and the amount of T present; the overall rate of reaction is given by Equation 9.7:

$$[Q]/[P] = k_R[T]/k_D \tag{9.5}$$

$$([P] + [Q])/[Q] = 1 + k_D/(k_R[T]) \tag{9.6}$$

$$-d[R]/dt = (k_D + k_R[T])[R] \tag{9.7}$$

$$RE = (d[R]/dt - d[R]/dt_{(T=0)})/d[R]/dt_{(T=0)}. \tag{9.8}$$

Rates are thus expected to show the first-order dependence on the concentration of T added, and the rate enhancement (RE, Equation 9.8) for a given concentration of T is easily shown to be the corresponding product ratio [Q]/[P]. Behaviour of this kind occurs, for example, in the solvolysis of 2-propyl tosylate in aqueous ethanol [22]. Products are those of nucleophilic substitution – a mixture of 2-propanol and 2-propyl ethyl ether. When the reaction is run in the presence of sodium azide, an excellent nucleophile, rates of disappearance of the tosylate are enhanced and 2-propyl azide is also formed. The relationships between amounts of azide product and rate enhancements for a series of different azide concentrations were accurately described by Equations 9.5–9.8.

Although the values of the constants differ, relationships of identical form hold for product ratios and overall rate when the reaction proceeds exclusively via the intermediate, i.e. when $k_D = k_R = 0$. Then, the starred terms in the rate equations vanish and further manipulation establishes Equations 9.9 and 9.10 as the relationships between product distribution and the amount of trapping agent present:

$$[Q]/[P] = k_1[T]/k_2 \qquad (9.9)$$
$$([P] + [Q])/[Q] = 1 + k_2/(k_1[T]). \qquad (9.10)$$

Furthermore, if destruction of the intermediate is rate limiting, i.e. $k_2 + k_1[T] \ll k_{-1}$, then the overall rate expression simplifies to Equation 9.11:

$$-\mathrm{d}[R]/\mathrm{d}t = k_1(k_2 + k_1[T])[R]/k_{-1}. \qquad (9.11)$$

Inspection of Equations 9.7 and 9.11 shows that they have the same form, so neither kinetics nor product analysis can test for the existence of an intermediate under these circumstances.

However, when the reaction proceeds exclusively via the intermediate, and $k_2 + k_1[T] \gg k_{-1}$, then the overall rate law reduces to the simple form of Equation 9.12 which is clearly distinguishable from both Equation 9.7 and Equation 9.11:

$$-\mathrm{d}[R]/\mathrm{d}t = k_1[R]. \qquad (9.12)$$

Rates should now be unaffected by the presence of the trapping agent, although we repeat the *caveat* that allowance has to be made for the effects of added trapping agents on the medium, especially when it is ionic. Product ratios in a particular reaction are expected to be constant over the course of the reaction, obeying Equations 9.9 and 9.10. Plots of $[Q]/[P]$ against $[T]$, or of the inverse mole fraction against $1/[T]$ for a series of reactions in which $[T]$ is varied, should yield linear plots with slopes of $k_1/k_2$ or $k_2/k_1$, respectively, with the corresponding intercepts of 0 and 1. Behaviour of this kind is found in the solvolysis of 2-adamantyl tosylate in aqueous ethanol. Experiments with added sodium azide, similar to those outlined earlier for 2-propyl tosylate, were performed [22]. In the case of 2-adamantyl tosylate, rate enhancements were found, as were small amounts of 2-adamanyl azide amongst the products. However, the rate enhancements were small, and, crucially, the fraction of alkyl azide produced was much smaller than that expected from Equation 9.6.

This is a useful and informative situation, and solvolytic experiments of this kind have a particular value if an absolute value for the second-order rate constant, $k_1$, for the reaction of the trap with the intermediate is known. In that case, an absolute value of the first-order rate constant, $k_2$, for the conversion of the intermediate into the solvent-derived product may be obtained, and hence an estimate of its lifetime under the reaction conditions. Measurements yielding values less than the vibrational limit ($1.7 \times 10^{-13}$ s at 25°C) indicate clearly that $I$ has no real lifetime and hence is not a viable intermediate, and an alternative mechanism is required. For non-solvolytic reactions in a solution where the forward reaction of the reactive intermediate (other than with $T$) is bimolecular/second order, its lifetime will be diffusion controlled and the limit is likely to be closer to $10^{-10}$ s (though dependent upon the concentration of its co-reactant).

Other combinations of reactions may occur, and useful information may be extracted from product ratios and rates, but reactions in which rates show a significant dependence on concentrations of $T$ are more difficult to unravel. If, for example, the pathway governed

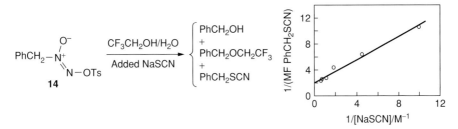

**Fig. 9.6**  Demonstration of a trappable intermediate in the solvolysis of benzyl azoxytosylate.

by $k_D$ operates avoiding any involvement of the trap, then the ratio of the products $[Q]/[P]$ is again expected to be proportional to $[T]$, so the situation is not differentiated from the simple case. The relationship between the slope of the plot and the rate constants is complex and shown in Equation 9.13. The plot of the inverse mole fraction of Q against $1/[T]$ is more revealing. If $(k_2 + k_l[T]) \gg k_{-1}$, the anticipated form of the relationship (Equation 9.14) shows that the plot should be linear, as in the simple case, but that the intercept should deviate from 1 by the ratio $k_D/k_1$. Rates should unaffected by the added trapping agent and the rate constant should equal $k_1 + k_D$:

$$[Q]/[P] = k_1 k_l[T]/(k_D + k_1 k_2) \tag{9.13}$$

$$([P] + [Q])/[Q] = 1 + k_D/k_1 + k_2(k_D + k_1)/(k_1 k_l[T]). \tag{9.14}$$

A situation of this type was encountered in the investigation of the solvolysis of benzyl azoxytosylate, **14** in Fig. 9.6, in aqueous trifluoroethanol [23]. In the absence of added nucleophilic reagents the products were benzyl alcohol and benzyl trifluoroethyl ether. Added excess sodium thiocyanate had minimal effect upon the rates, but the reaction then yielded substantial amounts of benzyl thiocyanate. The plot of the inverse mole fraction of benzyl thiocyanate against $1/[NaSCN]$, shown in Fig. 9.6, is linear with a non-zero intercept, consistent with the intermediacy of a species trappable by thiocyanate, and a *competing* pathway for the formation of ethereal and alcoholic products other than via this species.

If, on the other hand, a pathway involving a bimolecular reaction of a reactant and trap, governed by $k_R$, competes with the reaction via the intermediate, rates will again show the dependence on the added trapping agent. Expressions for the dependence of the product ratio on the added T show quadratic forms and are not reduced appreciably in complexity by the limits when $(k_2 + k_l[T]) \gg k_{-1}$ or when $(k_2 + k_l[T]) \ll k_{-1}$.

These trapping experiments have been illustrated with reactions involving heterolytic bonding changes and polar intermediates. The kinetic treatments are general, and applications to homolytic processes and radical traps are discussed in Chapter 10.

A variant of the trapping experiment, which physically excludes the possibility of the direct reaction of the trap with reactant, is found in the three-phase test [24]. This exploits the availability of an increasing range of polymeric solid supports (PS in Fig. 9.7) to which reactants and trap may be separately attached. The direct reaction between the solid-supported trap and solid-supported reactant is not anticipated, but control experiments to establish this are advisable. The formation of the trapping product (Q) when the polymer-supported reactant and polymer-supported trap are dispersed together in a solution under conditions where the suspected intermediate would be released from the reactant is then clear evidence

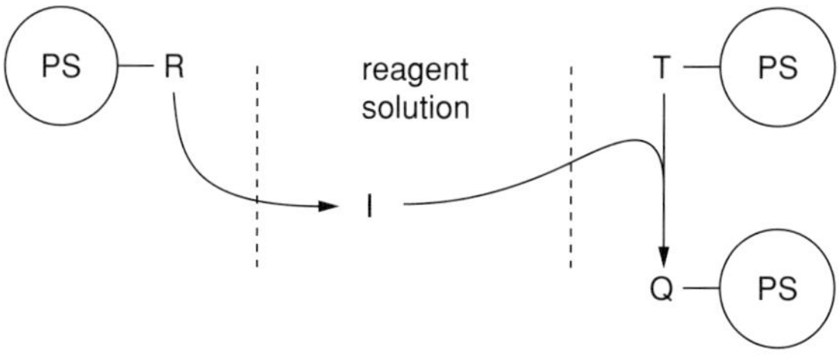

**Fig. 9.7** The basis of the three-phase test for an intermediate.

that this has indeed occurred, and the intermediate has diffused into solution, and then to the polymer with the trap. The technique is not applicable to very short-lived species.

Most recently, the test has been applied to examination of the mechanism of a heterogeneous Heck reaction, promoted by Pd on alumina [25]. In the presence of the solid catalyst, 4-iodobenzamide coupled efficiently with butyl acrylate yielding the cinnamate, and it was suspected that the catalytic agent was a soluble form of palladium released from and then recaptured by the alumina support. To test this, the amide was attached to a commercially available resin with suitable functionality, and the supported amide (15 in Scheme 9.10) was allowed to react with the acrylate and Pd on alumina. The same product, identified after release from the polymer by TFA treatment, was formed, and further experiments were able to narrow down the form of the soluble catalysing palladium species.

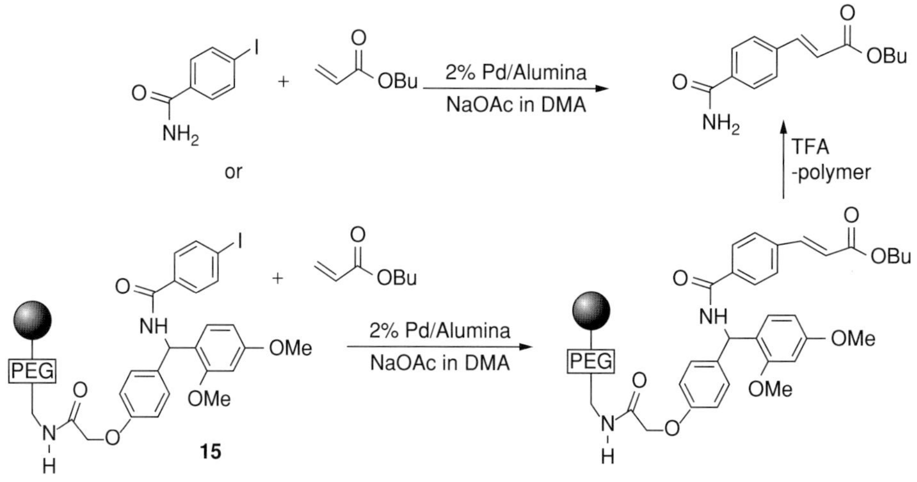

**Scheme 9.10** Application of a three-phase test to investigate the nature of the catalyst in a heterogeneous Heck reaction.

### 9.3.4 *Exploitation of stereochemistry*

Most organic molecules incorporate at least one stereogenic element (centre, plane or axis) [26] and may, therefore, exist in more than one stereoisomeric form. These stereoisomers, as far as mechanistic argument is concerned, are distinct species, and the kinetics and

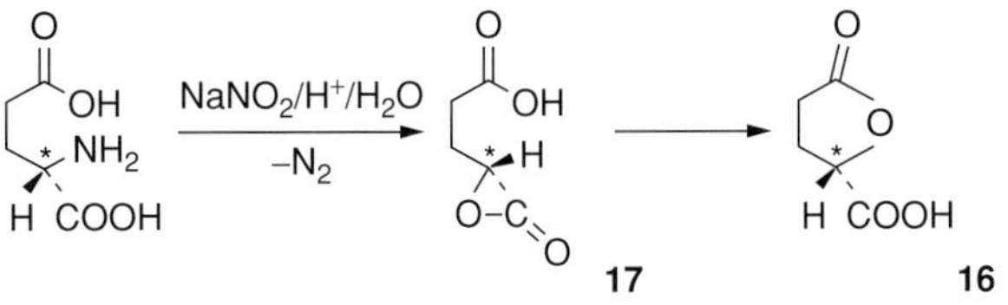

**Scheme 9.11**  Double displacement in the deamination of (*S*)-glutamic acid.

product distributions from the reaction of a particular stereoisomer can yield information about the existence and nature of an intermediate. Reaction schemes outlined in the earlier sections apply, and the key concept here is that *concerted reactions are stereospecific* [27]. The reaction of a particular stereoisomer in which a stereogenic element is retained can yield only one stereoisomeric product, or a set of products. Exclusive inversion of configuration in a nucleophilic displacement at saturated carbon, for example, indicates that the displacement is an $A_ND_N$ ($S_N2$) process. Other nucleophilic displacements which are stereospecific, but proceed with retention of configuration at the stereogenic centre, are, therefore, taken as evidence that the final product derives from a sequence of two stereospecific inversions. Such is believed to occur in the conversion of the (*S*)-glutamic acid into the lactone **16** in Scheme 9.11, a process in which the C—N bond at the stereogenic centre (*) has been replaced with net retention of configuration by the new C—O bond [28]. It was suggested that diazotisation of the amino group, and intramolecular displacement of molecular nitrogen by an oxygen of the nearer carboxylate group, yields the transient α-lactone (**17**), which suffers a second displacement at the stereogenic centre, again with inversion of configuration, by an oxygen of the more remote carboxyl group to complete the sequence.

Stereospecific behaviour in reactions at a stereogenic plane is observed in the addition of bromine to alkenes, Scheme 9.12 [29]. Alkenes with a very wide variety of *cis*-alkyl groups (**18**) react with bromine in methanol yielding only two of the three possible stereoisomeric bromides **19** and **20** (X = Br) – those of net *anti*-addition of a bromine molecule across the double bond. These might arise by two competing (and very awkward-looking) concerted additions of bromine across the double bond, or by the sequence shown where there is initial stereospecific formation of a non-accumulating intermediate cyclic bromonium ion, **21**, which is then converted into the stereoisomeric products by stereospecific opening of the three-membered ring by nucleophilic attack of a bromide ion at either of the carbons

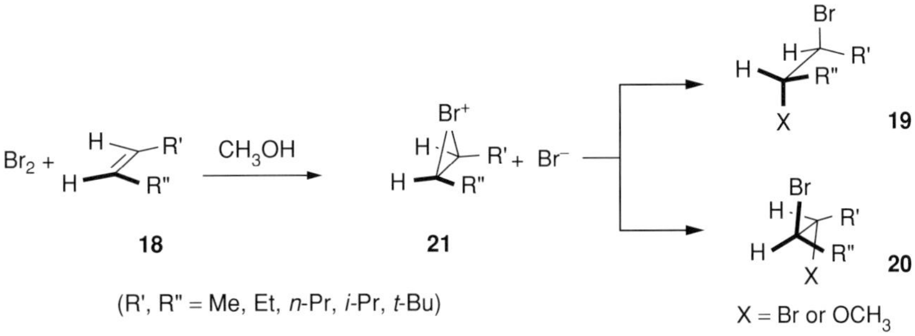

**Scheme 9.12**  Products and stereochemistry in the addition of bromine to alkenes.

of the ring. The latter interpretation is strongly supported by the concomitant formation of methoxybromides 19 and 20 (X = OCH$_3$), which can be accommodated by competing stereospecific nucleophilic ring opening of the intermediate 21 by a methanol molecule. Further internal consistency is provided by the stereochemical course of reactions of the corresponding *trans*-alkenes, which yield the third stereoisomeric dibromide, again the result of net *anti*-addition of bromine across the double bond.

Loss of stereochemical integrity during a chemical reaction is usually taken as good evidence that the products have been formed via an intermediate incapable of sustaining stereochemical information, but there are alternative explanations (competing concerted processes or equilibrations of stereoisomeric products under the reaction conditions) and these should be considered and excluded where possible by other experiments.

Most simply, loss of stereochemical integrity via an intermediate may occur because of its natural symmetry. Theoretical considerations and the study of carbocations under stable ion conditions, for example, show that, in the absence of steric constraints, the preferred arrangement of ligands around tri-coordinate cationic carbon is planar and trigonal, so that the cationic centre cannot itself be a stereogenic element, and some degree of racemisation during nucleophilic substitutions at chiral centres of optically active alkyl halides is key supporting evidence for the intermediacy of carbocations in those cases, formed by an initial D$_N$ process. Complete racemisation would indicate that the ion captured by the nucleophile is fully formed and symmetrically solvated. In practice, complete racemisation is exceptional, even in reactions where other evidence strongly supports the formation of a cationic intermediate (but see Chapter 7 for an account of the rather special *exo*-2-norbornyl system). Substantial amounts of retention or inversion may be observed and indicate product formation from an intermediate with high carbocation character (intimate or solvent-separated ion pairs, discussed in Chapter 2) but before all stereochemical information is lost.

Stereochemical integrity may also be lost when the reaction of a stereoisomer occurs via an intermediate which retains a stereogenic element, but whose bonding permits interconversion of stereoisomers faster than its conversion of stereoisomeric products. [2 + 2]-Cycloaddition of TCNE and *cis*-propenyl methyl ether [30] yields *cis*- and *trans*-adducts, 22 in Scheme 9.13, in ratios which depend on the solvent (84:16 in favour of the *cis*-adduct in acetonitrile). The dipolar 23 was proposed as an intermediate. The initial bonding destroys the double bond character between C1 and C2 of the enol ether reactant, and the much

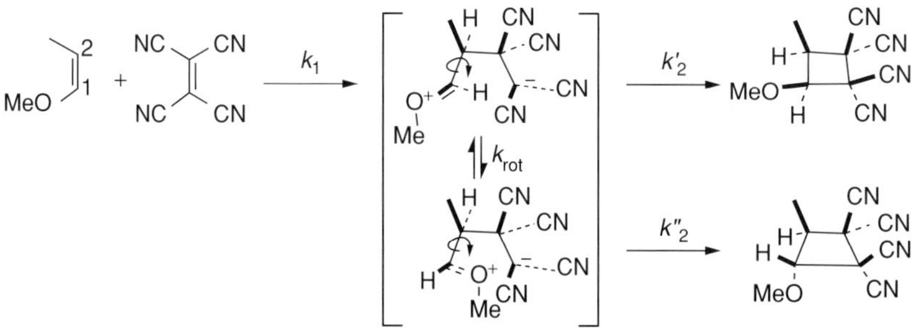

**Scheme 9.13** Loss of stereochemical integrity in cyclobutane formation from TCNE and *cis*-propenyl methyl ether.

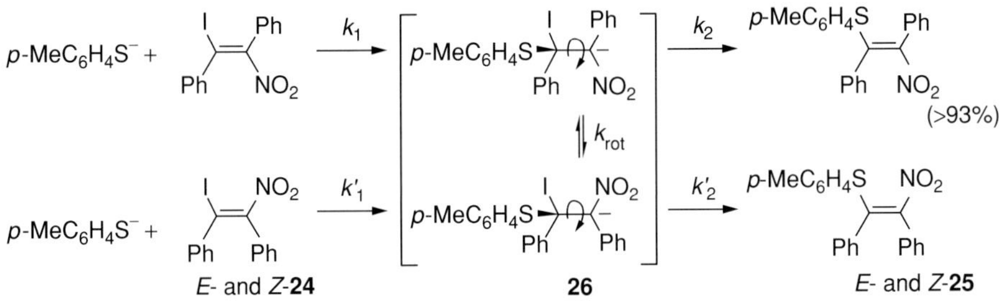

**Scheme 9.14** Stereoconvergence in nucleophilic substitutions at vinylic carbon.

faster rotation about the single bond in the intermediate competes with ring closure to give the isomeric cyclobutanes, i.e. $k_{rot} \approx k_2' \approx k_2''$. The suggestion that the polar intermediate (**23**) is involved is supported by separate trapping experiments with nucleophilic reagents, and by the variation of the isomer ratio with the polarity of the solvent. Isomer ratios when the reaction is carried out with *trans*-propenyl methyl ether (81:19 in favour of the *trans*-adduct in acetonitrile) are not equal to those from the *cis*-diastereoisomer, showing that the rotamers are not fully equilibrated.

Reactions which are apparently stereospecific occur in the nucleophilic displacement of vinylic iodide [31] in the electron-deficient alkenes *E*- and *Z*-**24** shown in Scheme 9.14. With ethanolic toluenethiolate, the sole detectable product from the reaction of *E*-**24** is *E*-**25**. However, *E*-**25** is also the sole detectable product from the reaction of *Z*-**24**. This stereoconvergence demands that the stereoisomers react through a common intermediate, and it was reasonably suggested that initial nucleophilic addition of the thiolate anion yields a resonance-stabilised carbanion (**26**) whose stereoisomerisation, again by rotation about a carbon–carbon single bond, is much faster than the loss of iodide to yield the substitution product ($k_{rot} \gg k_2'$).

In the examples of Schemes 9.13 and 9.14, formation of the equilibrating intermediate has been rate limiting. Any reversibility of intermediate formation would provide a pathway for interconversion of the stereoisomeric reactants.

## 9.3.5 Isotopic substitution in theory

The common elements of organic chemistry (C, H, N, O) occur naturally containing a high proportion of a single isotope. Other isotopes are available and can be used either at natural abundance or in specially synthesised substrates sometimes with high incorporation of isotopes at specific positions in studies to assist in the detection and characterisation of a transient intermediate. A useful resource here is the series, *Isotopes in Organic Chemistry*, edited by Buncel and Lee [32]. All isotopes, of course, are detectable in a molecule by their mass, and mass spectrometry remains the general tool for detection and quantification of isotopes in a molecule or a molecular fragment. As discussed later, vibrational modes are sensitive to mass effects, and IR spectra may reveal the presence of isotopes. Many isotopes, however, have additional useful properties which facilitate their detection. Some have non-zero spin and are detectable by their signals in an NMR spectrum or through coupling to other NMR-active nuclei. Isotopic substitution also induces small (intrinsic) chemical shift changes in signals from attached NMR active nuclei which may be exploited [33]. Some, notably $^{14}C$ and $^{3}H$, are radioactive ($\beta$-emitters, $t_{1/2} = 12.05$ and 1570 years, respectively) permitting detection at very low levels by scintillation counting. Tritium has an added benefit of being NMR active with a nuclear spin ($I = 1/2$) and a nuclear magnetic moment (5.1594) larger than that of $^{1}H$, so it is also detectable by NMR spectroscopy at low incorporations [34].

High precision isotope ratio mass spectrometry permits reliable measurement of the small changes in isotopic distributions which occur in the reactions of molecules with natural abundance incorporations, especially for experiments with isotopes of carbon, nitrogen or oxygen [35]. An alternative of growing importance is the use of NMR spectroscopy to monitor similar changes for $^{13}C$ and deuterium at natural abundance [36].

The theory governing the effects of isotopic substitution on equilibria and rates in reactions of isotopomeric species is well understood and is summarised here only to provide the framework for an understanding of the use of isotopes in the detection of intermediates [37]. Replacement of an atom by an isotope at a particular position in a molecule can have no effect on the electronic energy of the molecule and, for molecules with MW > 100, the fractional change in rotational and kinetic energies is so small that they may be neglected in most cases. The significant differences are in the zero-point energies of molecular vibrations since these are sensitive to the masses of the individual atoms in motion in that vibration. Since vibrations are largely localised, it is not unreasonable to consider the effects of isotopic substitution on a particular stretching or bending mode involving only two or three atoms. The stretching vibration of molecules $M–m_L$ and $M–m_H$ is represented in Fig. 9.8 where M is much heavier than m, and subscripts L and H refer to light and heavy isotopes of m. Arguments apply equally to other vibrational modes but, whatever mode is considered, it is customary to adopt the harmonic approximation so that the classical vibrational frequency ($\nu_{class}$) is given by the expressions shown, where $\kappa$ is the force constant for the vibration.

As microscopic systems, these oscillators have a zero-point energy (ZPE) equal to $\frac{1}{2}h\nu_{class}$ and the difference in zero-point energies for a particular vibration in isotopomers depends upon vibrational frequencies. These, in turn, depend on the force constant (the same for

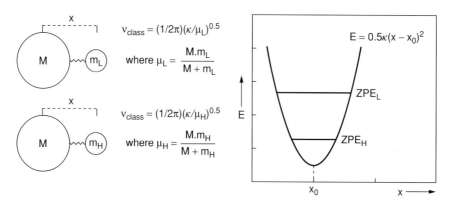

**Fig. 9.8** Vibrational behaviour of a pair of isotopomeric molecules; subscripts L and H refer to light and heavy isotopes, respectively.

both isotopes) and on the reduced masses, $\mu_H$ and $\mu_L$, Equation 9.15. Bonding changes affect the force constant but, obviously, not the reduced masses:

$$\Delta ZPE = ZPE_L - ZPE_H = 0.5\,h(\nu_L - \nu_H) = 0.5\,h\nu_L(1 - (\mu_L/\mu_H)^{1/2}). \tag{9.15}$$

The full effect is the sum of contributing terms from all vibrations of the molecule involving distinct motion of the isotopically substituted atom although, in practice, there is usually one major contributor. For isotopic substitution of hydrogen at tetracoordinate carbon by deuterium and tritium, for example, contributions from the bonds stretching ($\nu_L = 2850$ cm$^{-1}$) are 4.54 and 6.48 kJ mol$^{-1}$, respectively, and 2.15 and 3.15 kJ mol$^{-1}$ for angle bending ($\nu_L = 1350$ cm$^{-1}$). For substitution of heavier atoms, effects are much reduced, e.g. $^{12}$C by $^{13}$C with $\nu_L \approx 1000$ cm$^{-1}$ for a single bond stretch, the difference is only 0.22 kJ mol$^{-1}$.

In this semi-classical treatment, effects of isotopic substitution on equilibria and rates in chemical reactions reflect the changes in zero-point energy differences between (i) reactants (R) and products (P) – equilibrium isotope effects (EIE), or (ii) between reactants (R) and transition structures (TS) – kinetic isotope effects (KIE). In both cases, these changes reflect changes of force constants at the isotopically substituted site. When bonds are broken and made at the isotopic atom, the effects are described as *primary*; otherwise they are described as *secondary*.

The ratio of equilibrium constants (the EIE) for the reaction of light and heavy isotopomers is given by Equation 9.16, and these may be calculated with some accuracy from vibrational analysis of products and reactants. The simple zero-point energy picture demands that equilibrium isotope effects will be 'normal' ($K^L/K^H > 1$) in changes when bonding at the isotopic site is looser (lower force constant) in the product than in the reactant, and 'inverse' ($K^L/K^H < 1$) in changes when bonding at the isotopic site is tighter (increased force constant) in the product than in the reactant:

$$K^L/K^H = e^{(\Delta ZPE_R - \Delta ZPE_P)/RT}. \tag{9.16}$$

The ratio of rate constants (the KIE) for the reaction of light and heavy isotopomers is given by Equation 9.17:

$$k^L/k^H = e^{(\Delta ZPE_R - \Delta ZPE_{TS})/RT}. \tag{9.17}$$

Knowledge of structure and force constants of a transition structure, together with data for reactants, then also permits calculation of a kinetic isotope effect; however, one molecular vibration of the reactant species will become the reaction coordinate and so will not, therefore, be a proper vibration of the transition structure. In reactions involving transfer of an atom between covalently bonded sites, the reaction coordinate is dominated by stretching of the bond to the atom being transferred. Isotopic substitution of that atom will yield a maximum kinetic isotope effect when the vibration has no restoring force in the transition structure; $\Delta ZPE_{TS}$ then vanishes allowing the zero-point energy in the reactant, $\Delta ZPE_R$, to be fully expressed. For isotopic substitution of hydrogen at tetracoordinate carbon by deuterium and tritium, for example, rate ratios up to $k^H/k^D = 6.3$ and $k^H/k^T = 13.7$ at 25°C would be anticipated in reactions involving rate-limiting cleavage of the bond to the isotopically substituted atom, values calculated using the uncompensated loss of the zero-point energy differences (4.54 and 6.48 kJ mol$^{-1}$, respectively) calculated above for this vibration. Values close to these ratios, which may be measured by careful application of conventional kinetic methods, are commonly found in reactions involving transfer of hydrogen from carbon, and taken as evidence that the rate-limiting step does indeed involve cleavage of the C—H bond. Lower values are also common, and explicable either by residual restoring forces in the reacting C—H bond in the transition structure (i.e. incomplete C—H bond cleavage in the TS) or, in the extreme, by a mechanism in which the hydrogen transfer does not occur in the rate-limiting step.

A full calculation of a kinetic isotope effect requires estimates of force constants in the transition structure. These are available from ab initio PE calculations, but also from an empirical method based on the bond energy bond order (BEBO) assumption [38], which often provides more useful qualitative feedback for the experimentalist than high-level methods (see Chapter 7).

### 9.3.6  Isotopic substitution in practice

For isotopes other than those of hydrogen, equilibrium isotope effects are usually small (ca. <3%), and such isotopes are more often used in *labelling* studies with their concentrations at a particular position reflecting the probability of a mechanistic pathway, without consideration of the effects of isotopic substitution upon stability of isotopomers. Positional exchange in recovered reactants, for example, provides supporting evidence of intermediates and their nature, and Scheme 9.15 illustrates one dissociative process and one associative process.

In case (a), labelled secondary alkyl benzenesulfonates (alkyl = 2-adamantyl, 2-propyl, cyclopentyl, etc.) with 18–28% of $^{18}O$ in the sulfonyl group were partially solvolysed in a range of solvents (SOH), and recovered unreacted alkyl benzenesulfonates were then subject to reductive cleavage of the O—S bond of the sulfonate ester [39]. For water as solvent, the

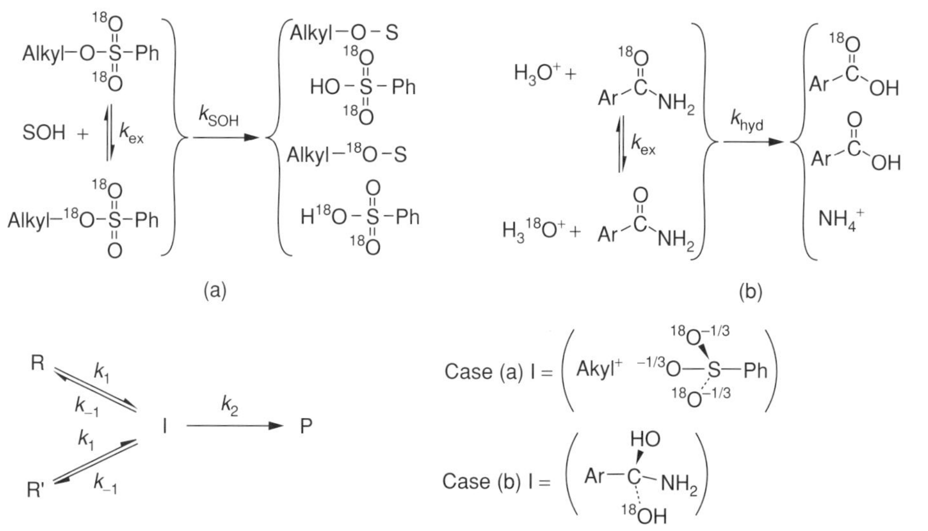

Scheme 9.15  Use of isotope exchange in the study of the mechanisms of nucleophilic substitution at alkyl and acyl carbon.

recovered alcohol and sulfonic acid (after methylation) could then be analysed by GC/MS to establish the extent of $^{18}O$ scrambling which had occurred in the alkyl benzenesulfonate, allowing the comparison of rates of hydrolysis ($k_{SOH}$) with rates of isotopic exchange ($k_{ex}$). In case (b), primary aromatic amides containing about 50% $^{18}O$ were hydrolysed in acidic or basic water [40]. Amide was recovered at known times and analysed mass spectrometrically to permit the comparison of the rate of hydrolysis with the rate of isotopic loss to solvent. In both cases, exchange competes with solvolysis/hydrolysis, demanding the transient formation of species which permit equilibration of the sulfonyl oxygens in the first case, and of oxygen atoms of the carbonyl group with a hydrolysing water molecule in the second. A range of additional experiments suggested that these intermediates lie on the solvolytic/hydrolytic pathways. Kinetic analyses of the reaction schemes, assuming steady-state intermediates, show that the minimum fraction of internal return in case (a), $k_{-1}/(k_{-1} + k_2)$, is given by $2k_{ex}/(2k_{ex} + k_{hyd})$, and the ratio $k_{-1}/k_2$ in case (b) can be identified with $2k_{ex}/k_{hyd}$.

Although heavy atom isotope equilibrium and kinetic isotope effects are small ($k_{12C}/k_{13C} < 1.05$ for breaking an isotopically substituted carbon–carbon bond) and are treated as negligible in labelling experiments of the kind described above, they have been used very effectively in mechanistic studies [35]. Double isotope effects may be used to probe the timing of bonding changes in reactions where bonds to the isotopically labelled sites are made or broken and an example is discussed in Chapter 11. Combinations of anticipated results are able to discriminate concerted from stepwise and, if the latter, which step is rate limiting. However, reliable measurement of the small effects is a specialist activity, requiring access to isotope ratio mass spectrometry [41].

Primary kinetic hydrogen isotope effects are usually large enough to be readily detected and measured, and their occurrence in any reaction involving bonding change to hydrogen is diagnostic for major changes in vibrational force constants at the isotopically substituted site in the rate-limiting step of the reaction. A large normal effect ($k^H/k^D \geq 3$ at $25°C$), therefore, is consistent with rate-limiting hydron transfer, or with coupling between the bonding changes at the transferring hydrogen and at other atoms. The absence of an appreciable effect indicates that the bonding changes at the isotopically labelled position occur either before or after the rate-limiting step in a multi-step process. In the former case, the isotopic substitution may still exert a much smaller secondary effect on the rate. In the latter, a small effect may arise from an equilibrium isotope effect on the formation of an intermediate. Either effect may be normal or inverse.

As an example, consider the reaction of $N$-phenyl or $N$-methyltriazolinedione with alkenes, Scheme 9.16 [42]. In aprotic solution, the products are formally those of an ene addition, with three major bonding changes linking reactants and products, one of which involves transfer of hydrogen from an allylic carbon to a nitrogen atom. Rate ratios, determined

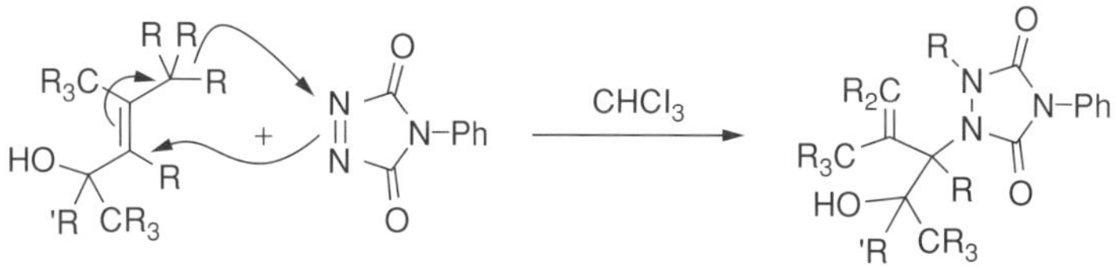

**27**        R = H or D, R' = H or CH$_3$

**Scheme 9.16**  Ene-addition of *N*-phenyltriazolinedione to isotopically labelled allylic alcohols.

by intermolecular competition at 25°C, for the isotopomeric tertiary and secondary alcohols 27 (R' = H and R' = CH$_3$), with R = H or D yielded $k^H/k^{D10}$ = 0.98 (±0.02) and 1.15 (±0.02), respectively. These results are not consistent with transfer of the allylic hydrogen in a rate-determining step (i.e. a single-step (concerted) mechanism) and require that the reaction occurs in at least two steps with the rate-determining step not involving the hydrogen transfer.

The possibility that the reaction involved formation of an intermediate by rapid reversible transfer of an allylic hydrogen in any form (H$^+$, H$^-$ or H$^\bullet$), followed by slower steps, was rejected on energetic grounds, and the proposal adopted included *rapid* hydrogen transfer *after* the rate-limiting step. Considerable information about the nature of the intermediate (or intermediates) in this type of reaction was gained also by examination of intramolecular kinetic isotope effects, measured by analysis of the products from reactions of the more selectively labelled 2,3-dimethyl-2-butenes, 28, 29 and 30 shown in Scheme 9.17. The product distributions (25°C) indicate clearly that there is discrimination between groups at the ends of the double bonds, but not between groups that have a *trans*-relationship on the double bond. These requirements are satisfied by an irreversibly formed transient dipolar aziridinium ion, 31, shown for the particular case of the *trans-d$_6$* isotopomer of 2,3-dimethylbutene, which may yield the ene addition product by [1,5]-hydrogen shift from carbon to nitrogen. The different ratios found in the reaction of the *trans-d$_6$* and *gem-d$_6$* isomers can be nicely reconciled with a value for the primary kinetic effect on the hydrogen transfer being 4.65 combined with normal secondary effects of 1.26 from $\beta$-deuterium atoms which operate to reinforce the primary effect in the case of *gem-d$_6$*, and to reduce it in the case of *trans-d$_6$*.

Consideration of the energetics of the sequence suggested that an intermediate like 31 might accumulate to observable concentrations in reactions of the triazolinedione with reactive alkenes whose structural features disfavour the [1,5]-hydrogen shift and, indeed,

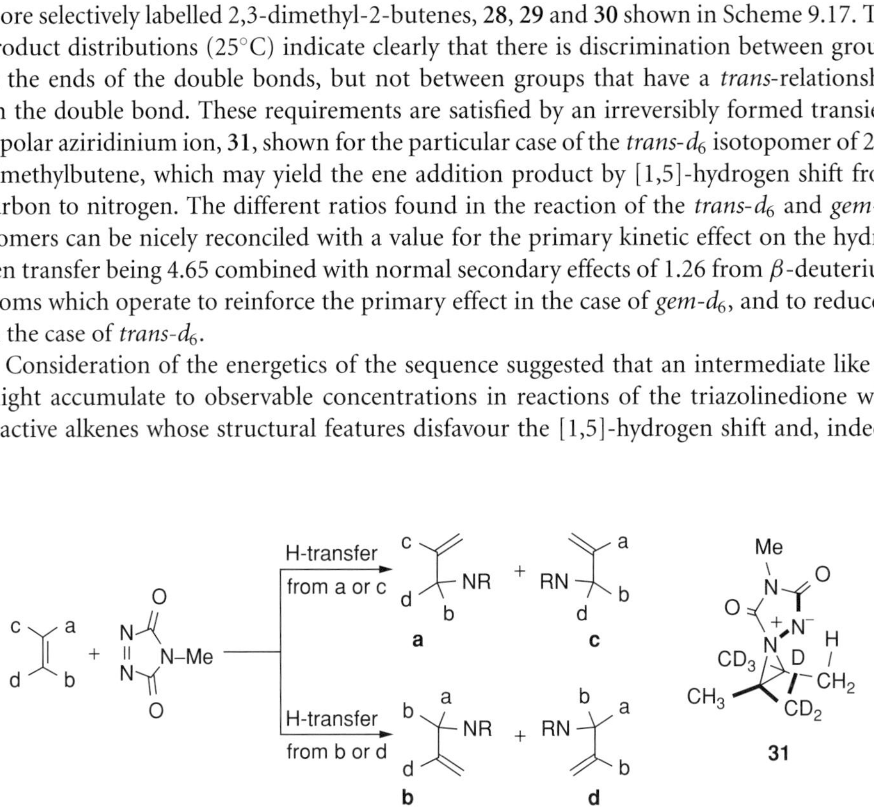

28: a = b = CH$_3$; c = d = CD$_3$ (*cis-d$_6$*)        $k^H/k^D$ = (a + b)/(c + d) = 1.08

29: a = d = CH$_3$; b = c = CD$_3$ (*trans-d$_6$*)        $k^H/k^D$ = (a + d)/(b + c) = 3.80

30: a = c = CH$_3$; b = d = CD$_3$ (*gem-d$_6$*)        $k^H/k^D$ = (a + c)/(b + d) = 5.70

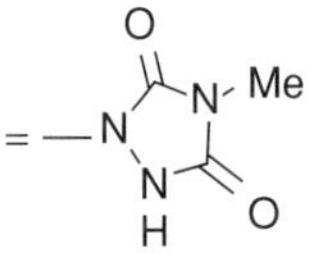

**Scheme 9.17** Product distributions from the reaction of isomeric d$_6$-2,3-dimethyl-2-butenes with *N*-methyltriazolidinedione. Products **a**, **b**, **c** and **d** arise form H or D transfer from sites a, b, c and d in the isomeric butenes.

such species have been observed spectroscopically in the reactions of biadamantylidene [43], *trans*-cycloheptene [44] and *trans*-cyclooctene [45].

Kinetic hydrogen isotope effects significantly larger than the range predicted by the semi-classical theory outlined above are relatively rare, but when they do occur, they have often been taken as evidence of enhanced quantum mechanical tunnelling by the lighter isotope [46]. Mechanisms involving branching in which products are formed from a common intermediate [47], however, can also generate large effects by quite reasonable combinations of microscopic rate constants and conventional primary isotope effects. Indeed, a number of reported large effects taken as evidence for the intervention of tunnelling have proved, upon re-examination, to be due to previously unrecognised branching from intermediates. Temperature dependences of kinetic isotope effects with substantial tunnel contributions are expected to be unusual, but this also can be mimicked by the behaviour of a reaction scheme involving branching [48].

Unusually large hydrogen kinetic isotope effects might, therefore, be better regarded in the first instance as diagnostic of a mechanism involving a transient intermediate, and the simplest case is illustrated in Scheme 9.18. Application of the steady-state assumption to the intermediate I yields the expressions shown for the pseudo-first-order rate constants for disappearance of the reactant R ($k_R$) and formation of products $P_A$ and $P_B$ ($k_{PA}$ and $k_{PB}$). The expressions for the kinetic isotope effects for formation of each product (superscripts L and H refer to light and heavy isotopes) are also shown.

For the formation of products, each KIE expression is the product of the separate kinetic isotope effects on the forward steps, multiplied by the ratio of sums of rate constants for destruction of the intermediate for heavy and light isotopomers. This ratio depends on the relative sizes of $k_A$ and $k_B$, and the isotope effects on each. For the disappearance of reactant (i.e. the overall reaction), the isotope effect is the product of three terms: (i) the kinetic isotope effect for intermediate formation, (ii) the ratio of sums of rate constants for the formation of products for heavy and light isotopomers and (iii) the same ratio of sums of rate constants for destruction of the intermediate for heavy and light isotopomers. The differences between these expressions ensure that competition between two processes which have different kinetic isotope effects, following a rate-limiting step which is also isotopically sensitive, will lead to an enhanced observed kinetic isotope effect for the conversion to the product proceeding through the competing step with the larger isotope effect, and an

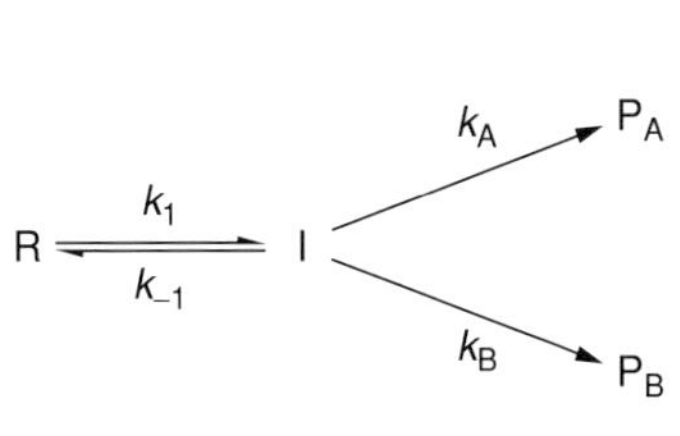

$$k_R = k_1 (k_A + k_B)/(k_{-1} + k_A + k_B)$$
$$k_{PA} = k_1 k_A /(k_{-1} + k_A + k_B)$$
$$k_{PB} = k_1 k_b /(k_{-1} + k_A + k_B)$$

$$k_R^L/k_R^H = (k_1^L/k_1^H)[(k_A^L + k_B^L)/(k_A^H + k_B^H)][(k_{-1}^H + k_A^H + k_B^H)/(k_{-1}^L + k_A^L + k_B^L)]$$
$$k_{PA}^L/k_{PA}^H = (k_1^L/k_1^H)(k_A^L/k_A^H)[(k_{-1}^H + k_A^H + k_B^H)/(k_{-1}^L + k_A^L + k_B^L)]$$
$$k_{PB}^L/k_{PB}^H = (k_1^L/k_1^H)(k_B^L/k_B^H)[(k_{-1}^H + k_A^H + k_B^H)/(k_{-1}^L + k_A^L + k_B^L)]$$

**Scheme 9.18**  Kinetic isotope effects in a branching mechanism.

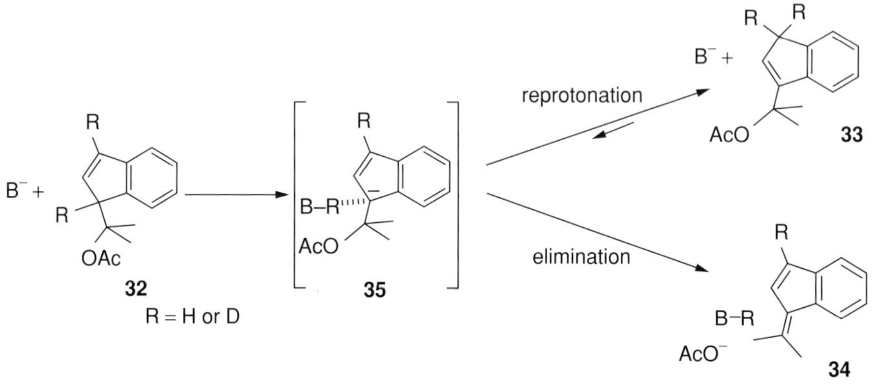

**Scheme 9.19**   Kinetic hydrogen isotope effects in competing rearrangement and elimination from 1-(2-acetoxy-2-propyl)indene.

attenuated isotope effect for the other, i.e. if $k_A^L/k_A^H > k_B^L/k_B^H$, then

$$k_{PA}^L/k_{PA}^H > k_R^L/k_R^H > k_{PB}^L/k_{PB}^H.$$

A situation of this kind has been characterised in the competing rearrangement of and elimination from 1-(2-acetoxy-2-propyl)indene (**32** in Scheme 9.19) induced by methanolic *p*-nitrophenoxide [49]. The deuterium isotope effect on disappearance of reactant was $k_H/k_D$ = 5.2 but, for the products, $k_H/k_D$ = 12.2 for the rearrangement to **33** and $k_H/k_D$ = 3.6 for the elimination to alkene **34**. At least one hydrogen-bonded carbanion, **35**, is required to account for these observations. Its reprotonation to yield the rearranged product was expected to be isotopically sensitive while the expulsion of acetate was not.

### 9.3.7   Linear free energy relationships

Linear free energy relationships describe the proportionality between rates and equilibria observed in many reactions of structurally related compounds. They have the general form of Equation 9.18 where $k_o$ and $k$ are rate constants of a parent compound and a second compound in the reaction under test, and $K_o$ and $K$ are the equilibrium constants for a (different) reversible reference reaction of compounds having the same structural relationship as the first pair:

$$\log k/k_o = r \log K/K_o. \tag{9.18}$$

In the Brønsted relationship, the reference equilibria are dissociations of protic acids, used directly as $\Delta pK_a$, and the constant of proportionality $r = \beta$ (or $\alpha$, see Chapter 11). In the Hammett relationship, acid dissociation constants are also used for the reference equilibria, but indirectly. The dissociation constants of substituted benzoic acids in water at 25°C are used to define a set of substituent parameters, $\sigma$, which are then used in the equation of correlation, and $r = \rho$ in Equation 9.18.

   Although introduced empirically, a detailed theoretical analysis of the shapes of computed PE surfaces suggests that such relationships are expected to describe the response of reactivity

$$\mathrm{R} \underset{k_{-1}}{\overset{k_1}{\rightleftharpoons}} \mathrm{I} \xrightarrow{k_2} \mathrm{P}$$

$$V = k_{obs}[\mathrm{R}] \text{ where } k_{obs} = k_1k_2/(k_2 + k_{-1})$$

$$k_1 = k_1^0\,10^{s\,r_1}$$

$$k_{-1} = k_{-1}^0\,10^{s\,r_{-1}} \quad \text{with } k_{-1}^0 = k_2^0$$

$$k_2 = k_2^0\,10^{s\,r_2} \quad k_{obs} = k_1^0\,10^{sr_1}/(1 + 10^{s(r_{-1} - r_2)})$$

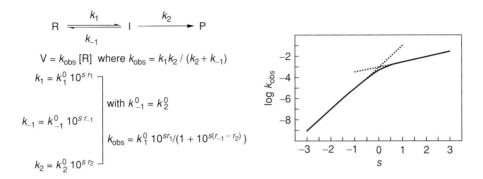

**Fig. 9.9** Application of Brønsted or Hammett relationships to a stepwise process. In the plot, $r_1 = 2$ and $\Delta r = (r_{-1} - r_2) = 1.5$.

to structural variation in a concerted reaction of a family of compounds. Where the reaction is not concerted, however, plots of observed rate constants against substituent parameters may show a downward curvature or discontinuity signalling the presence of an intermediate and a change in the rate-limiting step in response to the structural variation. The behaviour may be described quantitatively by applying the appropriate linear free energy relationship to each elementary step in the stepwise process. Fig. 9.9 shows the simplest case where the kinetic response (log $k_{obs}$) to structural variation in the reactant is mapped against either variation in acidity ($s = \Delta \mathrm{p}K_a$, the Brønsted relationship) or substituent constant ($s = \sigma$, a Hammett relationship), with response factors $r_i = \beta_i$ or $\rho_i$, respectively, for the $i$th step of the reaction. The form of the relationships is such that the plot of log $k_{obs}$ against $s$ comprises two straight lines whose slopes will be $r_1$ and $(r_1 - \Delta r)$ where $\Delta r = (r_{-1} - r_2)$. Setting $k_{-1}^0 = k_2^0$ ensures that the two lines intersect at $\Delta \mathrm{p}K_a$ or $s = 0$.

The principle here is general, but the absence of a break in a linear correlation does not exclude an intermediate. Values of $\Delta\beta$ or $\Delta\rho$ may be so small, or errors in determining the slopes of the components so large, that the break point is undetectable. Equally, prediction of the values of the parameters at which the break point occurs ($k_{-1} = k_2$) is not straightforward and it may lie outside the range of available reactants. The technique has been applied most often in the demonstration of non-accumulating intermediates in substitution processes, both at carbon and at other elements, in quasi-symmetrical reactions [50].

A recent example has been in an examination of the mechanism of alkaline cyclisations of uridine 3′-phosphate esters, Scheme 9.20 [51]. A plot of the logarithms of the second-order

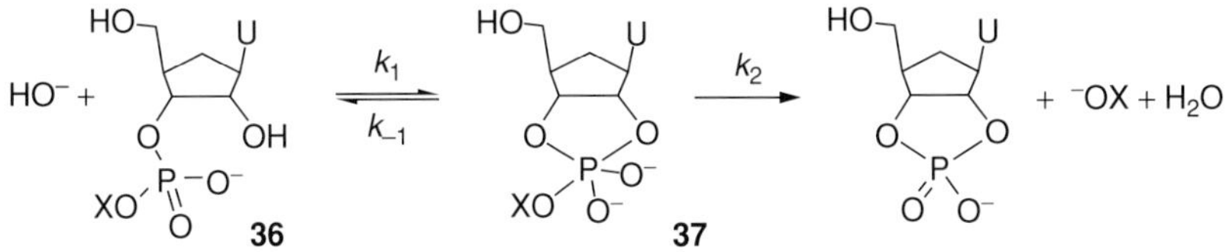

**Scheme 9.20** Stepwise cyclisation of uridine 3′-phosphate esters.

rate constants of the reaction of a series of esters, **36**, against the $pK_a$ of the conjugate acid of the leaving group (XOH) over an extended range ($8 < pK_a(HOX) < 17$) yields a downward convex plot, conforming to the behaviour outlined above, with $\beta_1 = -0.52$, $\Delta\beta = 0.82$ and $\Delta pK_a = 0$ when $pK_a(ROH) = 12.58$, close to the measured $pK_a$ of the 2'-hydroxyl group of uridine 3'-phosphate ethyl ester. The pentacoordinate oxyphosphorane di-anion species (**37**) is, therefore, indicated as a discrete intermediate in the cyclisation rather than a transition structure in a concerted process.

# References

1. (a) Special edition of *Chemical Reviews*, prefaced by Dantus, M. and Zewail, A.H. (2004) *Chemical Reviews*, **104**, 1717; (b) Polany, J.C. and Zewail, A.H. (1995) *Accounts of Chemical Research*, **28**, 119.
2. Benson, S. (1976) *Thermochemical Kinetics*. Wiley, New York.
3. Truhlar, D.G., Gao, J., Alhambra, C., Garcia-Viloca, M., Corchado, J., Sanchez, M.L. and Villa, J. (2002) *Accounts of Chemical Research*, **35**, 341.
4. Joseph Billo, E. (2001) *Excel for Chemists* (2nd edn). Wiley-VCH, New York.
5. Gampp, H., Maeder, M., Meyer, C.J. and Zuberbühler, A.D. (1985) *Talanta*, **32**, 95; Gampp, H., Maeder, M., Meyer, C.J. and Zuberbühler, A.D. (1985) *Talanta*, **32**, 257; Gampp, H., Maeder, M., Meyer, C.J. and Zuberbühler, A.D. (1985) *Talanta*, **32**, 1133; Gampp, H., Maeder, M., Meyer, C.J. and Zuberbühler, A.D. (1986) *Talanta*, **33**, 943.
6. Full details of the system appear in *The Report of the IUPAC Commission on Physical Organic Chemistry* (1989) *Journal of Pure and Applied Chemistry*, **61**, 23. For an abbreviated version, see Guthrie, R.D. and Jencks, W.P. (1989) *Accounts of Chemical Research*, **22**, 343.
7. Dewar, M.J.S. (1984) *Journal of the American Chemical Society*, **106**, 209.
8. Frank, I., Grimme, S. and Peyerinhoff, S.D. (1994) *Journal of the American Chemical Society*, **116**, 5949; Lechten, R., Breslow, R., Schmidt, A. and Turro, N. (1973) *Journal of the American Chemical Society*, **95**, 3025; Oth, J.F.M. (1968) *Receuil des Travaux Chimiques des Pays-bas*, **87**, 1185.
9. Crooks, J.E. (1975) In: E. Caldin and V. Gold (Eds) *Proton-Transfer Reactions*. Chapman & Hall, London, p. 153.
10. Moss, R.A., Platz, M.S. and Jones Jr., M. (Eds) (2004) *Reactive Intermediate Chemistry*. Wiley, Hoboken, NJ.
11. For microcalorimetry, see Thibblin, A. (2002) *Journal of Physical Organic Chemistry*, **15**, 233.
12. Mok, K.H., Nagashima, T., Day, I.J., Jones, J.A., Jones, C.J.V., Dobson, C.M. and Hore, P.J. (2003) *Journal of the American Chemical Society*, **125**, 12484; Hoetzli, S.D., Ropson, I. and Frieden, C. (1994) In: J.W. Crabb (Ed.) *Techniques in Protein Chemistry*. Academic Press, New York, p. 455; Couch, D.A., Haworth, O.W. and Moore, P. (1975) *Journal of Physics E: Scientific Instruments*, **8**, 831; McGarrity, J.F., Prodolliet, J. and Smyth, T. (1981) *Organic Magnetic Resonance*, **17**, 59.
13. Maryanoff, B.E., Reitz, A.B., Mutter, M.S., Inners, R.B. and Almond, H.R. (1985) *Journal of the American Chemical Society*, **107**, 1068. For reviews on the oxaphosphetane intermediates, see Maryanoff, B.E. and Reitz, A.B. (1989) *Chemical Reviews*, **89**, 863; Vedjes, E. and Petersen, M. (1996) In: V. Snieckus (Ed.) *Advances in Carbanion Chemistry* (vol 2). JAI Press, Greenwich.
14. Lopez-Ortiz, F., Lopez, J.G., Manzaneda, R., Alvarez, A., Isidro, J.P. and Perez, J. (2004) *Mini-Reviews in Organic Chemistry*, **1**, 65.
15. Crampton, M.R., Davis, A.B., Greenhalgh, C. and Stevens, J.A. (1989) *Journal of the Chemical Society, Perkin Transactions II*, 675.
16. For a recent comprehensive treatment see Helfferich, F.F. (2004) Kinetics of multistep reactions. In: N.J.B. Green (Ed.) *Comprehensive Chemical Kinetics* (vol 40). Eslevier, Amsterdam.

17. Leititch, J. and Sprintcheck, G. (1986) *Chemische Berichte*, **119**, 1640.

18. Schulman, J.M., Disch, R.L. and Sabio, M.L. (1984) *Journal of the American Chemical Society*, **106**, 7696; Ciganek, E. (1971) *Journal of the American Chemical Society*, **93**, 2207.

19. Guthrie, P.J., Cossar, J. and Klym, A. (1984) *Journal of the American Chemical Society*, **106**, 1351; Tapuhi, E. and Jencks, W.P. (1982) *Journal of the American Chemical Society*, **104**, 5785; Dubois, J.-E., El-Alaoui, M. and Toullec, J. (1981) *Journal of the American Chemical Society*, **103**, 5393.

20. Dias, E.L., Nguyen, S.B. and Grubbs, R.H. (1997) *Journal of the American Chemical Society*, **119**, 3887.

21. Keeffe, J.R. and Jencks, W.P. (1983) *Journal of the American Chemical Society*, **105**, 265.

22. Harris, J.M., Raber, D.J., Hall, R.E. and Schleyer, P.v.R. (1970) *Journal of the American Chemical Society*, **92**, 5730.

23. Maskill, H. and Jencks, W.P. (1987) *Journal of the American Chemical Society*, **109**, 2062.

24. Rebek Jr., J. (1979) *Tetrahedron*, **35**, 723.

25. Davies, I.W., Matty, L., Hughes, D.L. and Reider, P.J. (2001) *Journal of the American Chemical Society*, **123**, 10139.

26. For definitions of stereochemical terms, see Eliel, E.L., Whalen, S.H. and Mander, L.N. (1994) *Stereochemistry of Organic Compounds*. Wiley, New York, p. 1208. For definitions of specificity and selectivity, see Gold, V. (1979) *Pure and Applied Chemistry*, **51**, 1725.

27. For discussion of some fascinating exceptions, see Carpenter, B.K. (2004) Chapter 21 in: R.A. Moss, M.S. Platz and M. Jones Jr. (Eds) *Reactive Intermediate Chemistry*. Wiley, Hoboken NJ.

28. Gringore, O.H. and Rouessac, F.P. (1985) *Organic Syntheses*, **63**, 121; Markgraf, J., Hodge, D. and Howard, A. (1990) *Journal of Chemical Education*, **67**, 173.

29. Chretien, J.R., Coudert, J.-D. and Ruasse, M.-F. (1993) *Journal of Organic Chemistry*, **58**, 1917; Rouasse, M.F. (1993) *Advances in Physical Organic Chemistry*, **28**, 207.

30. Karle, I., Flippen, J., Huisgen, R. and Schug, R. (1975) *Journal of the American Chemical Society*, **97**, 5285; Huisgen, R. and Steiner, G. (1973) *Journal of the American Chemical Society*, **95**, 5054, 5055 and 5056.

31. Rappoport, Z. and Topol, A. (1989) *Journal of the American Chemical Society*, **111**, 5967; Park, P.K. and Ha, H.J. (1990) *Bulletin of the Chemical Society of Japan*, **63**, 3006.

32. Buncel, E. and Lee, C.C. (Eds) (1984–1992) *Isotopes in Organic Chemistry* (vols 1 to 8). Elsevier, Amsterdam.

33. Forsyth, D.A. (1984) Chapter 1 in: E. Buncel and C.C. Lee (Eds) *Isotopes in Organic Chemistry* (vol 6), Elsevier, Amsterdam.

34. Elvidge, J.A. and Jones, J.R. (1978) Chapter 1 in E. Buncel and C.C. Lee (Eds) *Isotopes in Organic Chemistry* (vol 4). Elsevier, Amsterdam.

35. O'Leary, M.H. (1980) *Methods in Enzymology*, **64**, 83.

36. Saettel, N.J., Wiest, O., Singleton, D.A. and Meyer, M.P. (2002) *Journal of the American Chemical Society*, **124**, 11552; Singleton, D.A. and Thomas, A.A. (1995) *Journal of the American Chemical Society*, **117**, 9357; Martin, M.L. and Martin, G.J. (1991) *NMR Basic Principles and Progress* (vol 23). Springer, Berlin, Chapter 1; Pascal, R.A., Baum, M.W., Wagner, C.K., Rodgers, L.R. and Huang, D.-S. (1986) *Journal of the American Chemical Society*, **108**, 6477.

37. Melander, L. and Saunders Jr., W.H. (1980) *Reaction Rates of Isotopic Molecules*. Wiley, New York.

38. Sims, L.B. (1986) Chapter 4 in: E. Buncel and C.C. Lee (Eds) *Isotope Effects in Organic Chemistry* (vol 6). Elsevier, Amsterdam.

39. Paradisi, C. and Bunnett, J.F. (1981) *Journal of the American Chemical Society*, **103**, 946.

40. Brown, R.S., Bennet, A.J. and Slebocka-Tilk, S. (1992) *Accounts of Chemical Research*, **25**, 481; Slebocka-Tilk, H., Bennet, A.J., Keillor, J.W., Brown, R.S., Guthrie, J.P. and Jodhan, A. (1990) *Journal of the American Chemical Society*, **112**, 8507.

41. Paneth, P. (1992) How to measure heavy atom isotope effects. Chapter 2 in: E. Buncel and W.J. Saunders Jr. (Eds) *Isotopes in Organic Chemistry* (vol 8). Elsevier, Amsterdam.

42. Reviewed recently by Vougioukalakis, G.V. and Orfanopoulos, M. (2005) *Synlett*, **5**, 713; Orfanop-
    polou, M.M., Simonou, I. and Foote, C.S. (1990) *Journal of the American Chemical Society*, **112**,
    3607; Seymour, C. and Greene, F.D. (1980) *Journal of the American Chemical Society*, **102**, 6384.
43. Nelson, S.F. and Kapp, D.L. (1985) *Journal of the American Chemical Society*, **107**, 5548.
44. Squillacote, M., Mooney, M. and De Felipis, J. (1990) *Journal of the American Chemical Society*,
    **112**, 5364.
45. Poon, T.H., Park, S., Elmes, S. and Foote, C.S. (1995) *Journal of the American Chemical Society*,
    **117**, 10468.
46. For recent discussion, see Romsberg, F.E. and Schowen, R.L. (2004) *Advances in Physical Organic
    Chemistry*, **39**, 27.
47. Thibblin, A. and Alhberg, P. (1989) *Chemical Society Reviews*, **18**, 209.
48. Thibblin, A. (1988) *Journal of Physical Organic Chemistry*, **1**, 161.
49. Thibblin, A. (1983) *Journal of the American Chemical Society*, **105**, 853.
50. Williams, A. (2003) *Free Energy Relationships in Organic and Bioorganic Chemistry*. Royal Society
    of Chemistry, Cambridge, p. 163.
51. Lomberg, H., Stromberg, R. and Williams, A. (2004) *Organic and Biomolecular Chemistry*, **2**, 2165.

# Chapter 10
# Investigation of Reactions Involving Radical Intermediates

*Fawaz Aldabbagh, W. Russell Bowman
and John M. D. Storey*

## 10.1 Background and introduction

There is a wealth of information on the kinetics and mechanisms of organic radical reactions which has been largely responsible for the vast expansion of synthetic methodology using reactions involving radical intermediates in recent years. In addition, many publications concerned primarily with synthetic reactions of radicals contain useful mechanistic information and discussion. Indeed, many synthetic procedures involving radical reactions are firmly based on mechanistic understanding, perhaps more so than in other areas of synthetic organic chemistry.

We refer the readers to a useful body of books and reviews in the bibliography which will prove helpful to investigators determining the mechanism of radical reactions. The early two-volume compendium edited by Kochi has much valuable information, even though 30 years old, and most modern texts on radicals provide excellent guidance to radical synthesis and mechanism. We shall not discuss stereochemistry explicitly which now forms an important part of the mechanisms of radical reactions except to note that excellent stereoselectivities can be obtained in radical reactions with a clear understanding of the mechanisms involved. Many concepts in radical polymerisations are equally applicable to small molecule reactions and we refer the reader to an excellent account on the subject by Moad and Solomon.

### 10.1.1 Radical intermediates

The first question of mechanism for radical reactions is to prove that radicals are in fact involved. Various techniques are available for gaining evidence for radical intermediates; the most useful and common is evidence of precedents in the literature. Many synthetic protocols with radical intermediates are well understood mechanistically and have been well reviewed. Likewise, the behaviour of radicals is often understood from their involvement in other reactions.

Once products and stoichiometry have been established from known starting materials, the following are some of the initial facets of mechanism to be considered.

(a) Does the reaction proceed by a chain or non-chain mechanism, and what are the radical initiators?
(b) Are redox steps involved?
(c) What are the structure, reactivity and polarity (nucleophilic or electrophilic) of the intermediate radicals in the reaction?
(d) What mechanistic steps do the radical intermediates undergo?
(e) How do reaction conditions affect the mechanism?

The techniques for determining mechanisms are similar to those used for other reactions, e.g. kinetic and thermodynamic measurements, and product analysis including stereochemical aspects. Electron spin resonance (ESR) spectroscopy, also called electron paramagnetic resonance (EPR) spectroscopy, is vital for determining the structures of compounds with unpaired electrons. Special equipment (ESR spectrometers) and experience are required and studies are best carried out in collaboration with ESR spectroscopy experts. Although radicals are normally present at very low concentrations, ESR spectroscopy is very sensitive and can be used to investigate radicals at concentrations as low as $10^{-8}$ M. If the lifetime of the radicals is too short to allow direct observation, as is the case with most synthetically useful radicals, there are several techniques which overcome the problem including the following.

(a) *Spin trapping*. A non-radical trap is added to capture the radical intermediate to yield a more stable radical which can then be studied by ESR spectroscopy. These traps are commonly nitroso compounds which react with radicals to yield nitroxyl (nitroxide) radicals (see later).
(b) *Continuous flow*. The reactants are mixed together in a flow system and passed continuously through the detection cell of the spectrometer.
(c) *Trapping in solid matrices*. The radicals are generated from reagents dissolved in solvents and frozen at low temperature so that solid matrices are formed. This reduces the thermal energy and mobility of the radical intermediates, and hence their reactivity, so their lifetimes are increased, which allows observation by ESR spectroscopy.

We shall not cover techniques (b) and (c) here (see the bibliography for sources of further information).

### 10.1.2    Some initial considerations of radical mechanisms and chapter overview

Radicals are normally very reactive, often reacting under diffusion control (second-order rate constant ca. $10^9$ dm$^3$ mol$^{-1}$ s$^{-1}$) with other species (radicals, neutral molecules or ions), Scheme 10.1. However, although radical–radical reactions normally have very high second-order rate constants, radical concentrations are generally very low. Therefore, radical–neutral molecule or radical–ion reactions with smaller rate constants take place preferentially because the concentrations of neutral molecules or ions can be very high. Radical

$$R^\bullet + R^{1\bullet} \xrightarrow{\phantom{xx}k_1\phantom{xx}} R-R^1$$

$$\frac{d[R-R^1]}{dt} = k_1[R^\bullet][R^{1\bullet}]$$

$$R^\bullet + AB \xrightarrow{\phantom{xx}k_2\phantom{xx}} R-A + B^\bullet$$

$$\frac{d[R-A]}{dt} = k_2[R^\bullet][A-B]$$

**Scheme 10.1**   Rates of second-order bimolecular free radical reactions.

reactions need to have second-order rate constants greater than about 50 dm$^3$ mol$^{-1}$ s$^{-1}$ to be significant.

The reactions of radicals with neutral molecules are normally dominant and each creates a new radical species. These reactions can be generally described by two overall mechanisms, $S_H2$ (substitution, homolytic, bimolecular) and addition to unsaturated bonds or its reverse (elimination to form unsaturated bonds), Scheme 10.2. $S_H2$ substitutions have some characteristics of the better known $S_N2$ substitutions and require approach by the attacking species at 180° to the bond about to be broken to facilitate maximum overlap of interacting orbitals. These are often referred to as *abstraction reactions* in which the centre of substitution is hydrogen (H-abstraction) or a halogen. Intramolecular abstraction ($S_Hi$) normally needs to be at least 1,5 (i.e. forming a six-membered ring transition structure) to allow an angle between bonds being made and broken of ca. 180°.

Radical addition can take place to almost every known unsaturated group including alkenes (Section 10.3), allenes, alkynes, arenes, carbonyls, thiocarbonyls (onto the $S$-atom), nitro groups, nitriles, isonitriles and imines. Of these, intramolecular additions are perhaps the most useful synthetically and are largely governed by Baldwin's rules. Large numbers of synthetic radical reactions are also intermolecular of which a large proportion involve polymeric radicals. The stereoselectivity of intramolecular addition, in particular 5-*exo* cyclisation, is defined by the 'chair-like' Beckwith transition structure. The reverse reaction is elimination, often termed $\beta$-scission. The direction of the reactions is determined by various factors, including thermodynamics, kinetics, bond strengths, concentration and other reaction conditions.

These radical–neutral molecule reactions tend to be chain reactions, which is an important facet of a large percentage of radical mechanisms. As with all chain mechanisms, there are initiation, propagation and termination steps. The efficiency of the chain is defined by the number of propagation cycles (chain length) and, in many reactions, termination becomes unimportant so is not commonly considered, except for conventional radical polymerisations or reactions in viscous solvents. Rate constants for propagation help to determine the synthetic utility and mechanism of radical reactions. Initiation is covered in Section 10.2.

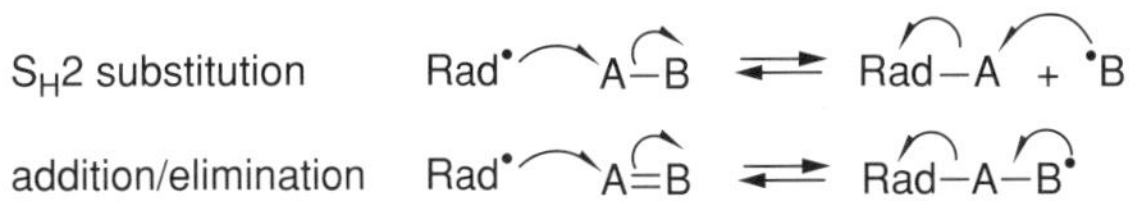

$S_H2$ substitution      $Rad^\bullet \frown A{-}B \rightleftharpoons Rad{-}A + {}^\bullet B$

addition/elimination     $Rad^\bullet \frown A{=}B \rightleftharpoons Rad{-}A{-}B^\bullet$

**Scheme 10.2**   Common free radical reactions.

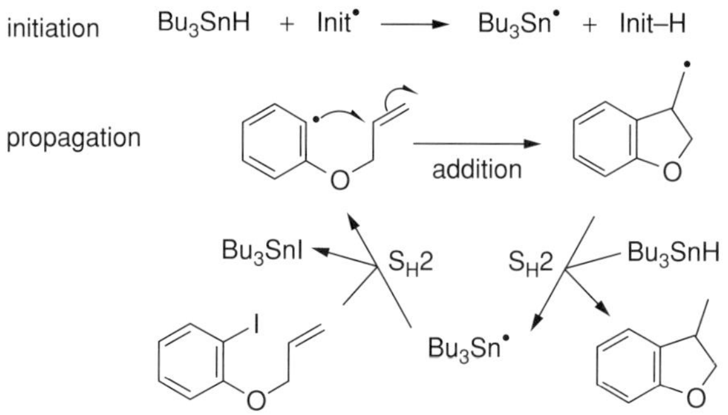


**Scheme 10.3** Example of a representative chain cyclisation using $Bu_3SnH$.

Termination steps can be caused by radical–radical combinations which remove radicals from the chain reaction. An example of a chain reaction using the commonly encountered tributyltin hydride ($Bu_3SnH$) to carry out a radical cyclisation is shown in Scheme 10.3 to illustrate $S_H2$ and addition reactions. The radicals act as the chain carriers, each reacting with a neutral molecule, and when a radical is removed from the chain by termination, the chain breaks down.

Other important mechanistic steps illustrated in Scheme 10.4 include reactions between radicals and ions (see Section 10.8) and redox reactions (see also Chapter 6).

$$R^\bullet + A^- \rightleftharpoons [R{-}A]^{\bullet -} \rightleftharpoons RA + e^- \quad \text{radical anion}$$

$$R^\bullet + Z^+ \rightleftharpoons [R{-}Z]^{\bullet +} \rightleftharpoons RZ - e^- \quad \text{radical cation}$$

$$R^\bullet - e^- \rightleftharpoons R^+$$

$$R^\bullet + e^- \rightleftharpoons R^-$$

**Scheme 10.4** Redox free radical reactions.

In the following sections, we deal with the main aspects of radical reactions and, in each section, examples of mechanistic determination are included. The selection is representative rather than exhaustive, and texts on radical chemistry should be consulted for greater detail. Key examples rather than detailed descriptions of mechanistic determination are presented.

## 10.2 Initiation

Initiation is the generation of the primary radical or initiator radical. The formation of propagating radicals in polymerisations by addition of initiator radicals to double bonds, so-called primary radical reactions, is discussed in Section 10.3. Initiation is essential for most radical reactions and therefore becomes a key but simple diagnostic tool in determining mechanism. The most basic test for a radical mechanism is to carry out blank reactions under identical conditions with all the reactants present except the initiator. A zero or very low yield of product in these blank reactions represents excellent evidence for a radical reaction.

Kinetics provides a quantitative understanding of the mechanism of initiation. Radicals are generated by any process that provides sufficient energy for homolytic bond cleavage, and these may be thermal, photochemical or redox reactions. This section focuses on initiation by the unimolecular, thermally induced, homolytic cleavage of covalent bonds. The rate of initiator decomposition, i.e. radical production ($R_{pr}$), increases with increasing delocalisation of the unpaired electron of the generated radical. A useful indication of the rate of decomposition is the half-life ($t_{1/2}$), which is the time taken for the initial concentration of the initiator to fall by half. As for any first-order process, $t_{1/2}$ is independent of the initial concentration and related to the decomposition rate constant ($k_{dec}$) by Equation 10.1. Polar solvents can marginally reduce the initiator half-life:

$$t_{1/2} = \frac{\ln 2}{k_{dec}} = \frac{0.693}{k_{dec}}. \tag{10.1}$$

The *Polymer Handbook* [1] lists $k_{dec}$ for a variety of initiators in different solvents at specified temperatures and, where the Arrhenius parameters ($A_{dec}$ and $E_{dec}$) are included, $k_{dec}$ may be calculated at any other temperature using the expression in Equation 10.2:

$$k_{dec} = A_{dec}e^{-E_{dec}/RT}. \tag{10.2}$$

The two most important classes of radical initiators are azo-compounds and peroxides (Fig. 10.1). The most commonly used azo-initiators are 2,2′-azobis(isobutyronitrile) (AIBN) and 1,1′-azobis(cyclohexane-1-carbonitrile) (ACN). The shorter half-life of AIBN ($t_{1/2} = 1.24$ h at 80°C in benzene) has led to international shipping restrictions and ACN ($t_{1/2} = 29.61$ h under the same conditions [1]) is becoming an increasingly valuable replacement. As explained in Section 10.4, however, this appreciable difference in decomposition rate means that ACN cannot replace AIBN as an initiator for all radical reactions. Furthermore, AIBN is soluble in a wider range of solvents, polar as well as non-polar (including alcohols, acetonitrile and benzene), compared with ACN which is restricted to use in non-polar solvents, such as benzene, toluene and cyclohexane.

Examples of commonly used peroxide initiators include the acyl peroxide, benzoyl peroxide (BPO) and the alkyl peroxide, *tert*-butyl peroxide (TBP), Fig. 10.1. Upon thermal homolysis of the O—O bond, the oxygen-centred radicals may undergo $\beta$-fragmentation. For example, the benzoyloxy (PhCOO$^\bullet$) radicals derived from BPO generate phenyl radicals and carbon dioxide. However, oxygen-centred radicals tend to be more reactive than carbon-centred ones and, in the presence of sufficient concentrations of substrate, readily undergo addition to unsaturated bonds and are likely to abstract labile hydrogen atoms owing to the high bond energy of the resultant O—H bond.

The rate of radical generation or production ($R_{pr}$) is related to the rate constant for initiator decomposition, $k_{dec}$, by Equation 10.3 where [I] is the instantaneous initiator concentration. The factor of 2 is included because two initiator or primary radicals are

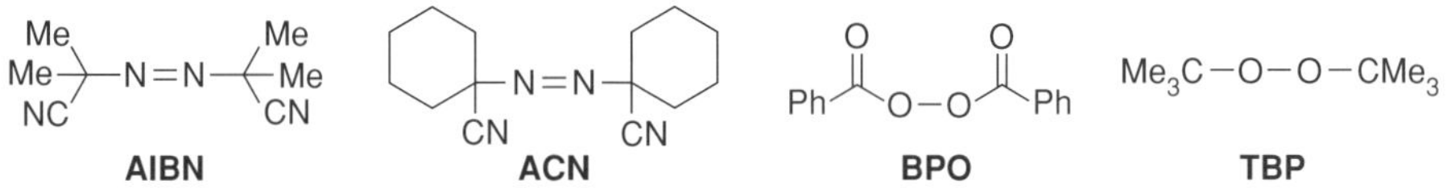

**Fig. 10.1** Commonly used free radical initiators.

generated upon homolysis of one molecule of the initiator, and the initiator efficiency term, $f$, is included because the initiator decomposition is not 100% efficient:

$$R_{pr} = 2 f k_{dec}[I]. \tag{10.3}$$

The value of $f$ will be less than unity when there are reactions of initiator radicals other than the required reaction, e.g. disproportionation. The result is some wastage of the initiator leading to $f$ values for azo-initiators of between 0.3 and 0.8 (which can be further reduced by viscous reaction conditions). For example, 2-cyanoisopropyl radicals, $(CH_3)_2(CN)C^\bullet$, derived from the decomposition of AIBN can disproportionate inside the solvent cage to give small amounts of methacrylonitrile, $CH_2{=}CCH_3(CN)$. Values of $f$ for peroxides are closer to unity owing to negligible solvent cage effects and the higher reactivity of oxygen-centred radicals. Furthermore, recombination of *tert*-butoxy radicals $(Me_3CO^\bullet)$ regenerates the initiator TBP, a process referred to as *cage-return*, and so has no effect on $f$ although it does reduce the rate constant for initiator decomposition, $k_{dec}$.

## 10.3   Radical addition to alkenes

Kinetics is used to investigate mechanisms of radical additions to alkenes. Outside the solvent cage, the initiator-derived radicals may undergo the desired bimolecular reaction with the substrate, or side reactions. When the substrate is an alkene, the exothermic intermolecular addition of the reactive radical $(R^\bullet)$ to the double bond results in the formation of two new single carbon–carbon bonds in place of the double bond. This reaction represents conversion of an initiator into a propagating radical in radical polymerisations, and is becoming increasingly important in a number of synthetically useful intermolecular small molecule reactions. The addition of $R^\bullet$ to monosubstituted and 1,1-disubstituted alkenes is nearly always at the unsubstituted carbon atom (tail addition), and thus is normally not affected by the individual steric demand of the alkene substituents. Equation 10.4 is the expression for the rate of addition $(R_i)$ of $R^\bullet$ to an alkene where [M] is the monomeric alkene concentration:

$$R_i = k_i[R^\bullet][M]. \tag{10.4}$$

The temperature dependence of the rate constants of radical addition $(k_i)$ is described by the Arrhenius equation (Section 10.2). At a given temperature, rate variations due to the effects of radical and substrate substituents are due to differences in the Arrhenius parameters, the frequency factor, $A_i$, and activation energy for addition, $E_i$. For polyatomic radicals, $A_i$ values span a narrow range of one to two orders of magnitude [$6.5 < \log(A_i/dm^3 \, mol^{-1} \, s^{-1}) < 8.5$] [2], which implies that large variations in $k_i$ are mainly due to variations in the activation energies, $E_i$. This is illustrated by the rate constants and Arrhenius parameters for the addition to ethene of methyl and halogen-substituted methyl radicals shown in Table 10.1.

Radical addition to alkenes is strongly influenced by polar effects, and reductive radical additions are nearly always anti-Markovnikov. Resonance stabilisation of the unpaired electron in $R^\bullet$ can decrease the rate of addition, while the stabilisation of the adduct radical

**Table 10.1**  Rate constants and Arrhenius parameters for the addition of methyl and halogen-substituted methyl radicals to ethene [2].

| Radical | $\log A_i$ | $E_i$ (kcal mol$^{-1}$) | $10^4\, k_i$ (dm$^3$ mol$^{-1}$ s$^{-1}$) at 164°C |
|---|---|---|---|
| $^\bullet$CH$_3$ | 8.5 | 7.7 | 5 |
| $^\bullet$CH$_2$F | 7.6 | 4.3 | 26 |
| $^\bullet$CCl$_3$ | 7.8 | 6.3 | 5 |
| $^\bullet$CF$_2$Br | 8.0 | 3.1 | 440 |
| $^\bullet$CF$_3$ | 8.3 | 2.0 | 5400 |

by the alkene substituent(s) can slightly increase the rate of addition. Although the methyl and benzyl radicals do not exhibit strong polar effects, the benzyl radical reacts 100–1000 times more slowly than methyl, much of which is due to resonance stabilisation of the benzyl radical [3]. Examples of strong polar effects include the addition of the electrophilic *tert*-butoxy radical (Me$_3$CO$^\bullet$) to styrene; this occurs five times faster than the addition to acrylonitrile, which possesses an electron-withdrawing nitrile (CN) substituent. The nucleophilic cyclohexyl radical, on the other hand, adds 8500 times faster to acrolein, which contains an electron-withdrawing and adduct-radical-stabilising aldehyde group, than to 1-hexene [4].

Non-activated double bonds, e.g. in the allylic disulfide **1** (Fig. 10.2) in which there are no substituents in conjugation with the double bond, require high initiator concentrations in order to achieve reasonable polymerisation rates. This indicates that competition between addition of initiator radicals (R$^\bullet$ = 2-cyanoisopropyl from AIBN) to the double bond of **1** and bimolecular side reactions (e.g. bimolecular initiator radical–initiator radical reactions outside the solvent cage with rate = $2k_t[\text{R}^\bullet]^2$ where $k_t$ is the second-order rate constant) cannot be neglected. To quantify this effect, [R$^\bullet$] was evaluated using the quadratic Equation 10.5 describing the steady-state approximation for R$^\bullet$ (i.e. the balance between the radical production and reaction). In Equation 10.5, [M]$_o$ is the initial monomer concentration, $k_i$ is as in Equation 10.4 (and approximately equal to the value for the addition of the cyanoisopropyl radical to 1-butene) [3] and $k_t = 10^9$ dm$^3$ mol$^{-1}$ s$^{-1}$; $f$ is assumed to be 0.5, which is typical for azo-initiators (Section 10.2). The value of $R_i$ for the cyanoisopropyl radicals and **1** was estimated to be less than $R_{pr}$ (Equation 10.3) by factors of 0.59, 0.79 and 0.96 at 50, 60 and 70°C, respectively, at the monomer and initiator concentrations used in benzene [5]:

$$2\,f\,k_{dec}[\text{I}]_o - k_i[\text{R}^\bullet][\text{M}]_o - 2k_t[\text{R}^\bullet]^2 = 0. \tag{10.5}$$

**Fig. 10.2**  2,2,4-Trimethyl-7-methylene-1,5-dithiacyclo-octane **1**.

## 10.4   Chain and non-chain reactions

A common task for chemists working in radical chemistry is to distinguish chain and non-chain radical reactions. The majority of synthetically useful radical reactions involve initiator radicals (Section 10.2) entering chain reactions, and the number of propagation steps dictates yield in small molecule reactions and molecular weight in polymerisation reactions.

In a radical chain reaction, only a sub-stoichiometric amount of initiator is normally required since radicals are continually being generated in the propagation steps. Consequently, chain reactions are normally represented as cycles that continually generate radicals and products. For example, a combination of $Bu_3SnH$ and AIBN, which is commonly used in radical reductions of R—X (X = halide, SePh, SPh) to R—H, and in reductive C—C bond-forming reactions, usually requires one equivalent of $Bu_3SnH$ but far less AIBN ($\approx 0.2$ equivalents). The 2-cyanoisopropyl radical from the AIBN generates $Bu_3Sn^\bullet$ by abstraction of the weakly bound hydrogen from $Bu_3SnH$ in the initiation step following AIBN breakdown. The $Bu_3Sn^\bullet$ now enters the propagation steps shown in Scheme 10.5. Since the $Bu_3Sn^\bullet$ radical is then continually regenerated upon reduction of the carbon-centred radical ($R^\bullet$) by $Bu_3SnH$ in the second propagation step, a sub-stoichiometric amount of AIBN is required. Obviously, at least one equivalent of $Bu_3SnH$ is needed since it provides the hydrogen in the actual reduction (the second propagation step) and $Bu_3Sn^\bullet$ is irreversibly converted into $Bu_3Sn$—X in the first step of propagation.

$$Bu_3Sn^\bullet \;+\; R-X \;\longrightarrow\; R^\bullet \;+\; Bu_3Sn-X$$

$$R^\bullet \;+\; Bu_3Sn-H \;\longrightarrow\; R-H \;+\; Bu_3Sn^\bullet$$

**Scheme 10.5**   Propagation steps in $Bu_3SnH$ reductions.

Chain reactions that require only a sub-stoichiometric amount of initiator thus have efficient chain carriers (e.g. $R^\bullet$, $Bu_3Sn^\bullet$), and loss of these radicals through termination reactions is insignificant compared with the number of propagation cycles. However, this does not apply to non-chain reactions and conventional radical polymerisations, where radical loss through irreversible terminations has to be balanced through continual generation of initiator radicals. Consequently, an initiator has to be chosen which has an appropriate half-life at the reaction temperature to generate radicals continually over the entire course of the reaction (see Section 10.2).

## 10.5   Nitroxides

Nitroxides ($T^\bullet$, general formula $R_2NO^\bullet$) have been used extensively to provide evidence for radical mechanisms through their ability to couple with (i.e. trap) reactive carbon-centred radicals ($R^\bullet$) to form alkoxyamines, R—T, e.g. Scheme 10.6. The elusive radical, $R^\bullet$, may thus be identified, characterised and quantified through analysis of the stable non-radical alkoxyamine by the usual spectroscopic techniques such as NMR and mass spectrometry.

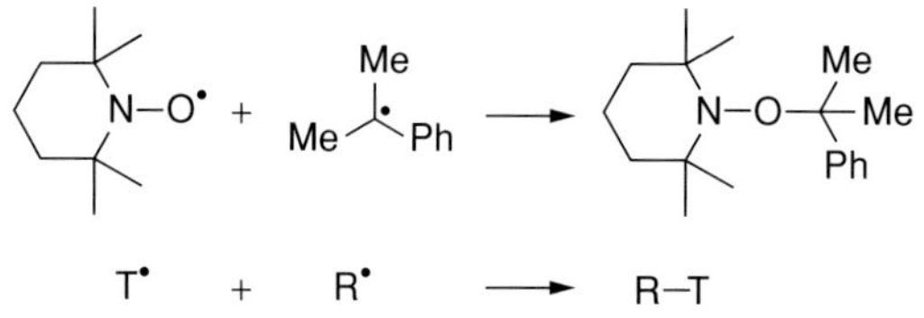

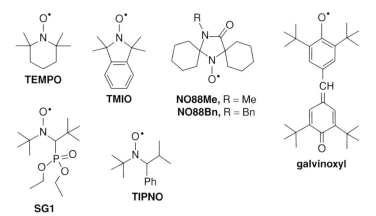

**Scheme 10.6**   Alkoxyamine formation by trapping of the cumyl radical by TEMPO.

## 10.5.1   Nitroxide-trapping experiments

Quantitative nitroxide-trapping experiments should be carried out under thoroughly de-oxygenated conditions since oxygen will act as a competitive radical trap for carbon-centred radicals. Consequently, several freeze-thaw degassing cycles using pressures $\leq 10^{-3}$ mm Hg are usually required.

The reactive radicals described in Sections 10.2–10.4 generally have half-lives of the order of $10^{-3}$ s at dilute concentrations of $10^{-6}$ mol dm$^{-3}$. In contrast, the nitroxides shown in Fig. 10.3 are observable red–orange compounds that can be stored under inert conditions for months or years in a laboratory. The primary reason for their stability is the resonance stabilisation (delocalisation) of the unpaired electron in the N—O$^\bullet$ bond, which is enhanced by the electronegativity of the atoms involved. The most stable nitroxides are fully substituted at the $\alpha$-positions by alkyl groups (e.g. TEMPO, TMIO, NO88 in Fig. 10.3), which prevents hydrogen abstraction and hence nitrone formation. Dilute solutions of TEMPO (2,2,6,6-tetramethyl-1-piperidinyloxy) and TMIO (1,1,3,3-tetramethylisoindoline-2-oxyl) are widely used to trap alkyl radicals, but the trapping of other types of radicals has proven less successful. Equation 10.6 gives the rate law for radical trapping by a nitroxide, where the rate of trapping or combination is $R_c$ and $k_c$ is the second-order rate constant:

$$R_c = k_c[R^\bullet][T^\bullet]. \tag{10.6}$$

**Fig. 10.3**   Persistent radicals including nitroxides.

TEMPO, which is commercially available, traps carbon-centred radicals with rate constants an order of magnitude lower than the diffusion-controlled limit in most organic solvents at $\leq 120°C$ (e.g. $k_c = 3.1 \times 10^8$ dm$^3$ mol$^{-1}$ s$^{-1}$ with benzyl radical at 50°C in *tert*-butylbenzene) [6], and somewhat more slowly if the radical is sterically congested (e.g. $k_c = 5.7 \times 10^7$ dm$^3$ mol$^{-1}$ s$^{-1}$ with cumyl radical under the same conditions, Scheme 10.6) [6]. Non-Arrhenius behaviour or non-temperature dependence has been observed for several radical coupling reactions [6, 7].

In trapping experiments, nitroxides will only trap carbon-centred radicals, and not oxygen-centred ones. This is particularly important since oxygen-centred radicals are often used as initiators (Section 10.2). The nitroxide should also not undergo other reactions, such as addition to double bonds or H-abstraction; this increases the probability that it will trap selectively carbon-centred radicals which act as chain carriers in many synthetically useful organic reactions, as propagating species in polymerisations and as reactive intermediates in biological pathways.

### 10.5.2 Alkoxyamine dissociation rate constant, $k_d$

Kinetics and ESR spectroscopy have become important tools for investigating the dissociation mechanism of alkoxyamines. Increased steric compression of the C—O bond of the alkoxyamine (R—T) through the presence of bulky substituents on the alkyl and nitroxide fragments leads to weakening of the bond; this promotes homolytic alkoxyamine dissociation. Thus, the alkoxyamine formed by the trapping of the cumyl radical by 1,1,3,3-tetraethylisoindoline-2-oxyl (Scheme 10.7) is more labile than the analogous alkoxyamine from the less congested tetramethyl analogue, TMIO (Fig. 10.3). Polar functional groups and polar solvents, as well as increased stability of the dissociated radicals, increase the rate constant, $k_d$, and hence the rate ($R_d$) of alkoxyamine homolysis (Equation 10.7):

$$R_d = k_d[R\text{—}T]. \tag{10.7}$$

Further, a dynamic equilibrium is established at higher temperatures between so-called active (R$^\bullet$ and T$^\bullet$) and dormant states (R—T), as shown in Scheme 10.7. The equilibrium

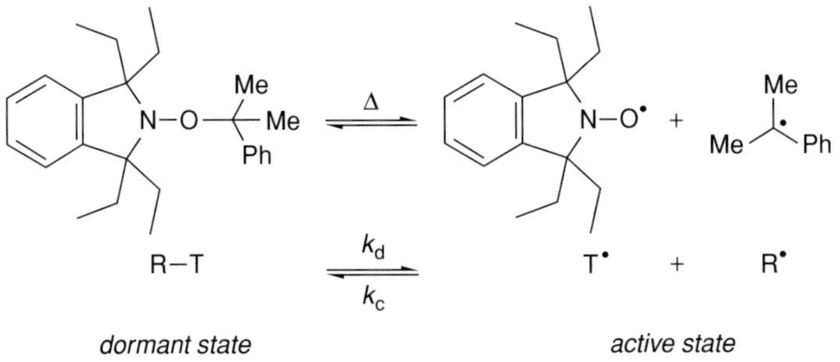

**Scheme 10.7** Reversible trapping of the cumyl radical by the nitroxide, 1,1,3,3-tetraethylisoindoline-2-oxyl.

constant between dormant and active states is $K$ in Equation 10.8:

$$K = \frac{[\text{T}^\bullet][\text{R}^\bullet]}{[\text{R--T}]}. \tag{10.8}$$

We may generalise that faster rates of alkoxyamine C—O bond cleavage correspond to slower cross-coupling (combination or nitroxide trapping) reactions, i.e. the rate constant for alkoxyamine dissociation, $k_d$, and coupling rate constant, $k_c$, exhibit opposite substituent steric effects. Consequently, greater steric congestion around the nitroxyl functionality results in a less efficient radical trap. However, this is not always the case as is observed for the low molecular weight polystyryl alkoxyamines of the imidazolidinone nitroxides, 2,5-bis(spirocyclohexyl)-3-methylimidazolidin-4-one-1-oxyl and the 3-benzyl analogue (NO88Me and NO88Bn in Fig. 10.3). These have $k_d$ and $k_c$ values at 120°C which are 4–8 and 25–33 times lower, respectively, than the reported values for the analogous polymeric alkoxyamine of TEMPO [7]. Steric congestion due to the 2,5-spirodicyclohexyl rings thus has a more profound effect on the coupling than on the dissociation reaction. Consequently, NO88 nitroxides are slower traps, but form stronger alkoxyamine C—O bonds, than TEMPO.

The rate constant for dissociation of an alkoxyamine, $k_d$ in Scheme 10.7, is most commonly obtained by techniques based on the addition of an efficient scavenger ($\text{S}^\bullet$) of thermally generated carbon-centred radicals ($\text{R}^\bullet$). $\text{S}^\bullet$ is present in large excess and traps $\text{R}^\bullet$, thereby preventing the recombination of $\text{R}^\bullet$ with $\text{T}^\bullet$, as shown in Scheme 10.8. Alkoxyamine dissociation now follows pseudo-first-order kinetics, and the relatively high concentration of $\text{S}^\bullet$ ensures that the only fate of $\text{R}^\bullet$ is to be trapped by $\text{S}^\bullet$. Scavengers ($\text{S}^\bullet$) used for this purpose include stable radicals such as galvinoxyl (2,6-di-*tert*-butyl-4-(3,5-di-*tert*-butyl-4-oxocyclohexa-2,5-dien-1-ylidene)phenoxyl) or nitroxides (e.g. TEMPO, TEMPO derivatives and TMIO), see Fig. 10.3. Obviously, in these so-called nitroxide-exchange experiments, the scavenger ($\text{S}^\bullet$) must be different from the nitroxide fragment ($\text{T}^\bullet$) of the alkoxyamine (R—T) under investigation. The rate of disappearance of R—T is equal to the rate of appearance of $\text{T}^\bullet$ and the new scavenger-containing alkoxyamine, R—S; if a nitroxide possessing a UV chromophore is used (e.g. TMIO), the reaction may be conveniently monitored by HPLC using a UV detector.

The most commonly used scavenging technique employs oxygen as $\text{S}^\bullet$, and $\text{T}^\bullet$ is monitored by ESR spectroscopy. This involves simply sealing ESR tubes without freeze-thaw degassing. The non-degassed reactions contain enough oxygen to scavenge irreversibly all thermally generated carbon-centred radicals ($\text{R}^\bullet$), thereby preventing the cross-coupling

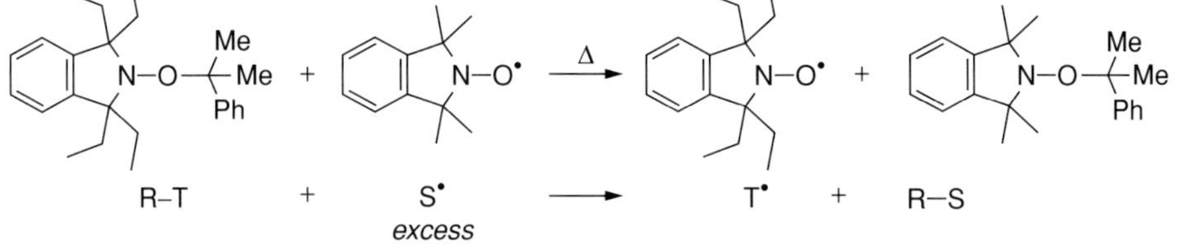

**Scheme 10.8** Measuring the alkoxyamine (R—T) dissociation rate using a nitroxide-exchange experiment where the scavenger $\text{S}^\bullet$ is TMIO.

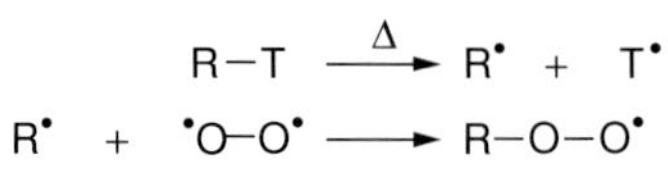

$$R-T \xrightarrow{\Delta} R^{\bullet} + T^{\bullet}$$
$$R^{\bullet} + {}^{\bullet}O-O^{\bullet} \longrightarrow R-O-O^{\bullet}$$

**Scheme 10.9**　Measuring an alkoxyamine dissociation rate by ESR using oxygen as the scavenger.

reaction (Scheme 10.9). Pseudo-first-order kinetics are now observed, as exemplified by the plots in Fig. 10.4 for the decomposition of the polystyryl alkoxyamine adduct (P—T) of the nitroxide NO88Bn [7]; the rate of P—T decomposition is equal to the rate of formation of T$^{\bullet}$, which was monitored by ESR spectroscopy. The linear relationship of $\ln([P-T]_o/[P-T])$ with time has the slope $k_d$ according to the first-order rate law of Equation 10.9. Such ESR spectral measurements are carried out using spectrometers with a variable temperature unit and are calibrated against known concentrations of nitroxide with benzene being the most commonly used reaction solvent:

$$\ln\left(\frac{[P-T]_o}{[P-T]}\right) = k_d t. \tag{10.9}$$

We can now plot the temperature dependence of $k_d$ using the Arrhenius equation given in Equation 10.10, see Fig. 10.5; $E_d$ (from the slope) and $A_d$ (from the extrapolated intercept

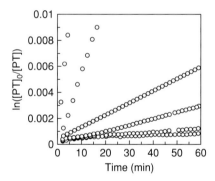

**Fig. 10.4**　First-order plots for the decomposition of polystyryl-NO88Bn (P—T) at 50, 60, 70, 80, 90 and 100°C [7].

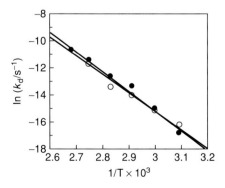

**Fig. 10.5**　Arrhenius plots for $k_d$ of the polystyryl alkoxyamine adducts (P—T) where T is NO88Me (●) and NO88Bn (o) [7].

not shown in Fig. 10.5) are useful parameters that enable us to estimate $k_d$ at any temperature. The values of $E_d$ are very similar for most alkoxyamine dissociations, and larger $k_d$ values are usually due to larger $A_d$ values as observed for nitroxides such as SG1 and TIPNO (see Sections 10.5.3 and 10.5.4):

$$\ln k_d = \ln A_d - \frac{E_d}{R}\left(\frac{1}{T}\right). \tag{10.10}$$

### 10.5.3    The persistent radical effect (PRE)

One of the consequences of a nitroxide ($T^\bullet$) not being able to couple with itself is the *persistent radical effect*, PRE. The PRE is observed in any reaction, involving either a small molecule or a polymeric species, where the reactive ($R^\bullet$ or $P^\bullet$) and persistent ($T^\bullet$) radicals are formed simultaneously from the same or different precursors. In the case of an alkoxyamine (R—T in Scheme 10.10), the precursor is the same. Upon thermal dissociation of R—T in an inert solvent, $R^\bullet$ can undergo two types of terminations – either rapid cross-coupling to reform R—T or self-termination at diffusion-controlled rates to form R—R; disproportionation may also occur to a minor extent. During a very short initial period, the concentrations of $R^\bullet$ and $T^\bullet$ will be the same but, soon after, the concentration of $R^\bullet$ will decrease by self-termination reactions, whereas that of $T^\bullet$ will increase. This causes a steady build-up of $T^\bullet$, which increases the rate of cross-coupling, i.e. the regeneration of the parent compound (R—T), and diminishes the role of self-termination. After an initial brief period, therefore, the concentration of $T^\bullet$ becomes effectively constant, which regulates the concentrations of $R^\bullet$ and R—T by making the level of irreversible self-termination small. In most quantitative experiments, [R—T] is assumed to be effectively equal to its initial concentration, $[R—T]_o$, with the PRE ensuring only a maximum of 2–3% loss of starting material. The PRE is now recognised as being the mechanism of control in nitroxide-mediated living/controlled radical polymerisations (Section 10.5.4), and has also been applied to the control of small molecule annulation reactions.

The use of this phenomenon to control carbon–carbon bond-forming reactions relies on $R^\bullet$ being rapidly converted into another transient radical which, in the case of a polymerisation, occurs by repetitive addition to a monomer double bond to give the propagating polymer radical, $P^\bullet$. Thus, the PRE prevents 'dead' polymer (P—P') formation and the dormant concentration of P—T remains effectively constant. It follows that the excess of $T^\bullet$ ensures that reversible termination and addition of $P^\bullet$ to monomer are dominant reactions allowing all polymer chains to grow practically simultaneously (Section 10.5.4).

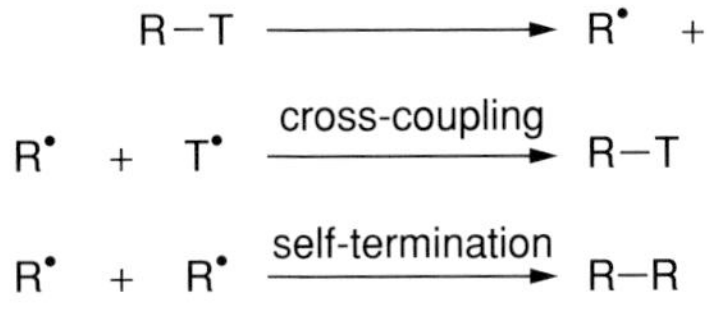

**Scheme 10.10**   The persistent radical effect (PRE) as a function of reversible and irreversible termination, i.e. cross-coupling and self-termination, respectively.

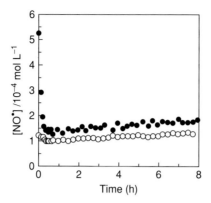

**Fig. 10.6**   Free nitroxide concentration as a function of time during P—T initiated bulk polymerisation of styrene at 120°C where T is NO88Me (●) and NO88Bn (o) [7].

Fig. 10.6 shows the continuous monitoring by ESR spectroscopy of the nitroxide concentration for the NO88Me- and NO88Bn-polystyryl alkoxyamine-initiated bulk polymerisation of styrene at 120°C under deoxygenated conditions. This allowed absolute nitroxide concentrations of NO88Me and NO88Bn to be estimated during the polymerisation. After an initial brief period, the nitroxide concentration, [T•], remained effectively constant throughout the reaction, and the higher free NO88Me concentration was due its higher $K$ value in Equation 10.8.

Scheme 10.11 shows a PRE-mediated 5-*exo-trig* radical cyclisation in which the controlled thermal formation of active radicals from the dormant alkoxyamine **2** is facilitated by steric compression of the alkoxyamine C—O bond by the bulky *N*-alkyl and *O*-alkyl groups [8]. Intramolecular H-bonding between a —CH$_2$—OH and the nitroxyl oxygen of the incipient nitroxide in a six-membered cyclic transition structure further facilitated the dissociation of **2**. After cyclisation, the resultant primary cyclopentylmethyl radical was trapped by the free nitroxide to form the new dormant isomerised alkoxyamine **3**, which is more stable than **2** since the *O*-alkyl is now primary. The same reaction using TEMPO as the nitroxide component did not work presumably because the C—O bond in the alkoxyamine precursor is much stronger.

Alkoxyamines have been promoted as alternative initiators to metal hydrides in radical cyclisations since metal hydride reactions often lack regioselectivity and produce mixtures of reduced uncyclised and cyclised products. Further, the nitroxide portion of the alkoxyamine

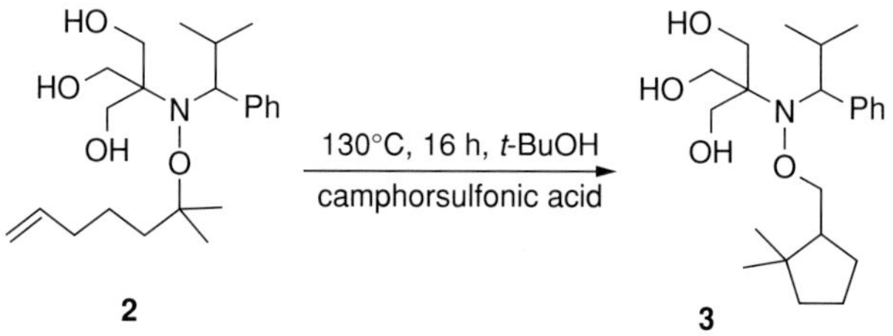

**Scheme 10.11**   Alkoxyamine-initiated 5-*exo* trig radical cyclisation.

product, both in small molecules (e.g. **3**) and polymer systems, can readily be reduced or converted into another functional group, or activated to mediate further co-polymerisation (Section 10.5.4).

## 10.5.4 Nitroxide-mediated living/controlled radical polymerisations (NMP)

Despite the ease and versatility of conventional radical polymerisation, the process is limited as far as building polymers with predictable molecular weights and well-defined architectures are concerned. This is because of diffusion-controlled random termination between growing radicals, which makes the lifetime of growing chains very short – in the range of 1 s. During this time, newly initiated radicals add to most of the monomer units prior to termination. Thus, macromolecular engineering is not feasible because it is difficult during 1 s to add a precise sequence of another monomer to form block copolymers. The solution to the problem is to perform the polymerisation in the presence of a *living reagent* (*mediator*) that will prevent termination reactions, and thereby prolong the life of propagating radicals. This process is now termed living/controlled radical polymerisation and allows the preparation of polymers with molecular weights which can be predicted by the [monomer]:[living reagent] ratio, and narrow molecular weight distributions can also be achieved. One of the most important procedures uses a nitroxide as the living reagent, i.e. *nitroxide-mediated living/controlled radical polymerisation* (NMP). Hawker and Fukuda, and their co-workers, have reviewed its synthetic and quantitative aspects, respectively [9]. NMP is based on the reversible trapping of polymeric radicals (P$^\bullet$) by nitroxides at temperatures usually higher than 100°C, thereby preventing coupling or irreversible termination reactions of these carbon-centred radicals. Therefore, P$^\bullet$ should be substituted for R$^\bullet$ in the equilibrium given in Scheme 10.7 to represent NMP (Scheme 10.12) and, in the active state, P$^\bullet$ propagates the chain.

In a conventional radical polymerisation, radicals have to be formed continuously at low concentrations in order to balance radical loss due to irreversible termination reactions. In contrast, in a living/controlled process, initiator decomposition must be fast as all polymer chains must start growing *simultaneously* to allow all chains to grow uniformly with conversion of monomer to polymer. A controlled polymerisation builds high molecular weight polymer much more slowly than a conventional radical polymerisation which achieves high molecular weights almost instantaneously. The resultant polymer chains are mostly capped by nitroxide, and are said to possess 'living ends'. This allows living polymers to be homolysed thermally in the presence of a second monomer to produce block copolymers, which

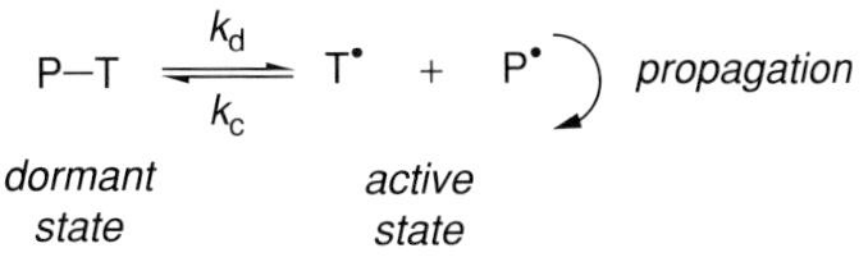

$$P\!-\!T \underset{k_c}{\overset{k_d}{\rightleftharpoons}} T^\bullet \;+\; P^\bullet \big) \; \textit{propagation}$$

*dormant*       *active*
*state*         *state*

**Scheme 10.12** Nitroxide (T$^\bullet$) as the living reagent in nitroxide-mediated living/controlled radical polymerisation (NMP).

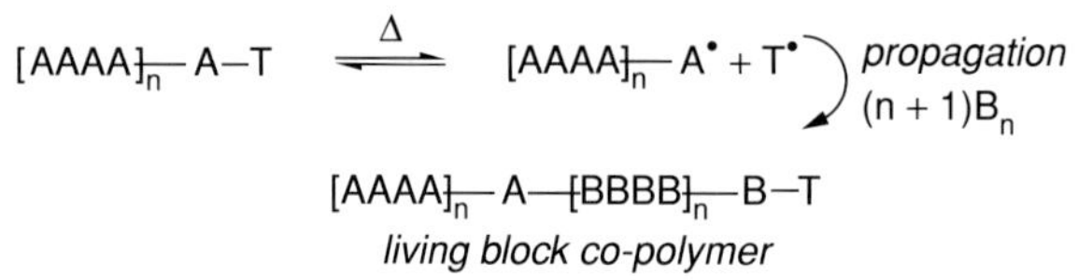

$$[AAAA]_n\!-\!A\!-\!T \;\; \overset{\Delta}{\rightleftharpoons} \;\; [AAAA]_n\!-\!A^{\bullet} + T^{\bullet} \quad \text{propagation} \atop (n+1)B_n$$

$$[AAAA]_n\!-\!A\!-\![BBBB]_n\!-\!B\!-\!T$$
*living block co-polymer*

**Scheme 10.13**  Block co-polymer formation using NMP.

cannot be prepared by conventional radical polymerisation. Scheme 10.13 illustrates block copolymer formation, where A and B are different monomers. We see, therefore, that living polymers can act as macro-alkoxyamine initiators for further polymerisations resulting in polymers with well-defined and often controlled complex architectures.

For a given monomer, a nitroxide which enables a rapid exchange between active and dormant states will give superior control or 'living' character and shorter polymerisation times. At present, acyclic nitroxides such as 2-[N-*tert*-butyl-N-(1-diethyloxyphosphoryl-2,2-di-methylpropyl)aminoxyl] and 2,2,5-trimethyl-4-phenyl-3-azahexane-3-aminoxyl (SG1 and TIPNO, Fig. 10.3) allow the controlled polymerisation of the widest range of monomers at the lowest possible temperatures. Presumably, steric congestion and electronic effects, which weaken the alkoxyamine (R—T) or macro-alkoxyamine (P—T) C—O bonds, are greater in open compared with cyclic nitroxides. In the absence of side reactions, optimal control requires $k_d$ and $k_c$ values between $10^{-3}$–$1$ s$^{-1}$ and $10^6$–$10^9$ dm$^3$ mol$^{-1}$ s$^{-1}$, respectively. Since, at polymerisation temperatures, most nitroxides exhibit $k_c$ values greater than $10^6$ dm$^3$ mol$^{-1}$ s$^{-1}$, it seems that $k_d$ is more important than $k_c$ in giving superior quality acyclic nitroxides in NMP and cyclisations (see Section 10.5.3). It follows that the rate of dissociation or activation of alkoxyamines is higher when T$^{\bullet}$ is SG1 or TIPNO than for cyclic nitroxides such as TEMPO, TMIO or NO88. This can be described in terms of the lifetime ($\tau$) of adduct P—T, defined by Equation 10.11. For example, the low molecular weight polystyryl adduct of SG1 has $\tau = 4.76$ min ($k_d = 0.21$ min$^{-1}$) whereas $\tau$ for the TEMPO analogue is three times greater at 120°C [10]. Since the recombination reaction is very fast in both cases, this enabled a greater number of activation–deactivation cycles during the course of a polymerisation reaction using SG1 compared with using TEMPO:

$$\text{lifetime }(\tau)\text{ of }\; \text{P—T} = \frac{1}{k_d}. \qquad (10.11)$$

## 10.6   Radical clock reactions

The ability to determine the rate constants of radical reactions is important both for mechanistic studies and when designing a synthetic sequence. In the past, ESR kinetics studies were commonly used to measure the rates of many radical processes with a fair degree of accuracy. More recently, other methods have found favour, particularly flash photolysis or pulse radiolysis in conjunction with high-speed, time-resolved analysis of reactions using UV–vis spectroscopy. Whilst these methods allow the rates of reactions to be investigated directly, they suffer from the major disadvantage of requiring large and expensive specialised equipment, a resource that is usually unavailable to most synthetic chemists.

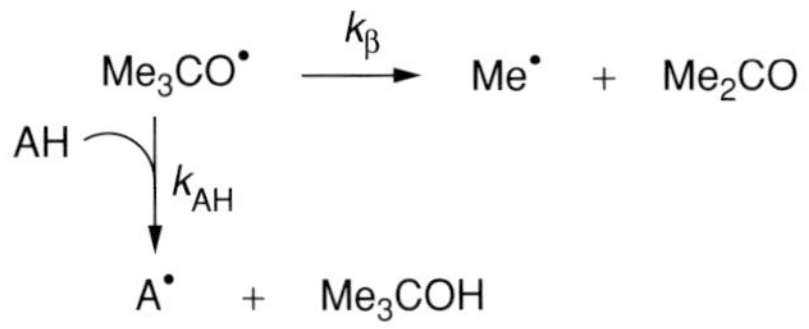

$$Me_3CO^\bullet \xrightarrow{\ k_\beta\ } Me^\bullet + Me_2CO$$

$$AH \searrow\ \Big\downarrow k_{AH}$$

$$A^\bullet + Me_3COH$$

**Scheme 10.14** $\beta$-Scission of the *tert*-butoxy radical for the measurement of the relative rates of hydrogen abstraction from organic compounds (AH).

To circumvent this problem, chemists have used competition methods to determine the rate constants of a number of reactions. These methods rely upon conducting a competition between a reaction (or rearrangement) of the known rate constant (measured as outlined above) with the reaction to be investigated; the unknown rate constant is simply obtained by product analysis (see Chapter 2). The reaction or rearrangement of the known rate constant is, therefore, acting as a timing device, commonly known as a *radical clock reaction*. In principle, any unimolecular radical reaction, fragmentation or rearrangement with a known rate constant can be used as a clock reaction to calibrate the rate of another competing reaction. The reader is directed to two excellent reviews covering methodology and containing abundant kinetic data [11, 12].

For ease of analysis, it is desirable that the clock reaction and the reaction to be investigated are irreversible. Similarly, in order to facilitate product analysis, it is desirable that the two competing processes give high conversions to products without the complication of side reactions.

The first-order $\beta$-scission of the *tert*-butoxy radical is one of the oldest radical clock reactions and has been used for over 50 years for the measurement of the relative rates of hydrogen abstraction from organic compounds (AH) in solution (Scheme 10.14). At low conversions, when the concentration of AH has not appreciably changed, the ratio of the rate constants for hydrogen atom abstraction, $k_{AH}$, and $\beta$-scission, $k_\beta$, can be determined simply by analysis for acetone and *tert*-butyl alcohol formation in the reaction. This is most conveniently achieved by gas chromatography:

$$\frac{k_{AH}}{k_\beta} = \frac{[Me_3COH]}{[AH]\,[Me_2CO]}. \tag{10.12}$$

If the clock reaction has been calibrated, i.e. if $k_\beta$ is known under the precise reaction conditions employed, then the absolute values of $k_{AH}$ (rate constant for hydrogen abstraction from the substrate AH) can be obtained using Equation 10.12. The relative rates of hydrogen abstraction for two different substrates AH and BH in separate reactions can also be determined using this simple reaction as long as the temperature, solvent system and concentrations remain the same for both reactions. It is then possible to compare $k_{AH}$ and $k_{BH}$ since $k_\beta$ remains constant throughout and can be eliminated from the expression.

It is clear that calibrated clock reactions with a wide range of rate constants are required for the method to be of significant value to the practising synthetic chemist. Furthermore, there needs to be a range of radical clock processes for each class of radical reaction. The range of clock reactions presently available covers the entire range of reaction rates that may be of interest, with the fastest having first-order rate constants exceeding $10^{11}$ s$^{-1}$ at room

**Table 10.2** A Selection of common radical clock reactions.

| Radical clock reaction | Rate constant, $k_{temp}$ (s$^{-1}$) |
|---|---|
| | $k_{25} = 0.14$ |
| | $k_{25} = 9 \times 10^2$ |
| | $k_{25} = 4 \times 10^3$ |
| | $k_{25} = 3 \times 10^5$ |
| | $k_{25} = 9 \times 10^6$ |
| | $k_{25} = 9.4 \times 10^7$ |
| | $k_{40} = 3 \times 10^8$ |
| | $k_{25} = 3 \times 10^{10}$ |
| | $k_{25} = 5 \times 10^{11}$ |

temperature. A selection of common radical clock processes covering a wide range of rate constants is shown in Table 10.2. Newcomb provides a more comprehensive list [12].

In addition to their use for the calibration of rates for radical reactions, radical clocks can be employed to distinguish between ionic and radical pathways. In the simplest embodiment of this idea, a suitable clock reaction that undergoes a known fast rearrangement with easily identifiable products is incorporated into the reaction system to be studied. This approach has been exploited in the pioneering work of Newcomb and co-workers in studies of the mechanism of cytochrome P450 oxidation reactions [13]. Newcomb has developed a range of ultrafast radical clocks able to detect radical species with lifetimes of 80–200 fs.

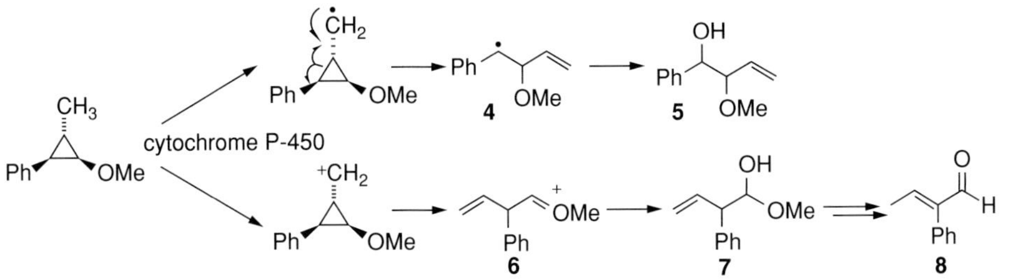

**Scheme 10.15**  Use of a radical clock to distinguish between radical and ionic mechanisms.

For example, if a methoxycyclopropylmethyl radical is generated as in Scheme 10.15, it would undergo a highly regioselective rearrangement to give a stable benzylic radical, **4**, which becomes oxidised to give, ultimately, alcohol **5**. However, if a methoxycyclopropylmethyl cation is involved, ring opening towards the methoxy group takes place to give the oxonium ion **6** which becomes hydrated to give an unstable hemiacetal **7**. Under the conditions of the reaction, **7** would be hydrolysed to give an unsaturated aldehyde which would isomerise to give the product, **8**. Since the rearrangements to **4** and **6** are effectively instantaneous, this provides a reliable mechanistic probe. In the cytochrome P450 studies, a radical mechanism was indicated.

In another use of the radical clock principle with the clock reaction incorporated within the substrate to be studied, Beckwith and Storey determined the rate constant for 5-*exo*-cyclisation of an aryl radical onto an acryloyl double bond, as depicted in Scheme 10.16 [14]. Since the rate constant for cyclisation onto the *O*-allyl double bond to give **9** is known ($k_{O\text{-allyl}} = 5 \times 10^8$ s$^{-1}$ at 25°C, see Scheme 10.3), the unknown rate constant, $k_{\text{acryloyl}}$, can be determined by Equation 10.13 simply by determining the ratio of the products **9** and **10**:

$$k_{\text{acryloyl}} = \frac{[\text{acryloyl cyclisation product}]}{[O\text{-allyl cyclisation product}]} \times k_{O\text{-allyl}}. \qquad (10.13)$$

An elegant example of the radical clock principle is illustrated in the investigations of Rychnovsky where a conformational radical clock is used [15]. This approach relies upon knowing the rate of racemisation of a radical formed at a centre which is originally configurationally pure and using this process as the 'clock'. The enantiomeric purity of the product from the reaction is, therefore, directly related to the lifetime of the radical.

In the study involving a tetrahydropyranyl radical illustrated in Scheme 10.17, Rychnovsky was attempting to establish whether cyclisation was from the radical or the anionic intermediate in the reductive lithiation of nitrile **11** to give **14**. The radical intermediate, **12**,

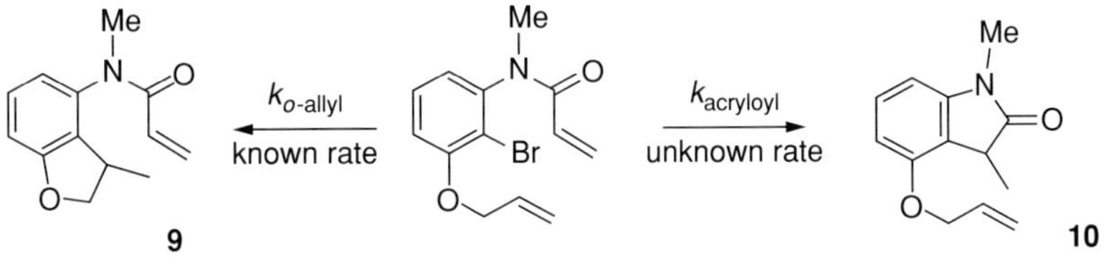

**Scheme 10.16**  Use of an internal radical clock.

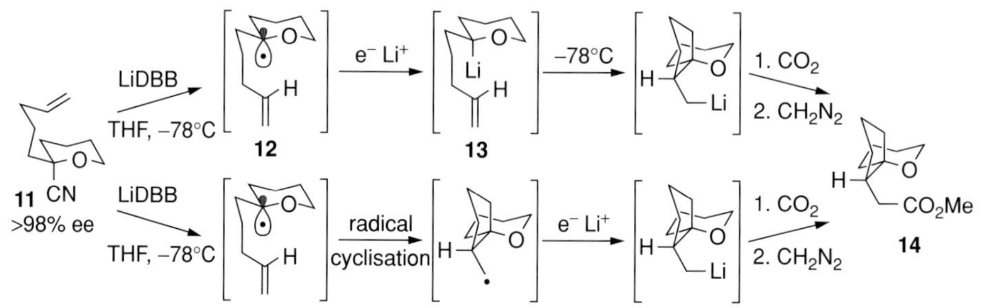

**Scheme 10.17**   Use of a conformational radical clock (LiDBB = lithium di-*tert*-butylbiphenylide).

is the only point in the reaction scheme where racemisation can occur, so the enantiomeric excess (*ee*) of the cyclised product is a direct measure of the lifetime of **12**. From essentially enantiomerically pure substrate, the enantiomeric composition of **9** was shown to correspond to 42% *ee*. From this, the half-life of the radical **12** was calculated to be $2.4 \times 10^{-7}$s from Equation 10.14 where $k_R$ is the known rate constant of its racemisation, $3.9 \times 10^6$ s$^{-1}$ at $-78°C$ [15]. The rate constant for cyclisation of a 5-hexenyl radical was estimated to be $4.0 \times 10^2$ s$^{-1}$ at this temperature, which is some five orders of magnitude too slow for cyclisation to take place to any appreciable extent within the lifetime of radical **12**. This represents strong evidence that the cyclisation is of the longer lived alkyl–lithium compound, **13**, in the upper route of Scheme 10.17.

$$t_{1/2} = \ln 2 / k_R \times (1 - ee) / ee \qquad (10.14)$$

There are numerous other examples of radical clock reactions in the literature used both for simple rate determinations to facilitate the quest for selectivity in synthesis and for more detailed probing of mechanistic pathways.

## 10.7   Homolytic aromatic substitution

Homolytic aromatic substitution is a valuable method for the substitution of arenes and heteroarenes, and has been reviewed recently by Studer and Bossart [16]. Both intramolecular [16] and intermolecular reactions [17] with arenes have become increasingly useful in synthesis. The intramolecular variant has received more attention with many elegant applications reported [18].

Scheme 10.18 illustrates the key features of the intramolecular reaction in general and also the specific radical intermediates in the mechanistic study [18]. After initial radical formation from a suitable halide precursor, in the vast majority of cases using Bu$_3$SnH and

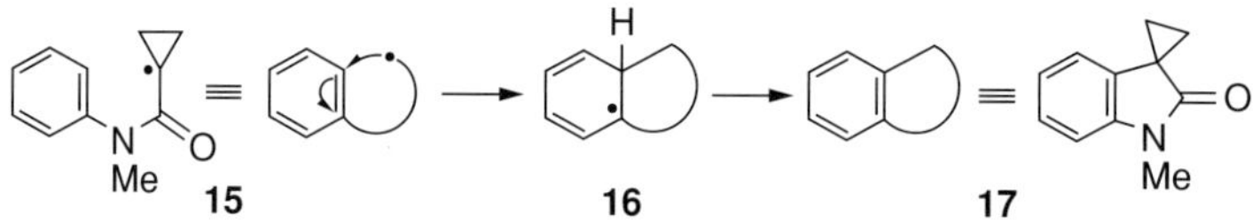

**Scheme 10.18**   Intramolecular homolytic aromatic substitution.

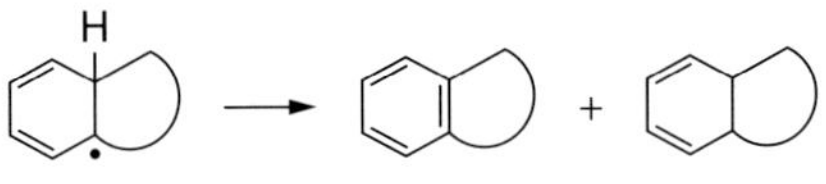

**Scheme 10.19**   Possible disproportionation route for **16** → **17** (see Scheme 10.18).

AIBN, the radical **15** can attack the aryl ring to give an intermediate cyclohexadienyl radical **16**; this is then converted to the fully aromatic product **17**. The feature of this reaction that is particularly noteworthy is that the conversion of **16** to **17** is formally an oxidation reaction, yet it takes place in the presence of $Bu_3SnH$, a reducing agent; various mechanistic possibilities could explain this anomaly. The approaches applied to investigate this reaction further illustrate some of the methods used to elucidate radical reactions generally and are applicable to other specific examples.

The most obvious of several mechanistic possibilities for the conversion of **16** to **17** is that **16** undergoes disproportionation to give **17** and the cyclohexadiene **18**, as illustrated in Scheme 10.19. Thorough NMR spectral analysis of the reaction mixture failed to detect any of the cyclohexadiene product **18**, and yields of the cyclised product **17**, measured with an internal standard, were greater than 50%. Consequently, this mechanistic possibility may be ruled out. Note that such NMR spectral analysis of reaction mixtures should, wherever possible, be undertaken prior to any chromatography or manipulations that could alter product ratios – an important point made in Chapter 2.

The next possibility investigated was that hydrogen transfer was from $Bu_3SnH$ to the $\pi$-radical **16**, followed by oxidation of **19** in the work-up to give the fully aromatised product **20** (Scheme 10.20). Since $Bu_3SnD$ is readily available, it provides a convenient means of investigating the interaction of the radical mediator in radical processes. If hydrogen transfer were occurring, some deuteriated cyclohexadiene **19** or aromatised product **20** would be expected. None of these products were observed by D-NMR spectroscopy. $^{13}C$ NMR spectral analysis also failed to detect any deuteriated products in the crude reaction mixture, as any deuteriation would lead to carbon signals appearing as triplets. Cyclohexadienes such as **18** and **19** were also known not to oxidise readily, which casts further doubt on their intervention as intermediates.

Other mechanistic possibilities that were considered involved the evolution of $H_2$ gas (Scheme 10.21) one of which invokes the reaction between $Bu_3SnH$ and radical **16**. Alternatively, $Bu_3SnH$ could act as a hydride donor which abstracts a proton from **16** to give the radical anion **21**; **21** could then undergo single electron transfer (SET) to a molecule of precursor halide ArRBr to propagate the chain – this would be a type of $S_{RN}1$ process (see Scheme 10.21). In order to test for both of these mechanistic possibilities, which involve the

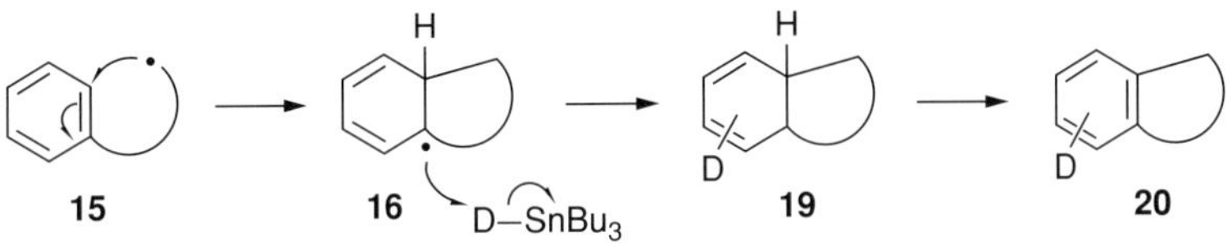

**Scheme 10.20**   Investigation of D-transfer from $Bu_3SnD$.

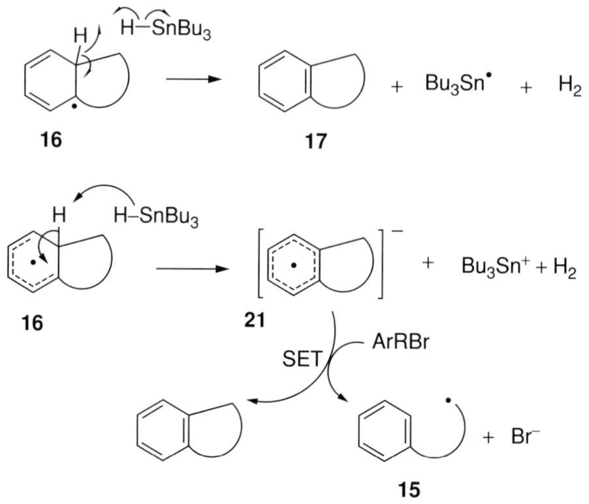

**Scheme 10.21**  Bu$_3$SnH-induced hydrogen atom or proton abstraction from **16** to yield H$_2$.

production of hydrogen gas, a substrate with all aryl protiums replaced with deuteriums was used. If either mechanistic process were occurring, HD would be expected from the reaction, which could be detected by quadrapole mass spectroscopy, Raman spectroscopy and D NMR spectroscopy in solution. No HD was detected, which ruled out these two mechanistic possibilities.

The role of the initiator was next investigated. Initial experiments were based upon the proposition that initiator fragments could be oxidising the cyclohexadienyl radical intermediate as outlined in Scheme 10.22 for **22** from labelled starting material. Again, deuterium labelling proved invaluable; D NMR spectroscopy revealed the formation of Me$_2$CDCN in 23%, and similar results were obtained when di-*tert*-butyl peroxide was used as the initiator in place of AIBN, giving *tert*-BuOD. The intervention of the initiator was, therefore, clear.

The need for stoichiometric amounts of initiator in many reactions involving radical addition to aryl systems has been reported as required by the mechanism in Scheme 10.22. In order to probe the generality of this phenomenon, we investigated cyclisation onto imidazoles, Scheme 10.23. $^1$H NMR spectral analysis of reaction mixtures containing an internal standard showed a clear relationship between the yield of cyclised product **23** and the amount of initiator used in the reaction. Less than one equivalent of initiator gave very low yields of product.

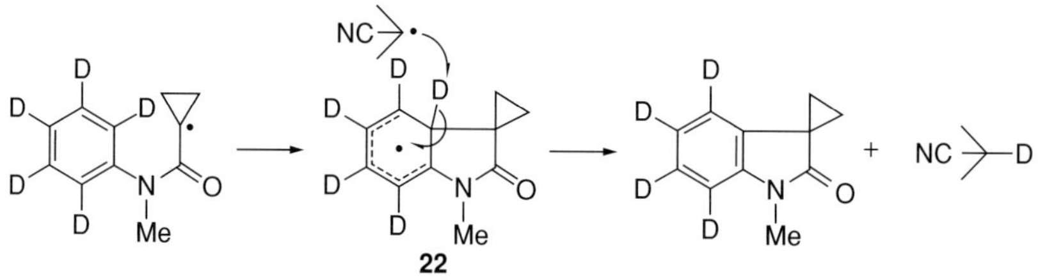

**Scheme 10.22**  Abstraction of deuterium from 15-d$_5$ by 2-cyanoprop-2-yl radicals.

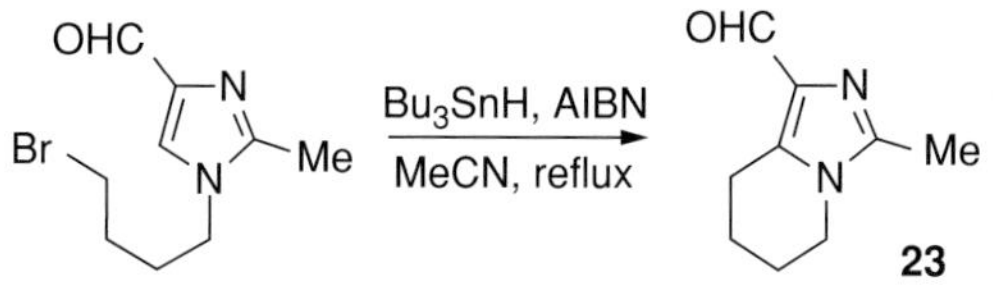

**Scheme 10.23** Radical cyclisation onto imidazoles.

Since only approximately 25% of the initiator in the experiment in Scheme 10.22 appeared as a deuteriated fragment, the fate of the other 75% required explanation. For every mole of initiator thermolysed, one mole of $N_2$ gas should be evolved. However, when the volume of gas from one of these reactions containing 1.2 equivalent of AIBN was measured using an equilibrated gas burette, only 0.3 equivalent of nitrogen was produced. This suggested that the initiator was not breaking down thermolytically in the normal manner, thus further implying that the role of the initiator in this reaction was more than that of a simple initiator. These results suggested that the initiator was acting as an oxidising agent through two pathways. A minor pathway, involving the initiator fragment formed after thermolysis, explained the occurrence of deuteriated initiator fragments. The major pathway accounting for the lack of $N_2$ and the dependence on stoichiometric amounts of the initiator is shown in Scheme 10.24. Thus, the AIBN abstracts a hydrogen atom from the cyclohexadienyl radical intermediate, and this semi-reduced form then abstracts a hydrogen from $Bu_3SnH$ to yield $Bu_3Sn^\bullet$ and the fully reduced form of AIBN.

If the mechanism in Scheme 10.24 is correct and cyclohexadienyl radicals do indeed react with the initiator, their presence should retard other radical processes. To test this, the reduction of 1-bromo-octane was investigated in the presence of **24** (Fig. 10.7), which readily undergoes cyclisation so should form intermediate cyclohexadienyl radicals. The initial reaction rates were measured between 2 and 12 minutes. During this period, less than 10% of the $Bu_3SnH$ in the reaction was consumed and so, to a fair approximation, the

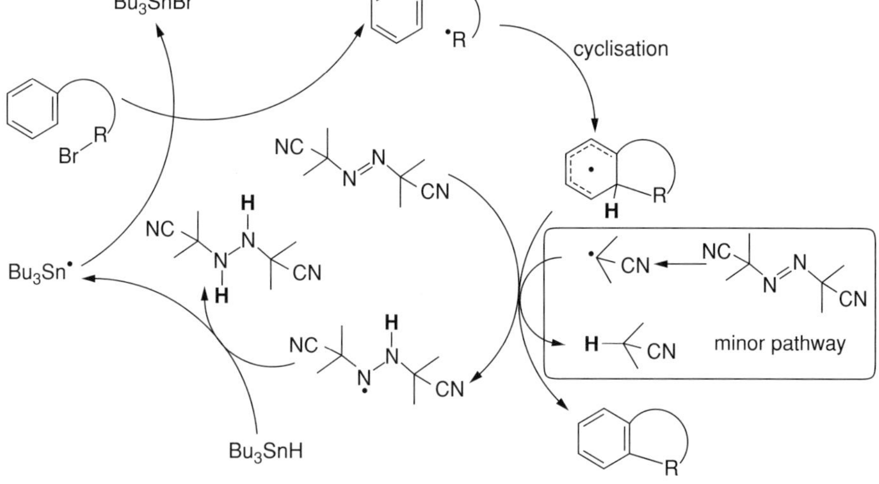

**Scheme 10.24** Dual role of AIBN in intramolecular homolytic aromatic substitution.

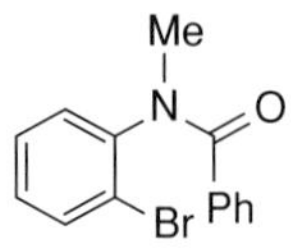

**Fig. 10.7**   Precursor for homolytic aromatic substitution studies.

concentration of $Bu_3SnH$ can be considered constant. In a series of experiments, mixtures of **24**, 1-bromo-octane, $Bu_3SnH$ and di-*tert*-butyl hyponitrite (BONNOB) as initiator in cyclohexane were heated at 45°C, and the progress of the reaction was monitored by GLC (a range of initiator concentrations was used in these experiments). These experiments showed that the initial rate of octane formation is proportional to the initial concentrations of 1-bromo-octane and initiator, and inversely proportional to the initial concentration of **24**, i.e. the reduction rate obeys the empirical rate expression in Equation 10.15. This result supports the notion that cyclohexadienyl radicals formed from **24** undergo termination with the initiator or other radicals in the solution:

$$\text{Rate of reduction} = k_{\text{reduction}}[\text{BONNOB}][\text{1-bromo-octane}]/[\mathbf{24}]. \qquad (10.15)$$

The methods involved in this study are typical of those used to investigate a radical mechanism, illustrating how it is necessary to use a number of investigative experiments to test reasonable working hypotheses. Then, having developed a reasonable model, supportive information can be obtained from a relatively straightforward kinetic analysis.

# 10.8   Redox reactions

A large number of radical reactions proceed by redox mechanisms. These all require electron transfer (ET), often termed single electron transfer (SET), between two species and electrochemical methods are very useful to determine details of the reactions (see Chapter 6). We shall consider two examples here – reduction with samarium di-iodide ($SmI_2$) and $S_{RN}1$ (substitution, radical-nucleophilic, unimolecular) reactions. The SET steps can proceed by inner-sphere or outer-sphere mechanisms as defined in Marcus theory [19, 20].

## 10.8.1   Reductions with samarium di-iodide, SmI₂

The elucidation of mechanisms of reactions of $SmI_2$ have involved polarography, kinetics, radical clocks and trapping techniques (radical cyclisation) [19, 20]. The reagent is able to reduce alkyl halides and ketones/aldehydes, as shown in Scheme 10.25, in non-chain radical reactions.

Studies using ESR spectroscopy have shown that the radical anions resulting from alkyl iodides and bromides are not stable and dissociative SET takes place, i.e. the cleavage of the aliphatic C—I and C—Br bonds is concerted with the SET. With aryl iodides and bromides, the radical anion does have a finite lifetime but then breaks down rapidly to aryl radicals. Evidence is required for each of the steps and, in particular, about whether radicals or

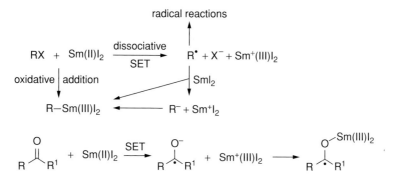

**Scheme 10.25**    Reactions of SmI₂ with alkyl halides and ketones.

organosamarium compounds (R–SmI₂) are the key intermediates. Such intermediates will facilitate non-radical reactions similar to those of Grignard reagents. Both alkyl and ketyl radicals have been trapped by radical cyclisation showing that direct oxidative addition is not a mechanism.

Cyclic voltammetry (see also Chapter 6) has been used to determine the reduction potential $E°$ (SmI₂-SmI₂$^+$) $= -1.41$ V in THF [19]. The rate-limiting step is the SET with a rapid equilibrium between Sm(III)I₂X and Sm$^+$I₂ plus halide. Using the measured value of $E°$ (SmI₂-SmI₂$^+$) and free energy plots, the SET step was shown to be outer sphere with benzyl bromide. The electronic interaction (orbital overlap) between SmI₂ and benzyl bromide amounts to only a few kcal mol$^{-1}$. In contrast, the reduction of acetophenone with SmI₂ to give the ketyl radical anion is clearly inner sphere with strong electronic interaction in the transition state.

Further measurements in THF and HMPA using cyclic voltammetry showed that complexation with HMPA increases the reductive power to $-2.05$ V [20], and the complex [SmI₂·(HMPA)₄] has been isolated and characterised. Comparison between the observed rate constant and that for SET (estimated using the Marcus equation) shows the minimal difference for reduction of alkyl radicals to yield the alkyl anion, i.e. minimal change of the complexation of the samarium species with HMPA [20]. After reduction of the alkyl radical to the anion, an HMPA molecule is displaced from the complex by the anion to yield the organosamarium species (R–SmI₂). These results indicate that the anion has a finite lifetime and that the alkyl radical does not react with SmI₂ to yield R–SmI₂ directly.

The elucidation of the mechanism of the samarium Barbier reaction has proved particularly taxing (Scheme 10.26) and represents a classic example of how to investigate radical mechanisms [21, 22]. To complicate the investigation of the mechanism, both alkyl halides and ketones can be reduced by SmI₂ (see above). Four feasible mechanisms can be postulated as shown in Scheme 10.26. Mechanisms 1 and 2 have reduction of the alkyl halide first to yield the alkyl radical, and mechanisms 3 and 4 have reduction of the ketone to the ketyl first.

Evidence of both types of potential intermediate in reduction by SmI₂, the alkyl radical and the ketyl radical **27**, has been provided by radical cyclisation reactions. Mechanism 4, which involves an S$_H$2 substitution, has been eliminated because optically active halides are completely racemised. The rate of addition of alkyl radicals to ketones is very slow ($<10^2$ dm$^3$ mol$^{-1}$ s$^{-1}$); the resulting alkoxy radicals (**26**) are very reactive and could not

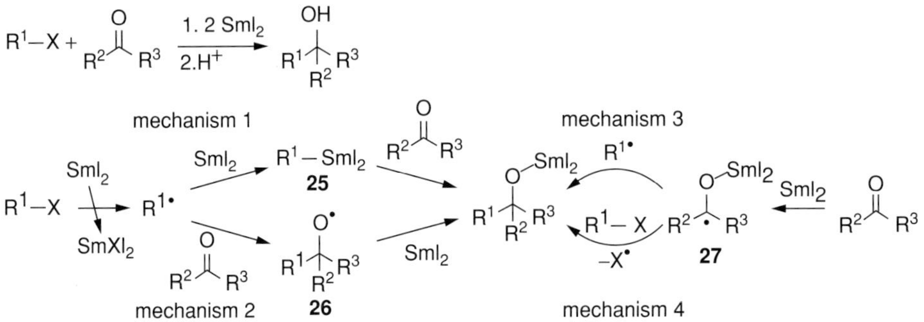

**Scheme 10.26**   The samarium Barbier reaction.

be trapped in radical cyclisation reactions showing that mechanism 2 is most unlikely. The combination of radicals ($R^{1\bullet}$ and ketyl **27**) in mechanism 3 would be very fast, probably at the diffusion limit. For this reaction to take place, both radicals ($R^{1\bullet}$ and ketyl **27**) would have to be formed at the same rate, but $R^{1\bullet}$ is very reactive and the ketyl **27** is stable. However, no self-coupling between the alkyl radicals was observed (see Section 10.1.2 and Scheme 10.1), which rules out mechanism 3. All evidence points towards mechanism 1 involving the organosamarium intermediate **25**.

Definitive evidence came from studies using aryl radical intermediates which cyclise extremely rapidly and are not reduced by $SmI_2$ (Scheme 10.27) [22]. Cyclisation of the aryl radical **28** is very fast ($k_c = 5 \times 10^8$ s$^{-1}$ at 25°C) allowing cyclisation prior to reduction by $SmI_2$. When the reaction was quenched with $D_2O$ at the end of the reaction, no deuterium was observed in the small amount of uncyclised **29** and E = 80% D in **30**; this shows that the cyclised primary radical is reduced by $SmI_2$, but the aryl radical is not. Electrophiles (e.g. $D_2O$, $I_2$, PhSeSePh, $Bu_3SnI$, aldehydes and ketones) can be added at the end of the reaction to yield **30** in high yield indicating that the organosamarium intermediate is relatively stable. The ratio of uncyclised products **29** to cyclised products **30** (3:97) is the same as is observed in $Bu_3SnH$ reductions (Scheme 10.3), indicating an aryl radical intermediate with no complexation with samarium. The result also shows that the rate of H-abstraction ($k_H$) from the THF solvent by **28** is fast, but not as fast as cyclisation.

The rate constant for reduction of primary alkyl radicals with $SmI_2$ has been determined using a radical clock (see Section 10.6) providing further information for understanding the mechanism [22]. The commonly used 5-hexenyl radical clock, where the rate constant for cyclisation is known ($k_c = 2.3 \times 10^5$ s$^{-1}$ at 20°C), was used to determine the rate constant

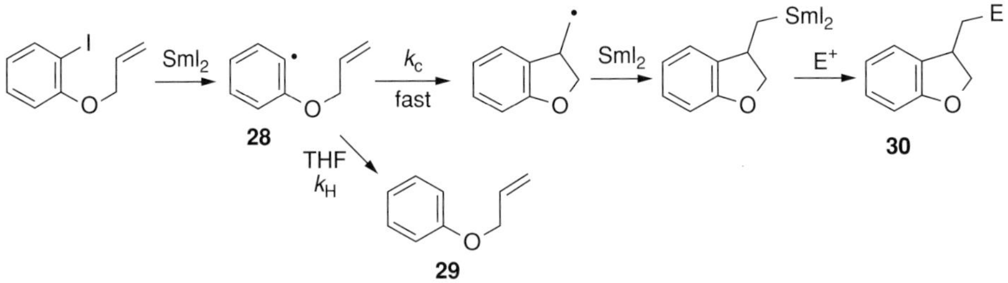

**Scheme 10.27**   Use of a radical clock to determine the rate of reduction of alkyl radicals with $SmI_2$.

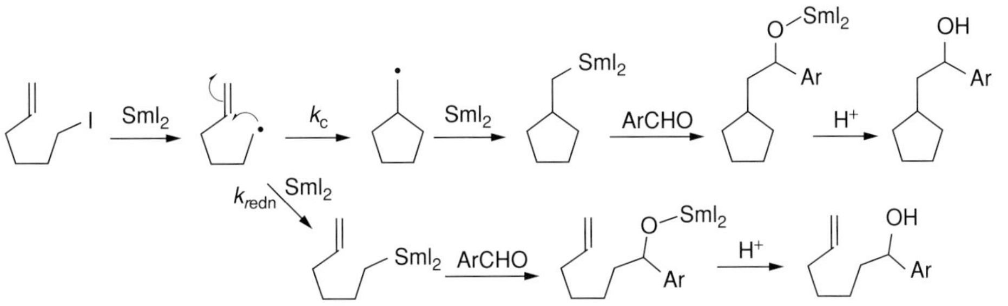

**Scheme 10.28** Use of the 5-hexenyl radical clock to determine $k_{redn}$ for primary alkyl radicals by $SmI_2$ in a samarium Barbier reaction (Ar $=$ $p$-methoxyphenyl).

of reduction, $k_{redn}$, by $SmI_2$ in the samarium Barbier reaction shown in Scheme 10.28; the result is $k_{redn} = 6 \times 10^6$ dm$^3$ mol$^{-1}$ s$^{-1}$. The rate constant of reduction for secondary and tertiary radicals ($<10^4$ dm$^3$ mol$^{-1}$ s$^{-1}$) is much slower as the corresponding anions are less stable than the primary alkyl anions. Therefore, primary radicals only undergo radical reactions if the rates are faster than reduction, whereas secondary and tertiary radicals will normally undergo radical reactions rather than reduction.

### 10.8.2 $S_{RN}1$ substitution

Substitutions by the $S_{RN}1$ mechanism (substitution, radical-nucleophilic, unimolecular) are a well-studied group of reactions which involve SET steps and radical anion intermediates (see Scheme 10.4). They have been elucidated for a range of precursors which include aryl, vinyl and bridgehead halides (i.e. halides which cannot undergo $S_N1$ or $S_N2$ mechanisms), and substituted nitro compounds. Studies of aryl halide reactions are discussed in Chapter 2. The methods used to determine the mechanisms of these reactions include inhibition and trapping studies, ESR spectroscopy, variation of the functional group and nucleophile reactivity coupled with product analysis, and the effect of solvent. We exemplify $S_{RN}1$ mechanistic studies with the reactions of $\alpha$-substituted nitroalkanes (Scheme 10.29) [23, 24].

The mechanism involves initiation and propagation steps. Most commonly, the reactions are initiated by SET between the nucleophile and the $\alpha$-nitroalkane (Equation 10.16). Strongly basic nucleophiles commonly undergo SET without photostimulation but weakly

$$R_2C(X)NO_2 + Nu^- \;\overset{SET}{\rightleftharpoons}\; [R_2C(X)NO_2]^{\bullet -} + Nu^{\bullet} \qquad (10.16)\;\text{Initiation}$$

$$[R_2C(X)NO_2]^{\bullet -} \;\rightleftharpoons\; R_2\overset{\bullet}{C}{-}NO_2 + X^- \qquad (10.17)$$

$$R_2\overset{\bullet}{C}{-}NO_2 + Nu^- \;\rightleftharpoons\; [R_2C(Nu)NO_2]^{\bullet -} \qquad (10.18) \quad \text{Propagation}$$

$$[R_2C(Nu)NO_2]^{\bullet -} + R_2C(X)NO_2 \;\overset{SET}{\rightleftharpoons}\; R_2C(Nu)NO_2 + [R_2C(X)NO_2]^{\bullet -} \qquad (10.19)$$

Summary: $R_2C(X)NO_2 + Nu^- \;\longrightarrow\; R_2C(Nu)NO_2 + X^-$

$X = I,\ Br,\ Cl,\ SCN,\ SR,\ SO_2R,\ NO_2,\ N_3$     $Nu = $ nitronates, enolates, $ArSO_2$, RS, $PO_3Et_2$, $N_3$

**Scheme 10.29** $S_{RN}1$ mechanism for reactions between nucleophiles and $\alpha$-substituted nitroalkanes.

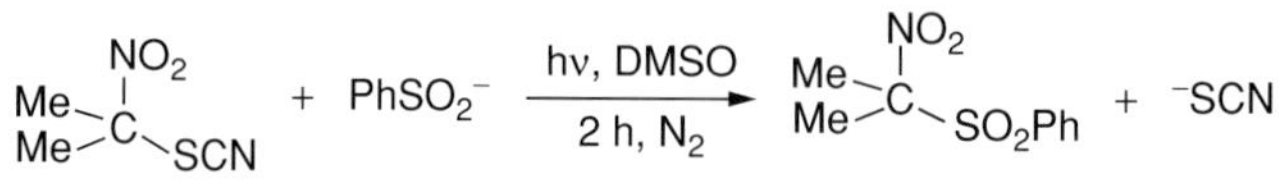

$$\underset{\substack{\text{Me}\\ \text{Me}}}{\overset{\text{NO}_2}{\underset{|}{\text{C}}}}{\text{SCN}} \quad + \quad \text{PhSO}_2^- \quad \xrightarrow[\text{2 h, N}_2]{\text{hv, DMSO}} \quad \underset{\substack{\text{Me}\\ \text{Me}}}{\overset{\text{NO}_2}{\underset{|}{\text{C}}}}{\text{SO}_2\text{Ph}} \quad + \quad {}^-\text{SCN}$$

| Conditions | % Yield | |
|---|---|---|
| | Me$_2$C(SO$_2$Ph)NO$_2$ | Me$_2$C(SCN)NO$_2$ |
| 2 h | 49 | 0 |
| 2 h, dark | 0 | 40 |
| 2 h, DNB (5 mol%) | 0 | 40 |
| 2 h, DTBN (10 mol%) | 0 | 35 |

**Scheme 10.30**  Inhibition studies of a typical S$_{RN}$1 reaction.

basic ones require photostimulation. The precursors, R$_2$C(X)NO$_2$, and the nucleophiles have been shown to form charge transfer (CT) complexes which absorb visible light and facilitate SET. In many reactions, a red CT complex forms on mixing and disappears when the reaction is complete. The three propagation steps (Equations 10.17–10.19) are reversible and depend on the stability of the intermediate radical anions.

The mechanism was deduced by a number of mechanistic arguments. As in most mechanistic determinations, the first conclusions come from eliminating other mechanisms. In this case, S$_N$2 substitution is not possible because of the steric hindrance at quaternary centres where substitution takes place. Likewise, an S$_N$1 mechanism can be ruled out because the intermediate, R$_2$(NO$_2$)C$^+$, would have a cationic centre on the carbon next to the strongly electron-withdrawing nitro group.

The key evidence came from inhibition and trapping studies showing a photolytic radical/radical anion chain mechanism. If radicals (R$^\bullet$) are present, they can be trapped by oxygen to give peroxyl radicals (R–OO$^\bullet$) which would break the chain thereby lowering product yields. Nitroxides and other radical traps such as di-*tert*-butylnitroxide (DTBN, (*t*-Bu)$_2$N–O$^\bullet$) and galvinoxyl (Fig. 10.3) combine with the R$_2$(NO$_2$)C$^\bullet$ radical intermediates. If radical anions are present, they can be intercepted by strong electron acceptors, e.g. oxygen and *p*-dinitrobenzene (DNB) which, on SET, yield superoxide and the radical anion of DNB, respectively. An example of inhibition studies is shown in Scheme 10.30; formation of the product is completely suppressed by DNB and DTBN, and large amounts of unaltered starting material are recovered [23, 24]. The inhibition studies are commonly less marked than in the example shown.

The use of ESR spectroscopy has provided crucial evidence for the first three steps of the mechanism and also the structure and reactivity of the intermediates. The intermediate radical anions are very unstable and have only been observed using ESR spectroscopy by being trapped in solid matrices of 2-methyltetrahydrofuran or CD$_3$OD at low temperature (liquid nitrogen cooling) [23, 24]. The nitro compounds readily undergo electron capture to give rise to radical anions which undergo dissociation when the temperature is raised to ca. −110°C to yield the intermediate radicals, R$_2$(NO$_2$)C$^\bullet$, Equation 10.17. Addition of nucleophiles to these radicals to yield new radical anions (Equation 10.18) has also been observed by using the same technique. The ESR spectral results indicate that the unpaired electrons of the radical anions occupy molecular orbitals derived from overlap of the nitro $\pi^*$ and C—X $\sigma^*$ components. The dissociation of [Me$_2$C(X)NO$_2$]$^{-\bullet}$ proceeds with smooth

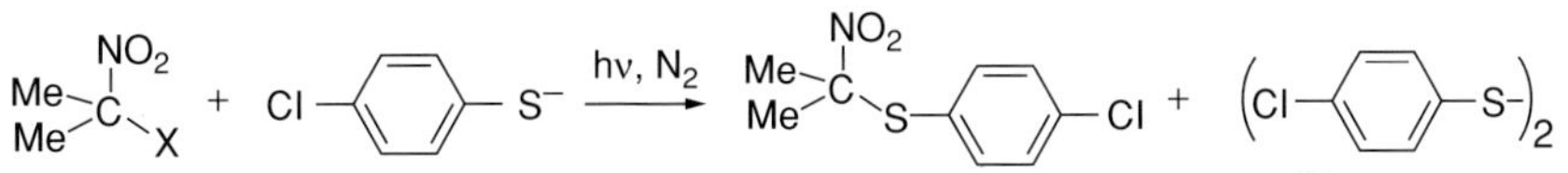

| X | Conditions | % Yield | |
|---|---|---|---|
| | | 31 | 32 |
| I | DMF, 30 min | 0 | 75 |
| Br | DMF, 4 h | 0 | 70 |
| Cl | DMSO, 2 h | 35 | 32 |
| SCN | DMSO, 2 h | 37 | 0 |
| NO$_2$ | DMF, 18 h | 69 | 0 |
| SO$_2$Ph | DMF, 5 h | 59 | 0 |

**Scheme 10.31**  Effect of the $\alpha$-substituent (X) in reactions between *p*-chlorophenylthiolate and 2-X-substituted 2-nitropropanes.

reorganisation of $\sigma^*$ and $\pi^*$ molecular orbitals in the required transition structure for loss of the anion (X$^-$).

More complex mechanisms are observed in the reactions between $\alpha$-substituted nitro-alkanes and thiolate anions. In dipolar aprotic solvents, both the expected substitution product and the disulfide are obtained (e.g. Scheme 10.31). The reactions provide a good example of changing functional group reactivity as a method of determining mechanism [23, 24]. Investigation using different functional groups (X) on the 2-substituted 2-nitropropane gave markedly different results. The most reactive radical anion intermediates (X = I, Br) yielded disulfides (32) and the more stable (X = NO$_2$, SO$_2$Ph) yielded S$_{RN}$1 products (31), see Scheme 10.31. A similar effect was observed by changing the substituent in the thiolates. For example, for X = NO$_2$, *o*-nitrophenylthiolate gave Me$_2$C(SR)NO$_2$ (55%) and no disul-fide, *p*-chlorophenylthiolate gave Me$_2$C(SR)NO$_2$ (69%) and no disulfide, phenylthiolate gave Me$_2$C(SR)NO$_2$ (32%) and disulfide (29%), and benzylthiolate gave no Me$_2$C(SR)NO$_2$ and disulfide (91%). Thus, as the thiolate becomes more nucleophilic, more disulfide is formed, and less S$_{RN}$1 product. Inhibition studies not only showed the expected reduction in the yield of the $\alpha$-nitrosulfides Me$_2$C(SR)NO$_2$ but also showed an increase in the yield of disulfide suggesting that the disulfide was formed by a non-chain non-photostimulated reaction.

In the initially proposed mechanism, the intermediate 2-nitro-2-propyl radicals, Me$_2$(NO$_2$)C$^\bullet$, undergo two reactions with the more basic (nucleophilic) thiolates: addition of thiolates leading to S$_{RN}$1 products (Equation 10.20 in Scheme 10.32), and SET to yield the nitro anion and thiyl radicals (RS$^\bullet$), Equation 10.21, which combine to give disulfide.

$$\text{Me}_2\overset{\bullet}{\text{C}}-\text{NO}_2 + \text{RS}^- \longrightarrow [\text{Me}_2\text{C(SR)NO}_2]^{\bullet-} \qquad (10.20)$$

$$\text{Me}_2\overset{\bullet}{\text{C}}-\text{NO}_2 + \text{RS}^- \xrightarrow{\text{SET}} \text{Me}_2\overset{-}{\text{C}}-\text{NO}_2 + \text{RS}^\bullet \qquad (10.21)$$

$$\text{RS}^\bullet + {}^\bullet\text{SR} \longrightarrow \text{RS}-\text{SR} \qquad (10.22)$$

**Scheme 10.32**  Intermediacy of 2-nitro-2-propyl radicals in reactions between $\alpha$-substituted 2-nitro-propanes and thiolate anions.

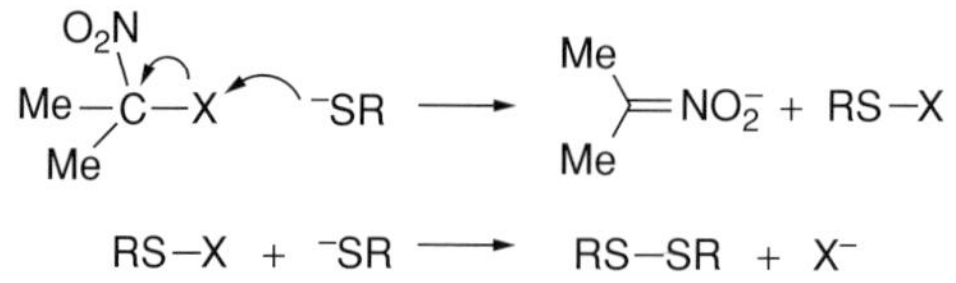

$$\begin{array}{c} \underset{\substack{\text{Me}-\text{C}-\text{X}\\|\\\text{Me}}}{\overset{\text{O}_2\text{N}}{|}} \quad {}^-\text{SR} \quad \longrightarrow \quad \underset{\text{Me}}{\overset{\text{Me}}{\Big\rangle}}{=}\text{NO}_2^- + \text{RS}-\text{X} \end{array}$$

$$\text{RS}-\text{X} + {}^-\text{SR} \quad \longrightarrow \quad \text{RS}-\text{SR} + \text{X}^-$$

**Scheme 10.33**  Abstraction of the $\alpha$-substituent from $\alpha$-substituted 2-nitropropanes by thiolate anion and formation of disulfide.

This mechanism involves radical anion intermediates but is not a chain reaction. Nitronate anions are formed in the reactions giving support to this mechanism. However, the nature of the $\alpha$-substituent X does not influence the partitioning between the two parallel routes, which argues strongly against this mechanism.

The evidence points towards competition between an $S_{RN}1$ substitution and a non-radical polar abstraction mechanism in which the nucleophilic thiolates attack the $\alpha$-substituents (Scheme 10.33). This mechanism is facilitated by the strong electron-withdrawing effect of the nitro group which promotes $S_N2$ substitution on the $\alpha$-substituent. This polar mechanism is favoured by strongly nucleophilic thiolates ($RS^-$ with $R = Bn > Ph > p\text{-Cl-C}_6\text{H}_4 > o\text{-}$ or $p\text{-NO}_2\text{-C}_6\text{H}_4$) and by $\alpha$-substituents which are readily polarisable and more weakly bonded ($I > Br > SCN > Cl > NO_2 > SO_2Ph$). We see, therefore, a competition between the SET ($S_{RN}1$) mechanism and the $S_N2$ substitution on the $\alpha$-substituent in which the resulting sulfenyl derivatives (RS–X) are very reactive and rapidly react further with more thiolate to yield the disulfide. Further evidence substantiates this assignment of mechanism; for example, sterically hindered thiolates yield isolable sulfenyl intermediates (e.g. $Ph_3CS–Cl$).

However, the competition between $S_{RN}1$ and polar abstraction mechanisms is complicated in certain reactions by the formation of disulfides which is inhibited by radical and radical anion traps, and requires photolysis [23, 24]. These results implicate a third possibility, the chain SET redox mechanism ($S_{ET}2$, i.e. substitution, electron transfer, bimolecular), Scheme 10.34. This alternative mechanism occurs when the intermediate radical anion can be intercepted by the thiolate (Equation 10.23) prior to the dissociation required in the $S_{RN}1$ mechanism (Equation 10.17 in Scheme 10.29). It becomes possible when either

(i)  the $\alpha$-substituent is less polarisable (and/or has a greater C—X bond strength), i.e. is a poorer nucleofuge as in the reaction between $Me_2C(SO_2Ph)NO_2$ and phenylthiolate, or

(ii)  the nucleophile is strongly nucleophilic, e.g. benzylthiolate in its reaction with $Me_2C(SCN)NO_2$.

$$Me_2C(X)NO_2 + RS^- \underset{}{\overset{\text{SET}}{\rightleftharpoons}} [Me_2C(X)NO_2]^{\bullet\,-} + RS^{\bullet} \qquad \text{Initiation}$$

$$[Me_2C(X)NO_2]^{\bullet\,-} + RS^- \underset{}{\overset{\text{SET}}{\rightleftharpoons}} Me_2C{=}NO_2^- + X^- + RS^{\bullet} \qquad (10.23)$$

$$RS^{\bullet} + RS^- \rightleftharpoons [RS{-}SR]^{\bullet\,-} \qquad (10.24)$$

$$Me_2C(X)NO_2 + [RS{-}SR]^{\bullet\,-} \rightleftharpoons [Me_2C(X)NO_2]^{\bullet\,-} + RS{-}SR \qquad (10.25)$$

Summary: $Me_2C(X)NO_2 + 2RS^- \longrightarrow Me_2C{=}NO_2^- + RS{-}SR + X^-$

**Scheme 10.34**  Chain SET redox mechanism ($S_{ET}2$) in reactions between $\alpha$-substituted $\alpha$-nitroalkanes and thiolate anions.

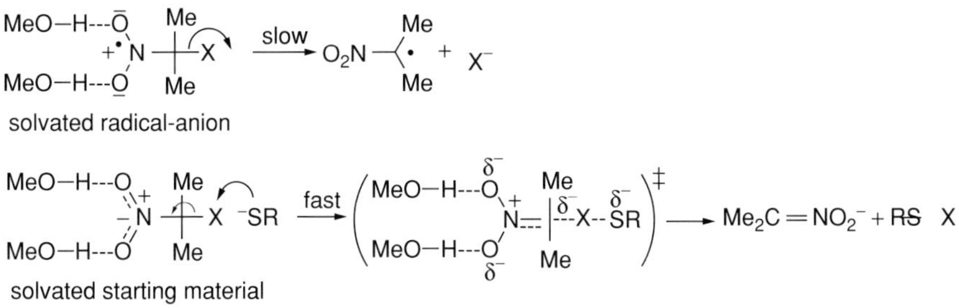

**Scheme 10.35** Solvation of the intermediate nitro radical anion and starting material by H-bonding in methanol.

The resulting thiyl radical (RS$^\bullet$) reacts with further thiolate to form a relatively stable disulphide radical anion (Equation 10.24) which in turn completes the chain reaction by SET to the starting material, $Me_2C(X)NO_2$ (Equation 10.25).

Study of reactions in different solvents often provides crucial mechanistic evidence. This is the case in the reaction between thiolates and $\alpha$-substituted nitroalkanes when changing from dipolar aprotic solvents to protic solvents. In contrast to the above reactions in dipolar aprotic solvents, where a mixture of disulfides and $S_{RN}1$ products is obtained, disulfides are exclusively formed in methanol with no $S_{RN}1$ products. Early studies had shown that $S_{RN}1$ substitution between $\alpha$-substituted nitroalkanes and anions such as nitronates and arylsulfinates takes place at a slower rate in methanol.

Solvation of thiolates is similarly low in both protic and dipolar aprotic solvents because of the size and polarisability of the large weakly basic sulfur atom, so is unlikely to contribute appreciably to the observed solvent effect. The intermediate nitro radical anion is stabilised by H-bonding in a manner which retards its dissociation in the $S_{RN}1$ mechanism (upper equation in Scheme 10.35). In contrast, the electron flow in the direct substitution at X (lower equation in Scheme 10.35) is such that solvation by methanol promotes the departure of the nucleofuge. In summary, protic solvation lowers the rate of the radical/radical anion reactions, but increases the rate of the polar abstraction yielding disulfide.

# Bibliography

Curran, D.P., Porter, N.A. and Giese, B. (1996) *Stereochemistry of Radical Reactions*. VCH, Weinheim.

Fossey, J., Lefort, D. and Sorba, J. (1995) *Free Radicals in Organic Chemistry*. Wiley, Chichester.

Perkins, M.J. (1994) *Radical Chemistry*. Ellis Horwood, New York.

Giese, B. (1986) *Radicals in Organic Synthesis: Formation of carbon–carbon Bonds*. Pergamon, Oxford.

Kochi, J.K. (Ed.) (1973) *Free Radicals*. Wiley, New York.

Moad, G. and Solomon, D.H. (1995) *The Chemistry of Free Radical Polymerization*. Pergamon, Bath.

Motherwell, W.B. and Crich, D. (1992) *Free Radical Chain Reactions in Organic Synthesis*. Academic Press, London.

Parsons, A.F. (2000) *An Introduction to Free Radical Chemistry*. Blackwell, Oxford.

Renaud, P. and Sibi, M.P. (Eds) (2001) *Radicals in Organic Synthesis* (vols 1 and 2). Wiley-VCH, Weinheim.

Zard, S.Z. (2003) *Radical Reactions in Organic Synthesis*. Oxford University Press, Oxford.

# References

1. Dixon, K.W. (1999) Decomposition rates of organic free radical initiators. In: J. Brandrup, E.H. Immergut and E.A. Grulke (Eds) *Polymer Handbook* (4th edn). Wiley, New York, p. 1–76.
2. Tedder, J.M. and Walton, J.C. (1976) *Accounts of Chemical Research*, **9**, 183–191.
3. Fischer, H. and Radom, L. (2001) *Angewandte Chemie (International Edition)*, **40**, 1340–1371.
4. Ghosez-Giese, A. and Giese, B. (1997) Factors influencing the addition of radicals to alkenes. In: K. Matyjaszewski (Ed.) *Controlled Radical Polymerization, ACS Symposium Series 685*. American Chemical Society, Washington, DC, p. 50–61.
5. Phelan, M., Aldabbagh, F., Zetterlund, P.B. and Yamada, B. (2005) *Macromolecules*, **38**, 2143–7.
6. Sobeck, J., Martschke, R. and Fischer, H. (2001) *Journal of the American Chemical Society*, **123**, 2849–57.
7. Dervan, P., Aldabbagh, F., Zetterlund, P.B. and Yamada, B. (2003) *Journal of Polymer Science, Part A (Polymer Chemistry)*, **41**, 327–34.
8. Studer, A. (2004) *Chemical Society Reviews*, **33**, 267–73.
9. Hawker, C.J., Bosman, A.W. and Harth, E. (2001) *Chemical Reviews*, **101**, 3661–88; Goto, A. and Fukuda, T. (2004) *Progress in Polymer Science*, **29**, 329–85.
10. Benoit, D., Grimaldi, S., Robin, S., Finet, J.-P., Tordo, P. and Gnanou, Y. (2002) *Journal of the American Chemical Society*, **122**, 5929–39.
11. Griller, D. and Ingold, K.U. (1980) *Accounts of Chemical Research*, **13**, 317–23.
12. Newcomb, M. (1993) *Tetrahedron*, **49**, 1151–76; Newcomb, M. (2001) Kinetics of radical reactions: radical clocks. In: P. Renaud and M.P. Sibi (Eds) *Radicals in Organic Synthesis* (vol 1). Wiley-VCH, Weinheim, pp. 317–36.
13. Newcomb, M. and Toy, P.H. (2000) *Accounts of Chemical Research*, **101**, 449–55.
14. Beckwith, A.L.J. and Storey, J.M.D., unpublished results.
15. Rychnovsky, S.D., Hata, T., Kim, A.I. and Buckmelter, A.J. (2001) *Organic Letters*, **3**, 807–10.
16. Studer, A. and Bossart, M. (2001) Homolytic aromatic substitution. In: P. Renaud and M.P. Sibi (Eds) *Radicals in Organic Synthesis* (vol 2). Wiley-VCH, Weinheim, pp. 62–80.
17. McLoughlin, P.T.F., Clyne, M.A. and Aldabbagh, F. (2004) *Tetrahedron*, **60**, 8065–71.
18. Beckwith, A.L.J., Bowry, V.W., Bowman, W.R., Mann, E., Parr, J. and Storey, J.M.D. (2004) *Angewandte Chemie (International Edition in English)*, **43**, 95–98. In particular, see the supporting information in this reference.
19. Enemaerke, R.J., Daasberg, K. and Skrydstrup, T. (1999) *Chemical Communications*, 343–4.
20. Shabangi, M., Kuhlman, M.L. and Flowers, R.A. (1999) *Organic Letters*, **1**, 2133–5.
21. Molander, G.A. (2001) Samarium(II) mediated radical reactions. In: P. Renaud and M.P. Sibi (Eds) *Radicals in Organic Synthesis* (vol 1). Wiley-VCH, Weinheim, pp. 153–82.
22. Curran, D.P., Fevig, T.L. and Totleben, M.J. (1992) *Synlett*, 943–61; Curran, D.P., Fevig, T.L., Jasperse, C.P. and Totleben, M.J. (1990) *Synlett*, 773–4.
23. Bowman, W.R. (1988) *Chemical Society Reviews*, **3**, 283–316.
24. Al-Khalil, S.I., Bowman, W.R., Gaitonde, K., Marley, M.A. and Richardson, G.D. (2001) *Journal of the Chemical Society, Perkin Transactions*, **2**, 1557–65, and references therein.

# Chapter 11
# Investigation of Catalysis by Acids, Bases, Other Small Molecules and Enzymes

*A. Williams*

## 11.1 Introduction

Speeding up the rate of a chemical reaction by catalysis is an economic advantage in industrial chemical processes. Provided catalysis is selective, nuisance by-products will occur to a smaller extent making work-up simpler while at the same time maintaining high overall yield in a shorter time. Simply increasing the temperature, whilst increasing the rate of the main reaction, would also increase the rate of formation of by-products. In biological systems, catalysis enables almost instantaneous response to changes; in some cases, enzyme catalysis is limited only by the diffusion together of reactants.

### 11.1.1 Definitions

A catalyst is a substance that participates in a chemical reaction and thereby increases the reaction rate but without a net change in the amount of that substance in the system. The catalyst is used and regenerated during each set of microscopic events leading from reactants to products, and the overall free energy of the reaction is unaltered by the presence of the catalyst which must, therefore, be in small amount relative to reactants. In radical chemistry, a catalyst can be the initiator of a chain reaction or a carrier (see Chapter 10).

A negative catalyst, inhibitor or stopper is a substance which decreases the rate of a reaction without causing a change in the free energy of the reaction (a topic not considered here). Autocatalysis occurs in a chemical reaction in which a product or an intermediate functions as a catalyst. Such catalysis is characterised by the existence of an induction period during the initial stages of the reaction (but, again, this is not considered here).

Catalysis provides a lower free energy pathway for a process by a mechanistic route different from that of the uncatalysed reaction. The study of catalysis, therefore, is an exercise in the elucidation of mechanisms. Distinction between mechanistic hypotheses is dominated by observing the effect on kinetics of defined alterations such as structural or concentration changes in the catalyst or substrate.

The following topics cover techniques for studying mechanisms of catalysis by proton transfer, small molecules and enzymes. These constitute very important areas of homogeneous catalysis and are paradigms for studies of related catalytic systems such as catalysis by metal ions, cyclodextrins, micelles and the currently burgeoning artificial enzyme systems. Catalysis by organometallic compounds is covered in the next chapter.

## 11.2 Catalysis by acids and bases

The influence of acids and bases on the speed and outcome of organic reactions has been known since the nineteenth century when organic chemistry was being developed and many fundamental synthetic routes were being discovered. The addition of either acid or base provides a very economical and easily handled procedure, and is probably one of the cheapest catalytic systems available to the industrial chemist. Acid–base catalysis is also ideal for reactions in water, a solvent with the lowest environmental impact for industrial processes. In the mid-1950s, it also became obvious that proton transfer, the fundamental reaction in acid–base catalysis, is involved as a facilitating process in many biological reactions. This realisation has given rise to theories of enzyme catalysis and, moreover, to attempts at the design and synthesis of relatively low molecular weight molecules, artificial enzymes, which have some of the properties of enzymes.

### 11.2.1 Experimental demonstration

Formulating credible mechanisms, i.e. accounting for all the electron movements and bonding changes in an acid- or base-catalysed reaction, is a difficult task although the framework on which the final mechanism is based is relatively easy to write. For this reason, and in the absence of evidence suggesting otherwise, a mechanism is conventionally illustrated as a series of steps as this enables the electron accounting to be most easily determined. This is illustrated by the mechanism for acid-catalysed acetal hydrolysis (Scheme 11.1a) shown

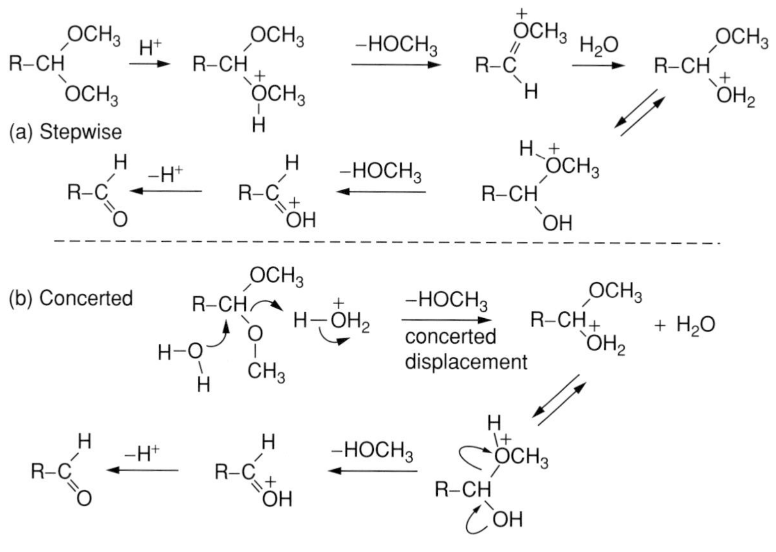

Scheme 11.1 Stepwise acid-catalysis mechanism for the hydrolysis of acetals (a) and the concerted alternative (b).

with oxocarbenium ion intermediates; these are extremely reactive in the presence of nucle-ophiles and in many cases do not exist as discrete molecules. The reaction could also involve proton transfer concerted with the heavy atom bond changes (Scheme 11.1b); where this is not proven, as in the reaction shown, the mechanism is conventionally written as stepwise.

The preliminary experiments in studying the hydrolysis of an organic compound such as aspirin [1] (Scheme 11.2) involve measuring the rate constant in buffered solutions and then extrapolating results to give the value at zero buffer concentration. The pH dependence of this rate constant (Fig. 11.1) will then be due only to the substrate and the solvent species. In water, these are simply water, hydroxide and hydronium ions; if the solvent is not water, these species are solvent, lyoxide and lyonium ions. These experiments will indicate whether or not any ionisable groups in the substrate are involved substantially in the catalysis. The pH profile can be interpreted initially by inspection. Thus, a *convex upward* curvature between two regions of dependence indicates either a change in rate-limiting step or the protonation or ionisation of a reactive species. A *concave upward* curvature indicates a change in the mechanism between two regions. The three linear sections A, B and C in Fig. 11.1, where log $k_{obs}$ changes with pH, correspond to terms $k_A$, $k_B$ and $k_C$ in Equation 11.1 where $k_{obs}$ is the experimental pseudo-first-order rate constant:

$$\text{rate} = k_{obs}\,[\text{aspirin}]$$
$$k_{obs} = k_A[\text{H}^+]/(1 + K_a/[\text{H}^+]) + (k_D[\text{H}^+] + k_B + k_C[\text{OH}^-])/(1 + [\text{H}^+]/K_a). \quad (11.1)$$

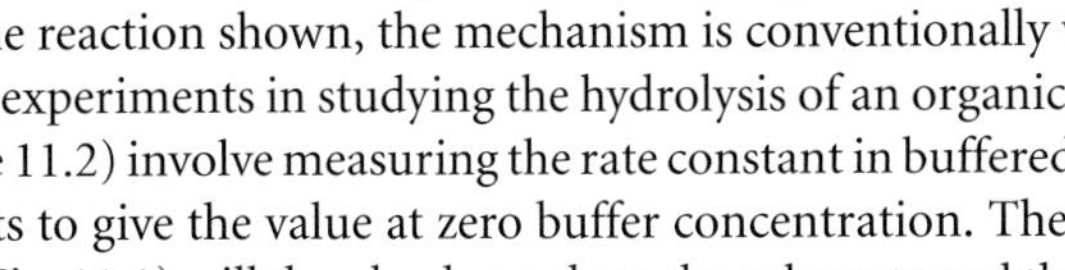

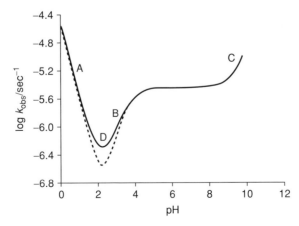

**Scheme 11.2**  Hydrolysis of aspirin.

Section A of Fig. 11.1 corresponds to the proton-catalysed hydrolysis pathway, section B corresponds to reaction of the carboxylate anion and C corresponds to the reaction of

**Fig. 11.1**  pH profile for the hydrolysis of aspirin at 25°C; the line is drawn from Equation 11.1 using experimental kinetic parameters from reference [1]; the dotted line is the curve calculated without inclusion of the $k_D$ term.

hydroxide ion with the anionic form of the aspirin. Scrutiny of the break at D between A and B indicates that the rates are larger than simply the sum of those from A and B (given by the dotted line in Fig. 11.1) and that a fourth term $k_D$ should be included; this is the hydrolysis of the neutral aspirin. Equation 11.1 then fits the pH profile, and each term ($k_A$ to $k_D$) corresponds to a separate mechanism, each one evident by sections A to D in the pH profile.

Equation 11.1 relates to the particular case of the catalysed hydrolysis of aspirin. More generally, we may write Equation 11.2 for the catalysed pseudo-first-order reaction (for example) of substrate HS:

$$\text{rate} = k_{obs}[\text{HS}]$$
$$k_{obs} = k_{solvent} + k_{HA}[\text{HA}] + k_B[\text{B}] + k_{HA.B}[\text{HA}][\text{B}]. \qquad (11.2)$$

Here, $k_{solvent}$ includes the solvent-induced reaction ($k_o$) plus terms for possible catalysis by $H_3O^+$ ($k_H$) and $OH^-$ ($k_{OH}$). These pathways can be investigated in the usual manner by measurements in the absence of general acids and bases. Catalysis by general acids (HA) and/or bases (B) is observed experimentally when the rate constant is plotted against the total concentration of general acid and general base at constant pH (Fig. 11.2A, where different possible outcomes are shown).

If there is no general acid or base catalysis, $k_{HA}$, $k_B$ and $k_{HA.B}$ are all equal to zero, so the horizontal line in Fig. 11.2A is observed, and the intercept on the y-axis corresponds to $k_o$ and $k_H$ or $k_{OH}$ paths (depending upon the pH). If $k_{HA.B}$ and either $k_{HA}$ or $k_B$ are zero, the linear plot corresponding to either general acid or general base is observed. Dissection of the catalysis into general acid or base catalysis can be understood by reference to Equation 11.2.

The slope of the linear plot in Fig. 11.2A, i.e. $k_{obs}/([\text{acid}]+[\text{base}])$, is plotted against the fraction of free base (FB) of the total added catalyst in Fig. 11.2B, where FB = [base]/([acid] + [base]). A linear plot from the origin with a positive slope (equal to $k_B$) indicates general base catalysis (rate = $k_B[\text{B}][\text{HS}]$). A linear plot with a negative slope (equal to $k_{HA}$) from FB = 1 on the x-axis indicates general acid catalysis (rate = $k_{HA}[\text{HA}][\text{HS}]$). When both general acid and general base catalysis occur in parallel mechanisms, there is a linear plot with a gradient equal to ($k_B - k_{HA}$), between finite values on the y-axis.

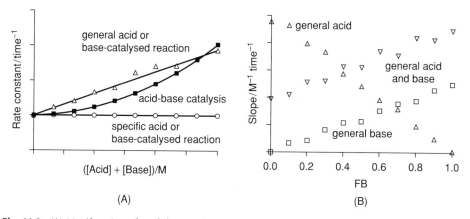

**Fig. 11.2** (A) Manifestation of acid, base and acid–base catalysis at constant pH; (B) determination of general acid and general base catalysis parameters {FB = [base]/([acid] + [base])}.

The non-linear plot in Fig. 11.2A indicates acid–base catalysis (i.e. when both acid and base are involved in the same catalytic mechanism with rate $= k_{HA.B}[HA][B][HS]$), and the $k_{HA.B}$ parameter can be determined from the slope of a plot of $k_{obs}/(1 - FB)([acid] + [base])^2$ versus FB (not shown).

It is possible that catalysis involves general acid, general base and general acid–base catalysis all together. Since $[B] = FB([HA] + [B])$, Equation 11.2 may be transformed into

$$(k_{obs} - k_{solvent})/([HA] + [B]) = k_{HA.B}FB(1 - FB)([HA] + [B]) + k_{HA} + (k_B - k_{HA})FB,$$

which shows that a plot of $(k_{obs} - k_{solvent})/([HA] + [B])$ against $([HA] + [B])$ will be a straight line with gradient $= k_{HA.B}FB(1 - FB)$ and intercept $= k_{HA} + (k_B - k_{HA})FB$. The intercepts from a set of experiments done at various pHs (i.e. at different FBs) are then plotted against FB to give a new straight line with gradient $(k_B - k_{HA})$ and intercept $k_{HA}$. All parameters are thus determined.

Except as discussed below, it is usually important that the concentration of the catalytic species in kinetic studies does not exceed about 0.1 molar because medium and/or specific solute effects may complicate the kinetics at higher solute concentrations (see Chapter 3).

### 11.2.2    Reaction flux and third-order terms

For confidence in the accuracy of $k_{HA.B}$, it is important that the curvature in Fig. 11.2A corresponds to an increase of two-fold or greater in the slope. Too often, the third-order term in Equation 11.2 ($k_{HA.B}$) is inaccurate because it derives from a very small fraction of the total reaction flux, and needs confirmation. Data for the enolisation of acetone are reproduced in Fig. 11.3, which shows that substantial differences between observed rate constants and calculated values *not* including the third-order term are seen only at high concentration of the acid–base pair [2]. Concentrations of acid–base pairs are between 0 and 2 M, and there are no major specific salt or solvent effects. This is important as the high

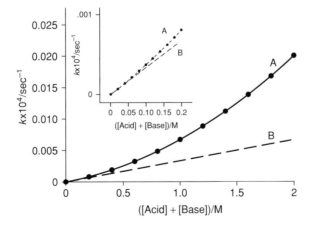

**Fig. 11.3**   Enolisation of acetone; data from reference [2]. (A) Calculated from the observed data. (B) Calculated from the kinetic parameters omitting the third-order term. The inset graph for a limited concentration range of acid and base shows only small curvature.

concentrations of catalyst necessary to observe curvature change the solvent composition substantially; in the case of acetate as catalyst at FB = 0.5, there is 1 M acetic acid at 2 M total concentration. The effect can be investigated by applying corrections derived from the effect of a similar but non-reactive solute such as acetonitrile. In the case of a given third-order term, the slopes of the corrected plots of $k_{obs}/[\text{catalyst}]$ against [catalyst] should be identical, and this is found to be so with the enolisation of cyclohexanone [2]. Hegarty and Jencks [3] showed that the observed third-order term for the enolisation of acetone has a slight dependence on the acidity of the conjugate acid of the acid–base pair for a range of carboxylic acids from chloroacetic to propionic acid. The regular range of values for the third-order term is consistent with it being a real term and not due to an artefact.

### 11.2.3   Brønsted equations

The graph of the logarithm of $k_B$ for a base-catalysed reaction against the logarithm of the equilibrium constant $K$ for the complete proton transfer from acidic substrate to catalyst is linear in the majority of cases, Equation 11.3. A corresponding relationship holds for acid-catalysed reactions (Equation 11.4). The slopes of these plots are classically denoted by the symbols $\alpha$ and $\beta$ for general acid and general base catalysis, respectively, and compare the change in free energy for the formation of the transition state for proton transfer with that of the complete proton transfer reaction for a range of substituted catalysts. It is invariably difficult to measure the equilibrium constant $K$ explicitly and, moreover, a change in $K$ is even more difficult to determine accurately. The deprotonation of acetone (Scheme 11.3) involves proton transfer to base and the slope of a plot of log $k_B$ versus $pK_a^{HB}$ is the same as versus log $K$ because $d\log K = dpK_a^{HB}$ ($K = K_a^{HS}/K_a^{HB}$ and $K_a^{HS}$ is constant; thus $d\log K = d\,pK_a^{HB}$ and $d\log k_B/d\log K = d\log k_B/d\,pK_a^{HB}$). Correspondingly, the slope of the plot of log $k_{HB}$ against $pK_a^{HB}$ for acid catalysis has a negative value:

$$\log k_B = \beta \log K + C \tag{11.3}$$

$$\log k_{HB} = \alpha \log K + D. \tag{11.4}$$

The values of the Brønsted coefficients $\alpha$ and $\beta$ for the reverse and forward reactions, respectively, may be deduced from Equations 11.3 and 11.4 to be related by the equation $\beta - \alpha = 1$. It should be noted that the plot of log $k_{HB}$ versus $pK_a^{HB}$ is the usual way of treating the data rather than the original one of plotting against log $K_a^{HB}$; in the latter case, the slope would be positive and the above relationship would change to $\beta + \alpha = 1$.

$$H_2O + CH_3COCH_3 \underset{}{\overset{K_a^{HS}}{\rightleftharpoons}} CH_3COCH_2^- + H_3O^+; \quad K_a^{HS} = [H_3O^+][S^-]/[HS]$$
$$(HS) \qquad\qquad\qquad (S^-)$$

$$B + CH_3COCH_3 \underset{k_{HB}}{\overset{k_B}{\rightleftharpoons}} CH_3COCH_2^- + HB^+; \quad K = [HB^+][S^-]/[HS][B]$$
$$(HS) \qquad\qquad\qquad (S^-)$$

$$HB^+ + H_2O \overset{K_a^{HB}}{\rightleftharpoons} B + H_3O^+; \quad K_a^{HB} = [H_3O^+][B]/[HB^+] \text{ hence } K = K_a^{HS}/K_a^{HB}$$

**Scheme 11.3**   Brønsted equilibria in the deprotonation of acetone (HS) to yield enolate anion (S⁻).

The Brønsted parameters measure the sensitivity of the reaction in one direction to polar substituent changes compared with that for the overall equilibrium. A Brønsted parameter is, therefore, a measure of the *charge distribution* in the transition structure relative to the reactant and product; the value of $\alpha$ should normally be between $-1$ and zero, and $\beta$ between $+1$ and zero. An $\alpha$ or $\beta$ of $-1$ or $+1$, respectively, indicates that the proton is completely transferred from the donor $HB^+$ to the substrate, or from the substrate to the acceptor base B, in the transition structure of the rate-limiting step.

### 11.2.3.1 Brønsted parameters close to $-1$, 0 and $+1$

It is generally difficult to obtain an accurate rate constant for general acid catalysis when the value of $\alpha$ is close to $-1$ or close to zero. This is because hydronium ion catalysis carries most of the reaction flux when $\alpha \sim -1$, which masks any catalysis by a general acid. When $\alpha \sim 0$, there is not much change in the rate constant as a function of the catalyst structure; therefore, solvent acting as an acid (at 55.5 M in the case of water) will take the major part of the flux, again masking the effect of a general acid. Consequently, under these respective conditions, changes in the general acid concentration will change the overall rate constants only marginally.

When $\alpha$ is intermediate in value, most of the flux is taken by the general acid HA, and general acid catalysis is easily observed. Table 11.1 [4] gives the percentage of the flux taken for given values of $\alpha$. Similar arguments can be made for base catalysis, i.e. general base catalysis is difficult to characterise when Brønsted $\beta$ values are close to 1 and 0.

**Table 11.1** Catalysis in 0.1 M HA and 0.1 M $A^-$ of reactions with $\alpha$ values of $-0.1$, $-0.5$ and $-1.0$ [4]. Percentages represent reaction flux taken by the acid catalyst indicated.

| Acid | $\alpha = -0.1$ | $\alpha = -0.5$ | $\alpha = -1.0$ |
|---|---|---|---|
| $H_2O$ | 98% | 0.01% | $5 \times 10^{-12}$% |
| HA | 2% | 96.4% | 0.2% |
| $H_3O^+$ | 0.002% | 3.6% | 99.8% |

## 11.2.4 Kinetic ambiguity

Kinetic ambiguity arises when different mechanisms lead to the same predicted rate equation, i.e. the mechanisms are kinetically equivalent (see Chapters 3 and 4). It is very widespread in reactions catalysed by acids and bases, but strategies have been developed to resolve some of the ambiguities.

### 11.2.4.1 Cross-correlation effects

General base catalysis of the reaction of a nucleophile (HNu) is kinetically equivalent to general acid catalysis of the reaction of the deprotonated nucleophile ($Nu^-$). A distinction can be made employing cross-correlation effects where the value of the Brønsted $\alpha$ is measured as a function of another parameter such as the nucleophilicity of the attacking nucleophile.

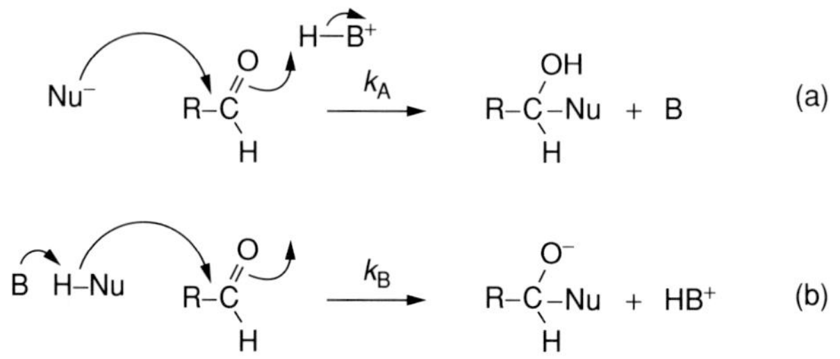

**Scheme 11.4** Catalysis of nucleophilic attack at an aldehyde by (a) general acid and (b) general base.

The technique is exemplified by consideration of the attack of several nucleophilic reagents on aldehydes where the values of $\alpha$ for catalysis by general acids become more negative as the nucleophilicity of the reagent decreases. Two mechanisms can be written for the reaction (Scheme 11.4) which are described by Equations 11.5a and 11.5b where $k_A$ is related to $k_B$ by Equation 11.6:

$$\text{rate (a)} = k_A[HB^+][Nu^-][\text{aldehyde}] \tag{11.5a}$$

$$\text{rate (b)} = k_B[B][HNu][\text{aldehyde}]. \tag{11.5b}$$

But

$$[H_3O^+] = K_a^{HB}[BH^+]/[B] = K_a^{HNu}[HNu]/[Nu^-],$$

so

$$\text{rate (a)} = k_A K_a^{HNu}[HNu][B][\text{aldehyde}]/K_a^{HB};$$

thus

$$k_B = k_A K_a^{HNu}/K_a^{HB}. \tag{11.6}$$

Consideration of the More O'Ferrall–Jencks diagram for the general acid-catalysed mechanism (Fig. 11.4A) shows that, on decreasing the nucleophilicity/basicity of $Nu^-$, i.e. reducing the energy of the states on the left-hand side ($\oplus$), the coordinates of the transition structure will move at right angles to the reaction coordinate towards the top-left corner (Thornton effect, i.e. towards the corner decreasing in energy) and along the reaction coordinate towards the top-right (Hammond effect, i.e. towards the corner increasing in energy). The resultant movement is towards a more negative $\alpha$ consistent with the observed result.

The general base catalysis mechanism, (b) in Scheme 11.4, has a More O'Ferrall–Jencks diagram (Fig. 11.4B) which shows that decreasing the nucleophilicity/basicity of the nucleophile (decreasing the energy, $\oplus$, of just the top-left corner) would cause movement of the transition structure coordinates towards the top-left. Resolving this into its components indicates an increase in $\beta$. The value of $\beta$ is related to that of $\alpha$ by $\alpha = \beta - 1$, so that the value of $\alpha$ should become less negative, which is *not* consistent with the observed results.

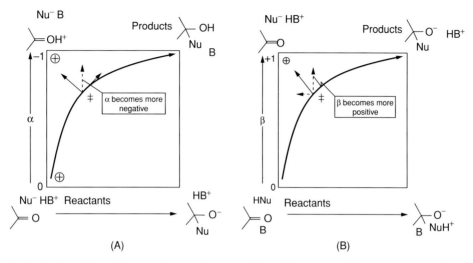

**Fig. 11.4** More O'Ferrall–Jencks diagrams for nucleophilic addition to an aldehyde. (A) For general acid catalysis. (B) For general base catalysis. The symbol ⊕ indicates an imposed decrease in energy of the state indicated.

## 11.2.4.2   The diffusion-controlled limit as a criterion of mechanism

A mechanism involving a transient intermediate can be excluded if it can be shown to require a rate greater than that for diffusion to support the observed rate of reaction. The general acid-catalysed formation of the intermediate in the formation of the semicarbazone from 4-nitrobenzaldehyde can be represented by two mechanisms (a) and (b) in Scheme 11.5. The mechanisms give rise to the rate laws of Equations 11.7 and 11.8:

$$\text{rate (a)} = k[\text{RNH}_2][{>}\text{C}{=}\text{O}][\text{AH}] \tag{11.7}$$
$$\text{rate (b)} = (k_1 K_a^{\text{HA}}/K_a)[\text{RNH}_2][{>}\text{C}{=}\text{O}][\text{AH}]. \tag{11.8}$$

In the case where AH = $H_3O^+$, the experimental result is $k_{\text{obs}} = 10^4$ $M^{-2}$ $s^{-1}$, which is a credible value for the third-order $k$ in Equation 11.7. However, since $K_a$ is known to be $2.8 \times 10^8$ M, $k_1$ in Equation 11.8 needs to be $2.8 \times 10^{12}$ $M^{-1}$ $s^{-1}$ to support the second mechanism in Scheme 11.5b with $A^- = H_2O$[5]. Since this value is larger than the likely diffusion-controlled rate constant ($7.4 \times 10^9$ $M^{-1}$ $s^{-1}$)[6], the second mechanism is ruled out. (The calculated rate constant is even larger than that of the fastest known

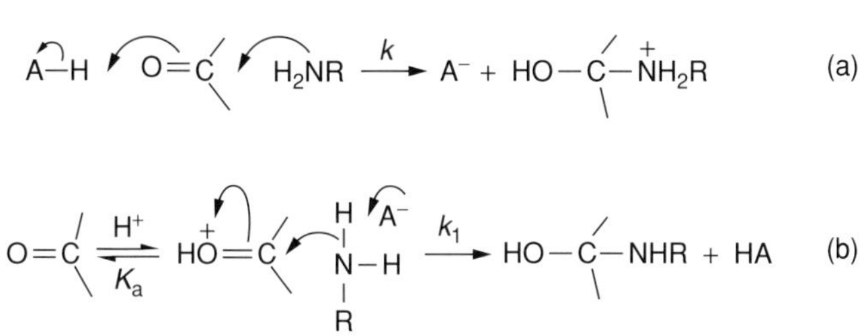

$$\text{A}{-}\text{H} \quad \text{O}{=}\text{C} \quad \text{H}_2\text{NR} \xrightarrow{k} \text{A}^- + \text{HO}{-}\text{C}{-}\overset{+}{\text{N}}\text{H}_2\text{R} \qquad \text{(a)}$$

$$\text{O}{=}\text{C} \underset{K_a}{\overset{\text{H}^+}{\rightleftharpoons}} \text{HO}{=}\overset{+}{\text{C}} \underset{\text{R}}{\overset{\text{H}}{\underset{|}{\text{N}{-}\text{H}}}} \xrightarrow{k_1} \text{HO}{-}\text{C}{-}\text{NHR} + \text{HA} \qquad \text{(b)}$$

**Scheme 11.5**   Distinction between mechanisms by the diffusion-controlled limit.

diffusion-controlled reaction in water, namely between hydroxide and hydronium ions, $1.4 \times 10^{11}$ M$^{-1}$ s$^{-1}$.) Similar arguments rule out the second mechanism when AH is a general acid.

### 11.2.4.3    Scatter in Brønsted plots

The absence of scatter in a Brønsted plot for a general base-catalysed reaction can imply that the reaction mechanism involves a rate-limiting proton transfer step. This is because proton transfer to the base in the reaction is closely similar to the equilibrium proton transfer to the base in the reaction which defines the p$K_a$ of the conjugate acid of that base. The observation of scatter, especially for sterically hindered bases (such as 2,6-dimethylpyridine), is evidence that nucleophilic catalysis is operating as opposed to general base catalysis.

### 11.2.4.4    Solvent kinetic isotope effects

Distinction between nucleophilic catalysis and mechanistic general base catalysis can be made because the latter involves rate-limiting proton transfer whereas the former does not. The solvent hydrogen isotope effect can be employed to demonstrate those mechanisms where the bonds to the transferring proton are being broken in the transition structure of the rate-limiting step. The proton in such cases is said to be *in flight* which would lead to a substantial primary isotope effect. In the case of C—H bond fission, a maximum value of $k_B^H/k_B^D \approx 7$ should result.

## 11.2.5    Demonstrating mechanisms of catalysis by proton transfer

Reaction via unstable acidic or basic intermediates can be promoted by proton transfer to a base or from an acid, respectively, thus giving rise to base or acid catalysis. The concept is illustrated for nucleophilic attack on carboxamide derivatives (Scheme 11.6). With base catalysis, deprotonation of the first-formed intermediate promotes the forward reaction, as does protonation of the first-formed intermediate with acid catalysis.

### 11.2.5.1    Stepwise proton transfer (trapping)

Proton transfer between simple electronegative atoms of simple acids and bases occurs in a series of steps [7]. Diffusion together of acid and base molecules is followed by proton transfer within the encounter complex and then diffusion apart (Eigen mechanism, Scheme 11.7). If proton transfer is thermodynamically favourable (p$K_a^{HS} <$ p$K_a^{HB}$), the rate-limiting step is diffusion together of the acid–base pair, the overall rate constant becomes $k_a$ and the slope of the Brønsted plot for invariant SH and a range of bases B is zero. If proton transfer is thermodynamically unfavourable (p$K_a^{HS} >$ p$K_a^{HB}$), the rate-limiting step is diffusion apart of the acid–base pair, $k_d$ in Scheme 11.7, and the overall rate constant is $(k_a/k_{-a})(k_1/k_{-1})k_d$. Since the diffusion terms are essentially invariant parameters, the rate constant is proportional to $k_1/k_{-1}$ which is $K_a^{HA}/K_a^{HB}$. Thus, a plot of log $k_B$ versus p$K_a^{HB}$ will have unit slope (p$K_a^{HS}$ is constant) and what is conventionally known as an Eigen plot is obtained, for example Fig. 11.5A. The reverse reaction where variant acid (HB) donates a proton to the invariant base (S$^-$) has a similar p$K$ dependence exemplified in (Fig. 11.5B). Both plots exhibit breaks at p$K_a^{HB} =$ p$K_a^{HS}$.

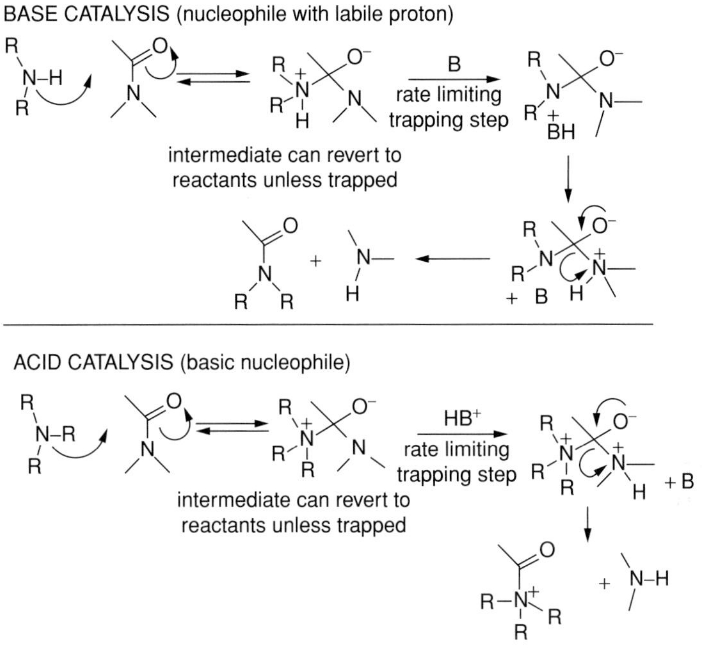

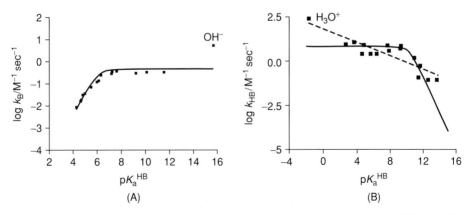

**Scheme 11.6**   Proton transfer in nucleophilic substitution in carboxamide derivatives.

$$B + HS \underset{k_{-a}}{\overset{k_a}{\rightleftharpoons}} [B.HS] \underset{k_{-1}}{\overset{k_1}{\rightleftharpoons}} [HB^+.S^-] \underset{k_{-d}}{\overset{k_d}{\rightleftharpoons}} HB^+ + S^-$$

**Scheme 11.7**   The Eigen mechanism for proton transfer from the substrate HS to the base B.

**Fig. 11.5**   (A) Eigen-type plot for the general base-catalysed reactions of 4-chlorobenzaldehyde with a substituted hydrazine [8]; the line is calculated from $\log k_B = -0.341^- \log(1 + 10^{6.09-pK_a})$. (B) Eigen-type plot for general acid catalysis in the reaction of cyanic acid (HNCO) with aniline, and an alternative interpretation [9]; the dashed line has a Brønsted slope of $-0.19$ and the break at $pK_a$ near 10 in the solid line is consistent with proton transfer to an addition intermediate ($PhNH_2^+CONH^-$).

The Eigen plots in Figs 11.5A and 11.5B are characterised, respectively, by a positive deviation of the point for hydroxide ion from the diffusion-limiting line, and a corresponding deviation of the point for the hydronium ion. These deviations are due to the much faster diffusion of these two ions compared with diffusion of non-lyate species in water. The diffusion limits are also subject to the effect of added non-reacting agents such as glycerol which slow the diffusion process by increasing the viscosity of the solution. The Eigen plot will only be observed if the $pK_a$ of the acid intermediate is larger than that of hydronium ion ($-1.7$) and less than that of water (15.7).

A further diagnostic device is to compare the experimental breakpoint in an Eigen plot with the *estimated* $pK_a$ of the postulated intermediate HS (it is unlikely that this $pK_a$ could be *measured* by conventional techniques because of its highly transient nature). For example, the plot in Fig. 11.5A for the general base-catalysed reactions of 4-chlorobenzaldehyde with a substituted hydrazine has a breakpoint at $pK_a^{HB} = 6.09$ which is larger than that estimated for the postulated tetrahedral intermediate ($pK_a^{HS} = 3.1$)[8]. This difference is greater than the likely uncertainty in the $pK_a$ estimation and led to the proposal of a preassociation mechanism (see Section 11.2.5.3) rather than a conventional Eigen-type proton transfer mechanism.

## 11.2.5.2   Stabilisation of intermediates by proton transfer

Reaction intermediates which are generated as a result of covalent bond formation or fission are often proton donors or acceptors. For example, the nucleophilic addition of water to the carbonyl carbon of an amide (Scheme 11.8) generates the tetrahedral intermediate, 1, which is unstable with respect to proton transfer to or from water, and with respect to breakdown back to reactants. Expulsion of water from 1 will occur much faster than expulsion of the unstable amine anion, $R'NH^-$. Removal of $H^+$ from the attacking water of 1, or proton addition to oxygen or nitrogen of 1, will generate relatively more stable intermediates although proton addition to the oxyanion of 1 requires proton transfer from the hydronium ion to give products. A rate increase will be observed if these proton transfers occur to or from added general acids or bases rather than solvent water.

The extent of catalysis depends critically upon the stability of the intermediate 1. If the rate of expulsion of $H_2O$ from 1 (rate constant $k_{-1}$) is slower than proton transfer to solvent water, the rate of formation of the intermediate (rate constant $k_1$) will be the rate-limiting step and no catalysis will be observed. The rate constant for protonation of the amine nitrogen of 1 by solvent water, $k_{HA}$ ($HA = H_2O$), depends on the basicity of the nitrogen and is given by $k_A K_w / K_a$, where $k_A$ represents the rate constant for diffusion-controlled abstraction of a proton by hydroxide ion, with a value of approximately $10^{10}$ $M^{-1}$ $s^{-1}$, and

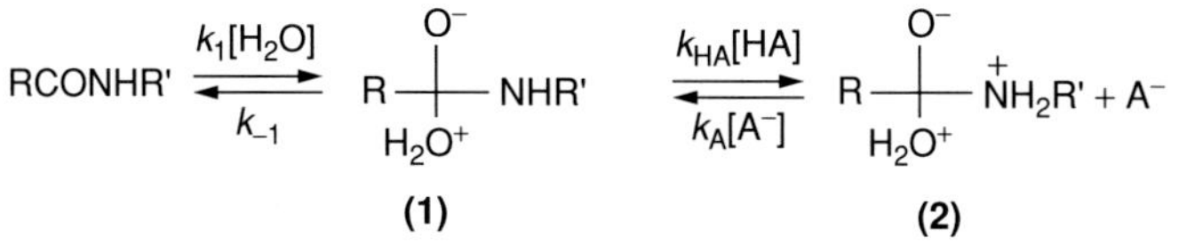

$$RCONHR' \underset{k_{-1}}{\overset{k_1[H_2O]}{\rightleftharpoons}} R\overset{O^-}{\underset{H_2O^+}{\vert}}NHR' \underset{k_A[A^-]}{\overset{k_{HA}[HA]}{\rightleftharpoons}} R\overset{O^-}{\underset{H_2O^+}{\vert}}\overset{+}{N}H_2R' + A^-$$

(1)                                     (2)

**Scheme 11.8**   Protonation as a catalytic process in the nucleophilic addition of water to the carbonyl of an amide.

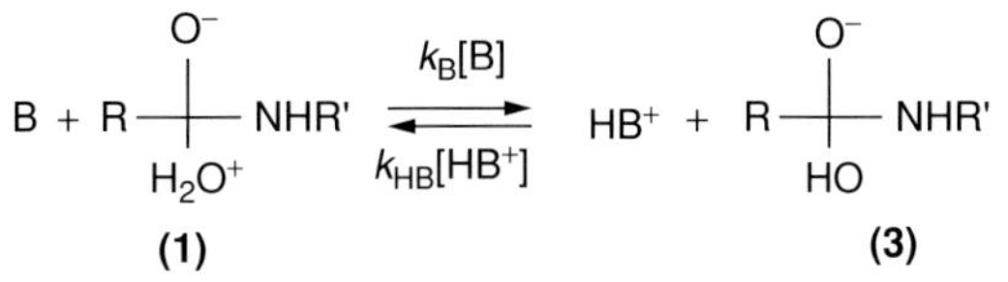

$$B + R \underset{H_2O^+}{\overset{O^-}{\vdash}} NHR' \underset{k_{HB}[HB^+]}{\overset{k_B[B]}{\rightleftharpoons}} HB^+ + R \underset{HO}{\overset{O^-}{\vdash}} NHR'$$

$$(1) \qquad\qquad\qquad\qquad (3)$$

**Scheme 11.9**   Deprotonation as a catalytic process in the nucleophilic addition of water to the carbonyl of an amide.

$K_a$ is the acid dissociation constant of the ammonium group of **2**. If the rate of expulsion of $H_2O$ from **1** to regenerate reactants is faster than (or similar to) the rate at which **1** is trapped by proton abstraction from water, the rate-limiting step occurs after the formation of intermediate **1**, and the addition of general acids (i.e. other proton donors) may increase the observed rate.

If HA is a stronger acid than the ammonium function of **2**, the rate constant for proton transfer to **2**, $k_{HA}$, will be for diffusion and the observed rate constant will be independent of the acidity of HA. On the other hand, if HA is a weaker acid than the ammonium function of **2**, the proton transfer from general acids, HA, to the nitrogen of **1** in Scheme 11.8 will be given by $k_A K_a^{HA}/K_a$, where $K_a^{HA}$ is the acid dissociation constant of HA and $k_A$ is the diffusion-controlled rate constant. The observed rate will now be dependent upon the acidity of the catalyst HA as described by a Brønsted correlation with slope equal to $-1$.

Intermediate **1** could also be stabilised by proton transfer from oxygen to give **3** in Scheme 11.9. The proton acceptor B could be solvent water or a general base catalyst. The reaction will only be catalysed if the rate of breakdown of **1** to regenerate reactants is faster than the rate of proton transfer. In this case, such catalysis would be independent of the base strength of the catalyst **B** as proton transfer would invariably be thermodynamically favourable and hence occur at the maximum diffusion-controlled rate. If proton transfer to solvent is thermodynamically favourable, such that proton donation to 55.5 M water is faster than to, say, 1 M added base, any observed catalysis by base must represent transition state stabilisation by hydrogen bonding, or a concerted mechanism.

It is of interest to examine the general case when a nucleofuge is displaced from a carbonyl group by a nucleophile with a labile hydrogen, HNu, which becomes much more acidic upon formation of the tetrahedral intermediate (Scheme 11.10). Catalysis by the general base **B** will be observed when the intermediate **4** breaks down to reactants faster than it transfers a proton to water. The rate constant for formation of **5**, which may or may not represent the overall rate constant of reaction, is given by $k_B[B]K$, where $K$ is the equilibrium constant for formation of **4** and $k_B$ is the rate constant of proton transfer from **4** to the catalyst **B**.

The rate constant for expulsion of $RNH_2$ to regenerate the reactants, $k_{-1}$, is between $10^6$ and $10^9$ s$^{-1}$ and much faster than that for proton transfer to water ($B = H_2O$), which is between $10^{-1}$ and $10^2$ s$^{-1}$. When the general base catalyst is a second molecule of amine

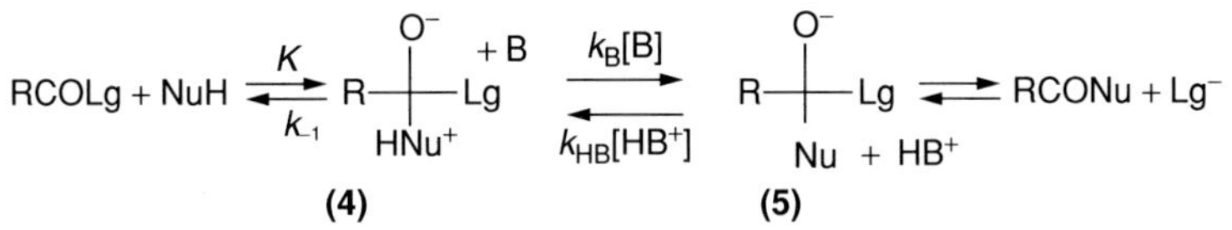

$$RCOLg + NuH \underset{k_{-1}}{\overset{K}{\rightleftharpoons}} R\underset{HNu^+}{\overset{O^-}{\vdash}}Lg \overset{+B}{\underset{k_{HB}[HB^+]}{\overset{k_B[B]}{\rightleftharpoons}}} R\underset{Nu + HB^+}{\overset{O^-}{\vdash}}Lg \rightleftharpoons RCONu + Lg^-$$

$$(4) \qquad\qquad\qquad\qquad (5)$$

**Scheme 11.10**   Nucleophilic substitution at a carbonyl by a nucleophile with a labile hydrogen.

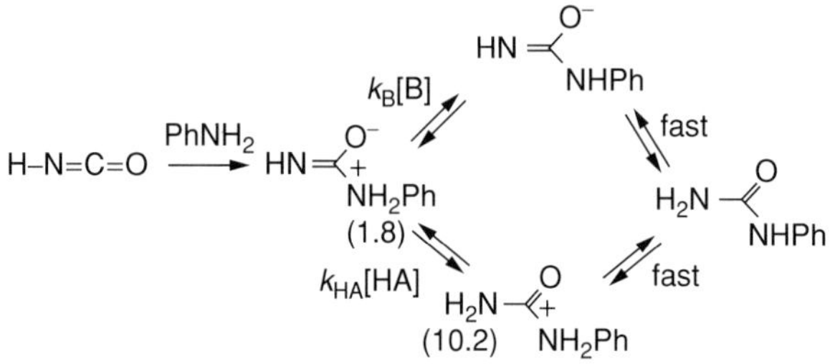

**Scheme 11.11**   General acid–base catalysis of the reaction of HNCO with aniline. The numbers in the parentheses below the functional groups are $pK_a$ values estimated for the functional groups.

($B = RNH_2$), proton transfer is much more efficient and gives a rate enhancement of about 1000-fold. The greater basicity of the amino group compared with water is offset to some extent by the solvent water concentration of 55.5 M, so the amine catalyst at 1 M is then only about 20 times more efficient than water. In general, rate enhancements of 10–100 are brought about by general acids and bases.

Brønsted plots with some scatter should always be examined carefully because they might conform with the Eigen equation. A classic case is the acid-catalysed reaction of aniline with HNCO which can be fitted to a shallow Brønsted line (slope $= -0.19$)[9]; however, the data are also consistent with an Eigen dependence (Fig. 11.5B) arising from a mechanism involving trapping of an addition intermediate by proton transfer (Scheme 11.11).

## 11.2.5.3   Preassociation

When the rate-limiting step of a reaction such as that of Scheme 11.10 is the diffusion-controlled encounter of two reagents, an enzyme may increase the rate simply by having the nucleophile and catalyst preassociated so that they do not have to diffuse through bulk solution before the reaction can occur. This situation is observed in simple intermolecular reactions when the rate of breakdown of the intermediate to regenerate reactants (rate constant $k_{-2}$, see Scheme 11.12) is faster than the rate of separation of the hydrogen-bonded tetrahedral intermediate and catalyst (**6.B**), with a rate constant between $10^{10}$ and $10^{11}$ s$^{-1}$, to give **6** and **B**. The formation of **6.B** is then forced to take place via a preliminary preassociation complex of reactants and catalyst (**Int** in Scheme 11.12).

The preassociation mechanism is more efficient than the trapping mechanism because it generates an intermediate which immediately reacts by an ultrafast proton transfer (in the pre-association complex, **Int**) and thus avoids the diffusion-controlled step bringing the catalyst and intermediate together. This mechanism is sometimes called a 'spectator' mechanism because, although the catalyst is present in the transition structure, it is not undergoing any transformation [10].

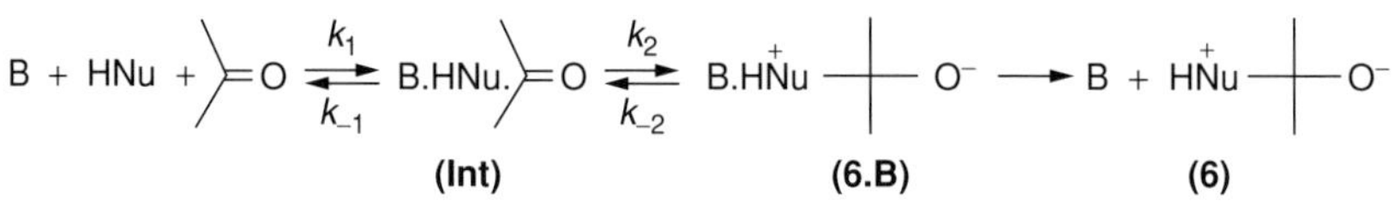

**Scheme 11.12**   The pre-association mechanism for addition of HNu to a carbonyl.

### 11.2.5.4   Concerted proton transfer

When a proposed intermediate is so unstable that it cannot exist, i.e. it would have a lifetime less than that of a bond vibration ($\sim 10^{-13}$ s), the reaction *must* proceed by a concerted mechanism (see reference [11], p. 5). The proton-transfer steps and other covalent bond-forming and bond-breaking processes are concerted but with varying degrees of coupling between their motions. However, it is still not clear whether a concerted mechanism can occur when the intermediates, which would be formed in an alternative stepwise mechanism, have significant lifetimes. This is an important question for reactions catalysed by enzymes because the nature of the intermediate itself will control whether the enzyme- and non-enzyme-catalysed mechanisms are forced to be similar if the *sole* criterion for a concerted mechanism is the stability of the intermediate.

If there is little coupling between the motions of bond making and bond-breaking because of unfavourable geometry and orbital overlap, a concerted mechanism can only occur when it is enforced, i.e. when there is no barrier for decomposition of the hypothetical intermediate. If there is an energetic advantage from coupling two steps of a reaction mechanism into one, the mechanism may become concerted. This is unlikely to occur if the energy barriers to bond making or bond-breaking steps are large. The barriers for proton transfer between electronegative atoms are usually smaller than those for carbon, so the advantage for coupling is more likely to arise in the former type of reaction.

### 11.2.5.5   Push–pull and bifunctional acid–base catalysis

In the 1950s, the *push–pull* mechanism whereby a proton is donated by one centre and received at another was considered as a major contributor to enzyme catalysis and to give rise to third-order terms in non-enzymic reactions. Third-order terms were observed in organic solvents but in water they are of minor importance because of the solvating power of this solvent for ions and polar centres.

In enzymes, the active site may possess acid and base groups intimately associated with the conjugate base and acid functions, respectively, of the complexed substrate; the push–pull mechanism is possible but might not be a driving force. The halogenation of acetone in the presence of aqueous solutions of carboxylic acid buffers exhibits the rate law of Equation 11.2 where the third-order term, although small, has been shown to be significant and due to bifunctional concerted acid–base catalysis (Scheme 11.13):

$$k_{\text{obs}} = k_{\text{solvent}} + k_{\text{HA}}[\text{HA}] + k_{\text{B}}[\text{B}] + k_{\text{HA.B}}[\text{HA}][\text{B}]. \qquad (11.2)$$

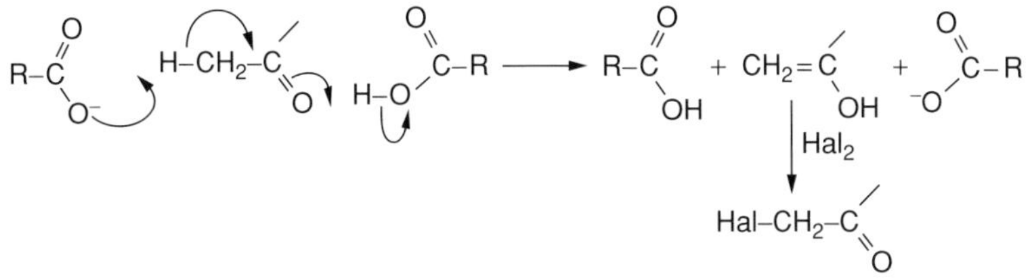

**Scheme 11.13**   Bifunctional acid–base catalysis in the halogenation of acetone via its enol.

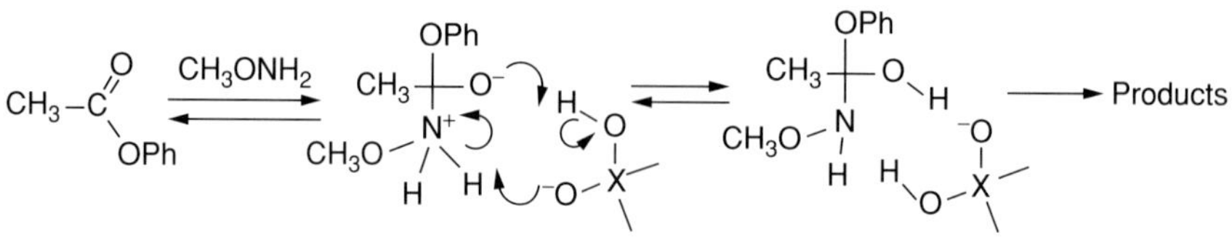

**Scheme 11.14** Bifunctional acid–base catalysis of the proton switch which traps the zwitterionic intermediate in methoxyaminolysis of phenyl acetate.

Push–pull acid–base catalysis has been proposed to account for the proton switch mechanism which occurs in the methoxyaminolysis of phenyl acetate (Scheme 11.14) where a bifunctional catalyst traps the zwitterionic intermediate. A requirement of efficient bifunctional catalysis is that the reaction should proceed through an unstable intermediate which has $pK_a$ values permitting conversion to the stable intermediate or product by two proton transfers after encounter with the bifunctional catalyst; the proton transfer with monofunctional catalysts should also be weak.

## 11.3 Nucleophilic and electrophilic catalysis

Nucleophilic and electrophilic catalysis occur when a nucleophile or electrophile reacts with the substrate to form an adduct which provides a more favourable alternative mechanism to that of the uncatalysed reaction. The intermediate can be formed as a transient species present in only a small concentration compared with the reactant or product, or it can build up to a measurable concentration. In this section, we exemplify the techniques used in their investigation using nucleophilic reactions. The same techniques can be used for reactions undergoing electrophilic catalysis, mutatis mutandis.

### 11.3.1 *Detection of intermediates*

The detection of intermediates depends upon the relative values of the rate constants for their formation and decay (see Chapter 4). An example is the imidazole-catalysed hydrolysis of aryl acetates where the concentration of acetylimidazole builds up and decays subsequently by hydrolysis. A stepwise process is manifestly obvious if, during a kinetic study, an intermediate species is observed to accumulate and then decay to give products (Fig. 11.6A)[12, 13]. The nature of the measuring device is not relevant to the argument but is likely to be spectroscopic (see Chapters 2 and 9). The direct observation of an intermediate depends on a build-up of its concentration to a measurable level and this requires that the decay to the product is relatively slow. The simplest possible case of a stepwise process is shown in Scheme 11.15 and this also happens to be one of the most generally applicable mechanisms (see Chapter 4).

The build-up of appreciable concentrations of the intermediate B depends on the initial reversible step being favourable; the rate of decay of the intermediate must be commensurate with its rate of formation, and the decay must also be slow enough to permit observation

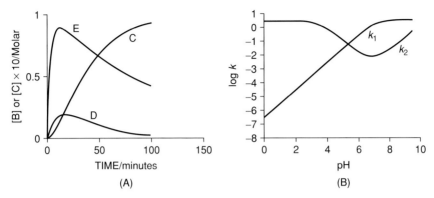

**Fig. 11.6**   (A) Spectroscopic detection of acetylimidazole in the imidazole-catalysed hydrolysis of 4-nitrophenyl acetate at pH 5 (D) and at pH 6 (E); curves are calculated from data in reference [12] and the curve for the acetate ion product (C) is for pH 5. (B) Reaction of imidazole with 4-nitrophenyl acetate ($k_1$) and the hydrolysis of acetyl imidazole ($k_2$); curves constructed from data in references [13] and [12].

[14]. The concentration of B in Scheme 11.15 is given by Equation 11.9:

$$[B] = A_o k_1 [\exp(-\lambda_1 t) - \exp(-\lambda_2 t)]/(\lambda_2 - \lambda_1) \qquad (11.9)$$

where

$$\lambda_1 = 0.5(k_1 + k_{-1} + k_2) - 0.5\{(k_1 + k_{-1} + k_2)^2 - 4k_1 k_2\}^{0.5}$$

and

$$\lambda_2 = 0.5(k_1 + k_{-1} + k_2) + 0.5\{(k_1 + k_{-1} + k_2)^2 - 4k_1 k_2\}^{0.5} \text{ (reference [15])},$$

and should increase to a value which can be measured by standard instrumental techniques (see Chapter 3).

Figure 11.6A illustrates the effect of a change in the ratio of $k_1/k_2$ for the imidazole-catalysed hydrolysis of 4-nitrophenyl acetate. When the pH is changed from 5 to 6, the maximal concentration of the intermediate increases. The pH profile in Fig. 11.6B provides a graphic illustration of the range of pH within which an intermediate can be detected; this is between the cross-over pH values of about 5 and 10 for these conditions. The range illustrated is for 0.1 M imidazole and could be extended because $k_1$ is dependent on the concentration of imidazole whereas $k_2$ is not.

The maximal concentration of a putative intermediate in the simplest mechanism (Scheme 11.15) may be obtained from estimates of the rate constants using linear free energy relationships. This concentration of the intermediate could then be assessed by a suitable analytical method and, provided there is confidence in the estimated rate constants, the non-observation of an intermediate would be good evidence for excluding a stepwise process. As far as we are aware, this direct procedure has not been achieved but it is relevant to studies of concertedness [11].

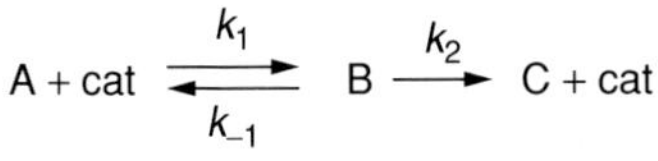

$$A + \text{cat} \underset{k_{-1}}{\overset{k_1}{\rightleftharpoons}} B \xrightarrow{k_2} C + \text{cat}$$

**Scheme 11.15**   Stepwise mechanism of nucleophilic catalysis.

### 11.3.2   Non-linear free energy relationships and transient intermediates

Intermediate B in Scheme 11.15 can be observed in principle but, if it undergoes rapid decay, it is usually expressed in concentrations very small compared with those of reactants and products, and it may be too dilute to be observed by instrumental techniques; it is then usually called a *transient* intermediate. Under these conditions, the Bodenstein steady-state hypothesis applies and the rate equation for Scheme 11.15 can be solved to give Equation 11.10 (see Chapter 4):

$$\text{rate} = k_{obs}[A][cat]$$
$$\text{where} \quad k_{obs} = k_1 k_2/(k_{-1} + k_2). \tag{11.10}$$

Equation 11.10 can be employed to demonstrate the presence of an intermediate if there is some factor such as substituent, concentration or other condition which can vary $k_{-1}$ and $k_2$ independently allowing the ratio $k_{-1}/k_2$ to vary from less than to greater than unity. Such behaviour is not usually predictable but, if a change in the rate-limiting step occurs, a plot of log $k_{obs}$ against (for example) a substituent parameter will be upwardly *convex* with the break at $k_{-1}/k_2 = 1$; this can be used to demonstrate the presence of an intermediate even when its concentration is below that for normal instrumental observation. An interesting example of this method related to acid–base catalysis is the hydroxide-catalysed elimination reaction of 4-nitrophenylethylpyridinium ions (Fig. 11.7A) which is interpreted to involve a carbanion intermediate (see Chapter 9 for the related base-induced elimination from *N*-(4-nitrophenylethyl)quinuclidinium ion)[16]. A free energy plot which is *concave* upwards is diagnostic of a mechanism with parallel pathways.

The individual steps in Scheme 11.15 should have linear Brønsted dependences where $\beta_n$ and $C_n$ are the general parameters corresponding to the equation for $k_n$. The overall

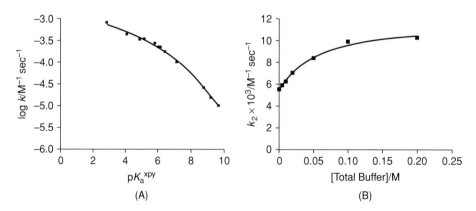

**Fig. 11.7** (A) E1cB mechanism for the elimination reactions of 4-nitrophenylethylpyridinium ions catalysed by OH⁻; deprotonation is rate limiting at p$K_a$'s below the breakpoint; data are from reference [16] and the line is calculated from Eqaution 11.11. (B) Effect of increasing 3-quinuclidinol buffer concentration on the rate constants for 2-methyl-3-thiosemicarbazone formation from 4-chlorobenzaldehyde at pH 11.10; results are consistent with a change in the rate-limiting step from proton transfer to formation of the intermediate; data from reference [8].

Brønsted dependence is expressed in Equation 11.11:

$$k_{obs} = (10^{\{\beta_1 pK_a + C_1\}})/(10^{\{(\beta_{-1} - \beta_2)pK_a + C_{-1} - C_2\}} + 1).$$     (11.11)

Equation 11.11 fits the data of Fig. 11.7A and the value of $(\beta_{-1} - \beta_2) = \Delta\beta$ refers to the difference in the slopes of the two linear portions of the correlation. A similar equation may be written for a Hammett correlation or any other free energy relationship including those for concentration changes (Fig. 11.7B) which might affect $k_2$ but not $k_{-1}$.

The absence of a convex break in a free energy relationship is not, in general, sound evidence for the absence of an intermediate, i.e. that the reaction has a single transition state and is concerted. This is because the results could refer to a linear portion of a free energy relationship with two linear parts with different slopes. If the position of the break could be predicted unequivocally, then the absence of a breakpoint would be evidence for the concerted path provided the results spanned the putative breakpoint. Such is the case in a substitution reaction when the leaving groups have the same general structure as the entering groups. An example is an acyl group transfer reaction where substituted phenolate ions displace 4-nitrophenolate ion from 4-nitrophenyl acetate. The breakpoint for the mechanism with a putative tetrahedral intermediate occurs when the rate constant for its formation is equal to that for its decomposition into products; this will be when the $pK_a$ of the conjugate acid of the entering phenolate ion equals that of 4-nitrophenol. The method, illustrated in Fig. 11.8, is called the *quasi-symmetrical technique* and is general for studying any displacement process.

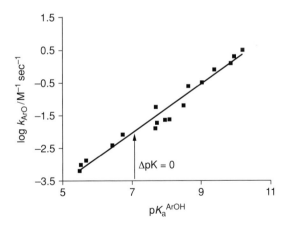

**Fig. 11.8**   The quasi-symmetrical reaction of phenolate ions with 4-nitrophenyl acetate.

## 11.4   Enzyme catalysis

Increases in rates of reaction effected by enzymes can be huge, but quantification of the rate enhancement, i.e. numerical comparison with the uncatalysed reaction rate, is seldom straightforward. Furthermore, the mechanism of catalysis of a given reaction by an enzyme may be very different from its catalysis by a non-enzyme. Of course, the main thrust of investigations of mechanisms of enzymic catalysis continues to be in biological and medicinal chemistry. However, the increasing use of enzymes and enzyme-like compounds in small molecule synthesis, for example, indicates the desirability of mechanistic investigations in such areas.

### 11.4.1   Technical applications

Enzymes provide catalytic pathways which are often superior to non-enzymic routes. The advantage of enzymes can overcome their disadvantages as relatively unstable species which work best in aqueous solution. In synthetic processes, the selectivity of the enzyme-catalysed process is not approached by chemical catalysis particularly regarding the enantioselective synthesis of chiral molecules. Furthermore, protective group chemistry is often not required in enzyme-catalysed reactions.

Enzyme catalysis can be employed to detect concentrations of metabolites such as blood sugar, urea, uric acid or ATP. The metabolite is often best reacted with a co-enzyme such as NADPH or pyridoxal phosphate which changes its UV–vis absorption spectrum upon reaction.

Enzymes can also be targets for drug design as they are often involved in significant metabolic pathways giving rise to a diseased condition. For example, familial hypercholesterolaemia is a level of cholesterol in the blood higher than normal. This level can be controlled by inhibiting the rate-limiting step in the formation of cholesterol, namely the reduction step catalysed by hydroxymethylglutaryl CoA reductase [17].

Inhibition can be reversible when it simply complexes at the active site preventing further catalysis. The active enzyme under these conditions can be recovered by dialysis. Another form of inhibition is the irreversible type where the active enzyme cannot be recovered by dialysis. A variant of this type of inhibition is *suicide* inhibition; a substrate of the enzyme reacts at the active site to yield an irreversible inhibitor which then reacts directly with groups at the active site [18]. A technique, 'in situ click chemistry', is related to that of suicide inhibition and involves 'click chemistry' components which complex at the active site of an enzyme and combine to form 'femtomolar' inhibitors. The technique can be used to synthesise inhibitors or by selection from a library of 'click chemistry' components to search structure space of the inhibitor for the drug target [19].

The technical applications referred to above can be studied using standard techniques to be described below; these differ in kind from those used in the study of catalysis by small molecules.

### 11.4.2   Enzyme assay

A fundamental problem associated with studies of enzyme mechanism is that even the purest samples of these catalysts contain relatively large amounts of impurities such as inactive protein and water. Moreover, samples are often solutions or suspensions of the enzyme in aqueous media and the basic analytical technique of accurate weighing is thus not appropriate to produce a standard solution of enzyme which meets the criteria normally demanded in physico-chemical studies of mechanism.

The assay of an enzyme solution, historically tied to a rate assay, depends on the correlation between the rate and a concentration (determined by some means, usually spectroscopic) of the purest possible sample of that enzyme. Since with even the most thoroughly characterised enzymes, such as $\alpha$-chymotrypsin, a substantial proportion of the matter is inactive enzyme and water (up to about 30% in some cases), the concentration of active enzyme from rate assays is liable to a wide margin of error.

## 11.4.3 Steady-state kinetics

Before we can discuss the measurement of active-site concentration, we need to consider the kinetics of the substrate reaction. The majority of kinetic studies of enzymes are carried out on systems described by Scheme 11.16 where all terms have their usual meanings and where the intermediates have come to a steady-state concentration; otherwise, studies of the kinetics of the pre-steady-state conditions usually require the use of specialist, fast reaction, equipment. The Michaelis–Menten equation, Equation 11.12, where all terms again have their usual meanings, can be derived from Scheme 11.16 when the system has reached a steady state; at this point the values of [ES] and [P] are still very much less than that of [S]:

$$\text{rate} = k_{cat}[E]_o[S]/(K_m + [S])$$
$$= V_{max}[S]/(K_m + [S]). \qquad (11.12)$$

The kinetics of most enzyme catalyses that have been investigated follow this rate law. The rates that are followed are normally over the first few percentages of the total reaction of the substrate. It is important that only initial rates be studied because enzymes tend to degrade readily, and the kinetics can also be modified by the accumulation of products (product inhibition). Michaelis–Menten kinetics are manifest in a plot of initial rate versus substrate concentration which is curved and comes to a maximum ($V_{max}$) at high substrate concentration (Fig. 11.9). The $V_{max}$ and $K_m$ terms can be derived from the initial rate data by fitting to Equation 11.12 using standard curve fitting software. Note that we are using enzyme concentration in these equations. Since a discrete volume of the enzyme solution is pipetted into the reaction, the relative concentrations can be determined quite accurately and the absence of an absolute concentration does not affect the validity of the Michaelis–Menten rate law. In order to obtain $k_{cat}$ accurately from $V_{max}$, it is necessary to possess an accurate value of $[E]_o$ which is the subject of the following discussion.

$$E + S \underset{}{\overset{K_m}{\rightleftharpoons}} ES \overset{k_{cat}}{\longrightarrow} E + P$$

**Scheme 11.16** Fundamental mechanism for the description of enzyme action.

## 11.4.3.1 Active-site titration

Titration of the intact active site obviates problems due to inactive protein which contribute to a false molarity. Active-site titrations of acyl group transfer enzymes such as $\alpha$-chymotrypsin utilise a substrate which has a good leaving group. This enables the build-up of an acyl enzyme intermediate which forms faster than it can degrade and results in

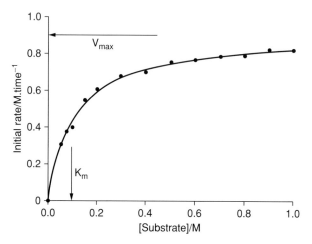

**Fig. 11.9**  Michaelis–Menten kinetics.

a so-called *initial burst* of leaving group prior to its steady-state release which is equivalent to the concentration of active sites in the solution. Let us consider the mechanism of α-chymotrypsin (Scheme 11.17) where the ester substrate (AcLg) complexes with the enzyme to give ES which decomposes to acyl enzyme (EAc) and leaving group; free enzyme is released on hydrolysis of the acyl enzyme (EAc).

A kinetic description of the reaction following Scheme 11.17 has been derived [20] and the release of LgH (or $Lg^-$) is given by Equation 11.13:

$$[Lg^-] = At + B(1 - e^{-bt}).  \tag{11.13}$$

In Equation 11.13, $A = \{k_2k_3[E]_o[S]_o/(k_2 + k_3)\}/\{[S]_o + k_3K_S/(k_2 + k_3)\}$, which has the form of a Michaelis–Menten equation, $B = [E]_o[S]_o\{k_2/(k_2 + k_3)\}^2/([S]_o + K_m(\text{apparent}))$, and $b$ is a composite rate constant describing the build-up of the acyl enzme intermediate (or, in the general case, the covalently bound enzyme intermediate). The non-linear plot of $[Lg^-]$ against time is shown in Fig. 11.10A for a typical substrate of α-chymotrypsin; extrapolation of the linear portion gives the intercept shown which allows evaluation of $B$.

When $[Lg^-]$ can be measured directly, it is clearly ideal if $k_3$ is relatively small and less than $k_2$; for such reactions, measurements can be made before the overall reaction is complete. Also, it is necessary to be able to measure $k_2$ in order that the equation for $B$ can be analysed. An ideal case would be a reaction where $k_3 = 0$; here, the substrate would essentially be an irreversible inhibitor and the value of $B$ becomes $[E]_o$.

If it is not possible to measure $[Lg^-]$ explicitly by analytical measurements, an irreversible inhibitor can be employed in conjunction with a rate assay using a substrate. Increasing

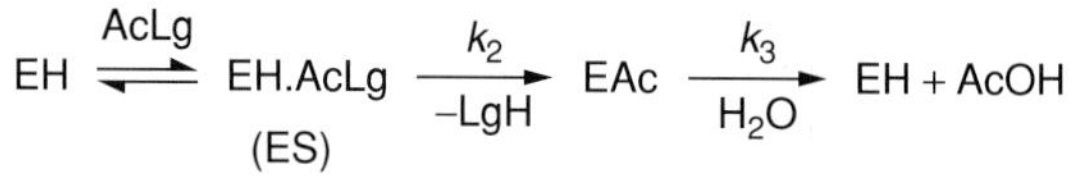

$$\text{EH} \;\underset{}{\overset{\text{AcLg}}{\rightleftharpoons}}\; \underset{\text{(ES)}}{\text{EH.AcLg}} \;\xrightarrow[-\text{LgH}]{k_2}\; \text{EAc} \;\xrightarrow[\text{H}_2\text{O}]{k_3}\; \text{EH} + \text{AcOH}$$

**Scheme 11.17**  The mechanism of catalysis by α-chymotrypsin.

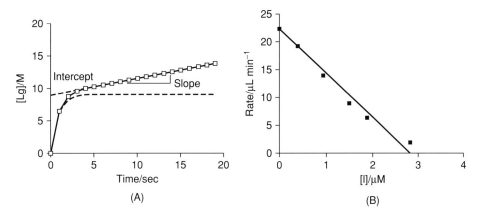

**Fig. 11.10** (A) Burst kinetics for release of the leaving group from a substrate of $\alpha$-chymotrypsin. (B) Titration of papain with the irreversible inhibitor 4-toluenesulphonamidomethyl chloromethyl ketone using methyl benzoylglycinate as substrate.

amounts of inhibitor at known concentrations are added to a given constant amount of enzyme assayed by a substrate. A plot of catalytic rate versus inhibitor concentration is linear and the intercept is at the inhibitor concentration equivalent to that of the enzyme active sites in the solution (Fig. 11.10B).

### 11.4.3.2   Active-site directed irreversible inhibitors

Associated with the problem of active-site titration is the question of the location of the active site in the three-dimensional structure of the protein. As a prelude to this investigation, a study is needed to indicate which amino acid residues in the overall peptide sequence are in the active site. The active site is defined as the location of the enzyme catalysis; thus, the substrate complexes at the active site prior to the catalytic process. Addition of a substrate will, therefore, protect the enzyme against reagents, such as inhibitors, which react at the active site. Of course, the active site may include amino acid residues from distant parts of the peptide chain; for example, both serine-195 and histidine-57 are in the active site of $\alpha$-chymotrypsin.

An experiment with an irreversible inhibitor should carry with it a control experiment involving the addition of a substrate; if the location of the reaction with inhibitor is at the active site, then the addition of a substrate will slow down the rate of inhibition. For example, the reactivity of papain (5 µM) with a 1.71 µM solution of 4-toluenesulphonylamidomethyl chloromethyl ketone suffers a drop of 1.68-fold when the substrate (methyl hippurate) is changed from 12.7 to 21.1 mM. The inhibitor which reacts covalently with the enzyme should carry either a radioactive or spectroscopic 'tag' which would enable the location of the altered amino acid to be determined in the sequence, and hence in the three-dimensional X-ray crystallographic map of the enzyme. An alternative approach is to design an inhibitor with groups (analogous to those attached to the substrate) which force it to bind at the active site (Scheme 11.18).

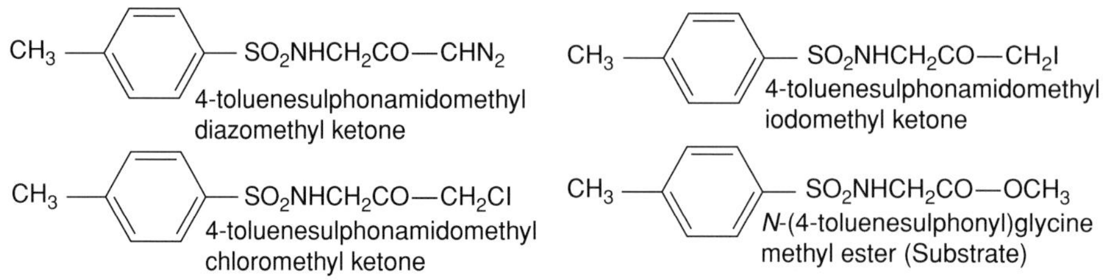

CH₃ —⟨benzene⟩— SO₂NHCH₂CO—CHN₂
4-toluenesulphonamidomethyl
diazomethyl ketone

CH₃ —⟨benzene⟩— SO₂NHCH₂CO—CH₂I
4-toluenesulphonamidomethyl
iodomethyl ketone

CH₃ —⟨benzene⟩— SO₂NHCH₂CO—CH₂Cl
4-toluenesulphonamidomethyl
chloromethyl ketone

CH₃ —⟨benzene⟩— SO₂NHCH₂CO—OCH₃
*N*-(4-toluenesulphonyl)glycine
methyl ester (Substrate)

**Scheme 11.18**   Active-site directed inhibitors analogous to the papain substrate, *N*-(4-toluenesulphonyl) glycine methyl ester.

Active-site directed inhibitors have reactivity with the enzyme greatly enhanced over that of non-specific inhibitors; thus phenacyl iodide inhibits papain 50-fold faster than iodoacetamide whereas the active-site directed inhibitor 4-toluenesulphonylamidomethyl chloromethyl ketone reacts some 650-fold faster. The enhanced rate is due to complexation of the inhibitor with the enzyme, and indicates that the inhibitor must be reacting at the active site.

### 11.4.4   Kinetic analysis

Historically, enzyme kinetics were visualised using the Lineweaver–Burk equation, Equation 11.14, where 1/rate is plotted against 1/[S] as seen in Fig. 11.11A:

$$1/\text{rate} = 1/V_{\max} + (K_m/V_{\max})/[S] \tag{11.14}$$

$$\text{rate} = V_{\max} - K_m \, \text{rate}/[S]. \tag{11.15}$$

The Michaelis–Menten equation can also be rearranged to give Equation 11.15, the Eadie–Hofstee equation, and the rate data plotted according to this, as shown in Fig. 11.11B, give a better graphical estimation of the Michaelis–Menten parameters than does the

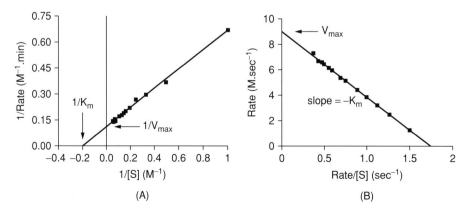

**Fig. 11.11**   (A) Lineweaver–Burk graphical analysis; (B) Eadie–Hofstee graphical analysis.

Lineweaver–Burk plot. None of the graphical procedures, however, yield reliable kinetic parameters, but are useful to obtain a qualitative feel for the kinetics of the system.

### 11.4.5 Reversible inhibitors

We have already dealt with the subject of *irreversible* inhibitors under enzyme titration and location of the active site (Section 11.4.3.2). The phenomenon of *reversible* inhibition involves simple complexation of the inhibitor with the enzyme at a site which modifies the reactivity of the enzyme catalysis.

$$\text{EI} \xrightleftharpoons{\text{I, } K_i} \text{E} + \text{S} \xrightleftharpoons{K_m} \text{ES} \xrightarrow{k_{cat}} \text{E} + \text{P}$$

**Scheme 11.19** Competitive inhibition.

Enzyme catalysis in the presence of competitive reversible inhibitors obeys the rate law in Equation 11.16, which can be derived from a mechanism (Scheme 11.19) where the inhibitor complexes reversibly at the active site, with equilibrium constant $K_i$, thus excluding the substrate:

$$\text{rate} = [\text{E}]_o [\text{S}] k_{cat} / \{ [\text{S}] + K_m (1 + [\text{I}]/K_i) \}. \tag{11.16}$$

Equation 11.16 is analogous to the Michaelis–Menten equation and changing [I] affects the apparent $K_m$ parameter but has no effect on $V_{max}$. The competitive inhibitor effect is observed graphically in families of Eadie–Hofstee plots where the lines intersect at $V_{max}$ on the $y$-axis, Fig. 11.12.

Non-competitive, uncompetitive and mixed inhibitions occur when both the inhibitor and the substrate bind simultaneously to the enzyme and do not compete for the same binding site as in competitive inhibition (Scheme 11.20). Non-competitive inhibition occurs when $K_m = K'_m$ (i.e. the dissociation constant of S from EIS is the same as that from ES) and

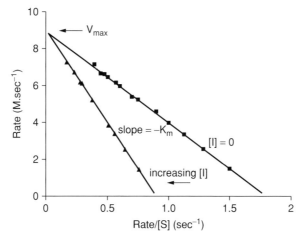

**Fig. 11.12** Eadie–Hofstee plots for competitive inhibition.

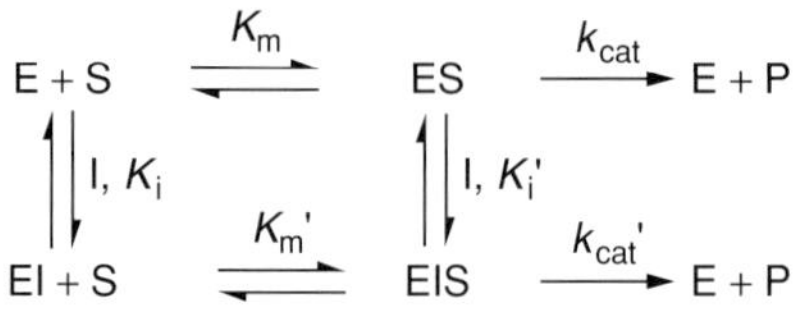

$$E + S \underset{K_m}{\rightleftharpoons} ES \xrightarrow{k_{cat}} E + P$$

$$\Big\updownarrow I, K_i \qquad \Big\updownarrow I, K_i'$$

$$EI + S \underset{K_m'}{\rightleftharpoons} EIS \xrightarrow{k_{cat}'} E + P$$

**Scheme 11.20**   General mechanism for reversible inhibitors.

where **EIS** does not react ($k_{cat}' = 0$). In this case, the rate law, Equation 11.17, is analogous to a Michaelis–Menten rate law:

$$\text{rate} = [E]_o[S]k_{cat}/\{(1 + [I]/K_i)([S] + K_m)\}. \tag{11.17}$$

Non-competitive inhibition is indicated by an Eadie–Hofstee plot with parallel lines (Fig. 11.13A).

Uncompetitive inhibition occurs when the inhibitor binds with ES but not with E ($1/K_i = 0$). The manifestation of this type of inhibition is an Eadie–Hofstee graph with lines intersecting on the rate/[S] axis (Fig. 11.13B). The general case where the parameters of Scheme 11.20 are finite and $K_m \neq K_m'$ and $k_{cat} \neq k_{cat}'$ is called *mixed inhibition*, and the lines of the Eadie–Hofstee plots do not intersect on either axis.

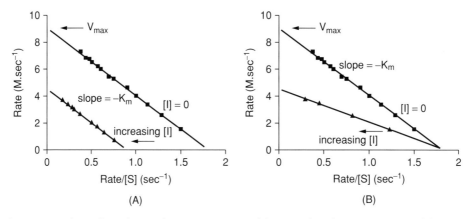

**Fig. 11.13**   Eadie–Hofstee plots (A) for noncompetitive inhibition and (B) for uncompetitive inhibition.

### 11.4.6    *Detection of covalently bound intermediates*

As well as complexing the substrate to the active site, many enzymes link covalently with the substrate, or a portion of it, to form an additional intermediate. Such intermediates occur in the action of enzymes as diverse as alkaline phosphatase (phosphoryl enzyme), serine and cysteine proteases (acyl enzymes), glycosidases (acylal enzymes) and aldolases.

#### 11.4.6.1    *Direct observation*

By virtue of being involved in a catalytic mechanism, a covalently bound intermediate is only transient. In such cases, it has often proved possible to change conditions so that decay of the intermediate is slow leading to an accumulation of intermediate which can then be detected by instrumental techniques.

The observation of burst kinetics is prima facie evidence for an intermediate; in the case of proteases, the substrate is modified so that it possesses a good leaving group rather than an amide, and a medium slightly more acidic than usual makes the decay of the acyl enzyme sufficiently slow for it to be observed. The liberation of the leaving group is monitored, and the burst kinetics are diagnostic of an intermediate lacking the leaving group.

Detection of the intermediate is possible if it has a spectrum sufficiently different from that of the enzyme. The cinnamoyl chymotrypsin intermediate is characterised by a UV maximum at 292 nm; the acyl papain intermediate *N*-benzoylaminothionacetyl papain has a UV maximum at 313 nm. The UV absorptions of the reactions catalysed by papain and chymotrypsin wax and wane in the presence of substrate giving rise to these intermediates.

### 11.4.6.2   *Structural variation*

Acyl-$\alpha$-chymotrypsins may be isolated if the acyl function can be made sufficiently unreactive by structural variation; several have been recrystallised at low pH, conditions which stabilise the acyl enzyme. The stabilised acyl enzymes can be characterised by physical methods and the first example of an X-ray crystallographic study of an enzyme intermediate is that of indolylacryloyl-$\alpha$-chymotrypsin [21].

The lysozyme mechanism (Scheme 11.21) has an *acylal* intermediate with a decay step which is fast compared with its rate of formation, so it is not normally possible to observe the intermediate. The acylal mechanism has recently been established by modifying both enzyme and substrate so that the intermediate can be isolated; in this case, the intermediate was stabilised so much that its X-ray crystallographic structure could be determined. Substrate A in Scheme 11.21 was employed [22] in which the 2-fluoro substituent reduces the reactivity at the C1 position so that any acylal enzyme formed from this substrate would have a

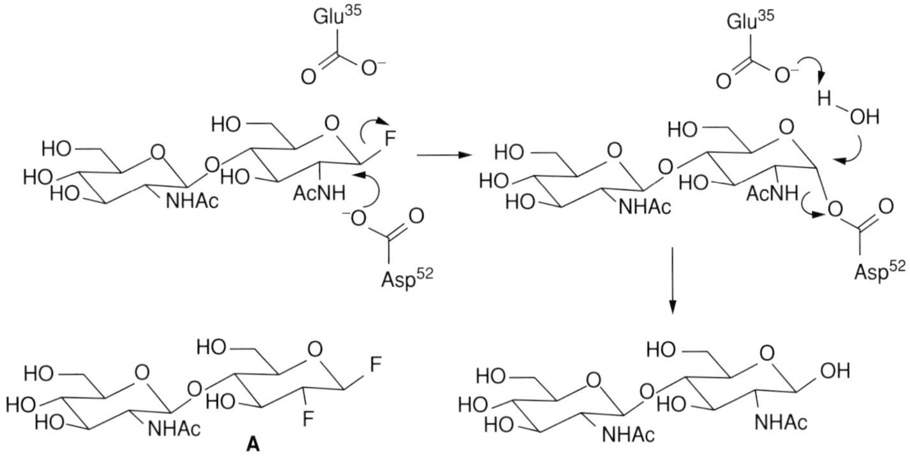

**Scheme 11.21**   Lysozyme mechanism for hydrolysis of the substrate $\beta$-cellobiosyl fluoride; A is a modified reactive substrate which yields a weakly reactive acylal intermediate.

good chance of accumulating and hence being observable. (In contrast, the fluorine at C1 ensures rapid glycylation of the enzyme.) In the case of lysozyme, the acylal intermediate was shown to be present by use of electrospray ionisation mass spectrometry. The intermediate hydrolyses slowly and, in order to provide structural information, it could be stabilised further by altering the sequence of the lysozyme by protein engineering to prevent loss of the Asp$^{52}$ nucleofuge (note that the intermediate is *not* an *acyl* enzyme). This can be done by changing glutamic acid-35 to glutamine-35 which cannot act as a base to assist nucleophilic attack by water. The X-ray crystallographic coordinates are at *pdb1h6m.ent* in the Brookhaven Protein Data Bank and clearly show an acylal structure bonded to the enzyme via the amino acid residue of aspartic acid-52.

### 11.4.6.3 Stereochemistry

Studies of the stereochemical course of the transglycylation reactions catalysed by lysozyme show retention of configuration at C1 in the product. This can be attributed most economically to the intervention of an acylal intermediate in the mechanism (see Scheme 11.21). A similar result is found in the alkaline phosphatase-catalysed solvolysis of isotopomeric phospho-esters which indicates a double inversion via a phospho-enzyme intermediate (Scheme 11.22).

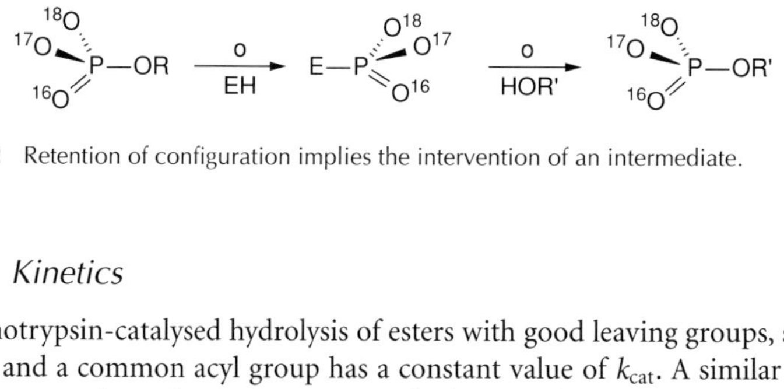

**Scheme 11.22** Retention of configuration implies the intervention of an intermediate.

### 11.4.6.4 Kinetics

The $\alpha$-chymotrypsin-catalysed hydrolysis of esters with good leaving groups, such as phenolate ions, and a common acyl group has a constant value of $k_{cat}$. A similar result holds for ester substrates of papain. Acyl groups attached to poor leaving groups such as peptides or weakly reactive esters have values of $k_{cat}$ lower than the constant values for active esters. These results are interpreted by the three-step mechanism involving an acyl enzyme intermediate (Scheme 11.23) where $k_{cat}$ is given by $k_2 k_3/(k_2 + k_3)$. In the case of reactive esters, $k_2 > k_3$ and thus $k_{cat} = k_3$, the function $k_3\{k_2/(k_3 + k_2)\}$ becomes increasingly less than $k_3$ as the value of $k_2$ becomes smaller.

These results are common with proteases and also for the hydrolysis of phosphate esters catalysed by alkaline phosphatase; in the latter case, there is a pitfall in that the constant value for $k_{cat}$ for phosphate esters is due to a rate-limiting conformational change in

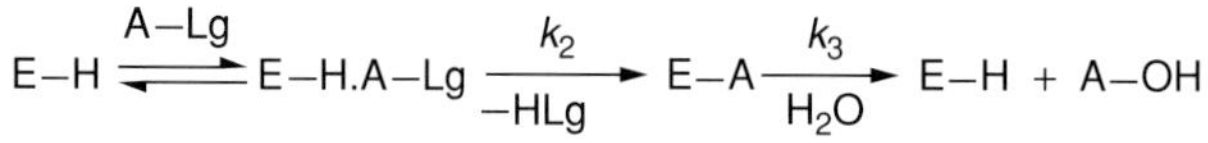

$$E-H \rightleftharpoons E-H.A-Lg \xrightarrow[-HLg]{k_2} E-A \xrightarrow[H_2O]{k_3} E-H + A-OH$$

**Scheme 11.23** Three-step mechanism for proteases (A = RCO−), phosphatases (A = −PO$_3^{2-}$) and glycosidases (A = glycosyl group).

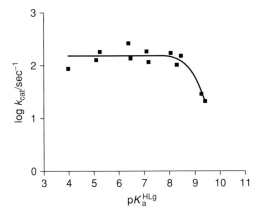

**Fig. 11.14** The value of $k_{cat}$ for substrates of $\beta$-glucosidase possessing a common glycosyl group and different leaving groups at C-1.

the enzyme substrate complex and not to a rate-limiting hydrolysis of phospho-enzyme. Figure 11.14 illustrates the effect of the leaving group ability on $k_{cat}$ for the catalytic effect of $\beta$-glucosidase.

### 11.4.6.5 Trapping

Trapping is a standard technique for demonstrating transient intermediates (see Chapter 9) and has been applied to studies of enzyme mechanisms. When the mechanism involves a transient intermediate which reacts with a nucleophilic trapping reagent as well as the normal reagent (water in hydrolysis reactions), then the product is diverted. If the amount of diverted product depends on the concentration of the trapping agent, whereas the overall rate of substrate consumption is constant, then the rate-limiting step must precede the product-forming step; this can only be consistent with a transient intermediate (Scheme 11.24).

An alternative type of trapping can be applied when group A in Scheme 11.24 is constant and the leaving group varies from active to relatively inactive. The trapping agent reacts with the common function (A) in the enzyme intermediate, and the rate will be independent of the nature of the leaving group. Thus, for a given concentration of trapping agent, the ratio of the hydrolysis product to the trapped product will be constant and independent of the leaving group structure. The classic example of this technique is $\alpha$-chymotrypsin reacting with benzoylglycine esters in the presence of a constant concentration of hydroxylamine as the trapping agent. The ratio of the hydroxamic acid product to the benzoylglycine acid product is independent of the nature of the leaving group, consistent with the mechanism involving a benzoylglycyl-$\alpha$-chymotrypsin intermediate (Scheme 11.24).

$$E{-}H \underset{}{\overset{A{-}Lg}{\rightleftharpoons}} E{-}H.A{-}Lg \overset{k_2}{\underset{-HLg}{\longrightarrow}} E{-}A \begin{array}{l} \overset{k_3}{\nearrow} A{-}OH + E{-}H \\ \phantom{xxx} H_2O \\ \underset{k_{trap}}{\searrow} \underset{A{-}Nu + E(H)}{Nu(H)} \end{array}$$

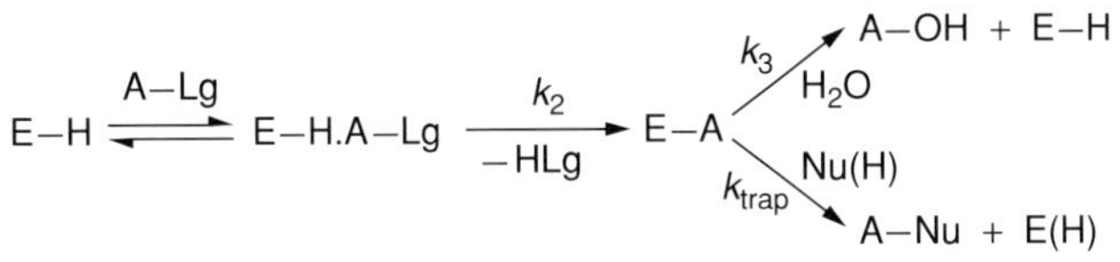

**Scheme 11.24** Trapping a transient acyl enzyme intermediate with a nucleophile, Nu(H).

## Bibliography

Bell, R.P. (1973) *The Proton in Chemistry* (2nd edn). Chapman & Hall, London.

Bender, M.L. (1971) *Mechanisms of Homogeneous Catalysis from Protons to Proteins*. Wiley-Interscience, New York.

Bernasconi, C.F. (Ed.) (1986) *Investigation of Rates and Mechanisms of Reactions* (4th edn). Wiley-Interscience, New York.

Bruice, T.C. and Benkovic, S.J. (1966) *Bioorganic Mechanisms* (vols 1 and 2). Benjamin, New York.

Espenson, J.H. (1995) *Chemical Kinetics and Reaction Mechanisms* (2nd edn). McGraw-Hill, New York.

Fersht, A.R. (1999) *Structure and Mechanism in Protein Science*. Freeman, New York.

Jencks, W.P. (1969) *Catalysis in Chemistry and Enzymology*. McGraw-Hill, New York.

Jencks, W.P. (1976) Enforced general acid-base catalysis of complex reactions in water. *Accounts of Chemical Research*, **9**, 425.

Maskill, H. (1985) *The Physical Basis of Organic Chemistry*. Oxford University Press, Oxford.

Page, M.I. and Williams, A. (1997) *Organic and Bio-Organic Mechanisms*. Addison-Wesley Longman, Harlow, Essex.

Price, N.C. and Stevens, L. (1999) *Fundamentals of Enzymology* (3rd edn). Oxford University Press, Oxford.

Roberts, S.M., Turner, N.J., Willetts, A.J. and Turner, M.K. (1995) *Introduction to Biocatalysis Using Enzymes and Micro-Organisms*. Cambridge University Press, Cambridge.

Williams, A. (2000) *Concerted Organic and Bio-Organic Mechanisms*. CRC Press, Boca Raton, FL.

Williams, A. (2003) In: I.T. Horvath (Ed.) *Encyclopedia of Catalysis* (vol 1). Wiley-Interscience, NJ, pp. 1–39.

Williams, A. (2003) *Free Energy Relationships in Organic and Bio-Organic Chemistry*. Royal Society of Chemistry, Cambridge.

## References

1. Garrett, E.R. (1957) *Journal of the American Chemical Society*, **79**, 3401.
2. Hand, E.S. and Jencks, W.P. (1975) *Journal of the American Chemical Society*, **97**, 6221.
3. Hegarty, A.F. and Jencks, W.P. (1975) *Journal of the American Chemical Society*, **97**, 7188.
4. Bell, R.P. (1941) *Acid-Base Catalysis*. Oxford University Press, Oxford, p. 94.
5. Kallen, R.G. and Jencks, W.P. (1966) *Journal of Biological Chemistry*, **241**, 5851.
6. Espenson, J.H. (1995) *Chemical Kinetics and Reaction Mechanisms* (2nd edn). McGraw-Hill, New York, p. 200.
7. Eigen, M. (1964) *Angewandte Chemie. (International Edition)*, **3**, 1.
8. Sayer, J.M. and Jencks, W.P. (1973) *Journal of the American Chemical Society*, **95**, 5637.
9. Williams, A. and Jencks, W.P. (1974) *Journal of the Chemical Society, Perkin Transactions* 2, 1760.
10. Jencks, W.P. (1976) *Accounts of Chemical Research*, **9**, 425.
11. Williams A. (2000) *Concerted Organic and Bio-Organic Mechanisms*. CRC Press, Boca Raton, FL.
12. Jencks, W.P. and Gilchrist, M. (1968) *Journal of the American Chemical Society*, **90**, 2620.
13. Jencks, W.P. and Carriuolo, J. (1959) *Journal of Biological Chemistry*, **234**, 1272.
14. Bernasconi, C.F., Killion, R.B., Fassberg, J. and Rappoport, Z. (1989) *Journal of the American Chemical Society*, **111**, 6862.
15. Espenson J.H. (1995) *Chemical Kinetics and Reaction Mechanisms* (2nd edn). McGraw-Hill, New York, p. 75.
16. Bunting, J.W. and Kanter, J.P. (1991) *Journal of the American Chemical Society*, **113**, 6950.
17. Endo, A. (1981) *Trends in Biochemical Sciences*, **6**, 10.

18. Silverman, A.B. and Abeles, R.H. (1976) *Biochemistry*, **15**, 4718; Abeles, R.H. and Maycock, A.L. (1976) *Accounts of Chemical Research*, **9**, 313.
19. Lewis, W.G., Green, L.G., Grynszpan, F., Radic, Z., Carlier, P.R., Taylor, P., Finn, M.G. and Sharpless, K.B. (2002) *Angewandte Chemie (International Edition)*, **41**, 1053.
20. Bender, M.L., Kézdy, F.J. and Wedler, F.C. (1967) *Journal of Chemical Education*, **44**, 84.
21. Henderson, R. (1970) *Journal of Molecular Biology*, **54**, 341.
22. Vocadlo, D.J., Davies, G.J., Laine, R., Withers, S.G. (2001) *Nature*, **412**, 835.

Chapter 12
# Catalysis by Organometallic Compounds

*Guy C. Lloyd-Jones*

## 12.1   Introduction

The use of organometallic complexes to facilitate or promote organic reactions has an extensive and rich history, but blossomed in the golden age of organometallic chemistry, the 1960s and 1970s. During this period, many serendipitous discoveries emerged as side reactions from systematic investigations into organometallic reactivity. Slowly but surely, the organic community began to realise that these 'exotic' organometallics offered unique opportunities. This interest in *application* fuelled the progressive transition away from stoichiometric use of organometallic complexes towards their use in catalytic quantities. The last three decades have witnessed enormous advances in organometallic catalysis: C—C and C—X bond forming (and cleaving) reactions under mild conditions, selective oxidations, metathesis, asymmetric catalysis and even applications on industrial scales. These remarkable transformations have empowered organic chemists, and the use of organometallic complexes and catalysts has become less and less the preserve of the pure organometallic chemist. As a corollary, many of the ensuing developments have been empirical and incremental – made by those whose interests lie in harnessing the power of organometallic catalysis for organic synthesis rather than in the organometallic complexes themselves. As a consequence, many mechanisms are 'working models' rather than 'established' and, in general, there is a paucity of thorough mechanistic work.

### 12.1.1   The challenges inherent in the investigation of organic reactions catalysed by organometallics

In any organic reaction that is genuinely catalytic in an organometallic compound, the sum of all the organometallic species present is substoichiometric. As a consequence, the low concentration of intermediates will always be an issue in the study of the mechanism. The problem is exacerbated by a number of features. For efficient 'catalytic turnover' (one turnover is one complete revolution through all stages of the catalytic cycle), all intermediates must be relatively unstable under the reaction conditions. This can make intermediates very hard to observe, let alone isolate. The catalytic cycle will often have one step that, for the bulk of the net macroscopic reaction, is 'turnover limiting' (the equivalent of the rate-limiting step in a linear reaction consisting of a series of elementary steps, see I, Fig. 12.1). The *active*

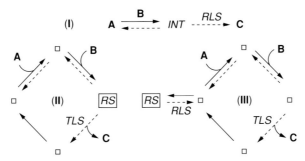

**Fig. 12.1**   Schematic representations of the reaction: A + B = C; **I** is a simple stoichiometric reaction, involving rate-limiting breakdown of an intermediate *INT*; **II** is a catalysed reaction in which there are four intermediates – the major concentration is the resting state (*RS*) located before turnover-limiting breakdown to give C; **III** is the same as **II** except that the resting state is now 'off cycle' and the release of this component into the catalytic cycle controls the global rate.

catalytic components will tend to accumulate at the reactant stage of the turnover-limiting step (the so-called resting state) – see the schematic representation of this in **II**, Fig. 12.1 – making other species even harder to locate.

Moreover, one or more species on the catalytic cycle are often in equilibrium with other complexes *off* the catalytic cycle. Such species may be more stable than any other, in which case the resting state is 'off-cycle' (see **III**, Fig. 12.1) and, thus, complexes observed may not be pertinent to the mechanism of turnover. Many organometallic complexes are rather sensitive towards air or moisture. When the complex is present in only catalytic quantities, seemingly low concentrations of impurities in substrates, solvent or head space (e.g. inert gas) can result in much of the organometallic component being irreversibly diverted to non-active species (termination). Moreover, the organometallic complex that is added to the reaction in the first place is very often a 'pro-catalyst' (an isolable, relatively stable complex that has to be 'activated' to enter the catalytic cycle under the reaction conditions). If the rate of activation is low, or the rate of termination competitive, then the concentration of all species on the cycle can be orders of magnitude lower than the concentration of the pro-catalyst added in the first place and, thus, the actual turnover rate is much higher than the net turnover rate based on the amount of pro-catalyst added. The above issues are compounded by the obvious but inescapable issue that the ideal catalyst system contains only a vanishingly small quantity of catalyst and that the turnover frequency is extremely high. As a consequence, the study of the organometallic catalysis falls broadly into two areas:

(i) the study of the macroscopic behaviour of the catalytic cycle, that is to say the 'input'/ 'output' (i.e. pro-catalyst(s), reactant(s), solvent(s)/products or side products) of the catalytic cycle is used *indirectly* to probe the mechanism, i.e. the cycle is treated as a 'back box'; and
(ii) the study of the microscopic behaviour of the catalytic cycle, that is the detection, isolation, diversion/trapping, independent generation or computation of intermediates, and the kinetics/thermodynamics associated with these species.

In most cases, empirical data arising from reaction development, as well as systematic and designed studies of the macroscopic behaviour (area (i) above), lead to the proposal of a 'working mechanism', based on analogy to related systems. This working mechanism, however crude, can be of significant utility in the optimisation of the reaction as well as in the development of new reactions. Studies in the second area are appreciably more demanding and are conducted much more rarely. Although in many cases, such studies are not feasible with the current analytical tools, e.g. due to the instability or low concentration of the intermediates, the catalytic cycle in some systems may be stalled at an observable intermediate by restricting access to a reactant. In such cases, significant insight may be obtained. In this chapter, examples of the study of both the macroscopic and the microscopic aspects of the mechanism of organometallic catalysis will be presented.

## 12.1.2   *Techniques used for the study of organometallic catalysis*

For the organometallic chemist, the process of treating mechanism as a hypothetical construct and then testing it is similar to, if not identical with, that used in mainstream physical organic chemistry, and the array of techniques available to do this are much the same. As outlined above, one is often limited to the study solely of the reagents going in and the organic products coming out (see Chapter 2). Nonetheless, under such conditions, standard techniques such as product and side-product identification, stereochemical studies, macroscopic kinetics and thermodynamics, effects of additives (e.g. as catalyst promoters or poisons), isotopic labelling, structure-reactivity relationships (e.g. Hammett and other free-energy relationships) as well as systematic variations in pro-catalyst, the solvent(s) and the pressure (particularly for reactions in which one or more of the reactants is a gas) can be used to great effect to investigate such systems.

In only relatively few cases, have *genuine* intermediates been observed, trapped or isolated in organic transformations catalysed by organometallic complexes. By and large, under such circumstances, nuclear magnetic resonance spectroscopy (NMR) and single crystal X-ray diffraction have proven to be the most valuable techniques, offering rich structural and dynamic data in the solid state and in solution (NMR only). Other techniques that have seen less widespread application include infrared spectroscopy (IR), X-ray absorption fluorescence spectroscopy (XAFS, giving information about the environment of the metal centre), electrochemical techniques (allowing the detailed and quantitative study of changes in oxidation states, see Chapter 6), electron spin spectroscopy (ESR, giving information about free radical species, see Chapter 10) and mass spectrometry (MS). It is notable that, with the recent developments in gentle ionisation techniques (lower energy and less destructive) particularly electrospray, which allows multiple ionisation, MS is seeing increasing application in the study of organometallic catalysis.

In addition to the analytical tools described above, computation has now become a significant component in the investigation of reaction mechanism (see Chapter 7). One major restriction is the complexity of the wave functions associated with the heavy nuclei which are typically encountered in reactions involving organometallic species. Nonetheless, density functional theory (DFT) using the B3LYP level of theory has made computational study of organometallics, or simplified models (e.g. with pared down ligand structures), a viable and important contributor to the debate. In the absence of the evolution of radically new analytical techniques, the ever-decreasing cost (in terms of both time and equipment) of computational techniques will result in escalating contributions from this area in the study of mechanism.

Despite the powerful analytical techniques that are available, one should always bear in mind that the hypothetical construct termed 'mechanism' cannot be proved, but only disproved; practical and computational investigations should be designed to *test* rather than *support* the current working model.

## 12.1.3   Choice of examples

The breadth of application of organometallic catalysis in organic synthesis continues to expand at an astonishing pace. Clearly, any discussion presented here cannot be widely representative, let alone comprehensive. A strong element of selection is necessary and, as such, a personal bias will influence the choice. Although this is inescapable, it is intended that the three examples presented in this chapter will exemplify the challenges, techniques and approaches that have been applied generally. Regarding the choice of examples, the formation of carbon–carbon bonds is arguably the most important fundamental process available to the synthetic organic chemist. Without the ability to concatenate, the generation of complex organic structures becomes limited to the modification of existing frameworks. The unique reactivity offered by organometallics has resulted in transition metal catalysis becoming the most important player in the arena of C—C bond forming reactions. Consequently, all three sections below concern examples of this type of process.

The potential to use single enantiomers of chiral ligands at a metal centre in the active species of a catalytic system to control stereoselectivity has resulted in significant effort being devoted to the development of *asymmetric* transition-metal-catalysed C—C bond forming reactions. Consequently, it will be no surprise that one of the example sections of this chapter concerns mechanistic studies related to this type of process – the Rh-catalysed addition of organoboron compounds to enones. In that section (Section 12.2), we see how classical NMR methods have allowed the identification of every major intermediate on a catalytic cycle. The second example (Section 12.3) involves a Pd-catalysed isomerisation reaction. Such *single-substrate* reactions present unique challenges to the mechanistic investigator. We shall see how a combination of isotopic labelling and kinetics allows insight through indirect methods. The final section (Section 12.4) examines the alkene metathesis reaction. We start with mechanistic studies from the very early days when alkene metathesis was more of a curiosity than a synthetic tool and see how the basic mechanism was established by statistical and kinetic analysis. We then move on some three decades to the point where developments in alkene metathesis offer a completely new dimension in the design of organic syntheses and see how advanced NMR techniques can be applied to determine the surprising origin of very different turnover rates arising from a pair of similar pro-catalysts.

Although the set of examples presented are unlikely to be *directly* related to the systems of immediate interest to the reader, it is hoped that the discussion will highlight the methodologies commonly applied, the various approaches that can be used to solve similar problems, as well as some of the difficulties encountered and the strategies used to surmount them. Above all, it is hoped that these case studies will provide both inspiration and guidance to readers in the design of their own experiments and in the interpretation of their results.

## 12.2    Use of a classical heteronuclear NMR method to study intermediates 'on cycle' directly: the Rh-catalysed asymmetric addition of organoboronic acids to enones

The ready preparation of enones ($\alpha,\beta$-unsaturated ketones or aldehydes) via Aldol processes has meant that the conjugate addition (or [1,4]-addition) of a nucleophile to enones has long been a favourite amongst synthetic organic chemists. The stoichiometric addition of organocuprates was the first major development in the use of transition metals to facilitate such processes. High selectivity for 1,4- over 1,2-addition made reagents such as 'Gilman cuprates' popular amongst the synthetic community and eventually led to the development of Cu catalysts for the stoichiometric addition of Grignard or organozinc reagents, via the cuprate.

In this section, we look at a related process, developed by Miyaura and Hayashi, that facilitates the addition of the much more readily stored and handled organoboron reagents, R-B(R')$_2$, to a broad range of pro-chiral enones with extremely high enantioselectivity. The advantage of organoboron reagents is that they allow a completely new level of functional group compatibility compared with Grignard or organozinc species. Consequently, a great variety of organoboronic acids and boronates have become commercially available. We shall see how the combination of $^{31}$P NMR and the three-component nature of the reaction (enone, organoboron reagent and water) has facilitated the detailed study of the mechanism by restricting access to reactants, thereby arresting the cycle. The investigation has led to the *informed* development of a much-improved pro-catalyst system, allowing reaction under milder conditions with higher selectivity. It has also led to a seminal development in chiral ligand architecture.

### 12.2.1    Background and introduction

The reaction of interest was first reported by Miyaura and co-workers in 1997 [1]. It was found that the addition of phenylboronic acid, a reagent pioneered for Pd-catalysed Suzuki–Miyaura cross-coupling, to simple $\alpha,\beta$-unsaturated ketones was catalysed by Rh(I) complexes.

The use of an aqueous solvent mixture was found to be essential for efficient catalytic turnover and the neutral conditions avoided complications arising from Aldol reactions. A series of phosphine ligands were screened and it was found that chelating diphosphines, such as 1,4-bisdiphenylphosphinobutane (DPPB), gave the best results. Using a pro-catalyst system comprising 3 mol% [Rh(acac)(CO)$_2$] plus 3 mol% DPPB in 86% aqueous MeOH, a broad range of enones and arylboronic acids were found to undergo smooth reactions. The only limitation was found to be with hindered boronic acids (e.g. with *ortho*-substituted aryl rings) where a competing process of protodeborylation (generation of Ar–H from Ar–B(OH)$_2$) reduced the yield. These results set the stage for development of an asymmetric process and, within a year, Hayashi and Miyaura had achieved this [2]. With cyclohexenone and PhB(OH)$_2$ as a benchmark test system, it was found that using the axially chiral ligand BINAP (2,2'-bisdiphenylphosphino-1,1'-binaphthalene) in place of DPPB resulted in very low reactivity (<2% conversion). However, changing from [Rh(acac)(CO)$_2$]

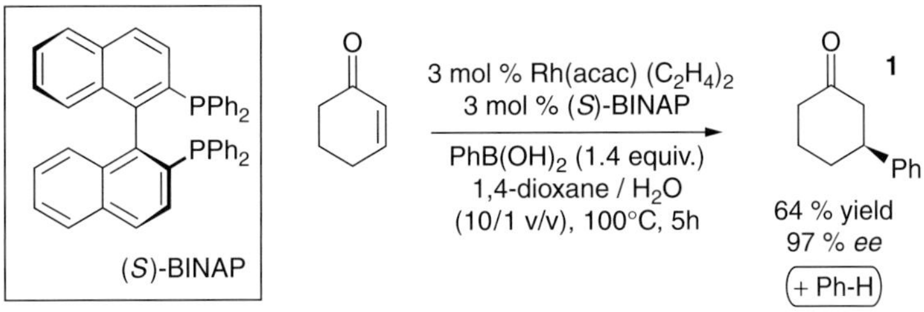

**Scheme 12.1** Highly enantioselective Rh-catalysed addition of phenylboronic acid to cyclohexenone with competing protodeborylation (which gives benzene).

to [Rh(acac)(ethene)$_2$], which bears the more readily displaceable ethene ligands in place of CO, resulted in better activity. Using this system, at 100°C in 1,4-dioxane/water with 1.4 equivalents of PhB(OH)$_2$, cyclohexenone was found to be phenylated to give 1 in 64% yield and with remarkable asymmetric induction (97% *ee*), Scheme 12.1.

The key to improving the reaction, in particular in finding a catalyst system that allows the reaction at lower temperatures to avoid competing protodeborylation, lay in a simple but highly effective study of the reaction mechanism by NMR. Many organometallic reactions utilise ligands in which phosphorus is a ligating atom, e.g. in phosphines, phosphites and phosphonites. $^{31}$P NMR provides an excellent tool for probing such species as the natural abundance of $^{31}$P, the NMR-active isotope of P, is close to 100%. Moreover, it has a reasonably high magnetogyric ratio affording workable sensitivity and the radio frequency for $^{31}$P is between those of $^1$H and $^{13}$C. So, provided that a pre-tuned or broad-band (tuneable) probe is available, a reasonably standard NMR set-up can be employed. A continuous series of 'hard' pulses at the $^1$H frequency are usually applied to avoid complications arising from scalar coupling of $^{31}$P with neighbouring $^1$H nuclei; this is denoted as $^{31}$P$\{^1$H$\}$NMR. The nuclear spin of $^{31}$P is 1/2, so relatively simple coupling patterns are observed and there is no quadrupolar broadening. The nucleus relaxes relatively efficiently and integration of peak area is a reliable method to determine the relative (or, with an internal standard, absolute) concentrations of the various species observed. In short, the $^{31}$P nucleus behaves in a manner essentially identical with that of the $^1$H nucleus with which most workers are more familiar, albeit with somewhat decreased sensitivity (ca. 7%). Thus, simple 1D as well as 2D experiments (COSY, EXSY, etc.), which are all within the scope of a standard modern NMR instrument, can be highly informative. In the reaction of interest here, not only are there two P atoms in the BINAP ligand, making it an ideal probe for the structure of intermediates, but also the Rh centre is NMR active. Like $^{31}$P, the $^{103}$Rh nucleus is of essentially 100% natural abundance and of spin 1/2; however, it is of low sensitivity (about 1/30 000 that of $^1$H) and at a frequency that is below the range accessible by most spectrometers equipped with a broad-band probe. Nonetheless, the scalar coupling between $^{31}$P and $^{103}$Rh ($^nJ_{PRh}$), as measured from the $^{31}$P$\{^1$H$\}$NMR spectrum, can be highly informative. Thus, the mere observation of $^1J_{PRh}$ coupling confirms the coordination of the BINAP to the Rh, and the coordination geometry and nature of other ligands bound to the Rh centre can be deduced from the magnitude of this coupling. In summary, study of the $^{31}$P$\{^1$H$\}$NMR spectra arising from the Rh-coordinated $C_2$-symmetric BINAP ligand (as well as 'free' or exchanging non-coordinated BINAP) during the reaction can give much information on the number of intermediates, their stoichiometry, chemical relationships and their symmetry and geometry, as well as indications regarding the other ligands that are also coordinated at the Rh centre.

### 12.2.2   The $^{31}P\{^1H\}$ NMR investigation of the Rh-catalysed asymmetric phenylation of cyclohexenone

As with most mechanistic investigations, the starting point is the consideration of possible working models using elementary steps that have precedent in other reactions. Based on the addition of cuprates to enones, Hayashi, Miyaura and co-workers had proposed a Rh–phenyl species (**2**) and an *O*-bound Rh enolate (**3**) as possible intermediates in the Rh-catalysed phenylation reaction [2], Scheme 12.2. They then developed independent syntheses of these intermediates, or close analogues.

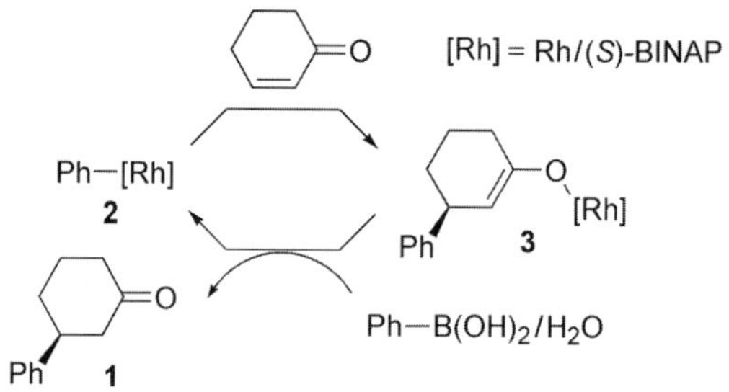

**Scheme 12.2**   The initial working model suggested by Hayashi and Miyaura et al. [2] for Rh-catalysed asymmetric phenylation of cyclohexenone.

By adaptation of a known synthetic route [3], the four-coordinate square planar phenyl–Rh complex **4** was prepared, Scheme 12.3 [4]. In addition to the Ph group, this complex bears three phosphine residues: one BINAP ligand and a PPh3. For enthalpic and entropic reasons, the latter is more labile than the BINAP. The $^{31}P\{^1H\}$ NMR spectrum of **4** displays characteristic signals comprising three sets of multiplets, each a double–double–doublet (*ddd*) arising from one $^1J_{PRh}$ and two $^2J_{PP}$. It is a characteristic of square planar complexes that coupling of a nucleus $X$ to a $^{31}P$ nucleus attached to the metal centre is much larger if $X$ and $^{31}P$ are *trans* related than when they are *cis*; this is much the same as the stereochemical

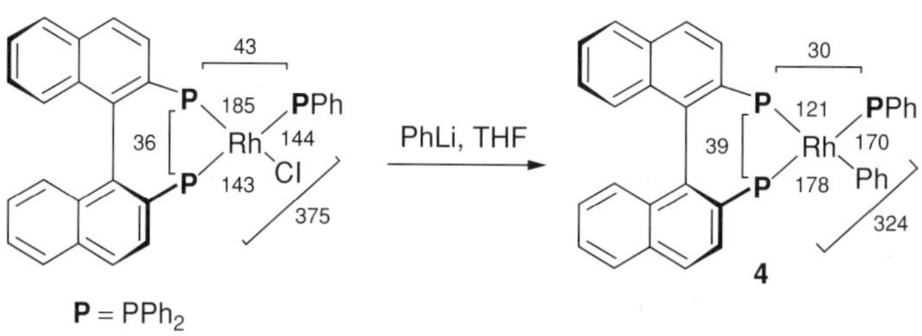

**Scheme 12.3**   Preparation of the phenylated Rh–BINAP complex, **4**; the $^2J_{PP}$ and $^1J_{PRh}$ values were determined by $^{31}P\{^1H\}$ NMR; note that the assignment of the *trans*-related phosphines is arbitrary.

dependence of $^3J_{HH}$ values in alkenes. Thus, the assignment of the three *ddd*'s is readily made, Scheme 12.3, with the *trans*-related $^{31}$P nuclei displaying the $^2J_{PP}$ values nearly an order of magnitude larger than the *cis*-related nucleis. It is interesting to note the changes in the $^1J_{PRh}$ values on exchange of Cl for Ph. These values are a good probe for (relative) bond orders, and the changes are strongly suggestive of a lengthening of the P—Rh bond *trans* to Ph and a shortening of the two *cis* P—Rh bonds.

Complex **4** was isolated and found to phenylate cyclohexenone to give **1** in 98.8% *ee* in the same solvent mixture (91% aqueous 1,4-dioxane) as the catalytic system and with the same sense of enantio-induction. More importantly, using the same conditions as in Scheme 12.1, but replacing [Rh(acac)(ethene)$_2$]/(*S*)-BINAP with either 3 mol% [Rh(acac)(*S*-BINAP)], or with 3 mol% **4**, gives **1** with the same *ee* and sense of induction, and in essentially identical yield. The *stoichiometric* phenylation by **4** and its efficacy as a pro-catalyst are good evidence that a Ph–[Rh] complex (cf. **2** in Scheme 12.2) is generated under the reaction conditions. Obviously, **4** bears a PPh$_3$ ligand that is not present under normal catalytic reactions conditions; however, the PPh$_3$ clearly has no effect on the enantioselectivity and is also expected, on enthalpic and entropic grounds, to be the most readily dissociable ligand in **4**.

The stoichiometric reaction of **4** with cyclohexenone in *anhydrous* THF was found to proceed readily at 25°C to afford essentially a *single* rhodium complex. The $^{31}$P{$^1$H}NMR spectrum of the product mixture showed that PPh$_3$ had been displaced (singlet at −4.5 ppm) and that the two BINAP phosphorus atoms were both coordinated to Rh but were inequivalent. The one $^2J_{PP}$ value (51 Hz) and the two $^1J_{PRh}$ values (205 and 215 Hz) were again indicative of a square planar rhodium complex. Addition of water at 25°C immediately liberated **1** (99% *ee*) which suggested that the complex was a rhodium enolate (**5**, Scheme 12.4). This conclusion was supported by an experiment in which addition of the potassium enolate of **1** (99% *ee*) to [(*S*)-BINAP)RhCl]$_2$ gave a product with an identical set of $^{31}$P{$^1$H}NMR signals, together with two other minor isomeric enolates (**6** and **6′**) whose $^{31}$P shifts were very similar and whose coupling constants ($^2J_{PP}$ and $^1J_{PRh}$) essentially identical with those of **5**. This is an important result as the enolisation of **1** can proceed at the methylene *distal* as well as that *proximal* to the phenylated centre, Scheme 12.4. The generation of distinct isomers (**6** and **6′**), together with the square planarity of the enolate complexes, is good evidence that these species are best described as bidentate $\pi$-oxo-allyl rhodium complexes (**5**, **6**; and **5′**, **6′**), as opposed to monodentate *O*- (cf. **3** in Scheme 12.2) or *C*-bound enolates (see the inset to Scheme 12.4). It is interesting to speculate that the phenylation most likely proceeds with *syn* stereochemistry such that the kinetic product will be **5′**. However, this can rapidly isomerise to **5** via the *O*-bound enolate, **3**. The observed ratios of isomeric complexes **5:5′** (>95:5) and **6:6′** (ca. 1:8) thus reflect the thermodynamic preference for the isomers in which Rh and Ph are on opposite faces of the planar $\pi$-oxo-allyl unit.

The co-product of the addition of water to the $\pi$-oxo-allyl rhodium complex, **5**, in THF at 25°C was suggested to be an (unobserved) hydroxy rhodium species which rapidly dimerises to give complex **7**, a square planar $\mu$-hydroxy dimer. This complex was easily identified by the high field shifts of the $\mu$-hydroxy protons (−1.9 ppm) and its symmetry (a single Rh-coupled doublet is observed at 55 ppm in the $^{31}$P{$^1$H}NMR spectrum). The assignment was confirmed by independent synthesis: [(*S*)-BINAP)Rh($\mu$-Cl)]$_2$ reacted cleanly with KOH in aqueous THF to give **7**, as did addition of (*S*)-BINAP to the conveniently prepared complex [Rh($\mu$-OH)(cod]$_2$ (cod = 1,5-cyclo-octadiene).

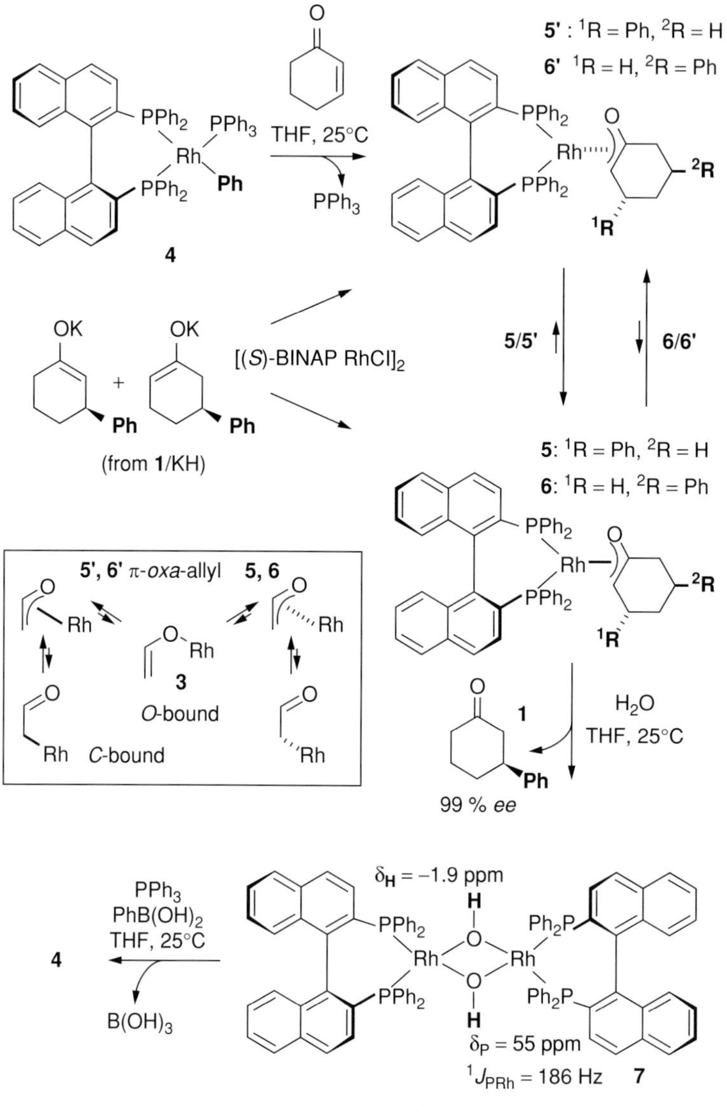

**Scheme 12.4** Stoichiometric stepwise exploration of the catalytic cycle in the asymmetric phenylation of cyclohexenone.

The final stage of this investigation was to study the reaction of the hydroxy dimer 7 with phenylboronic acid to test whether 7 is able to undergo efficient phenylation. The reaction with $PhB(OH)_2$ was found to proceed smoothly and, when conducted in the presence of $PPh_3$, gave the stable four-coordinate complex 4 which had already been characterised, Scheme 12.4. As 4 was already known to be effective as a pro-catalyst (see above), this process formally 'closed the circle' to constitute a very convincing and conclusive collection of experiments.

## 12.2.3 Summary and key outcomes from the mechanistic investigation

The key to the success of the investigation by Hayashi and co-workers lay in the stoichiometric stepwise exploration of the cycle, using the *monodentate* ligand PPh$_3$ to stabilise intermediate 4 and by restricting access, as required, to the three components (water, PhB(OH)$_2$, cyclohexenone). The astute reader will have noticed that *all* of the three stages of the cycle proceed readily at *ambient* temperature when conducted stoichiometrically in the manner outlined in Scheme 12.4. Yet, in the catalytic reaction, temperatures of over 80°C are required for reasonable turnover rates (at 60°C, the yield is <3%). At first, the two sets of results from catalytic versus stoichiometric conditions may seem at odds. However, on closer inspection, it becomes evident that the pro-catalyst system (Scheme 12.1) will release acetylacetone (acac-H, 9) into the reaction medium through phenylation of [Rh(acac)(BINAP)], 8 in Scheme 12.5. This is confirmed by the stoichiometric reaction and a key point to note is that this phenylation step is very slow compared with the reaction with the $\mu$-hydroxy dimer, 7. Indeed, it requires heating to 80°C for efficient conversion. Since, under the catalytic conditions, this process is simply a step *into* the catalytic cycle, one might then expect rapid catalysis to ensue. However, it was found that acac-H (9) reacts rapidly with the $\mu$-hydroxy dimer, 7, to regenerate 8 and this then shifts the resting state 'off-cycle' – analogous to III in Fig. 12.1.

Together with the NMR studies 'on-cycle', see above, this remarkable observation concerning the resting state then allows an *informed* redesign of the reaction conditions. By avoiding the generation of acac-H (9) in the medium, the resting state can be shifted 'on cycle' which facilitates faster turnover. Using the $\mu$-hydroxy dimer 7 as a pro-catalyst, the reaction can now be conducted at significantly lower temperatures. This allows higher enantioselectivity as well as higher yields to be achieved.

The reaction has one more secret to reveal before we move on. The careful work of Hayashi and co-workers led to the use of the $\mu$-hydroxy dimer, 7, as a pro-catalyst. Whilst

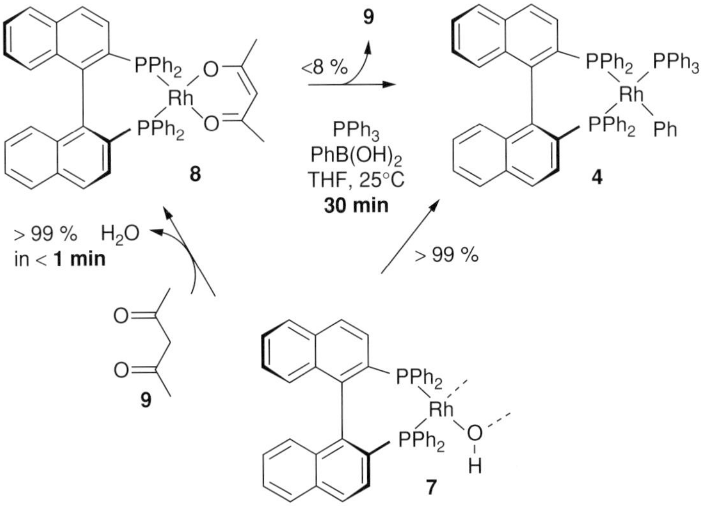

**Scheme 12.5** The slow phenylation of [Rh(acac)(BINAP)] (8) with release of acac-H (9), and its *rapid* regeneration from [Rh($\mu$-OH)(BINAP)]$_2$ (7) + 9.

this pre-formed isolated and pure complex gives outstanding results, it is often useful to be able to prepare catalysts in situ from the phosphine ligand and a suitable general Rh source as this allows new chiral ligands to be readily tested. The Japanese group found that preparing the catalyst in situ from BINAP + [Rh(cod)($\mu$-OH)]$_2$ gave high turnover rates but very poor enantioselectivity. This leads to two conclusions: (i) the rate of displacement of the *bidentate* 1,5-cyclo-octadiene (cod) ligand by BINAP is slow, and (ii) the [Rh(cod)($\mu$-OH)]$_2$ itself must be an active catalyst (giving racemic product). This was a surprising result as bidentate alkenes are often employed as readily displaceable ligands in transition metal pro-catalysts. On the basis of this observation, Hayashi and co-workers went on to report the truly seminal preparation and application of a chiral norbornadiene ligand in the phenylation reaction, which opened the door to a completely new family of chiral ligand architecture and design [5].

## 12.3 Kinetic and isotopic labelling studies using classical techniques to study intermediates 'on cycle' indirectly: the Pd-catalysed cyclo-isomerisation of dienes

There are rather few reactions that can be described as fully 'atom economical', i.e. when there are no co-products and all the atoms in the starting material(s) appear in the product(s). However, all isomerisation reactions necessarily fall into this category. The use of a transition metal to catalyse such a process with an appropriate substrate brings the possibility of effecting asymmetric isomerisation, a very efficient method to generate enantiomerically enriched products. Indeed, the asymmetric Rh-catalysed isomerisation of an allylamine to an enamine, which proceeds in over 96% *ee*, was scaled up a number of years ago for industrial production. The enamine product forms a multi-tonne feedstock for menthol and perfumery synthesis. In contrast, the cyclo-isomerisation of dienes, an equally atom-economical process that generates synthetically useful cyclic products, has seen relatively little development despite the reaction having been known for some 30 years.

### 12.3.1   *Background and introduction*

In this section, we look at Pd-catalysed cyclo-isomerisation of 1,6-dienes. This is a reaction that was first explored by Schmitz and Grigg and their co-workers in the late 1970s [6]. However, it has been revisited in the last few years with the aim of developing asymmetric cyclisation processes. The key to realising these goals undoubtedly rests on an appreciation of what 'control elements' one must build into the ligand architecture and thus an understanding of the reaction mechanism. However, in only one case so far has an 'on-cycle' intermediate been detected and, thus, nearly all of the evidence regarding mechanism has been accumulated *indirectly*. We shall see how the combined study of kinetics, pro-catalyst side products and isotopic labelling of substrates allows some insight into the basic mechanism and into the subtle stereochemical and regiochemical issues of this 'one-component' reaction. It is useful to contrast this isomerisation with the preceding study where there were three components and it was thus possible to arrest the cycle at various stages. This

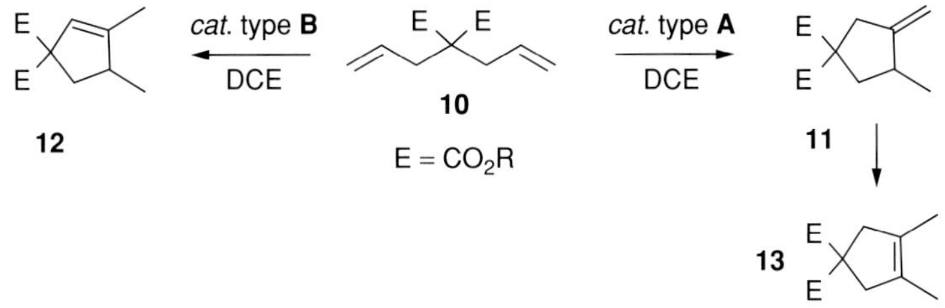

**Scheme 12.6** Pro-catalyst-dependent regioselectivity in the Pd-catalysed cyclo-isomerisation of 1,6-dienes; DCE = 1,2-dichloroethane (solvent); see the text for catalysts.

approach is not feasible with the one-component system described below, which severely limits what intermediates can be directly detected.

A fairly large range of palladium complexes have been reported to catalyse the cyclo-isomerisation of 1,6-dienes, usually giving rise to five-membered ring cyclo-isomers. The major product generated depends on the exact conditions and pro-catalyst employed, and five different mechanisms have been suggested. As we shall see from the two investigations outlined below, and consistent with the principle of parsimony (Occam's Razor), a single mechanism can account for all of the products. It is of note that one of the key points addressed in these studies was the mechanism by which the pro-catalyst is converted to the active catalyst. The two investigations that we shall review were conducted independently; however, they are complementary and arrive at essentially the same conclusions and will, therefore, be considered in parallel [7].

We shall consider reactions catalysed by two different types of pro-catalyst: the first (type A) employs Pd-allyl cations ($[Pd(allyl)(PCy_3)]^+/Et_3SiH$ or $[Pd(allyl)(MeCN)_2]^+$), and the second (type B) employs Pd-alkyl or chloro complexes ($[(phen)Pd(Me)(MeCN)]^+$, where phen = phenanthroline, and $[(RCN)_2PdCl_2]$). These two types of catalysts give very different products in the cyclo-isomerisation of typical 1,6-dienes such as the diallyl-malonates (**10**), Scheme 12.6. Since there is known to be a clear order of thermodynamic stability: **11** < **12** < **13**, with a difference of ca. 3–4 kcal mol$^{-1}$ between successive pairs, any isomerisation of *products* under the reaction conditions will tend towards production of **12** and **13** from **11**; and **13** from **12**. Clearly, when **11** is the major product (as with pro-catalysts of type A), it must be the kinetic product (see Chapter 2 for a discussion of kinetic and thermodynamic control of product distributions). However, when **12** is generated selectively, as it is with pro-catalysts of type B, there is the possibility that this is either generated by rapid (and selective) isomerisation of **11** or generated directly from **10**.

## 12.3.2   *Kinetic studies employing classical techniques*

The reactions employing the two types of catalysts (A and B) give very different results, but all proceed at ambient temperatures or slightly above. The substrate and products are stable and are readily separated by gas-chromatographic techniques. Moreover, the reactions proceed at a rate that is amenable to a classical approach: thus, reactions were conducted in the presence of an internal standard, samples were withdrawn periodically, inorganic species were removed by passage through a short plug of silica gel, and then the organic fraction was

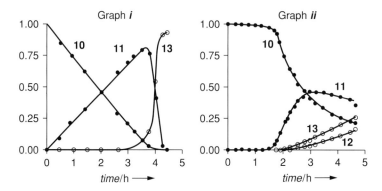

**Fig. 12.2** Evolution profiles for the cyclo-isomerisation of **10** catalysed by pro-catalysts of type **A**. Graph *i*, 5 mol% [Pd(allyl)(PCy$_3$)]$^+$/Et$_3$SiH, DCE, 40°C. Graph *ii*, 5 mol% [Pd(allyl)(MeCN)$_2$]$^+$, CHCl$_3$, 40°C.

analysed by GC. By such an approach, the kinetics were analysed with all four pro-catalyst systems. The outcomes are quite revealing.

With pro-catalysts of type **A**, the kinetic product is **11** which accumulates and then isomerises to **12** and **13**, Fig. 12.2. With the [Pd(allyl)(PCy$_3$)]$^+$/Et$_3$SiH pro-catalyst system (Fig. 12.2, graph *i*), the accumulation–depletion phenomenon is rather dramatic: after addition of pro-catalyst to **10**, the *exo*-cyclic alkene product **11** accumulates immediately and *linearly* with time (i.e. zero-order kinetics with $k_{obs} = 8.1 \times 10^{-5}$ M s$^{-1}$) until ca. 80% conversion. After this point, it very rapidly and selectively isomerises to **13**, at a rate that is at least six-fold faster than its initial generation ($k_{obs} = 5 \times 10^{-4}$ M s$^{-1}$). It should be noted that Et$_3$SiH acts as an activator; the reaction is very much slower in its absence and induction periods are observed when Et$_3$SiD is employed. Although unusual in appearance, the kinetics for the reaction are actually rather simple and readily explained: rapid induction of the allylpalladium pro-catalyst by Et$_3$SiH is followed by a prolonged phase (ca. 4 hours) during which the substrate (**10**) reacts rapidly with an intermediate 'on-cycle' to generate the catalyst resting state, which may be on- or off-cycle. Since Pd is catalytic and the concentration of the resting state becomes essentially steady state, the turnover-limiting *unimolecular* breakdown of the resting state gives rise to the pseudo-zero-order kinetics and there is a linear accumulation of **11**. As the concentration of **10** falls and that of the product (**11**) rises, the latter, which is less reactive towards the active catalyst than **10**, begins to compete and is converted by alkene isomerisation to **13**. The latter process is much more rapid (complete in approximately 30 minutes) because the resting state associated with this pathway is much less stable. In effect, **10** acts as a powerful inhibitor for the isomerisation of **11** to **13**. Indeed, so effective is **10** as an inhibitor that when *pure* **11** was exposed to the pro-catalyst system, it was quantitatively isomerised to **13** in less than a minute!

In contrast, with [Pd(allyl)(MeCN)$_2$]$^+$ (Fig. 12.2, graph *ii*), there is a rather variable induction period, a much less pronounced accumulation–depletion phase, non-linear (complex) evolution, and lower selectivity for **13** upon isomerisation of **11** (**13**:**12** ≈ 2:1). A major complication is the slow induction phase which gives rise to a 'trickle-feed' of active catalyst and, consequently, this system is far harder to analyse. Nonetheless, with both systems, the sole

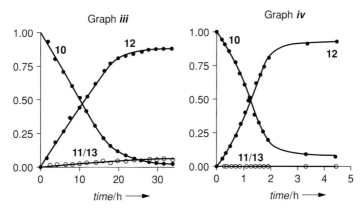

**Fig. 12.3** Evolution profiles for the cyclo-isomerisation of **10** catalysed by pro-catalysts of type **B**. Graph *iii*, 5 mol% [(phen)Pd(Me)(MeCN)]⁺, DCE, 40°C. Graph *iv*, 5 mol% [PdCl₂(*t*-BuCN)₂], DCE, 40°C.

kinetic product is **11** (as can be confirmed by material balance and by spiking the reactions with isotopically labelled samples of **11** and **12**) and the active catalyst for the conversion of **10** to **11** is also the species that then converts **11** to **12** and **13**.

With pro-catalysts of type **B**, the evolution profiles (Fig. 12.3) are very different; the product is **12** instead of **11** and this does not undergo the extreme accumulation–depletion cycles observed with pro-catalysts of type **A**, which give **11** as the initial product. With the [(phen)Pd(Me)(MeCN)]⁺ system, clean pseudo-zero-order kinetics are again observed for the first 75% of reaction ($k_{obs} = 7.1 \times 10^{-7}$ M s$^{-1}$) and the co-products (**11** and **13**) are formed at a level of ca 5% throughout the evolution (Fig. 12.3, graph *iii*).

With the neutral [(RCN)₂PdCl₂] pro-catalyst system (Fig. 12.3, graph *iv*), computer simulation of the kinetic data acquired with various initial pro-catalyst concentrations and substrate concentrations resulted in the conclusion that the turnover rates are controlled by substrate-induced 'trickle feed' catalyst generation, substrate concentration-dependent turnover and continuous catalyst termination. The active catalyst concentration is always low and, for a prolonged phase in the middle of the reaction, the net effect is to give rise to an *apparent* pseudo-zero-order kinetic profile. For both sets of data obtained with pro-catalysts of type **B** (Fig. 12.3), one could conceive that the kinetic product is **11**, but (unlike with type **A**) the isomerisation to **12** is extremely rapid such that **11** does not accumulate appreciably. Of course, in this event, one needs to explain why the isomerisation of **11** now proceeds to give **12** rather than **13**. With the [(phen)Pd(Me)(MeCN)]⁺ system, analysis of the relative concentrations of **11** and **13** as the conversion proceeds confirmed that the small amount of **11** that is generated isomerises predominantly to **13**. With the [(RCN)₂PdCl₂] pro-catalyst system, spiking the reaction with **11** results in strong inhibition and its slow isomerisation to **13**, not **12**.

Overall, on the basis of the kinetics and evolution profiles, one must thus conclude that, on changing from catalyst set **A** to **B**, the *kinetic* product (that emerging directly from the catalytic cycle) is switched essentially completely from **11** to **12**. But with both sets, any subsequent isomerisation of the product generates **13**, the thermodynamically most favoured product, Scheme 12.7.

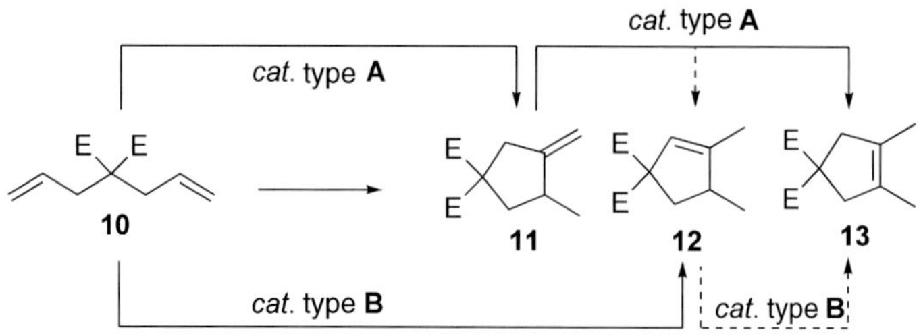

**Scheme 12.7** Conclusions from the studies of the kinetics of the Pd-catalysed cyclo-isomerisation of 1,6-diene **10** (E = CO$_2$R) to give **11**, **12** and **13**.

### 12.3.3   'Atom accounting' through isotopic labelling

Under none of the conditions (pro-catalysts of type **A** or **B**) had any 'on-cycle' intermediates been identified, or even detected, e.g. by NMR. Thus, the mechanisms had to be probed indirectly and since the cyclo-isomerisation process must, by its very nature, involve the generation of a C—C bond with concomitant H migration, deuterium labelling is an obvious technique. Both groups of researchers prepared a range of isotopically labelled samples of **10** and tested these under the catalytic reaction conditions. Scrambling compromised complete analysis in two of the systems, ironically those systems affording the clearest kinetic data. Nonetheless, the outcomes were consistent within types **A** and **B**. The key to the design of effective labelling experiments, where one is aiming simply to effect an 'atom accounting' type of analysis, is to consider all mechanistic possibilities and aim to eliminate those which cannot accommodate the results. Over the years, there have been four distinct mechanisms proposed for the palladium-catalysed cyclo-isomerisation of 1,6-dienes to give *exo*-methylenecyclopentanes (e.g. **10** → **11**) as illustrated in Scheme 12.8 and labelled **I**, **II**, **III** and **IV**.

By considering the H-migration origin/destination, one may distinguish **I**, **II** and **III/IV**. On this basis, experiments (i) and (ii) with a type **A** catalyst as shown in Scheme 12.9 eliminated mechanisms **I** and **II** from consideration; this left **III** and **IV** which were both fully consistent with the results. The outcome for (i) is obvious: the allylic hydrogens (see H$_b$ in mechanism **I**, Scheme 12.8) are not involved in the reaction. The outcome for (ii) is more subtle and relates to the stereochemistry attending *beta*-carbopalladation and *beta*-hydride elimination which are both known to proceed with *syn* stereochemistry. Thus, mechanism **II** which does not involve a *beta*-hydride elimination would not affect the alkene stereochemistry (see H$_c$ in **II**, Scheme 12.8), as was revealed by D-labelling, Scheme 12.9. In contrast, mechanisms **III** and **IV** should *reverse* the stereochemistry (see H$_c$ in **III** and **IV**, Scheme 12.8), as was observed.

Mechanisms **III** and **IV** both predict that H$_a$ in **10** (Scheme 12.8) migrates to the terminal carbon on the *opposite* allylic chain and becomes part of a methyl group. However, in mechanism **III** H$_a$ migrates intramolecularly whereas in **IV** the migration is intermolecular via **14**. The deployment of a mixed [13]C/D-labelled substrate (experiment (iii), Scheme 12.9) allows this distinction to be made as products are obtained in which D has transferred intermolecularly (see the right-hand product in experiment (iii), which demonstrates 'cross-over' of D).

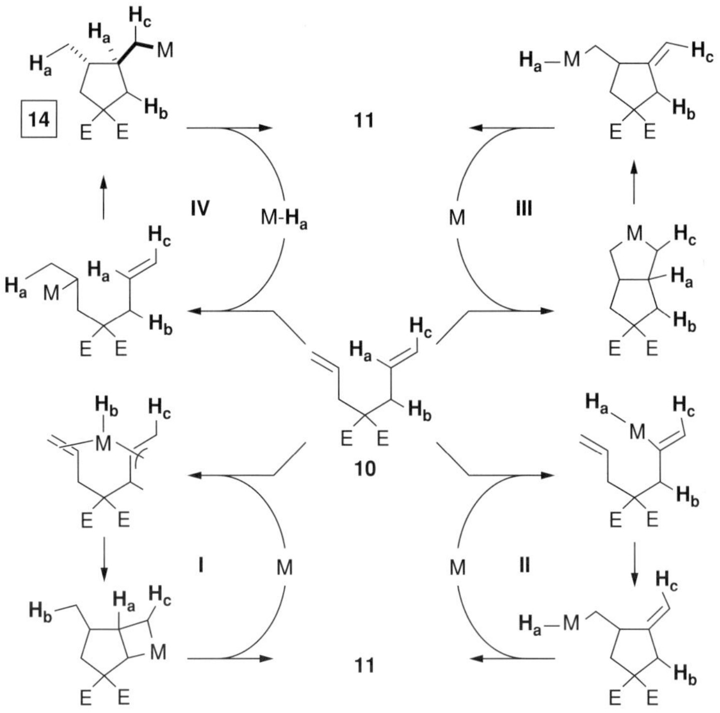

**Scheme 12.8** Mechanisms proposed and tested for the Pd-catalysed cyclo-isomerisation of 1,6-dienes (e.g. **10**) to give products of type **11**.

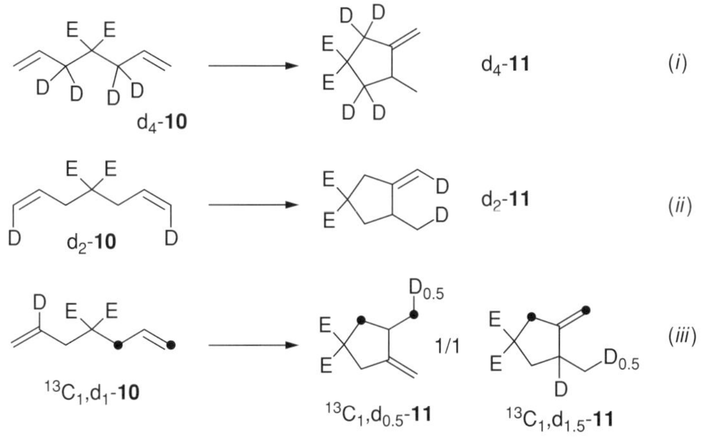

**Scheme 12.9** Three isotopic labelling experiments to distinguish mechanisms **I, II, III** and **IV** (see Scheme 12.8) for the Pd-catalysed cyclo-isomerisation of **10** to **11** with a type **A** catalyst.

Moving next to pro-catalysts of type **B**, which give **12** as the kinetic product, it is of note that all but one of the mechanisms previously proposed in the literature for the generation of **12** were based on **II–IV** to generate **11** as the *kinetic* product with a *subsequent* isomerisation to **13**. However, as was clear from the reaction kinetics (Fig. 12.3 and Scheme 12.7), these mechanisms cannot be operative because **11** is isomerised almost exclusively to **13**, and not **12**, under the reaction conditions. The sole mechanism that had been proposed earlier and is consistent with this later evidence is an adaptation of mechanism **I** in Scheme 12.8 in which a *beta*-hydride elimination at the palladacyclobutane stage proceeds with alternative regioselectivity to yield **12** after reductive elimination, Scheme 12.10. Two isotopic labelling experiments ((i) and (ii), Scheme 12.10), employing D to probe the two H-migrations ($H_a$ and $H_b$), gave results which are completely consistent with such a mechanism. Given that all the other proposed mechanisms are inconsistent with the kinetic/labelling results, it is tempting to conclude that the adaptation of mechanism **I** is correct. However, as ever, one must seek to test rather than support, and the salient issue here is that mechanism **I** is intramolecular. As above, this was tested by seeking evidence for cross-over products.

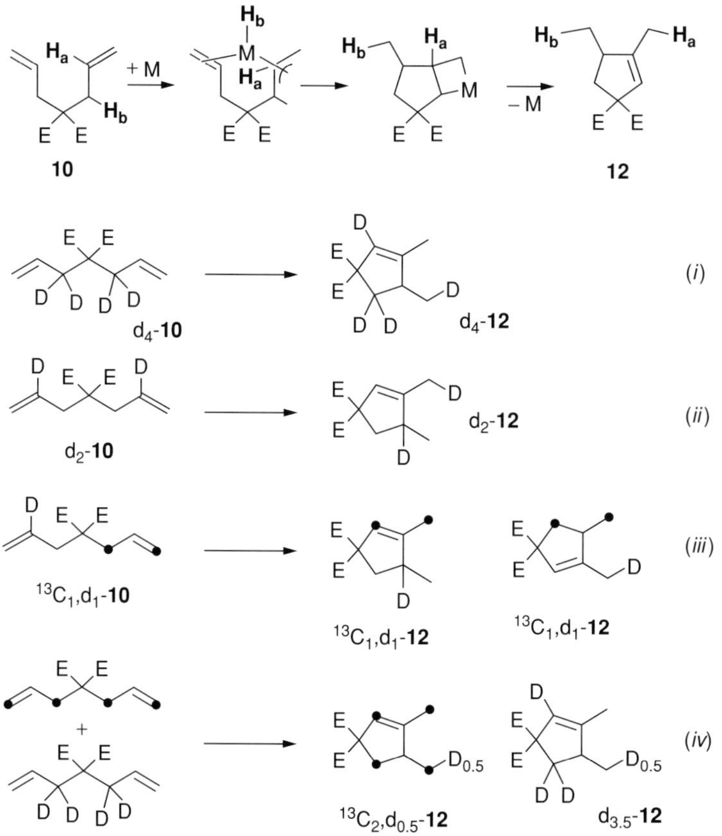

**Scheme 12.10**   Isotopic labelling experiments to investigate the mechanism of isomerisation of **10** to **12**, rather than **11**, as the kinetic product with a type **B** catalyst; (i), (ii) and (iii) are consistent with a modification of mechanism **I** in Scheme 12.8, but (iv) is not.

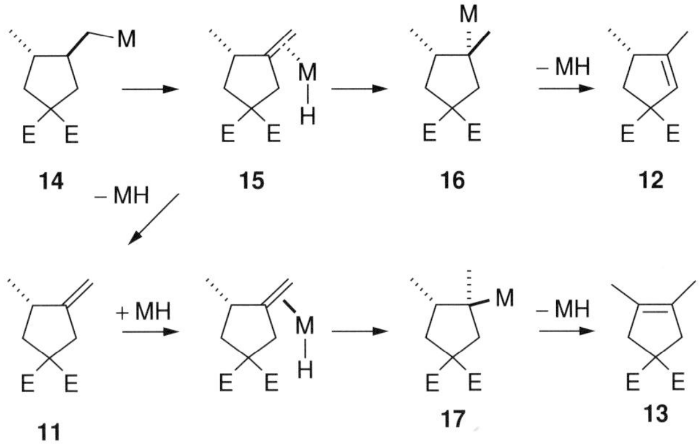

**Scheme 12.11** Modification of mechanism **IV** to allow for the generation of either **11** or **12** with a type **B** catalyst, depending on the propensity of intermediate **15** for dissociation or hydropalladation to give **16**.

These were indeed found, but they only arise from one of the two H-migrations. That is, $H_a$ migrates *intra*molecularly whilst $H_b$ migrates *inter*molecularly, as demonstrated (for example) by experiments (iii) and (iv) in Scheme 12.10. Clearly then, another mechanism is operative.

It is instructive at this juncture to consider two points concerning this reaction: first, the *inter*molecular H migration is suggestive of a Pd–H-based mechanism (as with pro-catalysts of type A) and, second, that addition of **11** to the reaction results in its isomerisation to **13**. This is exactly what would be predicted from reaction of **11** with a Pd–H intermediate: the Pd–H would add to the alkene from the face that is *anti* to the methyl group and the resulting sigma-alkyl–Pd complex (**17** in Scheme 12.11) would be expected to undergo selective *beta*-hydride elimination to generate the thermodynamically favoured **13**.

If one now considers **16**, the diastereoisomer of **17**, it is evident that, due to the *syn* relationship between the Pd and the methyl group, *syn beta*-hydride elimination can only proceed in two directions, one to generate **11** (via **15**) and one to generate **12**. Here, then, was solution to the *apparent* paradox: mechanism **IV** need merely be adapted so that *intramolecular* addition of the Pd–H to the alkene in intermediate **15** generates **16**, and thus **12**.

## 12.3.4 Observation of pro-catalyst activation processes by NMR spectroscopy

Having suggested, then, that pro-catalysts of types A and B both generate catalytic cycles involving Pd–H intermediates (**IV** + extension), evidence was needed for the origin of the Pd–H species. For one of the pro-catalysts from each type, the co-products of catalyst activation have been detected by NMR. However, for the $[Pd(allyl)(PCy_3)]^+$ and the $[(RCN)_2PdCl_2]$ pro-catalyst systems, nothing has yet been observed although, for the former, the $Et_3SiH$ activating reagent is well known to transfer hydride and there is an induction period when $Et_3SiD$ is employed. With the other two pro-catalyst systems, NMR studies of the activation process proved very informative. With $[Pd(allyl)(MeCN)_2]^+$, a relatively short-lived intermediate in very low concentration was detected by NMR. Since there is no

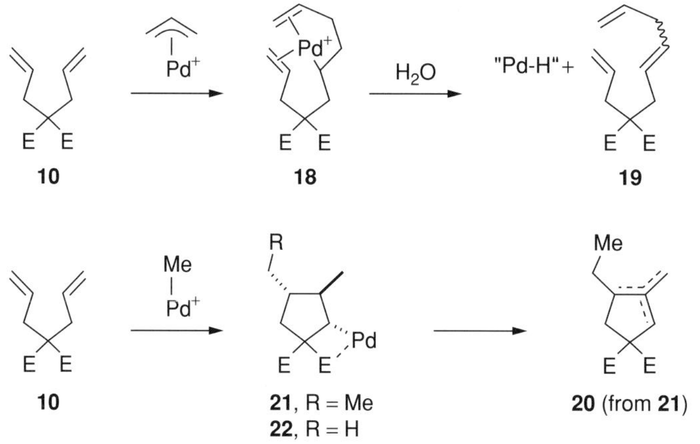

**Scheme 12.12**   Generation of the active catalyst (Pd-H) from two pro-catalysts.

external activator (such as Et$_3$SiH), it seemed reasonable that it is a substrate/pro-catalyst reaction that generates the Pd–H species. A series of $^1$H NMR studies of *stoichiometric* mixtures of isotopically labelled substrate and pro-catalyst facilitated its identification as **18** (Scheme 12.12), arising from allyl palladation of the diene. Consideration of the inefficiency of this process (the displacement of the nitrile ligands must be slow and the allyl palladation fast as the intermediate diene complex is not observed) led to an alternative preparation, and the isolation and full characterisation of this complex. Crucially, when **18** was tested to confirm that it can act as a genuine pro-catalyst for the cyclo-isomerisation of **10**, it then emerged that the variable induction periods observed in the original catalytic reaction arise from the requirement for traces of water to trigger the *beta*-hydride elimination in complex **18** to generate the active Pd–H and **19**, Scheme 12.12.

For the [(phen)Pd(Me)(MeCN)]$^+$ pro-catalyst system, within two turnovers of the initiation stage of the catalytic reaction, the methylated carbocycle **20** (as a mixture of isomers, Scheme 12.12) could be identified and correlated with the catalyst stoichiometry. Clearly, **20** arose from **10** and thus a stoichiometric activation was conducted, this time at lower temperature (25°C), and again a Pd complex (**21**) could be isolated and its structure confirmed by X-ray crystallography. On heating to 50°C in the presence of **10**, complex **21** liberated **20** and thereby acted as pro-catalyst. NMR analysis of the ensuing mixture undergoing turnover allowed the identification, on the basis of similarity to **21**, of complex **22**, which would then be the resting state of the catalytic cycle.

## 12.3.5   Summary and mechanistic conclusions

Compared with the previous example involving Rh-catalysed phenylation of cyclohexenone, the mechanistic investigation presented in this section involves a much less direct approach. This arises from the nature of the cyclo-isomerisation reaction, involving a single reactant and rather unstable intermediates. It is thus hard to manipulate the resting state of the catalytic cycle in order to facilitate direct study, e.g. by NMR, although Widenhoefer has made some progress in this direction. Nonetheless, much of the investigation relied on the indirect testing of proposed mechanisms by kinetic studies and isotopic labelling strategies. This requires that *all* prior mechanistic proposals for the reaction be tested. Obviously, once one has disproved all but one mechanism, it is tempting to be overly confident in assigning what remains. As was the case here, one must be cautious. The strength of the proposed Pd–H mechanism **IV** (which is fully consistent with the kinetic and labelling studies and is supported by detection of catalyst activation co-products) is that it obeys the principle of parsimony: one mechanism explains the results from both types of pro-catalyst (**A** and **B**).

## 12.4 Product distribution analysis, and kinetics determined by classical and advanced NMR techniques: the transition-metal-catalysed metathesis of alkenes

With the advances in pro-catalyst design that have been witnessed over the last decade or so, the transition-metal-catalysed alkene metathesis reaction has now become a practical procedure that can be utilised by the chemist at the bench. Undeniably, this has added a new dimension to the repertoire of synthetic organic chemistry as it facilitates disconnections that, pre-metathesis, simply would not have been considered. Take, for example, a macro-cyclic amide where the normal disconnection would be at the amide. Now, with the ready reduction of alkenes to alkanes, a ring-closing diene metathesis (RCM), followed by hydro-genation, becomes an alternative disconnection. And, when one considers that *any* of the C—C linkages could be established in such a manner, the power of the RCM disconnection becomes obvious.

In this section we shall see how the establishment of the currently accepted mechanism for the transition-metal-catalysed alkene metathesis required highly skilled experimental design in order to disprove the 'pairwise' quasi-cyclobutane and metallocyclopentane mechanisms originally proposed and defended. The identification of the currently accepted mechanism was absolutely crucial in facilitating the ensuing development of alkylidene complex pre-cursors as pro-catalysts. We then move forward some 30 years, to the point where the fine detail of the mechanism is explored using an advanced NMR technique to probe the origins of the different turnover rates engendered by two closely related pro-catalysts.

### 12.4.1 Background and introduction

The alkene metathesis reaction arose serendipitously from the exploration of transition-metal-catalysed alkene polymerisation. Due to the complexity of the polymeric products, the *metathetic* nature of the reaction seems to have been overlooked in early reports. However, in 1964, Banks and Bailey reported on what was described as the 'olefin disproportionation' of *acyclic* alkenes where exchange was evident due to the monomeric nature of the products [8]. The reaction was actually a *combination* of isomerisation and metathesis, leading to complex mixtures, but by 1966 Calderon and co-workers had reported on the preparation of a homogeneous W/Al-based catalyst system that effected extraordinarily rapid alkylidene

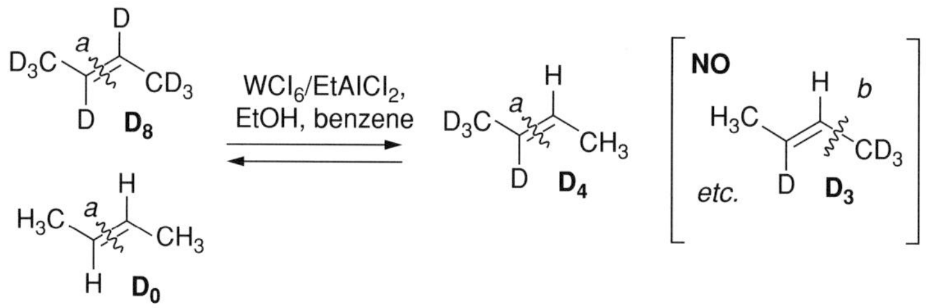

**Scheme 12.13** Metathesis of but-2-ene proceeds via alkylidene exchange (a) and not alkyl exchange (b).

exchange, with no double bond isomerisation [9]. The terms 'metathesis' and 'dismutation' were both used in the early literature, with the term 'metathesis' being widely adopted by ca. 1971. That it was genuine *alkylidene* metathesis (a), and not *alkyl* metathesis (b), in Scheme 12.13 was proven by study of the product mixture from equilibration of a mixture of but-2-ene and perdeutero-but-2-ene; only $D_0$, $D_4$ and $D_8$ mixtures were generated.

## 12.4.2 Early mechanistic proposals for the alkene metathesis reaction

The alkene metathesis reaction was unprecedented – such a non-catalysed concerted four-centred process is forbidden by the Woodward–Hoffmann rules – so new mechanisms were needed to account for the products. Experiments by Pettit showed that free cyclobutane itself was not involved: it was not converted to ethylene (<3%) under the reaction condition where ethylene underwent degenerate metathesis (>35%, indicated by experiments involving $D_1$-ethylene) [10]. Consequently, direct interconversion of the alkenes, via an intermediate complex (termed a 'quasi-cyclobutane', 'pseudo-cyclobutane' or 'adsorbed cyclobutane') generated from a bis-alkene complex was proposed, and a detailed molecular orbital description was presented to show how the orbital symmetry issue could be avoided, Scheme 12.14 (upper pathway) [10].

Although the quasi-cyclobutane mechanism accounts fully for the products of the metathesis of alkenes at equilibrium, see above, some workers were uncomfortable with the unconventional nature of the bonding in the proposed intermediate. An alternative mechanism which brings about the same net alkylidene exchange (lower pathway in Scheme 12.14)

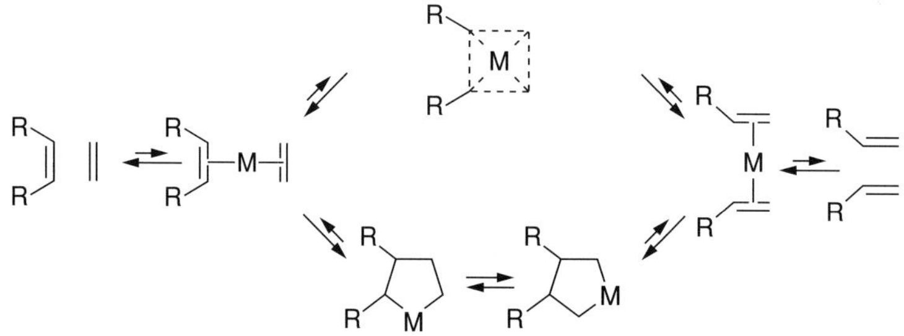

**Scheme 12.14** The 'quasi-cyclobutane' (upper) and metallocyclopentane (lower) pathways for metathesis; both are 'pairwise' mechanisms (see the text for full details).

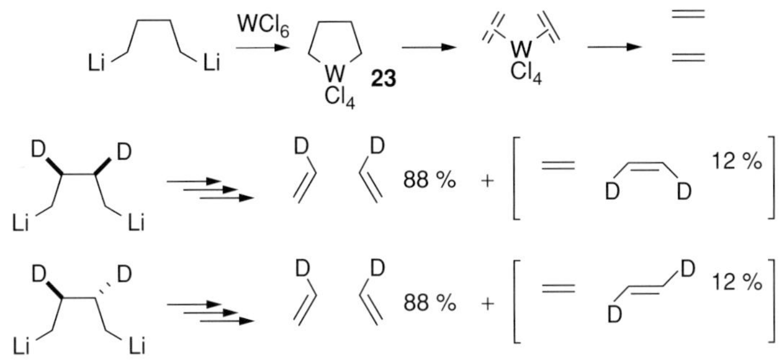

**Scheme 12.15** Evidence in support of the intermediacy of a metallocyclopentane as a pathway for metathesis.

was proposed by Grubbs [11]; in this, cyclometallation of two alkenes to form a metallo-cyclopentane intermediate is followed by a sigmatropic rearrangement and then a retro-cyclometallation. It was known from earlier work that the reaction of $WCl_6$ with 2BuLi generates a $WCl_4$ species which is active for metathesis. This was proposed to proceed via reductive elimination of octane from $WBu_2Cl_4$, generated from $BuLi/WCl_6$. To try to access a tetrachlorotungstacyclopentane (23), Grubbs and co-workers reacted $WCl_6$ with 1,4-dilithiobutane. This generated ethylene quantitatively, Scheme 12.15.

When 1,2-dideutero-1,4-dilithiobutane was employed, the product was found to be not just $D_1$-ethylene, but included 12% of a 1:1 mixture of $D_0$- and $D_2$-ethylene. This was taken as evidence that the tungstacyclopentane had rearranged. One might propose that secondary metathesis of the ethylene was responsible for these isotopologues: after all, $WCl_4$ that should be generated as a co-product is a metathesis catalyst. However, it was demonstrated that diethyl ether (the solvent in which 1,4-dilithiobutane was prepared) inhibits metathesis. Moreover, the *dl* and *meso* isomers of 1,2-dideutero-1,4-dilithiobutane gave *trans*- and *cis*-1,2-dideuteroethylene, respectively; subsequent metathesis would scramble the stereochemistry and, furthermore, the stereochemistry is fully consistent with retro-cyclometallation. Given this rather convincing experiment [11], the metallocyclopentane mechanism was surprisingly short lived.

## 12.4.3 Disproving the 'pairwise' mechanism for metathesis

In 1970, Chauvin and Hérisson presented a study of the co-metathesis of cycloalkene/alkene mixtures using a $WOCl_4/SnBu_4$ pro-catalyst mixture [12]. Whilst the *fully* quantitative analysis of the product mixtures was made complicated by the range of techniques that were required for the low, medium and high molecular weight products (mono alkenes, telomers and polymers), it became clear that product ratios were not consistent with what would be predicted by *either* mechanism in Scheme 12.14. The analysis and associated mechanistic interpretation were seminal and worthy of consideration in some detail here. The key point is that both mechanisms in Scheme 12.14 are *pairwise*, i.e. each turnover of the catalyst cycle involves two alkenes that undergo concerted alkylidene exchange. When a single alkene, e.g. pent-2-ene ($C_5$), is considered, the products of alkylidene metathesis

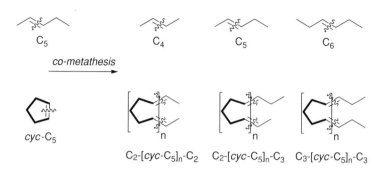

$C_5$        $C_4$        $C_5$        $C_6$

co-metathesis

cyc-$C_5$

$C_2$-[cyc-$C_5$]$_n$-$C_2$     $C_2$-[cyc-$C_5$]$_n$-$C_3$     $C_3$-[cyc-$C_5$]$_n$-$C_3$

**Scheme 12.16**   The co-metathesis of pent-2-ene and cyclopentene to yield mono acyclic alkenes (upper) and high molecular weight telomers and polymers (lower); double bond geometry is arbitrary.

are but-2-ene ($C_4$) and hex-3-ene ($C_6$), *irrespective* of the mechanism, and at equilibrium this will be an approximately 1:2:1 mixture of the $C_4$, $C_5$ and $C_6$ products, Scheme 12.16. Obviously, metathesis of cyclopentene will give polymers in which the monomer is the [(*cyc*-$C_5$)$_n$] unit. It is the cross-metathesis products, which are telomeric and formed from shorter strands of polymeric cyclopentene terminated by acyclic alkene units, that are the key to the analysis. The telomers are of the form ($C_{2,3}$)–[(*cyc*–$C_5$)$_n$]–($C_{2,3}$) where the end units are derived from the termination of a growing cycloalkene polymer through cross-metathesis with an acyclic alkene, thus, but-2-ene ($\rightarrow$ $C_2$ units), pent-2-ene ($\rightarrow$ $C_2$ and $C_3$ units) and hex-3-ene ($\rightarrow$ $C_3$ units).

Consider a telomer being formed from a cyclopentenyl polymer growing under the *pair-wise* mechanism (Scheme 12.14) with growth being curtailed by cross-metathesis under two extreme conditions: (i) with *only* pent-2-ene present ($C_4$:$C_5$:$C_6$ = 0:100:0) and (ii) with a *fully equilibrated* mixture of acyclic monoalkenes ($C_4$:$C_5$:$C_6$ = 1:2:1). Under condition (i), one would expect the formation of only hierarchical telomers ($n$ = 1, 2, 3, 4, 5, etc.) of the type ($C_2$)–[(*cyc*-$C_5$)$_n$]–($C_3$) as the pent-2-ene is split into a $C_2$ and a $C_3$ unit across the growing *cyclo* polyene. In contrast, under condition (ii), one would expect each hierarchical telomer to be formed in a 1:2:1 ratio of ($C_2$)–[(*cyc*-$C_5$)$_n$]–($C_2$):($C_2$)–[(*cyc*-$C_5$)$_n$]–($C_3$):($C_3$)–[(*cyc*-$C_5$)$_n$]–($C_3$), depending on whether there is cross-metathesis with $C_4$, $C_5$ or $C_6$ (ratio = 1:2:1). The outcome will thus depend on how quickly the pent-2-ene is equilibrated by homo-metathesis to yield the $C_4$, $C_5$ and $C_6$ mixture. Analysis of the rate of pent-2-ene homo-metathesis showed that it was not fast. Indeed, it proceeded at approximately the same rate as the telomerisation reaction. One would thus expect the telomer product early in the reaction to be essentially pure ($C_2$)–[(*cyc*-$C_5$)$_n$]–($C_3$) species. Then, as $C_4$ and $C_6$ increase in concentration relative to $C_5$, formation of the ($C_2$)–[(*cyc*-$C_5$)$_n$]–($C_2$) and ($C_3$)–[(*cyc*-$C_5$)$_n$]–($C_3$) telomers should increase proportionally. This was not found to be the case.

In fact, the ($C_2$)–[($C_5$)$_n$]–($C_2$) and ($C_3$)–[($C_5$)$_n$]–($C_3$) telomers were produced *in parallel* with the $C_4$ and $C_6$ alkenes, with the constant statistical telomer ratio of 1:2:1. The result was inconsistent with the pairwise mechanism and suggests that the two ends of the telomer arise from *two different* alkene units; thus, even in the early stages of the reaction, the pent-2-ene can generate *all three* telomer forms. In other words, a *sequential* rather than *pairwise* mechanism seems to have been operative. It was proposed, therefore, that the catalyst acts as a carrier for one half of the alkene and thereby allows sequential reactions of one alkene with

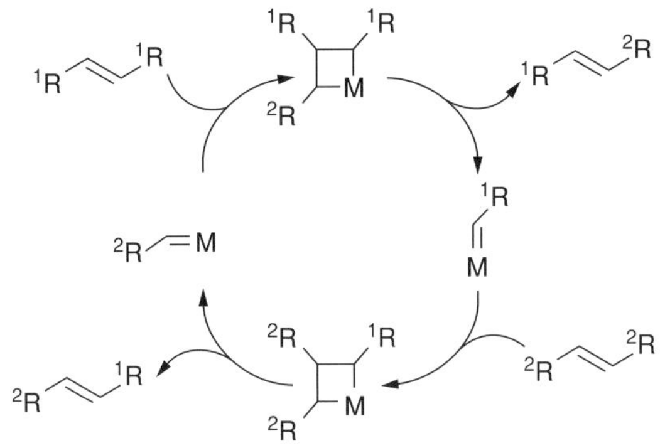

**Scheme 12.17** The *Chauvin mechanism* for alkene metathesis.

two other alkenes. A tungsten alkylidene, or 'carbenoid' complex of the form [W]=CHR, was proposed as the carrier and a chain reaction mechanism proposed, Scheme 12.17. This is now known as the *Chauvin mechanism* and remains accepted to this day across all alkylidene metathesis catalyst systems.

There then followed reports by Katz [13] and Grubbs [14] and their co-workers on studies that aimed to simplify and confirm the analysis. The key remaining issue was whether a modified pairwise mechanism, in which another alkene can coordinate to the metal and equilibrate with the product prior to product displacement, would *also* explain the appearance of the anomalous cross-over products *early* in the reaction evolution. However, a statistical kinetic analysis showed that for a 1:1 mixture of equally reactive alkenes, the *kinetic* ratio of cross-metathesis should be 1:1.6:1 for the *pairwise mechanism* and 1:2:1 for the *Chauvin mechanism*. Any equilibration (substrate or product) would, of course, cause an approach towards a statistical distribution (1:2:1) and thus allow no distinction between the mechanisms.

Clearly, an accurate determination of the ratios is required, without corruption of the data by equilibration. One tactic, of course, is to look only at products emerging in the first few turnovers before equilibration becomes an issue. However, this would be undesirable because of (i) the very low product concentrations at low conversion and (ii) the issues associated with potential complexities in the establishment of the genuine catalyst from the pro-catalyst at the start of the reaction. An alternative approach lay in monitoring all three processes (substrate equilibration, cross-metathesis and product equilibration) against time. If a smooth relationship could be determined, then extrapolating back to time zero allows the accurate deduction of the cross-metathesis ratio at this point. The demanding conditions involved a linear diene as the substrate whereupon RCM generates a cycloalkene which, by careful choice of the diene, is itself *inert* to metathesis, for example cyclohexene or phenanthrene. This ensures non-equilibrating conditions in at least one of the products, the other product being an acyclic alkene. Then, by using some form of label at the termini of the alkene units to generate two distinguishable but essentially identically reactive pairs of diene substrates, the stage was set to test the mechanism reliably.

A series of increasingly refined experiments were performed. The final example in the series is outlined in Scheme 12.18, where *E*/*Z* equilibration is used to monitor substrate (deca-2,8-diene) and product (but-2-ene) equilibration, and a pair of $D_3$-labelled methyl

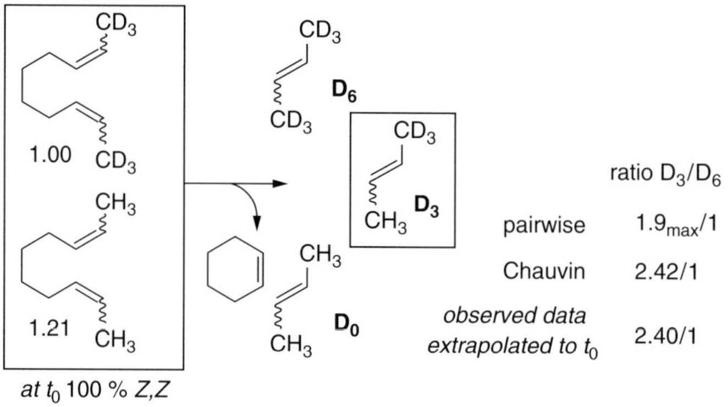

**Scheme 12.18** The co-RCM (ring-closing diene metathesis) of a 1.21:1.00 ratio of $D_0$ and 1,1,1,10,10,10-$D_6$-Z,Z-deca-2,8-diene; see the text for full details.

groups are used to determine cross-metathesis ratios at time zero by extrapolating back from data obtained at conversions up to 30%. At time zero, $D_3:D_6 = 2.4\,(\pm\,0.1)$ for the but-2-ene. The ratios calculated for the *pairwise* and the *Chauvin* mechanisms starting from the 1.21:1 ratio of $D_0/D_6$ Z,Z-deca-2,8-diene are $1$–$1.9_{max}$ and 2.42, respectively.

### 12.4.4  Mechanistic investigation of contemporary metathesis catalysts

After the Chauvin mechanism became established, the pathway to the development of new and improved catalysts became obvious: one needed to use isolable alkylidene complexes as direct precursors to 'on-cycle' intermediates. Over the following two decades or so, and predominantly through work by Schrock and Grubbs and their co-workers, catalyst design for metathesis underwent several 'quantum leaps'. One of the major problems with the early catalysts was functional group tolerance; essentially there was none, and only simple alkenes could be used. Ruthenium alkylidenes are amongst some of the more recent very functional-group-tolerant systems, e.g. the bis(tricyclohexylphosphine) pro-catalysts of the type $[Cl_2(PCy_3)_2Ru=CHR]$ (24) where (a) R = H, (b) R = CH=CPh$_2$ and (c) R = Ph, the so-called first generation (generation I) Grubbs catalysts. These catalytic systems have been the subject of a detailed mechanistic study by Grubbs, predominantly by analysis of the kinetics of the RCM of diethyl diallylmalonate, 10 in Scheme 12.19 [15].

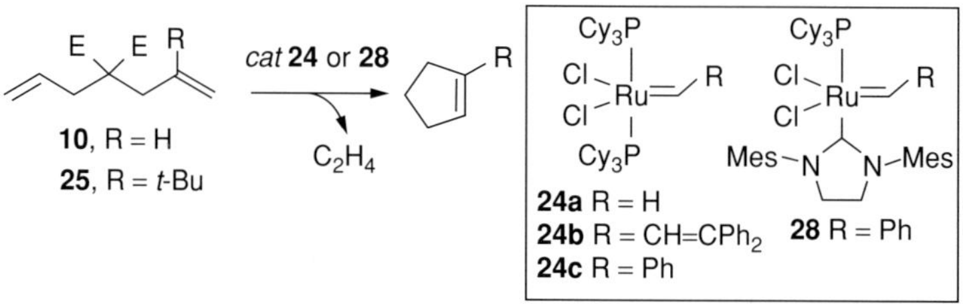

**Scheme 12.19** The RCM of dienes 10 and 25 by generation I (24) and generation II (28) pro-catalysts.

Reactions employing either **24a** or **24b** as pro-catalyst revealed a complex quasi-double-exponential dependence of the rate of RCM on the concentration of the diene (**10**) that defied simple analysis. When a reaction was performed in the presence of an excess of the phosphine ligand (up to 0.02 M excess of $Cy_3P$ with 0.02 M **24a**), the reaction was up to 20-fold slower. Moreover, the rate dependence became pseudo first order in diene and pseudo zero order in Ru, thus giving readily extractable $k_{obs}$ values from a standard linear 'log plot' of diene versus time. This transformation from complex to simple kinetics allowed determination of $k_{obs}$ over a series of excess $Cy_3P$ concentrations ([$Cy_3P$]$_{xs}$) and the establishment of an inverse relationship with $k_{obs} = (\{2.73 \times 10^{-5}/[Cy_3P]_{xs}\} + 5.27 \times 10^{-4})$ min$^{-1}$. There is, therefore, a first-order inhibitive effect arising from the excess of $Cy_3P$ and, because extrapolation gives a non-zero intercept on the $y$-axis ($k_{obs} = 5.27 \times 10^{-4}$ min$^{-1}$ when [$Cy_3P$] is infinite), there must be an additional process that *is not* inhibited by excess $Cy_3P$. Since the methylidene pro-catalyst (**24a**) does not need to undergo cross-metathesis to enter the catalytic cycle, there are no issues concerning induction rate. Analysis of the effect of pro-catalyst concentration on the pseudo-first-order rate constant, i.e. plotting $k_{obs}$ determined from a series of runs with different [**24a**]$_0$ values gave a good linear correlation and thus a first-order rate dependence on [Ru]. The $y$-intercept was, reassuringly, found to be zero: there should be no metathesis in the absence of added catalyst!

Returning to reactions conducted *without* added $Cy_3P$, four runs were performed with different initial concentrations of **24a** and the gradients of the quasi-double-exponential curves were determined at numerous points along the curves. The comparison of the relative rates (tangential gradients) between runs, at points that correspond to identical conversions (and thus diene concentrations), gave an *approximately* half-order relationship between the rate and [Ru]. Thus, as [**24a**]$_0$ was increased from 0.005 M to 0.010 M, 0.015 M and 0.02 M (concentration increases by factors of 2, 3 and 4), the rate was found to increase by factors of 1.33, 1.61 and 1.81. For a half-order relationship (rate $\propto$ [Ru]$^{0.5}$), the rate would have increased by factors of $2^{0.5}$, $3^{0.5}$ and $4^{0.5}$ (i.e. 1.41, 1.73 and 2.00). Such a relationship shows that there is a reversible ligand dissociation occurring from Ru, in which the dissociated form of the Ru complex undergoes turnover.

Operating within the framework of the Chauvin mechanism, the main consideration for the reaction mechanism is the order of events in terms of addition, loss and substitution of ligands around the ruthenium alkylidene centre. Additionally, there is a need for two pathways (see above), both being first order in diene, one with a first-order dependence on [Ru] and the other (which is inhibited by added $Cy_3P$) with a half-order dependence on [Ru]. From the analysis of the reaction kinetics and the empirical rate equation thus derived, the sequence of elementary steps via two pathways was proposed, one 'non-dissociative' (**I**) and the other 'dissociative' (**II**), as shown in Scheme 12.20. The *mechanism-derived* rate equation is also shown in the scheme and it can thus be seen how the constants '*A*' and '*B*' relate to elementary forward rate constants and equilibria in the proposed mechanism.

A key part in the analysis and proposal was 'electron accountancy' – on the basis of the usual propensity for the adoption of a noble gas outer shell configuration – and all of the proposed intermediates have 16- or 18-electron configurations except the metallocyclobutane (**26**) in pathway **II** which has 14. On this basis, one might expect that pathway **I**, the non-dissociative pathway, would predominate over pathway **II**. In the presence of a large excess of $Cy_3P$, which was used only to simplify the kinetic analysis, this is clearly the case. However, under the

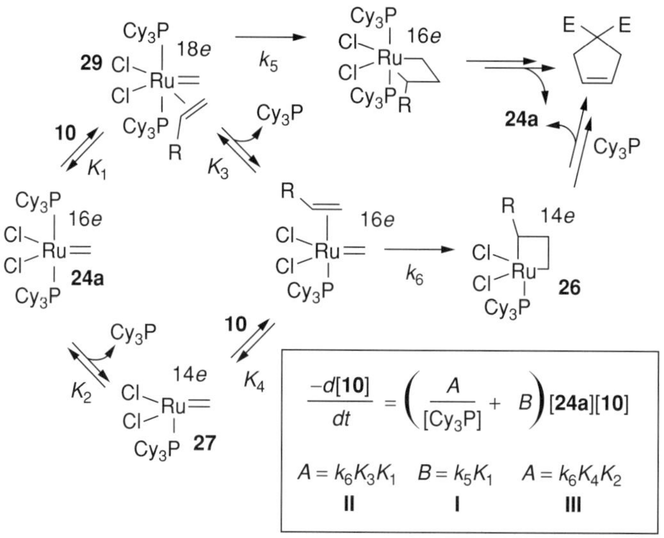

**Scheme 12.20** Pathways I ($K_1 k_5$) and II ($K_1 K_3 k_6$), the mechanism proposed for the RCM of **10** on the basis of the empirical rate equation and the '18-electron rule'; pathway III ($K_2 K_4 k_6$) involves two 14-electron intermediates (**26** and **27**) and was not considered viable in the initial analysis; inset: the mechanism-derived rate equation.

conditions of the *synthetic* reaction, where no excess Cy$_3$P is added, it was concluded from the relationship $k_{obs} = (\{2.73 \times 10^{-5}/[Cy_3P]_{xs}\} + 5.27 \times 10^{-4})$ min$^{-1}$, see above, that pathway **II** dominates by a factor of more than 20.

There is an alternative pathway to **II**, in which the phosphine dissociates *before* the alkene group coordinates: pathway **III**. On the basis of electron accountancy alone, this should be viewed as unfavourable as it involves two 14-electron intermediates (**26** and **27**). However, it should be noted that the *mechanism-derived* rate equation for reaction via pathways **I/III** rather than **I/II** would be *equally consistent* with the empirical rate equation.

After further investigation, and arising from advances in the application of *N*-heterocyclic carbenoids (NHC) as highly electron-donating ligands for transition metal complexes, 'second generation' (generation II) pro-catalysts of the form [Cl$_2$(PCy$_3$)(NHC)Ru=CHPh] (e.g. **28**, Scheme 12.19) were developed [16]. These pro-catalysts afford far more efficient catalytic systems, giving higher turnover rates and better catalyst longevity. The latter point can be ascribed to the bulk of the NHC providing steric inhibition of the bimolecular self-destruction of the ruthenium methylidene intermediate, this having been found to be the major catalyst decomposition pathway for the first generation system. The higher efficiency of the generation II system is exemplified neatly by the comparison of the RCM of the dienes **10** and **25** (Scheme 12.19). With 5 mol% of either **24c** or **28**, diene **10** is quantitatively converted in less than 10 minutes at 45°C. With **25**, a significantly more hindered alkene, the first generation pro-catalysts give no product at all over periods up to 60 minutes. In contrast, with second generation **28** the conversion is *quantitative*. The origin of this increased activity is surprising and was elucidated by a detailed kinetics investigation using advanced NMR techniques [17].

### 12.4.5 NMR studies of degenerate ligand exchange in generation I and generation II ruthenium alkylidene pro-catalysts for alkene metathesis

According to the mechanism for Ru-alkylidene-catalysed metathesis outlined in Scheme 12.20 (pathway II dominating by >95%), the increased reactivity of the second generation catalysts was suggested to originate from the highly electron-rich NHC ligand promoting more rapid $Cy_3P$ dissociation from the 18-electron alkene complex analogous to **29**. To study this without the complication of metathesis, the rates of degenerate exchange of $Cy_3P$ in complexes **28** and **24c** with excess $Cy_3P$ were studied by $^{31}P\{^1H\}$NMR in $d_8$-toluene, in effect using $Cy_3P$ as a surrogate for the alkene. The rate of exchange was found to be slow, having no measurable effect on the line-width of the $^{31}P\{^1H\}$NMR signals of either the complexed or free $Cy_3P$ up to 100°C. Consequently, one-dimensional 'magnetisation transfer' experiments were conducted. This technique allows reasonably accurate measurement of kinetics for site-exchange of NMR-active nuclei where the rate is too fast to measure by classical (laboratory or 'real' time frame) experiments, but too slow to have an impact on line shape [18].

Using this technique (inverting the spin of the free $Cy_3P$ and then studying magnetisation transfer into the Ru-bound $Cy_3P$), Grubbs and co-workers found that the rate constants for degenerate $Cy_3P$ exchange in **24c** and **28** were *unaffected* by the concentration of free $Cy_3P$ and large *positive* entropies of activation were found from Eyring-type analysis of the rate constants determined at different temperatures. These results conclusively established that the mechanism for exchange is not analogous to pathway II (Scheme 12.20). However, the results are fully consistent with a process in which a free $Cy_3P$ ligand occupies a vacant site generated by *prior* dissociation of a bound $Cy_3P$ ligand, analogous to pathway III (Scheme 12.20). What was equally surprising was that the rate constant for exchange was significantly *smaller* in the second generation pro-catalyst **28** than in the first generation **24c** (0.13 and 9.6 s$^{-1}$, respectively, at 80°C) – exactly the opposite to what had been expected based on their activity in metathesis.

In order to investigate this point more fully, the rates of reaction of the two complexes with ethyl vinyl ether (EVE) were studied. This alkene was chosen as it is rather reactive towards ruthenium alkylidene complexes and forms an inert alkoxyalkylidene product in an essentially irreversible manner. This alkene, therefore, should rapidly capture any nascent complex from which a $Cy_3P$ ligand has dissociated (**27** and **30** in Scheme 12.21). The two complexes displayed very different kinetics. The rate of reaction of the first generation pro-catalyst complex **24c** with EVE was found to be dependent on EVE concentration (over a range of 30–120 equivalents of EVE) and did not reach pseudo-first-order conditions

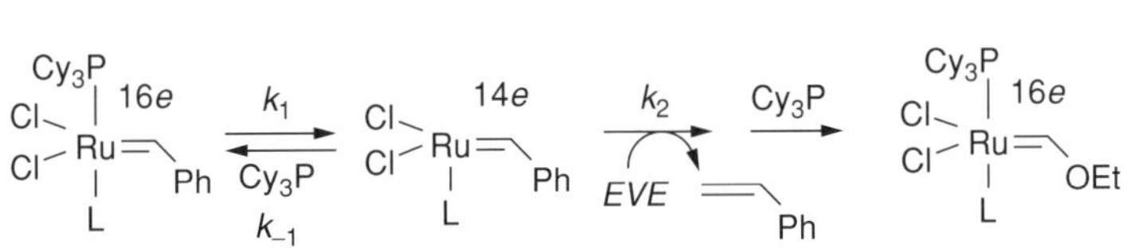

**Scheme 12.21** Contrasting kinetics for the irreversible reactions of first and second generation ruthenium alkylidene complexes **24c** and **28** with ethyl vinyl ether (EVE).

(i.e. $k_{obs}$ did not became independent of EVE concentration) until a *vast* excess of EVE was present. Using 5300 equivalents, and monitoring the kinetics by UV, the pseudo-first-order rate constant, $k_{obs}$, was determined as $180 \times 10^{-4}$ s$^{-1}$ at 20°C. By extrapolation of the Eyring analysis from the variable temperature magnetisation transfer experiments to 20°C, the rate constant of Cy$_3$P ligand exchange was predicted to be $160(\pm 20) \times 10^{-4}$ s$^{-1}$, in good agreement. This result shows that the reaction of the free Cy$_3$P (added or nascent from dissociation) with **27** is competitive with EVE.

In stark contrast, EVE reacted with the second generation complex **28** in a pseudo-zero-order manner with as little as ca. five equivalents of EVE. At 35°C, $k_{obs}$, the pseudo-first-order rate constant, was $4.6 \times 10^{-4}$ s$^{-1}$ and, again by extrapolation of the Eyring analysis to 35°C, the rate of Cy$_3$P ligand exchange was predicted to be $4(\pm 3) \times 10^{-4}$ s$^{-1}$. This excellent agreement shows that the rate-limiting step for the reaction with EVE is the Cy$_3$P ligand dissociation, and that capture of **30** by EVE is very rapid – much more so than the reaction of **27** with Cy$_3$P ($k_{-1} \ll k_2$).

By conducting further experiments in which the linear relationship between $k_{obs}$ and [EVE]/[Cy$_3$P] was determined under pseudo-first-order conditions in EVE, the rate ratios for $k_2/k_{-1}$ could be determined. The results are profound: for **24c**, $k_2/k_{-1} = 1.25$ (50°C); for **28**, $k_2/k_{-1} = 15\,300$ (37°C). This conclusively shows that although Cy$_3$P dissociation from the second generation pro-catalyst (**28**) to give **30** is some two orders of magnitude slower than **24c** → **27**, the reactive complex **30** effects many turnovers of metathesis before recapture by Cy$_3$P. In contrast, the reactive complex **27** from the first generation catalyst achieves only one or two turnovers before recomplexing Cy$_3$P and requiring a repeat initiation event.

### 12.4.6    *Summary and mechanistic conclusions*

Over a period of some 30 years, metathesis catalysis developed phenomenally, giving systems that now enjoy broad application. Early mechanistic experiments employed product distribution analysis to distinguish pairwise from sequential mechanisms, and so established a metal alkylidene complex as the propagating complex (the Chauvin mechanism). Most modern catalyst systems are now based on Mo (Schrock-type systems) or Ru (Grubbs-type systems). The joint award of the 2005 Nobel Prize in Chemistry to Chauvin, Schrock and Grubbs recognises the pioneering and widespread contributions that they have independently made to this still-developing field.

A detailed kinetic study on the first generation of the latter systems (e.g. **24c**) suggested a mechanism in which phosphine dissociation rate was crucial for efficient turnover. The emergence of second generation systems (e.g. **28**) in which one phosphine was replaced by an *N*-heterocyclic carbenoid ligand (NHC) gave pro-catalysts that were appreciably more active. Consideration of the mechanism established for **24c** led to the conclusion that the greater efficiency arises from the NCH inducing more rapid phosphine dissociation. Using advanced NMR techniques to measure these processes, together with a reactive alkene (ethyl vinyl ether, EVE), showed that this was not in fact the case. The outcome is more subtle: it is the competition between recomplexation of the Cy$_3$P ligand versus alkene complexation (and thus metathesis) that is the key issue, i.e. $k_2/k_{-1}$ in Scheme 12.21. This somewhat non-intuitive result will undoubtedly lead to further innovation in the field and demonstrates the level of insight that well-designed NMR kinetic experiments can provide.

# References

1. Sakai, M., Hayashi, H. and Miyaura, N.(1997) *Organometallics*, **16**, 4229.
2. Takaya, Y., Ogasawara, M., Hayashi, T., Sakai, M. and Miyaura, N. (1998) *Journal of the American Chemical Society*, **120**, 5579.
3. (a) Darensbourg, D.J., Grotsch, G., Wiegreffe, P. and Rheingold, A.L. (1987) *Inorganic Chemistry*, **26**, 3827; (b) Keim, W.J. (1968) *Organometallic Chemistry*, **14**, 179.
4. Hayashi, T., Takahashi, M., Takaya, Y. and Ogasawara, M. (2002) *Journal of the American Chemical Society*, **124**, 5052.
5. Hayashi, T., Ueyama, K., Tokunaga, N. and Yoshida, K. (2003) *Journal of the American Chemical Society*, **125**, 11508; see also Fischer, C., Defieber, C., Suzuki, T. and Carreira, E.M. (2004) *Journal of the American Chemical Society*, **126**, 1628.
6. (a) Schmitz, E., Heuck, U. and Habisch, D. (1976) *Journal für Praktische Chemie*, **318**, 471; (b) Grigg, R., Mitchell, T.R.B. and Ramasubbu, A. (1979) *Journal of the Chemical Society, Chemical Communication*, 669.
7. For leading references see (a) Widenhoefer, R.A. (2002) *Accounts of Chemical Research*, **35**, 905; (b) Lloyd-Jones, G.C. (2003) *Organic Biomolecular Chemistry*, **1**, 215.
8. Banks, R.L. and Bailey, G.C. (1964) *Industrial and Engineering Chemistry, Product Research and Development*, **3**, 170.
9. Calderon, N., Chen, H.Y. and Scott, K.W. (1967) *Tetrahedron Letters*, **8**, 3327; see also Calderon, N. (1972) *Accounts of Chemical Research*, **5**, 127.
10. Lewandos, G.S. and Pettit, R. (1971) *Tetrahedron Letters*, **12**, 789.
11. Grubbs, R.H. and Brunck, T.K. (1972) *Journal of the American Chemical Society*, **94**, 2538.
12. Hérisson, J.-L. and Chauvin, Y. (1970) *Makromolekulare Chemie*, **141**, 161.
13. Katz, T.J. and McGinnis, J. (1975) *Journal of the American Chemical Society*, **99**, 1903, and earlier work from Katz et al. cited therein.
14. Grubbs, R.H. and Hoppin, C.R. (1979) *Journal of the American Chemical Society*, **101**, 1499, and earlier work from Grubbs et al. cited therein.
15. Dias, E.L., Nguyen, S.T. and Grubbs, R.H. (1997) *Journal of the American Chemical Society*, **119**, 3887.
16. (a) Scholl, M., Ding, S., Lee, C.W. and Grubbs, R.H. (1999) *Organic Letters*, **1**, 953; (b) Huang, J., Stevens, E.D., Nolan, S.P. and Peterson, J.L. (1999) *Journal of the American Chemical Society*, **121**, 2674; (c) Weskamp, T., Kohl, F.J., Hieringer, W., Gliech, D. and Herrmann, W.A. (1999) *Angewandte Chemie (International Edition in English)*, **38**, 2416.
17. Sanford, M.S., Ulman, M. and Grubbs, R.H. (2001) *Journal of the American Chemical Society*, **123**, 749.
18. Orrell, K.G., Sik, V. and Stephenson, G. (1990) *Progress in Nuclear Magnetic Resonance Spectroscopy*, **22**, 141.

# Index

accelerating rate calorimetry (ARC), 200
acetal hydrolysis
   acid catalysed, 294
   concerted, 295
   stepwise, 295
acetaldehyde, 49, 95, 97, 175
   p$Ka$, 95
acetic anhydride, hydrolysis, 12, 72, 206, 213
acetohydroxamate, as a base in elimination, 241
acetone, acid-base catalysed halogenation, 307
acetylacetone, 333
acetylimidazole, 308
   in catalysed hydrolysis of 4-nitrophenyl
      acetate, 309
N-acetylimidazolium cation, 43
1-(2-acetoxy-2-propyl)indene, competing
      rearrangement and elimination, 256
acid catalysis, stepwise, 295
acid-base catalysis, bifunctional (push-pull),
      307
acid-base pre-equilibrium, 92
acidity, sigmoidal dependence of rate
      constants, 92
acrolein, addition of radicals, 267
acrylonitrile
   electrochemical reactions, 10
   electrohydrodimerisation, 128, 158
activated complex, 4, 50, 227
activation energy, 206
activation of organometallic catalyst, 325
activation parameters, 15
active and dormant states of alkoxyamines, 270
active-site-directed inhibitors, 315
active-site titration of enzyme, 313
   α-chymotrypsin, 315
   papain, 315
activity, 49, 83
activity coefficient, 49, 63
acyl enzyme intermediate, 313

acyl-α-chymotrypsin, 314
   isolation, 319
acylal enzymes, 318
acylal intermediate, 320
acylal-lysozyme intermediate, X-ray
      crystallography, 320
acylation
   of amines and phenols, 9
acylium cation, 14
acyl-oxygen cleavage, 25
adamantane, 30, 35
adamantanol, 4
adamantene, 36
adamantyl cation, 3
1-adamantyl derivatives, 32, 35
2-adamantyl tosylate, solvolysis, 244
addition as a basic reaction, 231
addition
   of bromine to alkenes, stereospecificity 247
   of methoxide to 1,4-dimethoxybenzene, 132
   of nucleophiles to carbonyl groups, spectator
      mechanism, 306
adiponitrile by electrohydrodimerisation of
      acrylonitrile, 10, 128, 158
σ-adduct, 237
Ag/AgCl electrode, 135
AIBN, 283
Albery-More O'Ferrall-Jencks diagrams, 14
Aldol reaction, 49, 95, 98, 328
aldolases, covalently bound intermediates, 318
aliphatic acid chlorides
   hydrolysis, 122
   substitution reactions, 105
alkaline cyclisation of uridine 3′-phosphate
      esters, 257
alkaline phosphatase mechanism, 320
   covalently bound intermediates, 318
alkanediazonium ion, 7
alkane-oxodiazonium ion, 7

alkene metathesis, 327
alkenes
addition of dihalocarbene, 9
addition of HBr, 18
reaction with *N*-methyltriazolinedione, 253
reaction with *N*-phenyltriazolinedione, 253
alkoxides, epimerisation, 25
alkoxyamines, 268
as initiators, 274
dissociation, 270
homolysis, 270
steric compression, 270
alkyl anions, 287
alkyl arenesulfonates, solvolysis, 90
alkyl halides
electrochemical reduction, 131
solvolysis, 90
alkyl metathesis, 344
alkyl radicals, 285
alkyl radical traps, 269
1,2-alkyl rearrangement, 22, 25
alkylation by PTC, 112
alkylidene metathesis, 344
allylamine, catalysed isomerisation to enamine, 334
AM1, 174
amination of halobenzenes, 26
amines
acylation, 9
oxidation by HOCl, 69
reaction with chloranil, 109
aminium radical cation, from *N*-phenylglycine, 70
amino acids, chlorination, 92
aminolysis, 38
aminyl radical, from *N*-phenylglycine, 70
amperometry, 143
analysis by NMR, GC, or HPLC, 33
anchimeric assistance, 188
$A_ND_N$ mechanism, 14
anharmonic vibration, 169
aniline by hydrogenation of nitrobenzene, 222
aniline, reaction with cyanic acid, 303, 306
annulation reactions, 273
anode, 127
anomeric effect, 71, 187
anthracene, electrochemical reduction, 10, 130
anthracene radical anion, 130, 141
protonation by phenol, 146, 153, 163
*anti*-addition of bromine to alkenes, 247
anti-aromaticity, 232
anti-Markovnikov addition, 18, 266
arenediazonium ion, 7
thermal decomposition, 75

arene-oxodiazonium ion, 7
aromatic stabilisation, 171
aromaticity, 168, 176, 232
Arrhenius equation, 61, 266, 272
Arrhenius frequency factor $(A)$, 266
Arrhenius parameters, 265
Arrhenius plot, 206
artificial enzymes, 294
aryl acetates, imidazole catalysed hydrolysis, 308
3-aryl-2-butyl sulfonates, 40
aryl cation, 29
aryl radical intermediate, 286
5-*exo*-cyclisation, 279
arylhydrazine, oxidation with oxygen, 116
aspirin, hydrolysis, 295
assay of enzyme concentration, 312
associative mechanism/process, 14, 231
asymmetric
induction, 15
isomerisation, 334
phenylation, 15
transition-metal-catalysed C—C bond formation, 327
asynchronous concerted mechanism, 14
atmospheric investigations, 76
atom abstraction reactions of radicals, 263
atom economical reaction, 334
atomic force microscopy (AFM), 123
atomisation energy, 171
attenuated total reflectance infrared spectroscopy (IR-ATR), 11, 200, 205
attenuated total reflectance, 235
autocatalysis, 52, 293
automated sensitivity analysis, 211, 217
azide clock, 13, 42
azide ion, 40
aziridinium ions as a transient intermediates, 254
1,1'-azobis(cyclohexane-1-carbonitrile) (ACN) 265
2,2'-azobis(isobutyronitrile) (AIBN) 265
azo-compounds as initiators, 265

B3LYP, 180, 326
Baldwin's rules, 263
base strength, rate dependence on, 98
basis set superposition error (BSSE), 190
basis sets, 181
Beer-Lambert law, 67, 205, 209
bell-shaped curve, 92
benzene
aromatic stabilisation, 171
nitration, 104

benzhydryl chloride, solvolysis and common ion rate effect, 241

benzoyl peroxide (BPO), 265

*N*-benzoylaminothionacetyl papain, 319

benzoylglycine esters as enzyme substrates, 321

benzyl azoxytosylate, solvolysis, 245

benzyl bromide, multiphase reaction with acetate, 115

benzyne intermediates, 20, 27, 242

β-carbopalladation, 338

β-hydride elimination, 338, 342

β-fragmentation, 265

β-scission, 263, 277

biadamantylidene, reaction with *N*-methyltriazolinedione, 255

bifunctional acid-base catalysis, 307

bimolecular mechanism/step, 1, 5, 49, 80, 86, 87

BINAP, 328

biphasic media, 63

biradical, 25

biradicaloid mechanisms, 174

1,4-bisdiphenylphosphinobutane (DPPB), 328

2,2′-bisdiphenylphosphino-1,1′-binaphthalene (BINAP), 328

2,5-bis(spirocyclohexyl)-3-benzylimidazolidin-4-one-1-oxyl (NO88Bn), 271

2,5-bis(spirocyclohexyl)-3-methylimidazolidin-4-one-1-oxyl (NO88Me), 271

bis(tricyclohexylphosphine) pro-catalysts, 348

block copolymers, 275

Bodenstein (steady state) approximation, 238, 243, 310

intermediate, 242

bond dissociation energy (BDE), 176

bond energy bond order (BEBO) assumption, 252

bond lengths, $r_0$, $r_e$, 169

Born-Oppenheimer approximation, 167

bornyl cinnamate, hydrodimerisation, 154

borohydride in stereoselective reduction, 242

boronates, 328

break points in linear free energy relationships, 257

1-bromoadamantane, 37

*p*-bromomethoxybenzene, 26

bromonium ion intermediate, 247

1-bromo-octane in radical reactions, 283

bromotrimethylbenzene, 28

Brønsted
  acids and bases, 13
  catalysis law, 6
  coefficient, 137
  equation in catalysis, 298

parameters in catalysis, 299

plots, scatter as criterion of nucleophilic catalysis, 302

relationship, 73, 256

Brookhaven protein data bank, 320

buffer
  concentration, 64
  ratio, 62
  solutions, 294

burst kinetics, 319
  in α-chymotrypsin catalysis, 315

Butler-Volmer equation, 137, 140

*tert*-butoxy radical, 266, 267, 277

*tert*-butyl cation, 35

2-[*N*-*tert*-butyl-*N*-(1-diethyloxyphosphoryl-2,2-dimethylpropyl)aminoxyl] (SGI), 276

*tert*-butyl hydroperoxide, 216

*tert*-butyl peroxide (TBP), 265

butyl acrylate in heterogeneous Heck reaction, 246

*n*-butyl formate, reaction with hydroxide, 109

*n*-butylamine, reaction with pyridyl ketenes, 8

$^{13}$C-NMR spectroscopy, 237, 281

$^{14}$C labelling, 27

cage return, 266

calorimetry, 11, 199

canonical orbitals, 173

capillary dilatometer, 74

carbanion, 182
  formation in PTC, 112
  in E1cB elimination, 310
  intermediates, 242
  trapping, 50

carbenium ion, 40
  nonclassical, 90

carbenoid complex, 347

carbocation, 90, 175
  intermediates, 34
  lifetime, 13, 42
  under stable ion conditions, 248

carbon monoxide, 104

carbon-centred radical, 268, 270

carbonyl, nucleophilic addition, 4, 304

carbonylation, 104

carboxamides,
  nucleophilic attack, 302
  proton-catalysed addition of water, 304
  proton transfer in nucleophilic substitution, 302

Cartesian coordinates, 170

catalysis, 13, 80, 234
  by cyclodextrins, 294
  by enzymes, 311

by metal ions, 294
by micelles, 294
by organometallics, 13
by proton transfer, demonstration of
    mechanism, 302
electrophilic, 308
experimental demonstration, 294
nucleophilic, 43, 308
pH-dependence, 294
catalyst, 21
  deactivation, 223
  definition, 293
  negative, 293
  poisons, 326
  promoters, 326
catalytic
  cracking, 104
  cycle, 325
  hydrogenation, 104
  turnover, 324
cathode, 127
cationic intermediates, 40
cell membranes, 104
$\beta$-cellobiosyl fluoride as substrate for
    lysozyme, 319
chain carriers, propagation cycles, 268
chain radical reactions, 262, 263, 268
chain SET redox mechanism ($S_{ET}2$), 290
chain termination, 264
chair-like (Beckwith) transition structure, 263
channel flow cell, 118
charge distribution, 299
charge transfer (CT) complex, 288
Chauvin mechanism, 347, 352
chemically induced dynamic nuclear
    polarisation (CIDNP), 13, 32
chiral ligands, 328
chiral norbornadiene ligand, 334
chloranil
  hydrolysis, 119
  reaction with amines, 109, 123
chloroarenes, reaction with fluoride, 109
4-chlorobenzaldehyde, substituted hydrazone
    formation, 304
chlorobenzene, amination, 27
1-chloronorbornane, 30
*p*-chlorobenzoyl chloride, 38
4-chlorophenol, 102
chlorotrimethylbenzene, 28
chromate, for oxidation, 100
chronoamperometry, 143
$\alpha$-chymotrypsin, 320
  active-site titration, 315
  cinnamoyl intermediate, 319
  mechanism of catalysis, 314

X-ray crystallography of
    indolylacryloyl-intermediate, 319
cinnamic esters, electrohydrodimerisation, 153
cinnamoyl-$\alpha$-chymotrypsin, 319
circular disc electrode, 160
*cis-trans* isomerisation of stilbene, 59
click chemistry and catalysis, 312
closed-shell singlet state, 177
collision theory, 7
combined determination of isothermal
    calorimetric and IR data, 211
combined glass electrodes, 73
co-metathesis, 345
common ion (rate depression) effect, 35, 43
  in solvolysis of benzhydryl chloride, 241
competitive inhibition of enzymes, 317
competitive reactions, 83
complex, enzyme-substrate, 313
computational chemistry, 6
concentration gradient, 108
concerted mechanism/reaction, 2, 6, 13, 25, 31,
    228, 232, 247
  degree of coupling, 14
  diagnosis from linear free energy
    relationships, 311
  enforced, 7, 307
concerted proton transfer, 307
conductimetry, 71
conductivity cell, AC, DC, 72
configuration interaction, 178
configuration state function (CSF), 178
confluence microreactor, 122
conformational
  change in alkaline phosphatase mechanism,
    320
  interconversions, 71, 235
  preferences, 187
  radical clock, 279
  space, 174
conjugate (1,4) addition, 328
consecutive reactions, 60, 81
contact ion pair, 35, 90
continuous
  on-line monitoring, 65
  static monitoring, 65
control elements in ligand architecture, 334
convex breaks, absence of as evidence for
    concerted mechanisms, 311
Coriolis coupling, 192
correlation analysis, application to solvent
    effects, 62
co-solvents, 63
Cottrell equation, 143
coulometry, 162, 163
Coulson-Fischer orbitals, 187

counter electrode, 133, 135
coupled cluster (CC) theory, 178
coupling of bond changes, 307
Cr(VI), 100, 101, 102
*o*-cresol, 102
cross coupling reactions involving radicals, 271, 273
cross-correlation effects, 299
cross-metathesis, 349
crossover experiments, 20, 26, 29, 231, 338, 340, 347
crystal structure predictions, 193
cul de sac
  mechanism, 236
  species, 237, 242
cumyl radical, 269
cuprates, addition to enones, 330
current density in electrochemistry, 138
curvature in linear free energy relationship
  as evidence for an intermediate, 257
curve-fitting software, 313
cyanic acid, reaction with aniline, 303, 306
2-cyanoisopropyl radical, 266, 267
2-cyanophenol, deprotonation by bicarbonate, 115
cyclic voltammetry, 138, 147, 285
3-*exo*-cyclisation, 263
5-*exo*-cyclisation of an aryl radical, 279
cycloaddition of TCNE and propenyl methyl ether, 248, 249
[2 + 2]-cycloaddition, 248
cyclodextrins as catalysts, 294
cycloheptatriene, reaction with maleic anhydride, 240
*trans*-cycloheptene, reaction with *N*-methyltriazolinedione, 255
cyclohexadienyl radical, 283
cyclohexenone, 328, 329
  asymmetric phenylation, 15
cyclohexyl radical, 267
cyclo-isomerisation, 334
1,5-cyclo-octadiene, 331, 334
*trans*-cyclo-octene, reaction with *N*-methyltriazolinedione, 255
cyclopentenyl polymer, 346
cyclopentylmethyl radical, 274
cyclopropane, 175, 181
cysteine proteases, covalently bound intermediates, 318
cytochrome P450 oxidation reactions, 278

data handling, 54
DDT, elimination of HCl, 74
deactivation of catalyst, 223
dead time after mixing reactants, 60

deamination, 7
Debye-Hückel limiting law, 63
decay time, 139
dediazoniation, 7
degenerate ligand exchange, 351
delocalisation, 176
density functional theory (DFT), 173, 179, 326
depolymerisation studies by dilatometry, 75
deprotonation of acetone, Brønsted relationships, 298
derivative cyclic voltammetry, 152
detection of intermediates, 308
deuterium-NMR spectroscopy, 281, 282
diallylmalonates, 335
diazoacetylpyridines, reaction to give pyridylketenes, 69
diazotization of amines, 109, 247
diborane, addition to alkenes, 19
2,5-di-*tert*-butyl-1,4-benzoquinone, sequential epoxidation, 216, 219, 221
2,6-di-*tert*-butyl-4-(3,5-di-*tert*-butyl-4-oxocyclohexa-2,5-dien-1-ylidene)phenoxyl (galvinoxyl), 271
2,6-di-*tert*-butylpyridine, 162
dication by oxidation of radical cation, 162
2,2-dichloronorbornane, 30
dichromate regeneration, 128
dielectric constant, 63
1,6-dienes, cyclo-isomerisation, 334
differential equations, 87
  for a rate law, 199, 230
differential method in kinetic analysis, 51
differential scanning calorimetry (DSC), 11, 200
differential thermal analysis (DTA), 11, 200
diffusion, effect of viscosity, 304
diffusion coefficient, 105, 139
diffusion controlled limit, 98
  criterion of catalytic mechanism, 301
  as rate-limiting step, 306
  reactions of radicals, 262
diffusion film/layer, 105, 106, 140, 155, 156
diffusion-controlled random termination, 275
digestive processes, 104
digital simulation, 142
dihalocarbene, addition to alkenes, 9
dihalocyclopropane formation, 9
dihydroanthracene, 10
dilatometry, 74
dimerisation, 10
1,4-dimethoxybenzene, 1,4-addition of methoxide, 132
*p, p'*-dimethylbenzhydryl chloride, 41, 42

2,4-dimethyl-3-ethylpyrrole
    oxidation, 162
    radical cation, dimerisation, 156
*N,N*-dimethyl-*p*-phenylenediamine, reaction
        with chloranil, 123
2,6-dimethylpyridine, 302
2,4-dimethylpyrrole, oxidation, 162, 164
*p*-dinitrobenzene (DNB) as an electron
        acceptor, 288
dinitrogen pentoxide, rate law for
        decomposition, 54
2,4-dinitrophenyl chloride, reaction with
        hydroxide, 237
diode array detector, 68
1,3-dioxane, anomeric effect, 187
2,2-diphenyl-1-picrylhydrazyl, 99
dipolar aprotic solvents
    in halex reactions, 109
    in radical reactions, 289, 291
disconnections, 15
discontinuity in linear free energy relationship
        as evidence for an intermediate, 257
disinfection of water, 91
dismutation, 344
dispersed liquid-liquid systems, 114
dispersed solids, 115
disproportionation, 266, 273
dissociated ion pair, 35, 90
dissociation energy, 175
dissociation of alkoxyamines, 270
dissociative electron transfer, 131
dissociative mechanism, 14, 231, 349
distribution coefficient, 104
di-*tert*-butyl
    hyponitrite, 284
    butyl peroxide, 282
di-*tert*-butyl nitroxide (DTBN), 288
donor-acceptor bonding, 185
dormant and active states of alkoxyamines, 270
dosing process, 208, 213
double beam spectrophotometers, 67
double inversion of configuration, 320
double potential step chronoamperometry
        (DPSCA), 145
dying of textiles, 104
dynamic magnetic resonance spectroscopy for
        kinetics, 71, 235

Eadie-Hofstee plots and enzyme catalysis, 316
    for non-competitive enzyme inhibition, 318
    for uncompetitive enzyme inhibition, 318
Eigen mechanism, 302, 306
    plots, 302
electroactive intermediates, 160
electroanalytical methods, 10, 132

electrochemical
    addition, 128
    cells, 132
    double layer, 138
electrochemical reactions classified, 128
    catalysis, 128
    cleavages, 128
    coupling and dimerisation, 128
    elimination, 128
    hydrogenation of acrylonitrile, 158
    processes and liquid-solid interfaces, 106
    substitution, 128
electrochemical transfer coefficient, 137
electrode potential, 132
electrohydrodimerization, 10
    of acrylonitrile, 158
    of cinnamic esters, 153
electrolysis cell, 139
electron acceptors, 288
electron accountancy, 349
electron correlation, 174, 192
electron correlation energy, 175
electron decomposition analysis (EDA), 186
electron density, 167, 173
electron spectroscopy, 192
electron spin resonance (ESR) spectroscopy,
        13, 262
electron transfer, 10, 71, 99, 127, 131, 140, 231,
        137, 284
    irreversible, 138
    quasi-reversible, 138
    reversible, 138
electron-donating ligand, 350
electronic energy, absolute, 170
electrophiles, 231, 232
    acid- and base-catalysed reactions with
        ketones, 240
electrophilic catalysis, 308
electrophilic radicals, 267
electrospray mass spectrometry, 320, 326
electrostatic potentials, 173
electrostatics, 168
electrosynthesis, 158
elementary rate constant, 84, 87
elementary reaction/step, 1, 4, 49, 81, 87
elimination, 231
empirical rate law, 7
enamine by catalysed isomerisation of an
        allylamine, 334
enantio-induction, 331
enantiomeric excess in a radical reaction, 280
enantiomerisation, 83
enantioselective synthesis using enzymes, 312
enantioselectivity, 328
encounter complex, 1, 63, 228

energy barrier to proton transfer, 307
energy decomposition analysis (EDA), 183
enforced concerted mechanism, 307
enol form of ketone, 240
enolate carbanion, 95
  from ketone, 240
  trapping, 49, 97
enolisation of acetone, third-order term, 297
enones, Rh-catalysed addition of organoboron
      compounds, 327
enthalpy
  activation of, 229
    determination, 69
  reaction of, 12, 15, 207
    determination, 69
entropy, 6
  activation of, 229, 351
    determination, 69
  reaction of, 15
    determination, 69
environment, 100
enzyme(s), 13
  active site, 307
  artificial, 294
  as targets for drug design, 312
  assay, 312
  catalysis, 311
  covalently bound intermediates, 318
  for determination of metabolite
      concentrations, 312
  general mechanism for reversible inhibition,
    318
  inhibitors, active-site directed, 315
  initial burst kinetics, 314
  kinetic analysis, 316
  kinetics and Eadie-Hofstee plots, 316
  kinetics and Lineweaver-Burk plots, 316
  mechanisms and stereochemistry, 320
  mixed inhibitors, 317
  non-competitive inhibitors, 317
  reversible inhibitors, 317
  steady-state kinetics, 313
  suicide inhibitors, 312
  structural variation and mechanisms, 319
  technical applications, 312
  trapping of transient intermediates, 321
  uncompetitive inhibitors, 317
enzyme-substrate complex, 313
epimerisation of alkoxides, 25
epoxidation of
    2,5-di-*tert*-butyl-1,4-benzoquinone,
    216, 219, 221
equilibrium, 48
  constant, 48, 58, 83
  control, 23
  isotope effect, 251

equivalent conductance, 72
ESR spectroscopy, 238, 270, 271, 274, 284, 288,
    326
$Et_3SiH$ as an activator, 336
ethane
  heat capacity, 183
  rotational barrier, 182
ethyl vinyl ether, 351
ethylene oxide, 175
  pyrolysis, 69
ethyltriphenylphosphonium bromide, 30
excited states, computation, 192
expanding drop methods, 121
extended Hückel theory, 174
extent of reaction, 46
extinction coefficient, 67
extraction of copper, 105, 112, 118, 123
extractive reaction, 105
Eyring-type analysis, 351, 352

familial hypercholesterolaemia, 312
fast reactions, 69
femtosecond
  gas phase studies, 7
  time scale, 67
ferrocene, oxidation, 161
ferrocene/ferrocenium redox couple, 135, 160
fibre optics, 235
Fick's law
  first, 105, 108, 140
  second law, 140
film thickness, 105
fine chemicals, 104
first-order reaction, 80
fission products, 104
flash photolysis, 70, 276
flow reactor, 60
fluid dynamics, 204
fluorimetry, 68, 238
force constant, 250
force field calculations, 172
formate esters, reaction with aqueous
    hydroxide, 114
Fourier transform (FT) [1]H NMR, 22
Fourier transform infrared (FTIR)
    spectroscopy, 11
Franck-Condon principle, 192
free energy diagram, 24
free energy
  of activation, 229
  of hydration, 190
  of reaction, 293
  of solvation, 190
  plots, 285
  relationships, concave/concave 294
freeze-thaw degassing, 268

full configuration interaction (FCI), 178
functional group
   intolerance, 348
   reactions, 13, 34

galvinoxyl, 271, 288
gas chromatography (GC), 21, 336
gas scrubbing, 104
gas-liquid reactions, 113
Gaussian energy distributions, 170
Gaussian-type orbital (GTO), 181
general acid catalysis, 4
general acid-base catalysis, determination of
   parameters, 296
general base catalysis, 95, 97
glass electrodes, 64
glucose
   anomeric effect, 187
   mutarotation, 73
(S)-glutamic acid, deamination, 247
glycine, chlorination, 92
glycosidase mechanisms, 318
glycylation of lysozyme, 320
Grignard reagents, 285, 328
group charges, 173
Grubbs catalysts, 348, 352
[1,5]-hydrogen shifts from carbon to nitrogen,
   254

H-transfer reactions, 71
H-abstraction, 286
halex reactions, 109
half-life, 265
half-order kinetic relationship, 349
halobenzenes, amination, 26
N-halodipeptides, decomposition, 68
halogen exchange (halex) reactions, 109
Hammett relationship, 256
Hammond effect, 300
Hammond postulate, 170
hanging mercury drop electrode, 134
harmonic approximation, 169, 250
Hartree-Fock (HF) Hamiltonian, 176
Hartree-Fock (HF) theory, 173, 175
HBr, addition to alkenes, 18
heat balance calorimeters, 202
heat capacity, 202
   of ethane, 183
heat conductivity, 204
heat flow calorimeters, 201
heat flow rate, 11
heat
   of formation, 170, 172, 174
   of mixing, 206, 213
heat-transfer coefficient, 201, 204
heavy atom isotope effects, 65

hemispherical diffusion in electrolysis, 140, 157
Henderson-Hasselbach equation, 61
Hessian matrix, 181, 185
N-heterocyclic carbenoid ligand (NHC)
hetero-Diels-Alder reactions of iminium ions,
   75
heterogeneous catalysis, 106
   hydrogenation of, 109, 222
heterogeneous electron transfer, 136, 141, 145
heterogeneous Heck reaction, 246
heterogeneous photocatalysis, 9
heterolysis, 12, 231
heteronuclear isotopic exchange reactions, 71
5-hexenyl radical clock, 286
5-hexenyl radical, 280
high field NMR spectroscopy, 22
high performance liquid chromatography
   (HPLC), 21
hindered rotation, 183
Hofmann reaction, 29, 32
Hohenberg and Kohn (HK) theorem, 179
homodesmotic reactions, 170
homogeneous catalysis, 293
homogeneous electron transfer, 141
homolysis, 12, 31, 231, 265
   of alkoxyamines, 270
homolytic aromatic substitution, 280
homo-metathesis, 346
HPLC, 21
Hund's rule, 177
hydrazine, oxidation with dispersed barium
   chromate, 115
hydrazone formation, general base catalysis,
   303
hydroboration, 19, 31
hydrocarbons, heats of formation, 172
hydrodimerisation, reductive, 128
hydrodynamics, 116
hydrogen atom abstraction, 277
hydrogen atom transfer, 98, 341
   from tributyltin hydride, 281
hydrogen bond donor/acceptor, 98
hydrogen bonding
   interactions, 174
   in transition state stabilisation, 305
hydrogen cyanide, 175
hydrogen isocyanide, 175
hydrogenation of nitrobenzene, 222
hydrogen-bonded complex, 99
hydronium ion, deviation from Eigen plot, 304
hydroxide ion, deviation from Eigen plot, 304
hydroxydiazenium oxide, 3
$\mu$-hydroxy dimer, 331
hydroxyl radical, 72
hydroxylamine as a trap for transient
   intermediates, 321

hydroxymethylene, 175
hydroxymethylglutaryl CoA reductase, 312
hyperconjugation, 183, 185
hypervalence, 231
hypochlorous acid, 91
  reaction with thiols, 69
hypovalence, 231

imaginary vibrational frequency, 170
imidazole, 43
imidazolidinone nitroxides, 271
iminium ions, in hetero-Diels-Alder reactions,
    75
indolylacryloyl-$\alpha$-chymotrypsin, X-ray
    crystallography, 319
induction period in autocatalysis, 293
in-film reactions, 116
infrared spectroscopy, 205, 326
inhibition, 52
  in $S_{RN}1$ substitutions 288
  competitive, 317
inhibitors, 21, 293, 336
  catalytic, 293
  irreversible, 315, 316
  mixed, 317
  non-competitive, 317
  reversible, 317
  suicide, 312
  uncompetitive, 317
initial burst enzyme kinetics, 314
initial rate, 51
  and enzyme kinetics, 313
initiation of radical chain reaction, 263,
    264
initiator efficiency factor, 266
initiator radical, 264, 268
inner sphere electron transfer, 127, 284
integral kinetic analysis, 199
integration by numerical methods, 208
interatomic interactions, 172
interface
  in multiphase systems, 105
  area per unit volume, 106
  deprotonation in PTC, 112
interfacial reaction, 105, 109
intermediate(s), 2, 6, 10, 12, 87, 227
  criteria for, 308
  estimation of p$K_a$ values as criterion of
      mechanism, 304
  stabilisation by proton transfer, 304
  trapping, 242
intermolecular
  interactions, 205
  reaction, 231
  rearrangement, 29

internal
  coordinates, 170
  energy, 168
  return, 90
    in solvolysis, 188
  standard, 22, 33
intimate ion pair, 90, 248
intramolecular
  reaction, 231
  abstraction reactions of radicals, 263
  homolytic aromatic substitution, 280
  hydrogen bonding, 274
  kinetic isotope effects, 254
  rearrangement, 29
intrinsic reaction coordinate, 227
inversion of configuration, 247
4-iodobenzamide in heterogeneous Heck
    reaction, 246
iodobenzene, amination, 27
ion pair, 35, 90
ion selective electrodes, 235
ionic strength, 39, 62, 63
ionisation, 35
ionisation-dissociation, 42
ionisation potential, 174
ion-selective electrodes, 73
IR-ATR spectroscopy, 205
irreversible
  electron transfer, 150
  inhibitors, 314, 315
  papain inhibitors, 316
isodesmic reactions, 170
isolation method, 56
isoleucine, chlorination, 92
isosbestic point, 68
isothermal
  calorimetry, 206
  infrared reaction data, 209
isotope-induced chemical shift changes,
    250
isotope ratio mass spectrometry, 253
isotopic labelling, 2, 20, 26, 29, 231, 326, 334,
    343
isotopic substitution, 249
isotopomers, 250, 252
  of 2,3-dimethylbutene, reaction with
      $N$-methyltriazolinedione, 254

jet reactor, 119

$K_m$ parameter, 313
ketones, acid- and base-catalysed reactions
    with electrophiles, 240
ketonisation of enols in buffer solutions, 70
ketyl radicals, 285

kinetic
  ambiguity/equivalence, 4, 7, 13, 79, 93, 299
  analysis of enzyme catalysis, 316
  control, 23, 84
  electrolyte effect, 63
  isotope effect, 251
  model, 79
  parameters from calorimetry, 208
Kitaura-Morokuma (MK) analysis, 186
$\alpha$-lactone as a transient intermediate, 247

laminar flow, 117
Langmuir adsorption, 122
Langmuir trough, 123
Laplacian, 185
laser flash photolysis, 8, 13, 70
leaving group (nucleofuge), 35
Lewis cell, 116
Lewis structures, 232
lifetime
  of adduct in NMP, 276
  of an intermediate, 307
ligation, 112
light pipes, 235
limitation of reaction, 32
limiting reactant, 11, 51, 56
line-shape analysis, 351
linear combination of atomic orbitals (LCAO),
    181
linear diffusion, 139
linear free energy relationships, 256
linear scaling methods, 193
linear semi-infinite diffusion, 157
linear sweep voltammetry, 138, 147
Lineweaver-Burk plots and enzyme catalysis,
    316
liquid-junction potentials, 64
liquid-liquid extraction kinetics, 116
liquid-solid reactions, 115
lithium diethylamide, 26
living ends in NMP, 275
living mechanisms, 13
living reagent/mediator, 275
localised orbitals, 173
London dispersion forces, 174
Löwdin population analysis, 185
LSV conditions in electrochemistry, 154
lyonium ion, 294
lyoxide ion, 294
lysozyme acylal intermediate, X-ray
    crystallography, 319
lysozyme mechanism, acylal intermediate, 319

macro-alkoxyamine initiators, 276
macrocyclic amide, 15, 343

magic acid, 30
magnetisation transfer, 351
maleic anhydride, reaction with
    cycloheptatriene, 240
many-body perturbation theory (MBPT), 176
Marcus theory, 284
Markovnikov addition, 18
mass spectrometry (MS), 3, 21, 26, 76, 249,
    326
mass transfer, 208
  coefficient, 105
  rate, 104, 106
mass transport, 9, 104, 140
  and current in electrochemistry, 139
mathematical modelling in multiphase
    systems, 122
mechanism as a hypothetical construct, 326
mechanism-derived rate equation, 350
mechanistic
  'cul de sac', 236
  nomenclature, 231
  rate constant, 4, 87, 90, 98, 100
  rate law, 79, 93, 95, 97, 99, 100
mediator in electrochemistry, 131
medium effect, 39
Menschutkin reaction, 72
mercury film electrode, 134
metal
  hydride initiators, 274
  ions as catalysts, 294
  recovery, 9
    by solvent extraction, 104
metathesis of alkenes, transition-metal
    catalysed, 15
methacrylonitrile, 266
methanolysis, 38
methionine, oxidation by HOCl, 57, 74
method of initial rates, 51
  and of integration, 53
methoxyaminolysis of phenyl acetate, proton
    switch mechanism, 308
*p*-methoxybenzoyl chloride, 37
methoxycyclopropylmethyl
  cation, 279
  radical, 279
methyl benzoylglycine as an enzyme substrate,
    315
methyl hippurate as an enzyme substrate, 315
methylbenzenediazonium ion, 31
methylene, 181
2,2'-methylene-bis(4-methyl-6-*tert*-
    butylphenol),
    99
*exo*-methylenecyclopentane, 338
methylidene pro-catalysts, 349

2-methyl-3-thiosemicarbazone formation catalysed by quinuclidinol, 310
*N*-methyltriazolinedione
 reaction with alkenes, 253
 reaction with isotopomers of 2,3-dimethylbutene, 254
1-methyl-2,3,5-triphenylpyrrole, 161
micelles as catalysts, 294
Michael reaction, 15
Michaelis-Menten
 equation, 316
 kinetics, 313, 314
 parameters ($K_m$ and $V_{max}$), 316
microelectrochemical measurements at expanding droplets, 121
microelectrode, 147
 techniques in multiphase systems, 122
microheterogeneous media, 63
mid-point potential, 149
minimum information requirements, 123
mixed inhibitors, 317
MM2 force field, 173
MNDO, 174
molar absorptivity, 67
molecular dynamics (MD), 7, 168
molecular
 mechanics, 172
 orbital, 173
 rotations and vibrations, 168
Møller-Plesset (MP) perturbation theory, 176
monodentate ligand, 333
Monte Carlo (MC) methods, 189
More O'Ferrall-Jencks diagrams, 14, 300
Mulliken
 charge, 173
 population analysis, 173, 185
multideterminant problems, 178
multi-mixing stopped-flow systems, 70
multiphase systems, 9, 104
multivariate evaluation techniques, 210
mutarotation of sucrose and glucose, 73
myeloperoxidase, 57, 68

natural bond orbital (NBO) analysis, 173, 183, 186
natural orbitals (NO), 186
natural population analysis, 186
natural product chemistry, 34
negative catalyst, 293
neighbouring group participation, 25, 34, 188
Nernst equation, 74, 137, 140, 147
Nernst-Planck equation, 139
nitramide
 acid- and base-catalysed decomposition, 6
 *aci*-form, 7

nitration
 of benzene, 104
 using nitronium tetrafluoroborate, 118
nitric oxide, 3
nitro radical anion, 291
$\alpha$-nitroalkane in $S_{RN}1$ substitution, 287
*m*-nitroaniline, 39
*o*-nitroaniline, 40
4-nitrobenzaldehyde, catalysed semicarbazone formation, 301
nitrobenzene, catalytic hydrogenation, 12, 222
*p*-nitrobenzoyl chloride, 38
nitronate anion, 289
nitronium tetrafluoroborate, use in nitration, 118
nitrophenyl acetate, imidazole catalysed hydrolysis, 309
4-nitrophenyl acetate, catalysed hydrolysis, 43
4-nitrophenyl acetate, reaction with phenolate ions, 311
*N*-(*p*-nitrophenylethyl)quinuclidinium, elimination of quinuclidine, 241
4-nitrophenylethylpyridinium ion, hydroxide catalysed elimination, 310
2-nitro-2-propyl radical, 289
nitroso compounds as spin traps, 262
nitrosodialkylhydroxylamines, 7
nitrosohydroxylamines, acid-catalysed decomposition, 3
nitrous acid, 7
nitrous oxide, 3
 rotational spectrum, 4
nitroxides, 268
nitroxide exchange experiments, 271
nitroxide mediated living/controlled radical polymerisations (NMP), 273, 275
nitroxyl (nitroxide) radical, 262
NMR and IR spectra, computation, 192
NMR spectroscopy, 235, 326, 350
no mechanism reaction, 1
NO88, 269, 276
NO88Bn, 271
NO88Me, 271
Nobel prize in chemistry, 352
noble gas outer shell electron configuration, 349
nomenclature of mechanisms, 231
nonaqueous solvents, 64
non-Arrhenius behaviour in radical coupling, 270
non-chain radical reactions, 268
nonclassical
 carbocation problem, 187
 structures, 174
non-competitive inhibitors, 317

non-dissociative pathway, 349
non-linear least-squares optimisation, 54, 209, 210
norbornan-2-ols by stereoselective reduction, 242
2-norbornanone, stereoselective reduction, 242
*endo*-2-norbornyl brosylate, 90
*exo*-2-norbornyl brosylate, 90
2-norbornyl cation, 187, 188
nuclear magnetic resonance (NMR) spectroscopy, 26
nuclear power industry, 9, 104
nuclear relaxation processes, 235
nucleofuge, 3
nucleophile, 231, 232
  catalysis, 43, 308
nucleophilic
  aromatic substitution, 28
  attack at aldehyde, general acid catalysis, 300
  attack at aldehyde, general base catalysis, 300
  capture, 90
  radicals, 267
  substitution reaction, 105
numerical integration, 142, 208
numerical methods, 87
$^{17}$O NMR spectroscopy, 67
$^{17}$O relaxation rates, 71
$^{18}$O labelling in solvolysis studies, 252
2-octyl sulfonates, 40

off cycle, 325, 333
olefin disproportionation, 343
olefin metathesis, catalysis by ruthenium methylidene, 241
one-component reaction, 334
on-cycle, 333
optical rotatory dispersion (ORD), 73
orbital relaxation terms, 186
orbital symmetry, 1, 12, 25, 31, 79, 344
order of reaction, 47
ordinary differential equations, 208, 210
organoboron reagents, 328
  Rh-catalysed addition to enones, 327
organoboronic acids, 328
organometallic
  catalysis, 13
  complexes, 324
organosamarium
  compounds, 285
  intermediates, 286
outer sphere electron transfer, 127, 284
overgrowth, 109
oxaphosphetane, 30, 237

oxidation, 91, 100
  by inorganic reagents, 127
  of amines by HOCl, 69
  of ferrocene, 157
  of inorganic complexes, 123
oxime ligand, 105, 112, 118
$\tau$-oxo-allyl rhodium complex, 331
oxocarbenium ion, 294
oxonium ion, 279
oxygen
  as a scavenger, 272
  as an electron acceptor, 288
  electrochemical reduction, 134
  interfacial mass transfer rate, 123
oxygen labelling, 3
oxygen-centred radical, 270, 265
$^{31}$P NMR, 237, 328, 329

palladacyclobutane, 340
palladium on alumina in heterogeneous Heck reaction, 246
papain, 315, 320
  active-site directed inhibitors, 316
  active-site titration, 315
  *N*-benzoylaminothionacetyl intermediate, 319
  competitive reactions, 60, 83, 98
parameterisation, 172, 174
parsimony principle (Occam's razor), 335
partial charges, 168
partition functions, 169
Pasteur, 32
patent laws, 20
Pauli (exclusion) principle, 183, 185
Pauli repulsion, 185
Pd-allyl cations, 335
Pd-catalysed isomerisations, 327
Pd-catalysed Suzuki-Miyaura cross-coupling, 328
Peltier calorimeters, 202
pentacoordinate oxyphosphorane di-anion as an intermediate, 258
pentane-$d_{12}$, 30
pericyclic reaction, 174, 232
periodic monitoring, 65
peroxides
  as initiators, 265
  in addition of HBr to alkenes, 18
peroxyl radicals, 288
persistent radical effect, 272, 269
pH
  control, 61
  measurement, 64
  profiles and catalysis, 295
pharmaceuticals, manufacturing, 104

phase boundary, 9
phase transfer catalysis (PTC), 9, 105, 110
phenacyl bromide, reaction with pyridine, 72
phenanthroline, 335
phenol(s)
    acylation, 9
    hydrogen atom transfer from, 98
    oxidation by Cr(VI), 100
phenyl acetate, methoxyaminolysis, 308
phenyl radical, 265
phenyl urea formation, general acid catalysis, 303
phenylation, asymmetric 15
phenylboronic acid, 328
phenylhydroxylamine, 12
*N*-phenyltriazolinedione, reaction with alkenes, 253
phenylurea formation, general acid-base catalysis, 306
phenylurea herbicides, photodegradation, 76
phosgene, solubility in water and hydrolysis, 121
phosphine ligands, 328
photochemical
    activation, 79
    initiation, 265
photodegradation of aniline derivatives, 76
photodissociation of *N*-tritylanilines, 71
photoinduced electron transfer, 127
photostimulation, 287
PM3, 174
polar abstraction mechanism, 290
polarimetry, 73, 83
polarisability, 173
polarisation, 176
polarography, 284
polycyclic arenes, 175
polymer reactions, 13
polymeric supports in three-phase test, 245
polymerisation
    by metathesis, 345
    studies by dilatometry, 75
    transition-metal-catalysed, 343
    viscosity changes during, 204
polystyryl alkoxyamines, 271
potassium
    bicarbonate, dissolution in DMF, 115
    enolate, 331
    fluoride, solubility in DMF, 118
potential energy hypersurface, 6, 51, 168, 192, 227
potential energy
    maxima, 227
    minima, 227
    profile, 227
    surface computations, 193

potentiometry, 73
potentiostat, 135
power-compensation calorimeters, 201
practical equilibrium constant, 49
preassociation, 306
pre-equilibria, 4, 87, 89, 90, 96, 100
pre-equilibrium approximation, 89
PRE-mediated 5-*exo*-trig radical cyclisation, 274
primary isotope effect, 251, 302
primary radical, 264
principal component analysis (PCA), 210
pro-catalyst, 325
    activation, 341
    generation I, 340
    generation II, 350
process development, 124, 198
process safety, 198
pro-chiral ketones, 328
product analysis, 2, 84, 91, 102, 164, 242, 277, 287, 326, 343
product equilibration, 347
product selectivity, 36
product-determining step, 35
propagation
    polymer radical of, 273
    species in radical polymerisations, 266, 270
    of radical chain reaction, 263
    steps of, 268
propene, 175
2-propyl sulfonates, 40
2-propyl tosylate, solvolysis, 243
protein engineering, 320
protic solvents, 291
protodeborylation, 328
proton
    exchange in secondary amines, 71
    switch mechanism, 308
proton transfer, 13, 95, 98, 100, 294
    concerted, 7, 294, 307
    energy barrier, 307
    in multiphase reactions, 112
    rate-limiting, 100, 302
pseudo-
    cyclobutane, 344
    equilibrium, 106
    first-order rate constant, 8
    first-order reaction, 38, 81, 100
    reference electrode, 135
    zero-order kinetics, 336, 337
pulse radiolysis, 13, 70, 276
pure spectra matrix, 209
push-pull acid-base catalysis, 307
pyridine, reaction with phenacyl bromide, 72
pyridyl diazomethyl ketones, 9

pyridyl ketenes
  capture by *n*-butylamine, 8
  from diazoacetylpyridines, 69
pyrroles, electrochemical oxidation, 160

quadrupole mass spectrometry, 282
quantum mechanical operator, 173
quasi-
  cyclobutane, 344
  reversible electron transfer, 149
  symmetrical reactions, 257, 311
quaternary salts
  ammonium salts, 111
  phosphonium salts, 111
quinuclidine, elimination from
    *N*-(*p*-nitrophenylethyl)quinuclidinium,
    241
quinuclidinol as base catalyst, 310

$r_{12}$-methods, 193
racemisation, 83, 90
  in nucleophilic substitution, 248
  of a radical, 279
radical(s), 231, 232
  addition to alkenes, 266
  chain, 28
  chain carriers, 264
  clock, 13
  clock reaction, 276
  clock reactions, table, 278
  coupling, 26
  hydrogen atom transfer to, 98
  in functional group reactions, 13
  inhibition, 28
  initiation, 28
  initiator, 234, 262
  intermediate, 12, 130, 261
  ion, 10, 130
  mediators, 13
  pairs, 26
  polymerisation, 268, 275
  reactions, 12, 261
  reductions, 268
  scavenger, 28
  trap, 262, 268
  trapping by oxygen, 288
radical anion, 284, 289
  intermediates, 287, 288, 289
  mechanism, 28
radical cation, 161
  from methoxylated benzenes, 71
  oxidation to dication, 162
radical-radical combination reactions,
  264
radioactive isotopes, 26
  in labelling studies, 250

Raman spectroscopy, 68, 282
rate coefficient, 47
rate constant, 47
  pH dependence, 294
rate equations, table, 53
rate law, 4
rate-determining/limiting step, 34, 49, 90, 92,
  324
  approximation, 87, 89
  change in, 242
rate-limiting conformational change, 320
rate-limiting diffusion control, 306
rate-product correlations, 38
reaction
  calorimetry, 11, 200
  coordinate, 6, 168, 170, 227
  dynamics, 192
  flux, 105, 297
  intermediate, 2, 20, 228
  map of, 14
  mechanism, 49
  model, 11, 198
  parameters, 198
  path, 168
  profile, 169
  rate, 46
reaction order, 80
  analysis in electrochemistry, 145, 153
  with respect to concentration, 51
  with respect to time, 52
reactive dyes, 104
reactive intermediate, 7, 25, 42, 82, 88, 169,
  228
rearrangements, 170
  intermolecular, 29
  intramolecular, 29
recovery of copper, 104
redox reaction, 100, 127
  initiation, 264
  in radical chemistry, 262, 264, 284
reduced mass, 251
reduction
  by inorganic reagents, 127
  potential, 285
reductive
  elimination, 345
  hydrodimerisation, 128
  lithiation of a nitrile, 279
reference electrode, 133, 135
refractive index, 235
regioselective rearrangement, 279
regioselectivity in polymerisation, 274
relativistic effects, 169, 192
relaxation methods for very fast reactions,
  70
residuals, 54

resonance stabilisation
  of benzene, 171
  of benzyl radicals, 267
  of carbanion, 249
  of nitroxide radicals, 269
respiration, 104
response factor, 22
retention of configuration, 25, 247
  with alkaline phosphatase, 320
  via double inversion, 320
retro-cyclometallation, 345
retro-synthetic analysis, 15
reverse phase HPLC, 33
reversible
  electron transfer, 147, 156
  first-order reactions, 83
  inhibitors, definition, 317
  one-electron process, 161
  reaction, 58, 60
  termination, 273
Rh-catalysed
  addition of organoboron compounds to
    enones, 327
  isomerisation, 334
  phenylation, 330
ring-closing diene metathesis (RCM), 15
rotated diffusion cell, 118
rotated disc reactor, 117
rotational barriers, 174
ruthenium alkylidenes, 348
ruthenium methylidene catalysis in olefin
    metathesis, 241

saddle point, 6, 50, 170, 227
salt bridge, 135
salt effects, specific, 297
samarium Barbier reaction, 285
samarium di-iodide, 284
sampling, 33
saturated calomel electrode (SCE), 134
saturation kinetics, 122, 241
scalar coupling, 329
scanning electrochemical microscopy (SECM),
    122
scavengers, 271
Schotten-Baumann reaction, 9
Schrock-type systems, 352
Schrödinger equation, 168, 175
scientific software package, 52
scintillation counting in tracer studies, 250
secondary
  alkyl benzenesulfonates, solvolysis, 252
  alkyl sulfonates, 40
  deuterium kinetic isotope effect, 65, 189
  isotope effect, 251
  metathesis, 345

second-order rate constant, 4
second-order reaction, 80
  worked example, 52
selectivity, 36, 85
  in synthesis, 280
self-consistent (SCF) theory, 173
self-coupling, 286
self-termination, 273
semi-batch operation mode, 208
semi-empirical methods, 173
semi-infinite linear diffusion, 155
sequential epoxidation of
    2,5-di-*tert*-butyl-1,4-benzoquinone,
    216, 219, 221
serendipity, 30, 324
serine proteases, covalently bound
    intermediates, 318
SET reaction, 287
SET($S_{RN}$) mechanism, 290
SG1, 272, 276
$S_H2$ mechanism, 263, 285
sigmatropic rearrangement, 345
signal:noise ratio, 67
simulated voltammograms, 154
single crystal X-ray diffraction, 326
single electron transfer (SET), 231, 281, 284
single peak evaluation, 210
single step process/reaction, 80, 228
single-substrate reactions, 327
site-exchange of NMR-active nuclei, 351
Slater-type orbital (STO), 181
$SmI_2.(HMPA)4$, 285
$S_N1$ mechanism, 32, 34, 41
$S_N2$ mechanism (ANDN), 14, 25, 40, 170, 288,
    290
sodium azide as a nucleophile, 243
sodium hypobromite, 29
solid matrix
  to trap radicals, 262
  to trap radical anions, 288
solvated electrons, 28
solvation
  by hydrogen bonding in methanol, 291
  of ions, 12
solvation effects, 62
solvation energy, 189
solvent effects
  specific, 297
  in $S_{RN}1$ substitutions 287
  on rate constants, 62
solvent extraction in metal recovery, 104
solvent kinetic isotope effects, 302
solvent-separated ion pair, 35, 37, 90, 248
solvolysis, 34, 40, 90, 105
  of 2-adamantyl tosylate, 244
  of benzyl azoxytosylate, 245

of 2-norbornyl derivatives, 188
of 2-propyl tosylate, 243
of secondary alkyl benzenesulfonates, 252
solvolytic rearrangement, 40
specific
    acid catalysis, 4
    base catalysis, 49, 97
    conductance, 72
    rotation, 73
spectator mechanism, 306
spectrophotometric techniques, 8
spin trapping, 262
spin-orbit coupling, 192
spreadsheet, 52
square
    planar complexes, 330
    planar Rh complex, 331
$S_{RN}1$ mechanism, 28, 281, 287, 290
stabilised acyl enzymes, 319
stable ion conditions for carbocations, 248
standard reduction potential, 74
standard state, 49
stationary points, 168, 227
statistical mechanics, 169
steady-state approximation (SSA), 83, 88, 90,
    95, 99, 255 267, 310
steady-state kinetics of enzyme action, 313
stepwise acid catalysis, 295
stepwise mechanism/process, 2, 6, 13, 232
stereochemistry, 25, 246
    and enzyme mechanisms, 320
stereoconvergence, 249
stereogenic centre/elements, 25, 246
stereoselectivity, 25
    in reduction of ketones, 242
    of intramolecular addition, 263
stereospecific reactions, 247
stereospecific ring-opening of bromonium
    ions, 247
steric compression in alkoxyamines, 270
steric hindrance, 168, 288
steric repulsion, 183
sterically hindered base, 302
Stevens rearrangement, 26
stilbene, *cis-trans* isomerisation, 59
stirred reactor, 60, 113
stoichiometry, 80
stopped-flow reactors, 69
stopper, catalytic, 293
strain, 168
strained rings, 175
structured layer model of electrochemical
    double layer, 138
structure-reactivity relationships, 326
styrene, polymerisation, 274
substituent parameters, 256

substitution, 231
    in arenes and heteroarenes, 280
sucrose, mutarotation, 73
suicide inhibitors, 312
sulfenyl
    chlorides from thiols, 69
    intermediates, 290
super acid media, 189
superoxide, 288
    from oxygen, 134
supporting electrolyte, 135
symmetry, 181
*syn beta*-hydride elimination, 338, 341
*syn* stereochemistry, 331
synchronous concerted mechanism, 2, 14

tail addition in polymerisation, 266
telomerisation, 346
telomers by metathesis, 345
temperature stability, 60, 61
TEMPO, 269, 276
termination, 325
    of radical chain reaction, 263
termolecular mechanism, 4
terpenes, 34
tetrabutylammonium hexafluorophosphate,
    135
tetrahedral intermediate, 12, 14
    in nucleophilic addition to carbonyl, 305
tetrahydrodicyclopentadiene, 30
1,1,3,3-tetramethylisoindoline-2-oxyl (TMIO),
    269, 270
2,3,4,5-tetraphenylpyrrole, oxidation, 160
thermal
    activation, 79
    initiation, 265
    process safety, 199
    reaction, 79
    runaway, 11
thermodynamic control, 23, 85
thermodynamic properties, calculated, 169
thermostatting device, 61
thiolate anions, 289
third-order
    rate constant, 4
    rate law, 97
    terms in catalysis, 297
Thornton effect, 300
three-electrode cell, 133
three-phase test, 245
tight ion pair, 35
time-resolved
    IR spectroscopy, 8, 69
    spectra, 68
TIPNO, 272, 276
TLC, use in kinetics, 75

TMIO, 269, 276
total wave function, 173
transglycylation, 320
transient/reactive intermediates, 7, 25, 42, 82, 88, 169, 228
  and non-linear free energy relationships, 310
  trapping in enzyme mechanisms, 321
transition metal catalysis, 13
transition state, 169, 229
transition state
  spectroscopy, 228
  stabilisation by hydrogen bonding, 305
  theory, 7, 50, 170, 229
transition structure, 4, 50, 168, 172, 227
  and Brønsted parameters, 299
  lifetime, 228
trapping mechanism, 42
  by nucleophiles, 8
  of intermediates, 242
  of transient intermediates in enzyme mechanisms, 321
trapping techniques, 284
2,2,6,6-tetramethyl-1-piperidinyloxy (TEMPO), 269
tetraphenylhydrazine, 28
tributyltin deuteride, 281
tributyltin hydride, 264, 268, 281
trickle-feed of catalyst, 336, 337
2,2,4-trimethyl-7-methylene-1,5-dithiacyclo-octane, 267
2,2,5-trimethyl-4-phenyl-3-azahexane-3-aminoxyl (TIPNO), 276
1,2,5-trimethylpyrrole, oxidation, 164
triple bonds, 175
Triton-B, 216
tungstacyclopentane, 345
tungsten alkylidene, 347
turbulence in stirred reactions, 60, 69
turnover limiting step, 324
two-compartment electrochemical cell, 132

ultrafast radical clock reactions, 278
ultramicroelectrode, 10, 121, 134, 139, 140, 155, 157
ultrasound, used for mixing, 115
uncompetitive enzyme inhibitors, 317
uncoupled bonding changes, 232
uniform electron gas model, 179
unimolecular process/reaction/step, 1, 49, 80, 81, 87, 228

unimolecular nucleophilic substitution ($S_N1$), 32, 34, 41
univariate evaluation technique, 210
unsaturated enones, conjugate (1,4) addition, 328
urethanes, 29
uridine 3′-phosphate esters, alkaline cyclisation, 257
UV chromophore, 271
UV-vis spectra, computation, 192
UV-vis spectrophotometry, 8, 67, 238, 276

valence bond (VB) theory, 183
valence
  electron models, 174
  electrons, 174
  isomerisations, 71
validation of computational methods, 182
variational theorem, 175
vibrational anharmonicities, 192
viscosity
  effect upon diffusion, 304
  effect on mixing, 60
  increase during polymerisation, 204
  rate dependence on, 98
vitamin B12, 32
$V_{max}$ parameter, 313
voltage sweep rate, 160
voltammetry, 147
voltammogram, 148
volume stability, 60

Walden inversion, 25
wave function, 173
weighting factors, 211
Wittig reaction, 30, 237
Woodward, 31
Woodward-Hoffmann rules, 344
working
  curve approach, 153
  electrode, 133, 134
  mechanism/model, 324, 326

X-ray
  absorption fluorescence spectroscopy (XAFS), 326
  crystallographic map of an enzyme, 315
  crystallography, 21, 342
  structures, 167

zero point energy, 176, 229, 250
zero-order kinetics, 100, 240, 336
zwitterionic intermediates, 308